ADVANCES IN CHEMICAL PHYSICS

VOLUME XCIX

Advances in CHEMICAL PHYSICS

Resonances, Instability, and Irreversibility

Edited by

I. PRIGOGINE

Center for Studies in Statistical Mechanics and Complex Systems
The University of Texas
Austin, Texas
and
International Solvay Institutes
Université Libre de Bruxelles
Brussels, Belgium

and

STUART A. RICE

Department of Chemistry
and
The James Franck Institute
The University of Chicago
Chicago, Illinois

VOLUME XCIX

AN INTERSCIENCE® PUBLICATION
JOHN WILEY & SONS
NEW YORK • CHICHESTER • BRISBANE • TORONTO • SINGAPORE • WEINHEIM

This text is printed on acid-free paper.

An Interscience® Publication

Library of Congress Catalog Number: 58-9935

ISBN 0-471-16526-3

Printed in the United States of America

10 9 8 7 6 5 4 3 2 1

CONTRIBUTORS TO VOLUME XCIX

I. Antoniou, International Solvay Institute for Physics and Chemistry, Brussels, and Theoretische Natuurkunde, Free University of Brussels, Brussels, Belgium

G. Bhamathi, Department of Physics and Center for Particle Physics, University of Texas, Austin, Texas

Erkki J. Brändas, Department of Quantum Chemistry, Uppsala University, Uppsala, Sweden

C. A. Chatzidimitriou-Dreismann, Iwan N. Stranski Institute for Physical and Theoretical Chemistry, Technical University of Berlin, Berlin, Germany

Charles B. Chiu, Department of Physics and Center for Particle Physics, University of Texas, Austin, Texas

E. Eisenberg, Department of Physics, Bar-Ilan University, Ramat-Gan, Israel

Pierre Gaspard, Service de Chimie Physique and Centre for Nonlinear Phenomena and Complex Systems, Université Libre de Bruxelles, Brussels, Belgium

L. P. Horwitz, Department of Physics, Bar-Ilan University, Ramat-Gan, and School of Physics, Raymond and Beverly Sackler Faculty of Exact Sciences, Tel-Aviv University, Ramat-Aviv, Israel

V. V. Kocharovsky, Institute of Applied Physics, Russian Academy of Science, Nizhny Novgorod, Russia

Vl. V. Kocharovsky, International Solvay Institute for Physics and Chemistry, Brussels, Belgium

T. Petrosky, Center for Studies in Statistical Mechanics and Complex Systems, University of Texas, Austin, Texas, and International Solvay Institutes, Université Libre de Bruxelles, Brussels, Belgium

I. Prigogine, Center for Studies in Statistical Mechanics and Complex Systems, University of Texas, Austin, Texas, and International Solvay Institutes, Université Libre de Bruxelles, Brussels, Belgium

Z. Suchanecki, International Solvay Institute for Physics and Chemistry, Brussels, Belgium, and Hugo Steinhaus Center and Institute of Mathematics, Wroclaw Technical University, Wroclaw, Poland

E. C. G. Sudarshan, Department of Physics and Center for Particle Physics, University of Texas, Austin, Texas

S. Tasaki, Institute for Fundamental Chemistry, Kyoto, Japan

PREFACE

All of chemistry deals with irreversible processes. Here we come immediately to one of the most important and also most controversial problems of modern physics: what are the dynamical roots of irreversibility? How can past and future be distinguished on the fundamental level of description? It is well known that both Newton's and Schrödinger's equations are time-reversible. Is then time an "illusion," as Einstein liked to repeat? In the past there have been many proposals to find an easy way out of this dilemma: additions of "friction" terms breaking the time symmetry, using trucation or coarse graining. But when do we need a new formulation at all?

There are indeed situations where the time-reversible laws of classical or quantum mechanics lead to predictions that are verified to the highest precision reached today in experimental sciences. On the other hand, truncation or coarse graining would make us, through our approximations, responsible for the existence of the breaking of the time symmetry. This is quite unsatisfactory. We would be the progenitors of time of evolution and not its children.

It is realized by an increasing number of physicists and chemists that the solution of the problem of irreversibility first formulated by Boltzmann about 125 years ago requires an extension of classical and quantum mechanics. The problem of irreversibility has many aspects. Non-equilibrium physics has shown the appearance of new spatiotemporal structures all associated with long-range correlations (the so-called "dissipative" structures). What is the dynamical origin of these correlations?

Simple examples of irreversible processes have been provided by the study of unstable quantum systems that began with Gamov's theory of alpha decay and Dirac's theory of spontaneous emission of radiation. But how to define decaying states (or unstable particles as well)? This is still a question of much debate. Irreversibility as manifest in the decay of unstable dynamical systems is only a special case of the problem of approach to equilibrium as described by processes such as diffusion, viscosity. . . .

It is interesting to recall Heisenberg's program at the origin of quantum theory. Observables should be associated with the eigenvalues of suitable operators. This idea has proved to be extraordinarily fruitful. But once we want to include irreversible processes, there appear difficulties. The operators corresponding to observables are supposed to be Hermitian operators. According to a well-known theorem, Hermitian operators have only real eigenvalues in the Hilbert space. But to include irreversibility, we need complex eigenvalues (the imaginary part being precisely related to irreversible processes). Therefore, the problem of irreversibility is basically connected to the problem of "extension" of the Hilbert space. This is a most fascinating field of modern mathematics, functional analysis.

There is at present no unanimity in choosing the way in which the extension of classical or quantum mechanics has to be performed to include the dynamical description of irreversible processes. There are a number of directions in which this problem is pursued. This volume of *Advances in Chemical Physics* presents a selection of these various directions.

It would go beyond the limits of this foreword to try to summarize the papers included in this volume. Let me only make a few remarks: the papers run from contributions close to experiment (C. A. Chatzidimitriou-Dreismann; P. Gaspard) to highly methematical presentations (E. Eisenberg and L. P. Horwitz). From the point of view of the subjects that are treated, we may distinguish three categories: classical systems with emphasis on chaos and dynamical instability (P. Gaspard); resonances and unstable quantum systems (E. J. Brandäs; V. V. and Vl. V. Kocharovsky and S. Tasaki; E. C. G. Sudarshan, Ch. B. Chiu, and G. Bhamathi); and the general problem of irreversibility (T. Petrosky and I. Prigogine; I. Antoniou and Z. Suchanecki).

We are at a most interesting moment in the history of science. Classical science emphasized equilibrium, stability, and time reversibility. Now we see instabilities, fluctuations, evolution on all levels of observations. This change of perspective requires new tools, new concepts. This volume invites the reader not to an enumeration of final achievements of contemporary science, but to an excursion to science in the making.

I. Prigogine

INTRODUCTION

Few of us can any longer keep up with the flood of scientific literature, even in specialized subfields. Any attempt to do more and be broadly educated with respect to a large domain of science has the appearance of tilting at windmills. Yet the synthesis of ideas drawn from different subjects into new, powerful, general concepts is as valuable as ever, and the desire to remain educated persists in all scientists. This series, *Advances in Chemical Physics*, is devoted to helping the reader obtain general information about a wide variety of topics in chemical physics, a field that we interpret very broadly. Our intent is to have experts present comprehensive analyses of subjects of interest and to encourage the expression of individual points of view. We hope that this approach to the presentation of an overview of a subject will both stimulate new research and serve as a personalized learning text for beginners in a field.

I. PRIGOGINE
STUART A. RICE

CONTENTS

ADVANCES IN CHEMICAL PHYSICS

VOLUME XCIX

THE LIOUVILLE SPACE EXTENSION OF QUANTUM MECHANICS

T. PETROSKY and I. PRIGOGINE

Center for Studies in Statistical Mechanics and Complex Systems, University of Texas, Austin, Texas, and International Solvay Institutes, Université Libre de Bruxelles, Brussels, Belgium

CONTENTS

* This is a review paper which also contains a number of new results. The reader should compare it with our paper "Poincaré Resonances and the Extension of Classical Mechanics" [1], which will be referred to as "*I*".

Advances in Chemical Physics, Volume XCIX, Edited by I. Prigogine and Stuart A. Rice.
ISBN 0-471-16526-3

ABSTRACT

Quantum mechanics can be formulated in terms of individual description (wave functions) or in terms of statistical ensembles whose time evolution is described by the Liouville–von Neumann equation. For classes of large nonintegrable Poincaré systems (LPS), the two descriptions are not equivalent. Thus, we must obtain a new formulation of quantum theory. This formulation is applicable to practically all dynamical systems studied in statistical mechanics that belong to this class. The first step is the extension of the Liouville operator L_H outside the Hilbert space to functions singular in their Fourier transforms. The eigenvalue problem for the Liouville operator L_H is then solved in this generalized function space. We obtain a complex, irreducible, spectral representation. Complex means that the eigenvalues are complex numbers, whose imaginary parts refer to various irreversible processes such as relaxation time and diffusion. Irreducible means that these representations cannot be implemented by trajectory or wave-function theory. As a result, the dynamical group splits into two semigroups. Moreover, the laws of dynamics take a new form because they have to be formulated on the statistical level. They express "possibilities" and not "certitudes." Several applications are presented. Our theoretical predictions have been verified by extensive computer simulations.

I. INTRODUCTION

Quantum mechanics has been remarkably successful in all its predictions. Yet discussions about its meaning and scope are as lively as ever. The basic assumption in quantum mechanics is that every problem can be solved at the level of probability amplitudes. But in order to attribute well-defined properties to matter, we need probabilities. Closely related to this issue is the basic duality we find in the formulation of quantum mechanics based on either the Schrödinger equation, which is time reversible and deterministic, or the reduction or collapse of the wave function, which leads from a single wave function to a "mixture" described by a density matrix. It seems that we would need our measurements to go from "potentialities" to "actualities" [2]. But, as emphasized by Bohr and Rosenfeld [3], measurements are *irreversible* processes, be it at the level of the apparatus or of our sensory mechanisms. This difficulty is common to classical and quantum mechanics and the solution we found is common to both [4,5].

It has been well known since the pioneering work of Gibbs and Einstein that dynamics can be formulated from two points of view—"individual" description in terms of wave functions Ψ (or trajectories for classical systems) or *statistical* description in terms of density matrices. It was always assumed that these two ways of description are equivalent, which was easy to verify for integrable systems. We have shown (e.g., see [1]) that this equivalence is destroyed for classes of systems called "large Poincaré systems" (LPS), which are nonintegrable systems (in the sense of Poincaré) in which the frequencies (or energies) vary continuously [6]. Large Poincaré systems play a fundamental role in both classical and quantum mechanics (examples are scattering, radiation damping, nonlinear field theory, and all the problems studied in nonequilibrium statistical mechanics). On the *statistical level*, new solutions have appeared which cannot be implemented by wave functions (or trajectories). The laws of dynamics take then on a new meaning, expressing "possibilities" and not "certitudes." Moreover, they incorporate time-symmetry breaking as they lead to a semigroup description.

The new statistical formulation of dynamics requires an appropriate mathematical formalism. Quantum mechanics was traditionally associated with Hilbert space. Because the evolution operators (the Hamiltonian for wave functions and the Liouvillian for density matrices) are hermitian operators, irreversibility requires the introduction of more general function spaces (so-called rigged Hilbert spaces or Gelfand spaces [7–10]) in which hermitian operators may have complex eigenvalues.

We have discussed the spectral decomposition of the "evolution" operator in these spaces in publications dealing with chaotic maps and classical Hamiltonian dynamics [1,11–22]. This paper is intended to give a self-contained presentation for quantum mechanics.

The physical reason for going beyond the Hilbert space is the same as in classical dynamics. When we expect irreversible processes to occur (such as during the approach to equilibrium), we deal with delocalized density matrices. We have then *persistent* interactions. As shown in Section IV, this leads to singular-density matrices which do not belong to the Hilbert space. This is already evident when we consider canonical equilibrium distribution. Independently of any "philosophical" considerations associated with the measurement process, it is clear that the inclusion of the approach to equilibrium requires an extension of the formulation of quantum mechanics. Historically, the discovery of quantum mechanics is closely related to the Rydberg–Ritz principle, according to which each spectral line is the *difference* between two energy levels. But this fundamental property does not extend to the imaginary part of the frequency related to relaxation processes. The characteristic time scales in the approach to equilibrium of quantum systems (e.g., of radiation interacting with matter) are not differences between two terms.

The need to go outside the Hilbert space in quantum mechanics for a continuous spectrum was recognized some years ago by various physicists and mathematicians, including Böhm and Gadella [8] (and in a slightly different form by Kumičák and Brändas [23] and also Sudarshan *et al.* [24]). The extension of the Hilbert space was based on the fundamental work of Schwartz [25] and Gelfand [26]. The physical motivation was to include decaying states as observed in the spectrum of the Hamiltonian (hence the name "Gamov vectors" used by Böhm and Gadella [8]). We present a short overview of the Gamov-state treatment of the simple Friedrichs model in Appendix H.

However, this generalization does not solve the "quantum paradox" associated with the collapse of the wave function and the problem of the approach to equilibrium (in the process of decay of an atomic state of the energy is only transferred to the electromagnetic field). To solve these problems we have to turn to the Liouville space and show that when we expect thermodynamic behavior (associated with the "thermodynamic limit" discussed in the text), we obtain new spectral decomposition which leads to *semigroups including irreversible processes.*

Situations in which our generalized formulation of quantum theory applies can easily be compared with numerical simulations. All of our theoretical predictions have been verified in this manner. Some of these predictions are very unexpected, such as the extension of scattering theory to finite times (in contrast with the usual *S*-matrix approach valid for asymptotic times). Our approach led to the discovery of new secular effects (see Appendix G and [27]). Moreover, we recovered all known results in nonequilibrium statistical mechanics (Pauli equation, generalized master equations, etc.).

Contrary to many existing proposals for reaching a better understanding of quantum theory, our approach is purely dynamical, carefully avoiding the introduction of speculative elements such as the many-world approach or environmental effects [28]. In short, we show that irreversibility is not due to our approximations but requires an extension of the basic laws of both classical and quantum dynamics. Traditional quantum mechanics is associated with operator calculus in Hilbert space. We have extended the operator calculus to more general spaces.

In Section II we give a short overview of our presentation. In the traditional formulations of both classical and quantum mechanics, the basic laws lead to dynamical groups in which past and future play symmetric roles. In our formulations, we obtain dynamical *semigroups*. The group description corresponds then only to an approximation which is valid, for example, when we can isolate a finite number of particles inside a large Poincaré system. Because our description includes irreversible processes, we can now describe measurement in dynamical terms. The quantum superposition principle is broken and the transition from "potentiality" to "actuality" is realized through the transition of a wave function (a "pure state") to a mixture described by the density matrix.

Starting with the group description we can derive two different semigroups, one in which equilibrium is reached in our future, the other in which it is reached in our past. How does it happen that we always find the same semigroup? Why is the world asymmetric in time? The universe is a nonequilibrium system; this is an experimental fact (in an equilibrium universe, life would be impossible). Now our approach shows that nonequilibrium systems are described by singular-density matrices. As we show in Section XIII, this has an important consequence. Once the system is described by one semigroup, it is *outside the range of the other semigroup* (the inversion of the semigroup leads to divergences due to products of distributions; see Section XIV). Even if we assume that initially, that is, at "the big bang," the density matrix is in the domain of the two semigroups, this is no longer true once dissipative processes begin to play. In more physical terms, we cannot proceed to "velocity (or time) inversions" for the system as a whole. However, this does not eliminate velocity inversion for a finite number of particles embedded in the large system.

The nature of the initial conditions for our universe is related to the formulation of the early evolutionary stages of our universe. If we assume that the birth of our universe corresponds to a kind of phase transition from the quantum vacuum in which we have only virtual particles, the matter is itself the result of irreversible processes. But these considerations are outside the aim of this chapter. We hope to investigate this problem in a separate paper.

II. OVERVIEW

We consider mainly systems of interacting particles and briefly mention anharmonic lattices. We concentrate our study on $\mathcal{N}$-body systems with Hamiltonians of the form (with short-range interactions)

$$H(q, p) = H_0 + \lambda V = \sum_{i=1}^{\mathcal{N}} \frac{\mathbf{p}_i^2}{2m_i} + \lambda \sum_{j>i}^{\mathcal{N}} V(|\mathbf{q}_i - \mathbf{q}_j|) \tag{2.1}$$

where λ is the coupling constant; q is an $\mathcal{N}$-component vector, that is, $q \equiv (\mathbf{q}_1, \mathbf{q}_2, \ldots, \mathbf{q}_{\mathcal{N}})$ with three-dimensional vectors $\mathbf{q}_j$; and p is an $\mathcal{N}$-component momentum operator, that is, $p \equiv (\mathbf{p}_1, \mathbf{p}_2, \ldots, \mathbf{p}_{\mathcal{N}})$ with $\mathbf{p}_j = -i\hbar\partial/\partial\mathbf{q}_j$. In this chapter, we assume for simplicity that the system consists of distinguishable particles, so that we do not consider quantum statistics. The extension to nondistinguishable particles (Fermions or Bosons) can be performed in terms of the second quantization formalism (see Appendix L and also [29–31]). The system is enclosed in a large box with volume L^3. These are the systems studied in equilibrium and nonequilibrium statistical mechanics. In the large limit $L^3 \to \infty$, we obtain LPS. The number of particles $\mathcal{N}$ may be finite or infinite. We are especially interested in the *thermodynamic limit*[1]

$$\mathcal{N} \to \infty \quad \text{and} \quad L^3 \to \infty \qquad \text{with } c = \mathcal{N}/L^3 = \text{finite} \tag{2.2}$$

The statistical description in quantum mechanics is expressed by the Liouville–von Neumann equation for the density matrices [4,29,32]:

$$i\frac{\partial}{\partial t}\rho(t) = L_H\,\rho(t) \tag{2.3}$$

Here

$$L_H\,\rho \equiv \frac{1}{\hbar}(H\rho - \rho H) \tag{2.4}$$

is the commutater of ρ with the Hamiltonian H.

In Section III we present a survey of the Liouville formalism. To emphasize the parallelism between classical and quantum mechanics we also introduce the well-known "Wigner representation."

We show how to extend the Liouville–von Neumann operator for LPS to a class of functions outside the Hilbert space. The class of functions of interest to us has a simple physical meaning because it includes equilibrium distributions (which are functions of the Hamiltonian). These functions are characterized by well-defined singularities in the momentum representation. We have to distinguish between ensembles "localized" in space, such as wave packets, and "nonlocal" ensembles. Associated with a finite number of particles, localized distributions ρ describe *transient interactions* (free "in" and "out" states), as studied in S-matrix theory. In contrast, nonlocal ensembles describe *persistent interactions*, as studied in statistical mechan-

[1] The characteristic feature of thermodynamics is the existence of extensive and intensive variables. We shall see that this imposes strict conditions on the thermodynamic limit [Eq. (2.2)].

ics, characterized by singularities in the momentum representation (see Section III).

Our extended spectral representation of L_H for singular functions is presented in Section IV. It has quite remarkable features, exhibiting "non-Schrödinger" behavior. There appear indeed *diffusive effects* associated with collision operators Θ of the Pauli type, familiar from phenomenological theories. These contributions result from the coupling of dynamical "events" through Poincaré resonances. The eigenvalues of L_H in this extended functional space are complex, implying time-symmetry breaking, and the eigenfunctions are not implementable by wave functions. Therefore, the basic object of our approach is associated with density matrices.

The new non-Schrödinger effects lead to the construction of nonunitary transformation operators Λ, which intertwine L_H and the collision operators Θ. The collision operators are dissipative operators. This generalizes the unitary transformations that lead to integrable systems from L_H to L_0, the Liouvillian corresponding to H_0 (Section V). The complex spectral representation also leads to subdynamics, which corresponds to an extension of the kinetic theory to all correlation spaces [33–45]. In our previous work, subdynamics has been constructed using an ansatz for the analytic continuation (the so-called $i\epsilon$ rule) (e.g., [45]). We may now derive subdynamics directly from the complex spectral representation. Using our nonunitary transformation, we can transform the Liouville–von Neumann equation for ρ into an infinite set of "kinetic equations" (see Section VI). We also obtain a new formulation of the Heisenberg equations of motion for the evolution of observables which makes explicit the role of dissipative processes. As the result of breaking the time symmetry, we can easily construct Lyapounov functions, which are dynamical analogies of the "$\mathscr{H}$-functions" usually derived through phenomenological assumptions (Section VII). The basic new element is that the $\mathscr{H}$-functions are constructed using the nonunitary transformation Λ of the density matrix which already has a broken time symmetry. The existence of $\mathscr{H}$-functions, which are microscopic models of entropy, is the direct consequence of this broken time symmetry. For weak coupling systems, we recover the well-known Boltzmann–Gibbs expressions of the $\mathscr{H}$-functions.

Our theory leads to a simple description of the irreversible approach to equilibrium as described in terms of the *flow of correlations*. In time, this flow leads to the building up of correlations involving an increasing number of particles. We show that causality is preserved in this process.

Nonequilibrium systems have two fundamental properties, as shown in Section XI. First, Poincaré resonances lead to *long-range correlations*. This is in agreement with macroscopic nonequilibrium theory, which shows that nonequilibrium leads to new forms of coherence [5]. Correlations are also associated with singular functions, an essential point in the stability of the dynamical semigroup (see also Section XIV).

It is very interesting to compare our theory to the well-known scattering theory based on the Lippmann–Schwinger equations. We show that the intertwining relations between L_H and Θ lead to a nonlinear extension of the Lippmann–Schwinger type equations, well known from quantum scattering theory (see Section IX). When dissipative effects are neglected, the nonlinear terms vanish and we come back to the usual Lippmann–Schwinger equations.

Still, our equations differ even in this case from the Lippmann–Schwinger equation by our analytic continuation of the propagators. There appears to be a degeneracy for LPS. The existence of Poincaré resonances[2] even in the nondissipative limit leads to a new spectral decomposition of the Liouvillian in addition to the usual spectral representations in terms of "advanced" or "retarded" solutions. It is this spectral representation that can be extended to include dissipative effects for LPS, whereas the usual spectral representation would lead to divergences.

Our approach reduces to the usual theory in simple situations. Therefore, we have to discuss the conditions under which the non-Schrödinger effects that appear in our spectral representation of L_H can be observed. This is done in Sections VIII–XII. The appearance of our new effects essentially depends on the type of density matrices (associated or not with singularities in the momentum representation) and on the number of particles ($\mathcal{N}$ finite or $\mathcal{N} \to \infty$ as in the thermodynamic limit).

For $\mathcal{N}$ finite and localized regular-density matrices all dissipative effects disappear. These systems are integrable while presenting Poincaré resonances. This is quite remarkable because these systems are "not integrable" in traditional Poincaré terminology. The situation changes dramatically when we consider persistent interactions associated with density matrices which are singular in their momentum representation (Sections X and XI). Of special importance is the thermodynamic limit. When we apply our spectral representation to this class of density matrices, we recover all results derived in nonequilibrium statistical mechanics [4] (such as the Pauli master equation, quantum Boltzmann equation, generalized master equations, etc.). This shows that dissipative processes are part of the exact dynamical description when we consider LPS and extend the functional space to include density matrices which are singular in their momentum representation.

[2] Here, we have in mind the Poincaré resonances expressed by the frequency (or "energy") conservation between the initial and final states such as $\delta(\omega_i - \omega_f)$ in the usual S-matrix theory. The Poincaré resonances appear already for repulsive interactions. These resonances are not related to "resonance poles" associated with the so-called resonance scattering.

To illustrate our theory, we present a number of examples, including persistent potential scattering, three-body scattering, the perfect Lorentz gas, the Friedrichs model (with both a regular and a singular distribution of fields) for matter–field interacting systems, and anharmonic lattices as a model of field–field interacting systems. The examples are presented in the appendix. We outline only the situations we considered and the results of our theory. Details of calculations have been published in separate papers or are being prepared for publication.

In contrast with many alternative "derivations" of existing quantum theory, our approach leads to a new formulation of quantum physics which avoids the basic difficulties that arise from the need to introduce "the collapse" of the wave function without being able to associate to it a realistic, dynamical formulation. Moreover, it extends the scope of quantum theory to include irreversible processes, which form an essential part of science, so as to include the chemistry as well as biology.

III. THE LIOUVILLIAN FORMALISM

We denote the eigenstates of H_0 by $|p\rangle$:

$$H_0|p\rangle = \hbar\omega_p|p\rangle \tag{3.1}$$

where

$$\hbar\omega_p \equiv \sum_{j=1}^{\mathcal{N}} \hbar\omega_{p_j} = \sum_{j=1} \frac{\mathbf{p}_j^2}{2m_j} \tag{3.2}$$

The system is enclosed in a large box of volume L^3. We impose the usual periodic boundary conditions. Then the spectrum of the unperturbed momentum is *discrete* and is given by (for $\Delta k \equiv 2\pi/L$ and with integer vector $\mathbf{n}_j$)

$$\mathbf{p}_j = \mathbf{n}_j \hbar\Delta k \tag{3.3}$$

We have a complete set of normalized eigenstates of H_0 for (3.1):

$$\sum_{p^{\mathcal{N}}} |p\rangle\langle p| = 1, \qquad \langle p|p'\rangle = \delta^{kr}(p - p') \equiv \prod_{j=1}^{N} \delta^{kr}(\mathbf{p}_j - \mathbf{p}_j') \tag{3.4}$$

where $\delta^{kr}(\mathbf{p} - \mathbf{p}') = \delta_{p_x, p_x'}\delta_{p_y, p_y'}\delta_{p_z, p_z'}$ is a product of Kronecker's delta. We use the term "wave-function space" for the space spanned by the set of these eigenstates, to distinguish it from the Liouville space introduced later. In the limit of large volumes $L^3 \to \infty$, we obtain a *continuous* spectrum and a LPS. In this limit we have

$$\frac{1}{\Omega_\hbar} \sum_{\mathbf{p}_j} \to \int d\mathbf{p}_j, \qquad \delta_{\Omega_\hbar}(\mathbf{p}_j) \equiv \Omega_\hbar\, \delta^{kr}(\mathbf{p}_j) \to \delta(\mathbf{p}_j) \tag{3.5}$$

Here

$$\Omega_\hbar \equiv \Omega/\hbar^3 \qquad \text{with } \Omega \equiv (L/2\pi)^3 \tag{3.6}$$

and $\delta_{\Omega_\hbar}(\mathbf{p}_j)$ is the "weighted" Kronecker's delta and $\delta(\mathbf{p}) = \delta(p_x)\delta(p_y)\delta(p_z)$ is a product of Dirac's delta function.

Associated with the box normalization, there exists a characteristic time scale $t_B \sim m/(|\mathbf{p}|L)$ which is the crossing time from one side of the box to the other with a typical momentum $\mathbf{p}$ of particles. Our formulation for LPS is applicable in the large volume limit for time scales such that

$$t \ll t_B \tag{3.7}$$

The evolution of the system is governed by the Liouville–von Neumann equation (2.3) for the density matrix ρ. The formal solution of the Liouville equation is

$$\rho(t) = \mathcal{U}(t)\rho(0) \tag{3.8}$$

with

$$\mathcal{U}(t) = e^{-iL_H t} \tag{3.9}$$

where $\mathcal{U}(t)$ is the evolution operator.

We may consider situations where the density matrices are localized in the configuration space. An example is a pure state $\rho = |\Psi\rangle\langle\Psi|$ which consists of a localized wave packet in a few-body system. This is the case where the usual S-matrix theory applies. The interaction is *transient*. As the wave function Ψ are localized in configuration space, there is a well-defined norm for the wave function:

$$\|\Psi\| = \left(\int d^{\mathcal{N}} q\, |\Psi(q)|^2 \right)^{1/2} < \infty \tag{3.10}$$

For this situation, the Liouville–von Neumann equation does not introduce any new features. If we can integrate the Schrödinger equation, we can solve the Liouville–von Neumann equation and vice versa.

Usually, one equips the "Liouville space" with a Hilbert space structure. In this space, a scalar product of the linear operators A and B acting on wave functions are defined as a Schmidt inner-product:

$$\langle\langle A \mid B \rangle\rangle \equiv \mathrm{Tr}(A^{+}B) \tag{3.11}$$

and their Hilbert norms by

$$\|A\| \equiv \sqrt{\langle\langle A \mid A \rangle\rangle} \tag{3.12}$$

where A^{+} is hermitian conjugate operator of A in the wave-function space. We have introduced Dirac's "bra" and "ket" notations, which are analogous to the wave-function space. The Hilbert norm for the example above of the density matrix associated with the wave pocket is given by

$$\|\rho\| = \int d^{\mathcal{N}}q \, |\Psi(q)|^2 < \infty \tag{3.13}$$

In the Liouville space one can introduce operators acting on density matrices. We call these operators "superoperators" in case it is necessary to emphasize the difference from operators in the wave-function space. We can then introduce the *adjoint* superoperator $\mathcal{Q}^{\dagger}$ of the superoperator $\mathcal{Q}$ through the relation

$$\langle\langle A_\nu \mid \mathcal{Q}^{\dagger} \mid A_\mu \rangle\rangle = \langle\langle A_\mu \mid \mathcal{Q} \mid A_\nu \rangle\rangle^{c.c.} \tag{3.14}$$

where *c.c.* denotes the complex conjugation and $\{A_\nu\}$ is a complete orthogonal basis of the Liouville space [see Eq. (3.28) for an example of the basis]. Here we have introduced the notation "†" and distinguished the adjoint operation denoted by "+", such as defined in Eq. (3.11) in the wave-function space. We have, as usual,

$$(|A\rangle\rangle\langle\langle B|)^{\dagger} = |B\rangle\rangle\langle\langle A| \tag{3.15}$$

We can then define hermitian superoperators that satisfy $\mathcal{Q}^{\dagger} = \mathcal{Q}$ as well as unitary superoperators. The Liouvillian L_H is an example of the hermitian superoperators and $\mathcal{U}(t)$ is unitary in the Liouville space. Thus, as long as we remain in Hilbert space, the eigenvalues w of L_H are real and the eigenvalues $\exp[-iwt]$ of $\mathcal{U}(t)$ are of modulo one. In short, the density matrix oscillates in time and there is no place for irreversible processes. To obtain irreversible processes associated to complex eigenvalues of L_H, we need to go out of the Hilbert space (this is a necessary condition).

In this Chapter we consider classes of density matrices outside the Hilbert space. The simplest example is a plane wave with a momentum p_0 which is normalized by the Dirac delta function

$$\rho = |p_0)(p_0| \tag{3.16}$$

where we introduce the new notation

$$|p) \equiv \Omega_\hbar^{\mathcal{N}/2}|p\rangle \tag{3.17}$$

which satisfies [cf. Eq. (3.4)]

$$\frac{1}{\Omega_\hbar}\sum_{p^{\mathcal{N}}}|p)(p| = 1, \qquad (p|p') = \delta_{\Omega_\hbar}(p - p') \tag{3.18}$$

Then, Eq. (3.16) leads to

$$\rho(p, p') \equiv (p|\rho|p') = \delta_{\Omega_\hbar}(p - p')\delta_{\Omega_\hbar}(p - p_0) \tag{3.19}$$

The Hilbert norm is then proportional to $[\delta_{\Omega_\hbar}(0)]^{2\mathcal{N}} = \Omega_\hbar^{2\mathcal{N}}$ which diverges in the limit of large volumes.

Among superoperators, we have factorizable superoperators $A \times B$ defined by

$$(A \times B)\rho = A\rho B \tag{3.20}$$

where A and B are linear operators in the wave-function space. We have [see Eq. (3.14)]

$$\begin{aligned}\langle\langle f|(A \times B)^\dagger|g\rangle\rangle &= \langle\langle g|(A \times B)|f\rangle\rangle^{c.c.} = [\mathrm{Tr}(g^+ A f B)]^{c.c.} \\ &= \mathrm{Tr}(B^+ f^+ A^+ g) = \mathrm{Tr}(f^+ A^+ g B^+) \\ &= \langle\langle f|(A^+ \times B^+)|g\rangle\rangle \end{aligned}\tag{3.21}$$

Hence we have

$$(A \times B)^\dagger = A^+ \times B^+ \tag{3.22}$$

The Liouvillian is then written as

$$L_H = \frac{1}{\hbar}(H \times 1 - 1 \times H) \tag{3.23}$$

which is indeed a hermitian operator. Corresponding to Eq. (2.1), we can decompose the Liouvillian into an unperturbed part $L_0 \equiv L_{H_0}$ and an interaction L_V:

$$L_H = L_0 + \lambda L_V \tag{3.24}$$

We denote dyadic operators $|p\rangle\langle p'|$ generated by the eigenstates $|p\rangle$ and $\langle p'|$ of H_0 by

$$|p; p'\rangle\!\rangle \equiv |p\rangle\langle p'| \tag{3.25}$$

They are eigenstates of L_0,

$$L_0|p; p'\rangle\!\rangle = \frac{1}{\hbar}[H_0|p\rangle\langle p'| - |p\rangle\langle p'|H_0] = w_{pp'}|p; p'\rangle\!\rangle \tag{3.26}$$

where

$$w_{pp'} \equiv \omega_p - \omega_{p'} \tag{3.27}$$

They form a complete orthonornal set for the Liouville space:

$$\begin{gathered}\sum_{p\mathcal{N}}\sum_{p'\mathcal{N}}|p; p'\rangle\!\rangle\langle\!\langle p; p'| = \sum_{p\mathcal{N}}\sum_{p'\mathcal{N}}|p\rangle\langle p| \times |p'\rangle\langle p'| = 1\\ \langle\!\langle p; p'|p''; p'''\rangle\!\rangle = \delta^{kr}(p - p'')\delta^{kr}(p''' - p')\end{gathered} \tag{3.28}$$

In this notation, we have a simple expression for matrix elements of an operator A acting on wave functions:

$$A_{pp'} = \langle p|A|p'\rangle = \langle\!\langle p; p'|A\rangle\!\rangle \tag{3.29}$$

The matrix element of the perturbed Liouvillian is then given by

$$\langle\!\langle p; p'|L_V|p''; p'''\rangle\!\rangle = \frac{1}{\hbar}[V_{pp''}\delta^{kr}(p''' - p') - \delta^{kr}(p - p'')V_{p'''p'}] \tag{3.30}$$

To emphasize the parallelism between classical and quantum mechanics, it is convenient to use the "Wigner function representation" for density matrices. The Wigner representation of linear operators A is defined by

$$A^W(Q, P) \equiv \sum_{k^{\mathcal{N}}} e^{ik \cdot Q} \langle\langle k, P | A \rangle\rangle = \sum_{k^{\mathcal{N}}} e^{ik \cdot Q} \left\langle P + \frac{\hbar}{2} k \middle| A \middle| P - \frac{\hbar}{2} k \right\rangle \tag{3.31}$$

where the "wave vector" k is given as $k = (\mathbf{k}_1, \ldots, \mathbf{k}_{\mathcal{N}})$ with $\mathbf{k}_j = \mathbf{n}_j \Delta k$ [see Eq. (3.3)], and $k \cdot Q \equiv \mathbf{k}_1 \cdot \mathbf{Q}_1 + \cdots + \mathbf{k}_{\mathcal{N}} \cdot \mathbf{Q}_{\mathcal{N}}$. Here we have introduced new notation (note the comma instead of the semicolon):

$$| k, P \rangle\rangle \equiv | p; p' \rangle\rangle = \left| P + \frac{\hbar}{2} k; P - \frac{\hbar}{2} k \right\rangle\rangle \tag{3.32}$$

where

$$P \equiv \tfrac{1}{2}(p + p'), \qquad \hbar k \equiv p - p' \tag{3.33}$$

In the Wigner representation, diagonal elements of the density matrices in the momentum representation correspond to components with vanishing wave vector $k = 0$.

The Wigner representation of the Hamiltoniam (2.1) is then given by

$$H^W(Q, P) = \sum_{i=1}^{\mathcal{N}} \frac{\mathbf{P}_i^2}{2m_i} + \lambda \sum_{j>i}^{\mathcal{N}} V(|\mathbf{Q}_i - \mathbf{Q}_j|) \tag{3.34}$$

where $\mathbf{P}_i$ and $\mathbf{Q}_i$ are c-number, and

$$V(|\mathbf{Q}|) = \frac{1}{\Omega} \sum_{\mathbf{k}} V_{|\mathbf{k}|}\, e^{i\mathbf{k} \cdot \mathbf{Q}} \tag{3.35}$$

We have put

$$\left\langle \mathbf{P}_i + \frac{\hbar}{2}\mathbf{k}, \mathbf{P}_j - \frac{\hbar}{2}\mathbf{k}, \{P\}^{\mathcal{N}-2} \middle| V \middle| \mathbf{P}_i - \frac{\hbar}{2}\mathbf{k}, \mathbf{P}_j + \frac{\hbar}{2}\mathbf{k}, \{P\}^{\mathcal{N}-2} \right\rangle = \frac{1}{\Omega} V_{|\mathbf{k}|} \tag{3.36}$$

where $\{P\}^{\mathcal{N}-2}$ denote a set of momenta excluding the momenta of particles i and j. We assume that $V_{|\mathbf{k}|}$ does not depend on L^3 for the large-volume limit. Moreover, we assume that $V_0 = 0$, that is,[3]

$$\int d\mathbf{Q}V(|\mathbf{Q}|) = 0 \tag{3.37}$$

For the eigenstates of L_0 we have

$$L_0 |k, P\rangle\rangle = (k \cdot v) |k, P\rangle\rangle \tag{3.38}$$

where

$$(k \cdot v) = \mathbf{k}_1 \cdot \mathbf{v}_1 + \cdots + \mathbf{k}_{\mathcal{N}} \cdot \mathbf{v}_{\mathcal{N}} \tag{3.39}$$

and $v \equiv (\mathbf{P}_1/m_1, \ldots, \mathbf{P}_{\mathcal{N}}/m_{\mathcal{N}})$. As in the classical case the appropriate variables are the "wave vectors" k and the momenta P.

Corresponding to Eq. (3.28), we have

$$\sum_{k\mathcal{N}} \sum_{P\mathcal{N}} |k, P\rangle\rangle\langle\langle k, P| = 1, \qquad \langle\langle k, P | k', P'\rangle\rangle = \delta^{kr}(k - k')\delta^{kr}(P - P') \tag{3.40}$$

We also introduce states $|Q, P\rangle\rangle$ representation of the eigenfunctions $|k, P\rangle\rangle$. As in classical dynamics

$$\langle\langle k, P' | Q, P\rangle\rangle = L^{-3\mathcal{N}/2} e^{-ik \cdot Q} \delta_{\Omega_\hbar}(P - P') \tag{3.41}$$

We have

$$\begin{aligned}\langle\langle Q, P | Q', P'\rangle\rangle &= \sum_{k\mathcal{N}} \sum_{P''\mathcal{N}} \langle\langle Q, P | k, P''\rangle\rangle\langle\langle k, P'' | Q', P'\rangle\rangle \\ &= \delta(Q - Q')\delta_{\Omega_\hbar}(P - P')\end{aligned} \tag{3.42}$$

and

$$\frac{1}{\Omega_\hbar^{\mathcal{N}}} \sum_{P\mathcal{N}} \int d^{\mathcal{N}}Q |Q, P\rangle\rangle\langle\langle Q, P| = 1 \tag{3.43}$$

Note the difference in the normalization on the "momentum variables" P using the weighted Kronecker delta in Eq. (3.42) from that with the usual

[3] If this is not the case, we redefine the unperturbed Hamiltonian by incorporating the element V_0 into H_0.

Kronecker delta in Eq. (3.40). With this new normalization, we recover the same expressions for Eqs. (3.42) and (3.43) as in classical dynamics (see *I*).

In terms of states $|Q, P\rangle\rangle$, we can write the Wigner representation of the linear operators A as

$$A^W(Q, P) = \langle\langle Q, P | A \rangle\rangle \tag{3.44}$$

For density matrices, we have (for $\Omega \to \infty$):

$$|\rho(t)\rangle\rangle = \int d^{\mathcal{N}}Q \int d^{\mathcal{N}}P \, |Q, P\rangle\rangle \rho^W(Q, P, t) \tag{3.45}$$

We assume that the distribution function vanishes quickly enough for large values of "momentum:"

$$\lim_{|P| \to \infty} \rho^W(Q, P) \to 0 \tag{3.46}$$

However, we do not in general impose a similar condition for the "coordinate variables" Q because we are interested not only in density matrices local in space, but also in nonlocal density matrices as considered in typical situations in statistical mechanics.

Similarly to (3.45), we have for observables M (for $\Omega \to \infty$):

$$\langle\langle M | = \int d^{\mathcal{N}}Q \int d^{\mathcal{N}}P M^W(Q, P) \langle\langle Q, P |$$

The evolution of the observables is given by

$$\langle\langle M(t) | = \langle\langle M | \mathcal{U}(t) \tag{3.47}$$

They satisfy Heisenberg's equation of motion [cf. Eq. (2.3)]:

$$i \frac{\partial}{\partial t} | M(t) \rangle\rangle = -L_H | M(t) \rangle\rangle \tag{3.48}$$

Then the expectation value of M is given by

$$\langle M \rangle_t = \langle\langle M(0) | \rho(t) \rangle\rangle = \langle\langle M(t) | \rho(0) \rangle\rangle = \int d^{\mathcal{N}}Q \int d^{\mathcal{N}}P M^W(Q, P) \rho^W(Q, P, t) \tag{3.49}$$

We therefore obtain a similar expression for the expectation value as in classical dynamics (see *I*).

In the Wigner representation the nonvanishing matrix elements of the interaction Liouvillian are [see Eqs. (3.30) and (3.36)]

$$\langle\!\langle \mathbf{k}'_j, \mathbf{k}'_n, \{k\}^{N-2}, P' \,|\, L_V \,|\, \mathbf{k}_j, \mathbf{k}_n, \{k\}^{N-2}, P \rangle\!\rangle$$
$$= -\frac{1}{\Omega} \sum_{\mathbf{k}} \delta^{kr}(\mathbf{k}_j - \mathbf{k}'_j + \mathbf{k}) \delta^{kr}(\mathbf{k}_n - \mathbf{k}'_n - \mathbf{k}) V_{|\mathbf{k}|} [\partial_{jn}^{\hbar \mathbf{k}/2} \delta^{kr}(P - P')] \quad (3.50)$$

with

$$\partial_{jn}^{\mathbf{a}} \equiv \frac{1}{\hbar} (\eta_j^{\mathbf{a}} \eta_n^{-\mathbf{a}} - \eta_j^{-\mathbf{a}} \eta_n^{\mathbf{a}}) \quad (3.51)$$

and $\eta_j^{\mathbf{a}}$ is the "displacement operator" defined by

$$\eta_j^{\mathbf{a}} f(\mathbf{P}_j) = f(\mathbf{P}_j + \mathbf{a}) \quad (3.52)$$

We have conservation of momentum:

$$\mathbf{P}'_j + \mathbf{P}'_n = \mathbf{P}_j + \mathbf{P}_n \quad (3.53)$$

In addition, we have the conservation law of wave vectors:

$$\mathbf{k}'_j + \mathbf{k}'_n = \mathbf{k}_j + \mathbf{k}_n \quad (3.54)$$

Except the two indices $\mathbf{k}_j$ and $\mathbf{k}_n$, all indices k in Eq. (3.50) keep their values. These are direct consequences of the assumption of binary interactions and of invariance in respect to translation.

In the large-volume limit, the displacement operator can be expressed as

$$\eta_j^{\mathbf{a}} = e^{\mathbf{a} \cdot \partial/\partial \mathbf{P}_j} \quad (3.55)$$

and

$$\partial_{jn}^{\hbar \mathbf{k}/2} = \frac{1}{\hbar} (e^{\hbar \mathbf{k} \cdot \mathbf{d}_{jn}/2} - e^{-\hbar \mathbf{k} \cdot \mathbf{d}_{jn}/2}) \quad (3.56)$$

where

$$\mathbf{d}_{jn} \equiv \frac{\partial}{\partial \mathbf{P}_j} - \frac{\partial}{\partial \mathbf{P}_n} \tag{3.57}$$

We note that

$$\int d\mathbf{P}_j\, g(\mathbf{P}_j) \eta_j^{\mathbf{a}} f(\mathbf{P}_j) = \int d\mathbf{P}_j\, f(\mathbf{P}_j) \eta_j^{-\mathbf{a}} g(\mathbf{P}_j) \tag{3.58}$$

where we have assumed that $g(\mathbf{P}_j) \to 0$ and $f(\mathbf{P}_j) \to 0$ for $|\mathbf{P}_j| \to \infty$.

We have (for $\hbar \to 0$)

$$\partial_{jn}^{\hbar\mathbf{k}/2} \to \mathbf{k} \cdot \mathbf{d}_{jn} \tag{3.59}$$

Therefore, the only difference between the matrix element (3.50) and the corresponding classical expression is the replacement of the classical differential operator by a displacement operator (see *I*). This corresponds to the fact that energy transfers are infinitesimal in the classical limit while they are finite in the quantum case.

Acting with the displacement operators in Eq. (3.50), one can easily verify the relation

$$\int d^{\mathcal{N}}P \langle\langle 0, P | L_V | \rho \rangle\rangle = \int d^{\mathcal{N}}P \sum_{j>i}^{\mathcal{N}} \frac{1}{\Omega} \sum_{\mathbf{k}} V_{|\mathbf{k}|} \left[\rho\left(\mathbf{P}_i + \frac{\mathbf{k}}{2}, \mathbf{P}_j - \frac{\mathbf{k}}{2}, \{P\}^{\mathcal{N}-2} \right) - \rho\left(\mathbf{P}_i - \frac{\mathbf{k}}{2}, \mathbf{P}_j + \frac{\mathbf{k}}{2}, \{P\}^{\mathcal{N}-2} \right) \right] = 0 \tag{3.60}$$

while (for $\partial f / \partial P_j \neq 0$ where $1 \leq j \leq s \leq \mathcal{N}$)

$$\int d^{\mathcal{N}}P\, f(\mathbf{P}_1, \ldots, \mathbf{P}_s) \langle\langle 0, P | L_V | \rho \rangle\rangle \neq 0 \tag{3.61}$$

The nonvanishing contribution comes only from the terms "connected" to the labeled particles 1 to *s*. All "disconnected" terms vanish. This is a general property of reduced quantities associated with the labeled particles 1 to *s*. Thanks to this property, the reduced quantities give finite contribution in the thermodynamic limit [see Eq. (5.8) and [4]].

So far, we have considered the Hamiltonian (2.1). Using a similar formulation, we can include potential scattering of a single particle corresponding to the interaction $V = V(|\mathbf{q}_1|)$. The matrix element (3.50) then reduces to

$$\langle\langle \mathbf{k}'_1, \mathbf{P}'_1 | L_V | \mathbf{k}_1, \mathbf{P}_1 \rangle\rangle = -\frac{1}{\Omega} \sum_{\mathbf{k}} \delta^{kr}(\mathbf{k}_1 - \mathbf{k}'_1 + \mathbf{k}) V_{|\mathbf{k}|} [\eta_1^{\hbar \mathbf{k}/2} \delta^{kr}(\mathbf{P}_1 - \mathbf{P}'_1)] \tag{3.62}$$

Before going to the next section, let us make a few comments on the Wigner functions. In general, $\rho^W(Q, P)$ may have negative values, so that we cannot consider them as probability densities. However, reduced distribution function depending only on momenta $\varphi_{\mathcal{N}}(P)$ [or only on coordinates $n_{\mathcal{N}}(Q)$] are probability distribution functions, because they are given by diagonal components of the density matrices;

$$\varphi_{\mathcal{N}}(\mathbf{P}_1, \ldots, \mathbf{P}_{\mathcal{N}}) = \int d^{\mathcal{N}}Q \rho^W(Q, P, t) = \langle P | \rho(t) | P \rangle \tag{3.63}$$

Moreover, the essential features of the extension of quantum mechanics for LPS are based on the singularities of the Fourier components of the Wigner distribution functions, which are momentum representation of the density matrices. In other words, we could as well present our extension of quantum mechanics in terms of the density matrices in momentum representation without referring to the Wigner distribution functions. The advantage of the Wigner distribution functions is the parallelism between quantum mechanics and classical mechanics (see *I*). In this way, it becomes clear that the origin of irreversibility for LPS is common to classical and quantum mechanics.

IV. SINGULAR FOURIER EXPANSIONS AND STATISTICAL DISTRIBUTION FUNCTIONS

The statistical description of dynamics includes ensembles localized in space as well as nonlocal ensembles. Let us first consider *local ensembles*, that is, $\langle Q | \rho | Q \rangle \to 0$ in the limit $|\mathbf{Q}_j| \to \infty$. In terms of the Wigner distribution function, we consider the *Fourier expansion*

$$\rho^W(Q, P, t) = \sum_{k^{\mathcal{N}}} \sum_{P'^{\mathcal{N}}} \langle\langle Q, P | k, P' \rangle\rangle \langle\langle k, P' | \rho(t) \rangle\rangle = \frac{1}{L^{3\mathcal{N}}} \sum_{k^{\mathcal{N}}} e^{ik \cdot Q} \tilde{\rho}_k(P, t) \tag{4.1}$$

where [note the volume dependence in Eq. (3.41)]

$$\tilde{\rho}_k(P, t) = L^{3\mathcal{N}/2} \langle\langle k, P \mid \rho(t) \rangle\rangle \tag{4.2}$$

and we assume the coefficients $\tilde{\rho}_k(P)$ do not depend on the volume in the limit $\Omega \to \infty$. Hereafter, we use the unit

$$\hbar = 1 \tag{4.3}$$

The distribution function (4.1) is normalizable in the sense of absolute integrable functions (i.e., L_1 functions) as

$$\int d^{\mathcal{N}}Q \int d^{\mathcal{N}}P \rho^W(Q, P) = \int d^{\mathcal{N}}P \tilde{\rho}_0(P) = 1 \tag{4.4}$$

The characteristic feature of local ensembles is that the diagonal elements of the density matrix in the momentum representation are "regular." In other words, the Wigner distribution functions have a "regular" Fourier expansion; that is, the Fourier component of the distribution function with $k = 0$ has the same volume dependence $L^{-3\mathcal{N}}$ as the Fourier components with nonvanishing k. This is in contrast to nonlocal ensembles, where $\langle Q \mid \rho \mid Q \rangle \neq 0$ in the limit $|\mathbf{Q}_j| \to \infty$. As we shall see, "singularities" appear in the Fourier components. To emphasize this fact and to distinguish Eq. (4.1) from the Fourier coefficients $\rho_k(P)$ for nonlocal ensembles, we put the tilde on the coefficients ($\tilde{\rho}_k$) for the local ensembles.

The Hilbert space norm of the distributions is given by (for $\Omega \to \infty$)

$$\langle\langle \rho \mid \rho \rangle\rangle = \frac{1}{L^{6\mathcal{N}}} \sum_{k\mathcal{N}} \sum_{P\mathcal{N}} |\tilde{\rho}_k(P)|^2 \to \frac{1}{(2\pi)^{6\mathcal{N}}} \int d^{\mathcal{N}}k \int d^{\mathcal{N}}P \, |\tilde{\rho}_k(P)|^2 \tag{4.5}$$

Hence, there exists a Hilbert norm for square integrable functions and for $\mathcal{N}$ finite (note that the 2π factors can be absorbed in the definition of $\tilde{\rho}_k$).

On the other hand, statistical mechanics (equilibrium and nonequilibrium) deals mainly with nonlocal distributions, such as canonical distribution function which admits no Hilbert norm. Before discussing this point, we introduce, as in equilibrium, reduced distribution functions ρ_s, referring to s particles.

$$\rho_s(\mathbf{Q}_1, \ldots, \mathbf{Q}_s, \mathbf{P}_1, \ldots, \mathbf{P}_s) = \int d^{\mathcal{N}-s}Q d^{\mathcal{N}-s}P \rho^W(Q, P, t) \tag{4.6}$$

where $\int d^{\mathcal{N}-s}Q \equiv \int d\mathbf{Q}_{s+1} \cdots \int d\mathbf{Q}_{\mathcal{N}}$, and so on.

Distribution functions that refer to specified particles are called *specific* distribution functions [4]. In general, we are more interested in distribution functions that concern only the distribution of momenta or the coordinates of s particles regardless of which particles they are. This distribution function, which we call f_s, is found by multiplying ρ_s by the factor $\mathcal{N}!/(\mathcal{N} - s)!$ This is the number of possible ways in which a sequence of s particles can be chosen out of $\mathcal{N}$. Therefore,

$$\begin{aligned} f_s(\mathbf{Q}_1, \ldots, \mathbf{Q}_s, \mathbf{P}_1, \ldots, \mathbf{P}_s) &= \frac{\mathcal{N}!}{(\mathcal{N} - s)!} \rho_s(\mathbf{Q}_1, \ldots, \mathbf{Q}_s, \mathbf{P}_1, \ldots, \mathbf{P}_s) \\ &= \frac{\mathcal{N}!}{(\mathcal{N} - s)!} \int d^{\mathcal{N}-s}Q d^{\mathcal{N}-s}P \rho^W(Q, P) \end{aligned} \tag{4.7}$$

We also use distribution function φ_s in momentum space and n_s in coordinate space, as defined by

$$\begin{aligned} \varphi_s(\mathbf{P}_1, \ldots, \mathbf{P}_s) &= \frac{(\mathcal{N} - s)!}{\mathcal{N}!} \int d^sQ f_s \\ n_s(\mathbf{Q}_1, \ldots, \mathbf{Q}_s) &= \int d^sP f_s \end{aligned} \tag{4.8}$$

As mentioned in Eq. (3.63), f_s can be negative, but φ_s and n_s are true probability densities because they depend only on P or Q. The reduced distribution functions f_s, φ_s, and n_s are called *generic* distribution functions to distinguish them from the specific distribution functions [4]. In the following discussion we use specific distribution functions whenever it is necessary to specify the coordinates and momenta of each particle.

In general, statistical mechanics deals with situations where there are no asymptotic free "in" and "out" states. The interactions are *persistent*. As mentioned, this requires the use of nonlocal distribution functions. For this case, density matrices have "delta-function singularities" in momentum representation, or equivalently in the Fourier expansion in their Wigner representation [4]. For example, let us consider the reduced number-density in space given by $n_1(\mathbf{Q}) = c + h(\mathbf{Q})$, where c is the concentration as defined in

Eq. (2.2) and $h(\mathbf{Q})$ is the "inhomogenuity," that is, the deviation from a uniform distribution. We assume that h is an absolutely integrable function. In the Fourier representation we have

$$n_1(\mathbf{Q}) = \frac{1}{\Omega} \sum_{\mathbf{k}} [c\delta_\Omega(\mathbf{k}) + h_{\mathbf{k}}]\, e^{i\mathbf{k}\cdot\mathbf{Q}} \tag{4.9}$$

where $\delta_\Omega(\mathbf{k}) = \Omega\delta^{kr}(\mathbf{k})$ [see Eq. (3.5)]. In the limit of large volume, the uniform part has a delta-function singularity at $\mathbf{k} = 0$.

We note that kinetic energy is a nonlocal quantity in space; therefore, the Hamiltonian has a delta-function singularity for the diagonal component in the momentum representation:

$$\langle P|H|P'\rangle = \frac{1}{\Omega} \sum_{j>i}^{\mathcal{N}} \left[\frac{\mathbf{P}_i^2}{2m_i}\, \delta_\Omega(\mathbf{P}_i - \mathbf{P}_i')\delta_i^{kr}(P - P') \right.$$
$$\left. + V_{|\mathbf{P}_i - \mathbf{P}_i'|}\, \delta^{kr}(\mathbf{P}_i + \mathbf{P}_j - \mathbf{P}_i' - \mathbf{P}_j')\delta_{ij}^{kr}(P - P') \right] \tag{4.10}$$

where $\delta_i^{kr}(P)$ is a product of $\mathcal{N} - 1$ Kronecker's delta which excludes the particle i, and $\delta_{ij}^{kr}(P)$ a product of $\mathcal{N} - 2$ Kronecker's delta which excludes the particles i and j, and so on:

$$\delta_i^{kr}(P) = \prod_{r \neq i}^{\mathcal{N}} \delta(\mathbf{P}_r), \qquad \delta_{ij}^{kr}(P) = \prod_{r \neq i,\, j}^{\mathcal{N}} \delta(\mathbf{P}_r) \tag{4.11}$$

In the Wigner representation, this leads to a delta-function singularity in its Fourier expansion [cf. Eq. (4.9)]:

$$H^W(Q, P) = \frac{1}{\Omega} \sum_{\mathbf{k}} \sum_{i}^{\mathcal{N}} \left[\frac{\mathbf{P}_i^2}{2m_i}\, \delta_\Omega(\mathbf{k}) + \lambda \sum_{j(>i)}^{\mathcal{N}} V_k\, e^{-i\mathbf{k}\cdot\mathbf{Q}_j} \right] e^{i\mathbf{k}\cdot\mathbf{Q}_i} \tag{4.12}$$

Let us now show that this leads to delta-function singularities for equilibrium distributions which are functions of the Hamiltonian (for $\Omega \to \infty$):

$$\rho^{eq}(Q, P) = \frac{f(H_0 + \lambda V)}{\int d^{\mathcal{N}}Q d^{\mathcal{N}}P f(H_0 + \lambda V)} \tag{4.13}$$

We assume the normalization

$$\int d^{\mathcal{N}} P f(H_0) = 1 \tag{4.14}$$

Then, using Eq. (3.37), we have the power series expansion in the coupling constant λ for the Hamiltonian (2.1):

$$\begin{aligned}\rho^{eq} = \frac{1}{L^{3\mathcal{N}}} \Bigg[1 &+ \frac{1}{2!} \lambda \sum_{i,j}^{\mathcal{N}} V_{ij} \frac{\partial}{\partial H_0} \\ &+ \frac{\lambda^2}{2} \left(\frac{1}{2!} \sum_{i,j}^{\mathcal{N}} V_{ij} V_{ij} + \frac{1}{3!} \sum_{i,j,m}^{\mathcal{N}} V_{ij} V_{jm} + \frac{1}{4!} \sum_{i,j,m,n}^{\mathcal{N}} V_{ij} V_{mn} \right) \frac{\partial^2}{\partial H_0^2} \\ &- \frac{\lambda^2}{2L^{3\mathcal{N}}} \int d^{\mathcal{N}} Q V^2 \int d^{\mathcal{N}} P \left(\frac{\partial^2 f(H_0)}{\partial H_0^2} \right) + \lambda^3 \cdots \Bigg] f(H_0)\end{aligned} \tag{4.15}$$

where we have written explicitly the particle indices, such as $V_{ij} = V(|\mathbf{Q}_i - \mathbf{Q}_j|)$. Different particle indices i, j denote different particles. Let us consider the canonical distribution function (for systems with the same mass $m_i = m$ of the particles):

$$f(H^W) = \left(\frac{\beta}{2\pi m} \right)^{3\mathcal{N}/2} \langle\langle Q, P | e^{-\beta H} \rangle\rangle = \left(\frac{\beta}{2\pi m} \right)^{3\mathcal{N}/2} e^{-\beta H^W} \tag{4.16}$$

with $\beta = (\mathbf{K}T)^{-1}$ and $\mathbf{K}$ is the Boltzmann constant. We then obtain [with Eq. (4.12)]

$$\begin{aligned}\rho^{eq}(Q, P) = \frac{1}{L^{3\mathcal{N}}} \left(\frac{\beta}{2\pi m} \right)^{3\mathcal{N}/2} e^{-\beta H_0} \Bigg[&(1 + \lambda^2 \cdots) \\ &+ \frac{1}{2!\Omega} \sum_{i,j} \sum_{\mathbf{k}} e^{i\mathbf{k} \cdot (\mathbf{Q}_i - \mathbf{Q}_j)} \left(-\lambda \beta V_{|\mathbf{k}|} + \lambda^2 \beta^2 \frac{1}{\Omega} \sum_{\mathbf{k}'} V_{|\mathbf{k}'|} V_{|\mathbf{k}' - \mathbf{k}|} + \lambda^3 \cdots \right) \\ &+ \frac{1}{3!\Omega^2} \sum_{i,j,n} \sum_{\mathbf{k}} \sum_{\mathbf{k}'} e^{i(\mathbf{k}+\mathbf{k}') \cdot \mathbf{Q}_i - \mathbf{k} \cdot \mathbf{Q}_j - \mathbf{k}' \cdot \mathbf{Q}_n} (\lambda^2 \beta^2 V_{|\mathbf{k}|} V_{|\mathbf{k}'|} + \lambda^3 \cdots) + \cdots \Bigg]\end{aligned} \tag{4.17}$$

Here, the first term in the bracket does not depend on the coordinates, so that this term is associated with a Fourier coefficient that has only vanishing wave vectors $k = 0$. The second term corresponds to contributions that have nonvanishing wave vectors $\mathbf{k}_i = \mathbf{k}$ and $\mathbf{k}_j = -\mathbf{k}$ for only two particles, i and j, and so on. Because the Hamiltonian is translational invariant, the equilibrium distribution is "homogenous" in space (it is invariant when $\mathbf{Q}_j \Rightarrow \mathbf{Q}_j + \mathbf{a}$ for all j, then the total wave vector vanishes, i.e., $\mathbf{k}_i + \mathbf{k}_j + \cdots = 0$).

The remarkable feature of the equilibrium distribution is that ρ^{eq} can be decomposed into the "vacuum of correlations" [i.e., the first term in the bracket of Eq. (4.17)], binary correlations (the second term), ternary correlations (the third term), and so on. Moreover, we see the appearance of *delta-function singularities*, as in Eq. (4.9). The existence of this expansion ensures the existence of both "extensive variables" (e.g., the average energy) and of "intensive variables" (i.e., reduced variables) depending on a finite number of particles. We obtain also a "cluster expansion" of the density matrix ρ in terms of correlation functions which have a finite range of correlations in the thermodynamic limit [4,29].

In our previous work in nonequilibrium statistical mechanics, we used the class of ensembles that corresponds to a natural generalization of the canonical distribution [4,29] [see Eq. (4.11)]:

$$\begin{aligned}\rho^W(Q, P) = \frac{1}{L^{3\mathcal{N}}} \sum_{k^{\mathcal{N}}} e^{ik\cdot Q} \Bigg[& \rho_0(\,|\,P)\delta^{kr}(k) \\ & + \frac{1}{\Omega} \sum_{j>i}^{\mathcal{N}} \rho_{\mathbf{k}_i, -\mathbf{k}_i}(\mathbf{P}_i, \mathbf{P}_j \,|\, P^{\mathcal{N}-2}) \delta^{kr}(\mathbf{k}_i + \mathbf{k}_j) \delta^{kr}_{ij}(k) \\ & + \frac{1}{\Omega^2} \sum_{n>j>i}^{\mathcal{N}} \rho_{\mathbf{k}_i, \mathbf{k}_j, \mathbf{k}_n}(\mathbf{P}_i, \mathbf{P}_j, \mathbf{P}_n \,|\, P^{\mathcal{N}-3}) \\ & \times \delta^{kr}(\mathbf{k}_i + \mathbf{k}_j + \mathbf{k}_n) \delta^{kr}_{ijn}(k) + \cdots \\ & + \frac{1}{\Omega} \sum_{j}^{\mathcal{N}} \rho'_{\mathbf{k}_j}(\mathbf{P}_j \,|\, P^{\mathcal{N}-1}) \delta^{kr}_j(k) \\ & + \frac{1}{\Omega^2} \sum_{j>i}^{\mathcal{N}} \rho'_{\mathbf{k}_i, \mathbf{k}_j}(\mathbf{P}_i, \mathbf{P}_j \,|\, P^{\mathcal{N}-2}) \delta^{kr}_{ij}(k) + \cdots \Bigg]\end{aligned} \tag{4.18}$$

Here, as in Eq. (4.17) we have decomposed the Fourier components according to the number of nonvanishing elements $\mathbf{k}_j$ in the set of wave vector $k = (\mathbf{k}_1, \ldots, \mathbf{k}_{\mathcal{N}})$, that is, the number of particles that have off-diagonal elements in momentum representation. In the expression $\rho_{\mathbf{k}_i, \mathbf{k}_j, \ldots}(\mathbf{P}_i, \mathbf{P}_j, \ldots \,|\, P^{\mathcal{N}-r})$, the momentum arguments on the left side of bar denote the particle i with a nonvanishing wave vector $\mathbf{k}_i$, the particle j with $\mathbf{k}_j$, and so on; the arguments on the right side of the bar denote the remaining particles that have zero wave vectors and are therefore uniformly distributed [29]. We assume that $\rho_{\mathbf{k}_i, \mathbf{k}_j, \ldots}$ and $\rho'_{\mathbf{k}_i, \mathbf{k}_j, \ldots}$ do not depend on the volume Ω. Moreover, we assume that their dependence on the wave vectors is smooth. The coefficients $\rho_{\mathbf{k}_i, \mathbf{k}_j, \ldots}$ are associated with the homogeneous components of the distribution function in space (i.e., the component with the total wave vector vanishes, i.e., $\mathbf{k}_i + \mathbf{k}_j + \cdots = 0$), while the coefficients $\rho'_{\mathbf{k}_i, \mathbf{k}_j, \ldots}$ are associated with the "inhomogeneous" components (with $\mathbf{k}_i + \mathbf{k}_j + \cdots \neq 0$). Because of the conservation law of wave vectors (3.54), the homogeneous components evolve independently from the inhomogeneous components.

To emphasize the difference in volume dependence from that of local ensembles such as Eq. (4.1), we have introduced the notations $\rho_{\mathbf{k}_i, \mathbf{k}_j, \ldots}$ and $\rho'_{\mathbf{k}_i, \mathbf{k}_j, \ldots}$ instead of $\tilde{\rho}_{\mathbf{k}_i, \mathbf{k}_j, \ldots}$, which was used for the Fourier coefficients in Eq. (4.18).

This class of distribution functions expresses the existence of extensive and intensive variables even outside equilibrium. It leads to an extension of the cluster expansion in terms of the correlation functions in nonequilibrium statistical mechanics; that is, the coefficients $\rho_0(P)$, $\rho_{\mathbf{k}_i, \mathbf{k}_j}(P)$, $\rho_{\mathbf{k}_i, \mathbf{k}_j, \mathbf{k}_n}(P)$, and so on are the Fourier components of the momentum distribution functions (which corresponds to the "vacuum of correlations"), of the binary correlations, of the ternary correlations, and so on [4,29]. As we have seen, interactions leads to transitions from one set of wave vectors to another. This corresponds to a "dynamics of correlations" [4].

Our expansion in Eq. (4.18) depends strongly on the distinguishability of particles. To incorporate quantum statistics in the Wigner representation for the variables P and Q, we need a significant improvement of the decomposition in the right-hand side of Eq. (4.18). This can be achieved by introducing the concept of the "contraction" of the Wigner functions (see [29] and also [30]). There is, however, an alternative method for quantum statistics, which is based on the second quantization formulation and uses the Wigner representation in the occupation number representation. We illustrate this alternative approach in Appendix K, when we discuss anharmonic solids.

A characteristic feature of the distributions (4.18) is that all reduced quantities are well defined. For example, the expectation value of $\mathbf{Q}_1 - \mathbf{Q}_2$ is given by (in the thermodynamic limit)

$$\begin{aligned} \langle \mathbf{Q}_1 - \mathbf{Q}_2 \rangle &= \int d^{\mathcal{N}}Q \int d^{\mathcal{N}}P (\mathbf{Q}_1 - \mathbf{Q}_2) \rho^W(Q, P) \\ &= -i \left[\frac{\partial}{\partial \mathbf{k}} \int d^{\mathcal{N}}P \rho_{\mathbf{k}, -\mathbf{k}}(\mathbf{P}_1, \mathbf{P}_2 \mid P^{\mathcal{N}-2}) \right]_{\mathbf{k}=0} \end{aligned} \tag{4.19}$$

Assuming a finite range of correlation, this quantity is finite.

An important aspect of this class of distribution functions is its stability during the time evolution. Dynamics of correlation leaves the form (4.18) invariant. For example, let us assume that the system is initially in the vacuum of correlation. Then, using Eq. (3.50), we have

$$\begin{aligned}\rho_{\mathbf{k}_j, -\mathbf{k}_j}(\mathbf{P}_j, \mathbf{P}_n | P^{\mathcal{N}-2}) &= L^{3\mathcal{N}/2} \langle\langle \mathbf{k}_j, -\mathbf{k}_j, \{0\}^{\mathcal{N}-2}, P | \lambda L_V | \rho \rangle\rangle \\ &= L^{3\mathcal{N}/2} \int dP' \langle\langle \mathbf{k}_j, -\mathbf{k}_j, \{0\}^{\mathcal{N}-2}, P | \lambda L_V | 0, P' \rangle\rangle \\ &\quad \times \langle\langle 0, P' | \rho \rangle\rangle \\ &= -\frac{1}{L^{3\mathcal{N}}} \frac{1}{\Omega} \lambda V_{|\mathbf{k}_j|} \partial_{jn}^{\mathbf{k}_j/2} \rho_0(|P) \end{aligned} \tag{4.20}$$

where the volume factor $L^{3\mathcal{N}/2}$ comes from the relation between Eq. (3.41) and the Fourier expansion in Eq. (4.18) [cf. Eqs. (4.1) and (4.2)]. Equation (4.20) gives the same volume dependence for the binary correlation (i.e., the second term) as in Eq. (4.18). One can extend this result to all orders of λ and to all Fourier components (see [4,29] for more details). This is quite remarkable. Indeed, this is the only class of distribution functions stable, in this sense, in the thermodynamic limit (see Sections VI and XIII).

We note that the distribution function (4.18) satisfies the normalization condition (4.4) in L_1. In contrast, the Hilbert space norm of Eq. (4.18) vanishes as in the thermodynamic limit:

$$\begin{aligned}\langle\langle \rho | \rho \rangle\rangle = \frac{1}{L^{3\mathcal{N}}} \Bigg(&\int d^{\mathcal{N}} P |\rho_0(|P)|^2 \\ &+ \frac{1}{\Omega^2} \sum_j^{\mathcal{N}} \sum_{\mathbf{k}} \int d^{\mathcal{N}} P |\rho'_{\mathbf{k}}(\mathbf{P}_j | P^{\mathcal{N}-1})|^2 + \cdots \Bigg) \to 0 \end{aligned} \tag{4.21}$$

Hence, distributions of this class *do not* belong to the Hilbert space.

We note that observables M, which depend on a reduced number r ($<\mathcal{N}$) of coordinates, have a delta-function singularity in their Fourier expansion as

$$M^W(\mathbf{Q}_1, \ldots, \mathbf{Q}_r, \mathbf{P}_1, \ldots, \mathbf{P}_r) = \frac{1}{\Omega^{\mathcal{N}}} \sum_{k^{\mathcal{N}}} e^{ik \cdot Q} M_k(\mathbf{P}_1, \ldots, \mathbf{P}_r) \delta_\Omega(\mathbf{k}_{r+1}) \cdots \delta_\Omega(\mathbf{k}_{\mathcal{N}}) \tag{4.22}$$

Hence, these observables also do not belong to the Hilbert space.

To investigate the time evolution of the class of these singular functions, it is convenient to introduce projection operator $P_a^{(\nu)}$ which extracts single eigenmodes of the unperturbed Liouvillian,[4]

$$
\begin{aligned}
P^{(0)} &= \sum_{P^{\mathcal{N}}} |k, P\rangle\!\rangle\langle\!\langle k, P|\, \delta^{kr}(k), \\
P_{ij}^{(k_i,\,-k_i)} &= \sum_{P^{\mathcal{N}-2}} |k, P\rangle\!\rangle\langle\!\langle k, P|\, \delta_{\mathbf{k}_i+\mathbf{k}_j,\,0}\, \delta_{ij}^{kr}(k), \qquad \cdots \\
P_j^{(k_j)} &= \sum_{P^{\mathcal{N}-1}} |k, P\rangle\!\rangle\langle\!\langle k, P|\, \bar{\delta}_{\mathbf{k}_j} \delta_j^{kr}(k), \\
P_{ij}^{(k_i,\,k_j)} &= \sum_{P^{\mathcal{N}-2}} |k, P\rangle\!\rangle\langle\!\langle k, P|\, \bar{\delta}_{\mathbf{k}_i+\mathbf{k}_j}\, \delta_{ij}^{kr}(k), \qquad \cdots
\end{aligned}
\tag{4.23}
$$

where

$$
\bar{\delta}_{\mathbf{k}_j} \equiv 1 - \delta_{\mathbf{k}_j,\,0} \tag{4.24}
$$

The index a in $P_a^{(\nu)}$ denotes the particles associated with nonvanishing wave vectors; the index ν denotes the value of their wave vectors. The summation of momenta P^{N-r} is taken over only $N - r$ particles with vanishing wave vectors. The projection operators in the first and second lines in Eq. (4.23) extract the homogeneous components ($\mathbf{k}_1 + \mathbf{k}_2 + \cdots = 0$) of the density matrices, while the projection operators in the third and fourth lines extract the inhomogeneous components ($\mathbf{k}_1 + \mathbf{k}_2 + \cdots \neq 0$).

Note that any observables that are diagonal in momentum representation lie in the vacuum of correlation subspace $P^{(0)}$. For example, the bra-state for the momentum $\hat{\mathbf{P}}_j$ defined by

$$
\langle\!\langle \hat{\mathbf{P}}_j | \equiv \sum_{p^{\mathcal{N}}} \mathbf{p}_j \langle\!\langle p; p | = \sum_{P^{\mathcal{N}}} \mathbf{P}_j \langle\!\langle 0, P | \tag{4.25}
$$

satisfies

$$
\langle\!\langle \hat{\mathbf{P}}_j | = \langle\!\langle \hat{\mathbf{P}}_j | P^{(0)} \tag{4.26}
$$

The Wigner representation of the momentum is given as a c-number by

$$
\hat{\mathbf{P}}_j^W(Q, P) = \langle\!\langle \hat{\mathbf{P}}_j | Q, P \rangle\!\rangle = \mathbf{P}_j \tag{4.27}
$$

[4] In the limit of large volume, $P^{(0)}$ is defined at a single point $k = 0$ in the continuous variable k. This leads to a delicate mathematical problem in the formulation using continuous variables from the beginning [46]. The box normalization allows us not only to deal with the thermodynamic limit, but also to deal consistently with the vacuum of correlations.

For the unperturbed system, all components of $\mathbf{P}_j = (P_{jx}, P_{jy}, P_{jz})$ are invariants of motion. This leads to $3\mathcal{N}$ independent invariants of motion. For interacting systems, some of the invariants of motion will be destroyed, owing to the resonances under certain conditions (see Sections IX–XI). Therefore, even for quantum mechanics, we can introduce a classification into "integrable systems" and "nonintegrable systems" in the sense of Poincaré as in the case of classical systems (see *I*).

The projection operators $P_a^{(\nu)}$ commute with the unperturbed Liouvillian,

$$L_0 P_a^{(\nu)} = (k \cdot v) P_a^{(\nu)} = P_a^{(\nu)} L_0 \tag{4.28}$$

Moreover

$$\sum_a \sum_\nu P_a^{(\nu)} = 1, \qquad P_a^{(\nu)} P_b^{(\mu)} = P_a^{(\nu)} \delta_{\nu,\mu} \delta_{a,b} \tag{4.29}$$

To shorten the notation, we have not written the summation sign and Kronecker's delta for the momenta [cf. Eq. (3.40)]. We also introduce the projection operators $Q_a^{(\nu)}$:

$$Q_a^{(\nu)} = 1 - P_a^{(\nu)} \tag{4.30}$$

which are orthogonal to $P_a^{(\nu)}$, that is,

$$P_a^{(\nu)} Q_a^{(\nu)} = Q_a^{(\nu)} P_a^{(\nu)} = 0 \tag{4.31}$$

We note that [see Eqs. (3.50) and (3.51) with $\mathbf{k} = 0$]

$$P_a^{(\nu)} L_V P_a^{(\nu)} = 0 \tag{4.32}$$

In the following discussion, we often use the notation

$$P^{(\nu)} = |\nu\rangle\rangle\langle\langle\nu| \tag{4.33}$$

as well as w_ν for the eigenvalue of L_0, where we have omitted the particle index a. Then, the spectral decomposition of L_0 is

$$L_0 = \sum_\nu |\nu\rangle\rangle w_\nu \langle\langle\nu| \tag{4.34}$$

We now come to the main problem, the study of the spectral representation of L_H in the extended function space which includes the canonical distribution.

V. COMPLEX SPECTRAL REPRESENTATIONS

For nonintegrable systems, the spectral decomposition of the Liouvillian corresponding to the Hamiltonian (2.1) in Hilbert space is generally *not known*. In contrast, we obtain the solution of the eigenvalue problem for the Liouvillian for the class of density matrices with singularities in the momentum representation. Because these density matrices have no Hilbert space norm (4.21), we have to extend the Liouville operator outside the Hilbert space. This has already been done in the case of deterministic chaos [17–22]. This extension is quite natural, because the class of density matrices we consider includes the equilibrium distributions. In this extended function space, the Liouvillian has "complex" eigenvalues. That means that time symmetry is broken. We may therefore expect that this complex spectral representation allows us to describe irreversible processes such as the approach to equilibrium. Moreover, the complex part of the eigenvalues generally does not satisfy the Rydberg–Ritz principle, that is, it is not the difference of two numbers. Hence the Liouvillian is no longer a commutator in the extended function space. For the special case with no singular components of the density matrix in the momentum representation, we recover the usual spectral representation in the Hilbert space.

We consider the eigenvalue problem (see *I* and [11])

$$L_H | F_\alpha^{(\nu)}(\lambda) \rangle\!\rangle = Z_\alpha^{(\nu)} | F_\alpha^{(\nu)}(\lambda) \rangle\!\rangle \tag{5.1}$$

with the boundary condition

$$| F_\alpha^{(\nu)}(\lambda) \rangle\!\rangle \to P^{(\nu)} | F_\alpha^{(\nu)}(0) \rangle\!\rangle \qquad \text{for } \lambda \to 0 \tag{5.2}$$

The indices α (together with ν) are the parameters characterizing the eigenfunctions. The class of eigenfunctions we consider [see Eq. (5.7)] has precisely the form of the singular distribution functions (4.18).

As we shall show, the eigenvalues $Z_\alpha^{(\nu)}$ are generally complex numbers. The time evolution of LPS splits into two semigroups. For the semigroup corresponding to $t > 0$, the eigenstates are associated with the eigenvalues with Im $Z_\alpha^{(\nu)} \le 0$, and equilibrium is approached in our future for $t \to +\infty$; for the other semigroup, the eigenvalues are the complex conjugate of $Z_\alpha^{(\nu)}$ and equilibrium is reached in our past.[5] In Section XIV, we show that if the universe, as a whole, is in one semigroup, one cannot prepare initial conditions which belong to the other semigroup. All irreversible processes have

[5] The definite sign of the imaginary part of $Z_\alpha^{(\nu)}$ is related to the singularity of the resolvent operator of L_H. See the discussion in *I*.

the same time orientation. Therefore, to be self-consistent, we have to choose the semigroup oriented toward our future.

For complex eigenvalues, the left eigenstates of L_H are not the hermitian conjugates of the right eigenstates. Let us denote the left eigenstates corresponding to the same eigenvalue $Z_\alpha^{(\nu)}$ by $\langle\langle \tilde{F}_\alpha^{(\nu)} |$, that is,

$$\langle\langle \tilde{F}_\alpha^{(\nu)}(\lambda) | L_H = \langle\langle \tilde{F}_\alpha^{(\nu)}(\lambda) | Z_\alpha^{(\nu)} \tag{5.3}$$

again with the boundary condition

$$\langle\langle \tilde{F}_\alpha^{(\nu)}(\lambda) | \to \langle\langle \tilde{F}_\alpha^{(\nu)}(0) | P^{(\nu)} \qquad \text{for } \lambda \to 0 \tag{5.4}$$

We assume the biorthogonality and bicompleteness relations:

$$\langle\langle \tilde{F}_\alpha^{(\nu)} | F_\beta^{(\mu)} \rangle\rangle = \delta_{\nu,\mu}\delta_{\alpha,\beta}, \qquad \sum_\nu \sum_\alpha | F_\alpha^{(\nu)} \rangle\rangle \langle\langle \tilde{F}_\alpha^{(\nu)} | = 1 \tag{5.5}$$

We also assume that the eigenstates of the Liouvillian are not degenerate for the different indices of ν and α. The biorthogonality relation is the direct consequence of the assumption of the nondegeneracy. This assumption, as well as the bicompleteness of the eigenstates, should be verified for each specific Hamiltonian.[6]

Moreover, we assume that the Liouvillian is diagonalizable in the extended function space:

$$L_H = \sum_\nu \sum_\alpha | F_\alpha^{(\nu)} \rangle\rangle Z_\alpha^{(\nu)} \langle\langle \tilde{F}_\alpha^{(\nu)} | \tag{5.6}$$

In this chapter, we do not consider more general situations which may lead to Jordan blocks (see [22,42]). We consider eigenfunctions that have the structure (4.18). We limit ourselves to homogeneous situations where the eigenfunctions are translationally invariant. As noticed [see Eq. (4.18)], the homogeneous components are independent of the inhomogeneous components. In the following argument, if we replace $P^{(\nu)}$ by the projection operators corresponding to the inhomogeneous components, the construction of eigenstates associated with the inhomogeneous situation is straightforward, and we do not repeat the calculations. Therefore, we shall study the eigenvalue problem for functions characterized by the singular Fourier

[6] The proof of the biorthogonality and bicompleteness for the complex spectral representation for the Friedrichs model as well as potential scattering can be found in our earlier papers [10,11,15].

expansions (abbreviating the argument λ):

$$
\begin{aligned}
&\langle\langle Q, P \mid F_\alpha^{(\nu)}\rangle\rangle \\
&= \frac{1}{L^{3N/2}} \sum_{k^N} e^{ik\cdot Q}\Bigg[F_0^{(\nu)}(P, \alpha)\delta^{kr}(k) + \frac{1}{\Omega}\sum_{j>i}^{N} F_{\mathbf{k}_i, \mathbf{k}_j}^{(\nu)}(P, \alpha)\delta^{kr}(\mathbf{k}_i + \mathbf{k}_j)\delta_{ij}^{kr}(k) \\
&\qquad + \frac{1}{\Omega^2}\sum_{n>j>i}^{N} F_{\mathbf{k}_i, \mathbf{k}_j, \mathbf{k}_n}^{(\nu)}(P, \alpha)\delta^{kr}(\mathbf{k}_i + \mathbf{k}_j + \mathbf{k}_n)\delta_{ijn}^{kr}(k) + \cdots \Bigg]
\end{aligned} \tag{5.7}
$$

and similar expansions for $\langle\langle \tilde{F}_\alpha^{(\nu)} \mid Q, P\rangle\rangle$.

We assume that the Fourier coefficients $F_{\mathbf{k}_i, \mathbf{k}_j, \ldots}^{(\nu)}$ and $\tilde{F}_{\mathbf{k}_i, \mathbf{k}_j, \ldots}^{(\nu)}$ do not depend on the volume in the limit of the large volumes $\Omega \to \infty$. Here $F_0^{(\nu)}$ corresponds to the vacuum of correlations, $F_{\mathbf{k}_i, -\mathbf{k}_i}^{(\nu)}$ to binary correlations, and so on, such as in Eq. (4.18), and so do for $\tilde{F}^{(\nu)}$.

Note that the eigenstates $\mid F_\alpha^{(\nu)}\rangle\rangle$ for $\lambda \neq 0$ contain components in the range of all projection operators $P^{(\nu)}$. We call $P^{(\nu)}$. We call $P^{(\nu)} \mid F_\alpha^{(\nu)}\rangle\rangle$ the "privileged" components of $\mid F_\alpha^{(\nu)}\rangle\rangle$.

We formulate the eigenvalue problem for an arbitrary number $\mathcal{N}$ of particles, including $\mathcal{N} \to \infty$. For this case, special care is necessary, as the perturbed Liouvillian L_V contains $\mathcal{N}^2$ terms involving all pairs of particles j and n [see Eq. (3.50)]. We therefore take the inner product of the eigenvalue equation (5.1) with observables (4.22) that depend on an arbitrary but *finite* number of particles:

$$
\langle\langle M \mid L_H \mid F_\alpha^{(\nu)}\rangle\rangle = Z_\alpha^{(\nu)} \langle\langle M \mid F_\alpha^{(\nu)}\rangle\rangle \tag{5.8}
$$

Then only terms connected to the labeled particles that appear in M give nonvanishing contributions, and all disconnected terms vanish [see Eqs. (3.60) and (3.61)]. This operation thus reduces the number of pairs and leads to a finite contribution in the thermodynamic limit (2.2). In this sense, observables play the role of "test functions" in functional analysis. In our discussion of the eigenvalue problem, our formulas should always be understood in terms of Eq. (5.8). We return to this problem in Sections VIII and XI (see also [4]).

Let us now display the results of the eigenvalue problem. The method is identical to that for classical mechanics in *I*. For that reason we give only the results and refer to Appendix A for more detail. Applying the projection operators $P^{(\nu)}$ and $Q^{(\nu)}$ in Eq. (4.30) to Eq. (5.1), we derive first the equations:

$$
P^{(\nu)} L_H (P^{(\nu)} \mid F_\alpha^{(\nu)}\rangle\rangle + Q^{(\nu)} \mid F_\alpha^{(\nu)}\rangle\rangle) = Z_\alpha^{(\nu)} P^{(\nu)} \mid F_\alpha^{(\nu)}\rangle\rangle \tag{5.9a}
$$

$$Q^{(\nu)}L_H(P^{(\nu)}|F_\alpha^{(\nu)}\rangle\rangle + Q^{(\nu)}|F_\alpha^{(\nu)}\rangle\rangle) = Z_\alpha^{(\nu)}Q^{(\nu)}|F_\alpha^{(\nu)}\rangle\rangle \tag{5.9b}$$

Equation (5.9b) leads to

$$(Z_\alpha^{(\nu)} - Q^{(\nu)}L_H Q^{(\nu)})Q^{(\nu)}|F_\alpha^{(\nu)}\rangle\rangle = Q^{(\nu)}\lambda L_V P^{(\nu)}|F_\alpha^{(\nu)}\rangle\rangle \tag{5.10}$$

This leads to $Q^{(\nu)}|F_\alpha^{(\nu)}\rangle\rangle$ as a functional of $P^{(\nu)}|F_\alpha^{(\nu)}\rangle\rangle$:

$$Q^{(\nu)}|F_\alpha^{(\nu)}\rangle\rangle = \mathscr{C}^{(\nu)}(Z_\alpha^{(\nu)})P^{(\nu)}|F_\alpha^{(\nu)}\rangle\rangle \tag{5.11}$$

where

$$\mathscr{C}^{(\nu)}(z) = \frac{-1}{Q^{(\nu)}L_H Q^{(\nu)} - z} Q^{(\nu)}\lambda L_V P^{(\nu)} \tag{5.12}$$

If the geometrical series converges, we have

$$\mathscr{C}^{(\nu)}(z) = \sum_{n=0}^{\infty} \left(\frac{-1}{L_0 - z} Q^{(\nu)}\lambda L_V Q^{(\nu)}\right)^n \frac{-1}{L_0 - z} Q^{(\nu)}\lambda L_V P^{(\nu)} \tag{5.13}$$

Care has to be taken in the analytic continuations of z to obtain a consistent description of a semigroup oriented toward the future; this is discussed later in more detail. The operator $\mathscr{C}^{(\nu)}(z)$ is called the "creation-of-correlations operator" (or "creation operator"). Graphically, the creation (of correlation) processes can be represented as in Fig. 1 (see [4] for the definition of the diagram). In the series expansion of λ, $\mathscr{C}^{(\nu)}(z)$ starts with the first order.

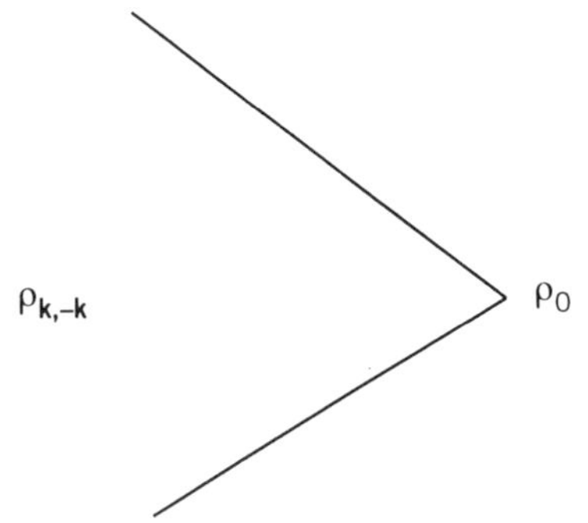

Figure 1. Graphical representation of the creation-of-correlations operator $\mathscr{C}^{(\nu)}(z)$ for $\nu = 0$. The lowest-order contribution in the series expansion by λ (i.e., λ order).

The creation operator describes *irreducible off-diagonal transitions* from a $P^{(\nu)}$ subspace to orthogonal states in $Q^{(\nu)}$ subspace:

$$\mathscr{C}^{(\nu)}(z) = Q^{(\nu)}\mathscr{C}^{(\nu)}(z)P^{(\nu)} \tag{5.14}$$

Here, "irreducible" means that all intermediate states in $\mathscr{C}^{(\nu)}(z)$ are in the $Q^{(\nu)}$ subspace which is orthogonal to the initial states in $P^{(\nu)}$ [see Eq. (5.13)].

Substituting Eq. (5.11) into Eq. (5.9a), we obtain

$$\psi^{(\nu)}(Z_\alpha^{(\nu)})\,|\,u_\alpha^{(\nu)}\rangle\!\rangle = Z_\alpha^{(\nu)}\,|\,u_\alpha^{(\nu)}\rangle\!\rangle \tag{5.15}$$

where $\psi^{(\nu)}(z)$ are the "collision operators" and are defined by [4]:

$$\psi^{(\nu)}(z) \equiv L_0\, P^{(\nu)} + \lambda P^{(\nu)} L_V\, \mathscr{C}^{(\nu)}(z)P^{(\nu)} \tag{5.16}$$

The eigenfunctions $|\,u_\alpha^{(\nu)}\rangle\!\rangle$ are the "privileged components" of $|\,F_\alpha^{(\nu)}\rangle\!\rangle$, as

$$|\,u_\alpha^{(\nu)}\rangle\!\rangle \equiv P^{(\nu)}\,|\,F_\alpha^{(\nu)}\rangle\!\rangle \tag{5.17}$$

The collision operator $\psi^{(\nu)}(z)$ describes *irreducible diagonal transitions* between two states in the same $P^{(\nu)}$ subspace:

$$\psi^{(\nu)}(z) = P^{(\nu)}\psi^{(\nu)}(z)P^{(\nu)} \tag{5.18}$$

Substituting the expansion (5.12) into (5.16), we see that the effects of the interaction in $\psi^{(\nu)}(z)$ begin with the second order in the series expansion of λ. For example, this process may be represented graphically as in Fig. 2.

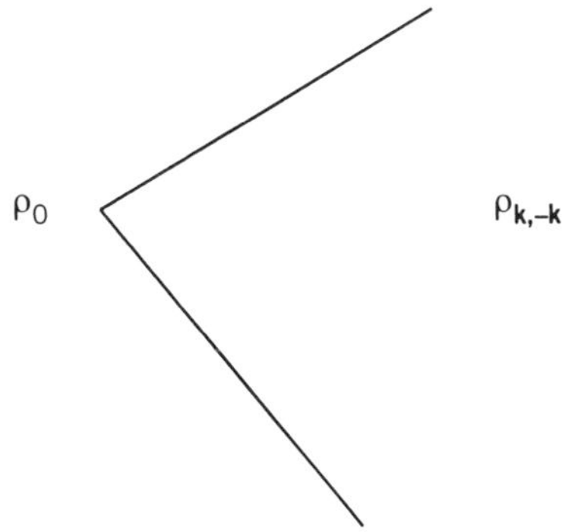

Figure 2. Graphical representation of the collision operator $\psi^{(\nu)}(z)$ for $\nu = 0$. The lowest-order contribution in the series expansion by λ (i.e., λ order).

The collision operators are dissipative operators, and they are the central objects of nonequilibrium statistical mechanics [4,29]. The relations described above show that the Liouville operator shares the same eigenvalues with these dissipative operators in the extended function space. The eigenvalues of the dissipative operators are generally complex numbers. This is consistent with our assumption for $Z_\alpha^{(\nu)}$. Moreover, the privileged components are the right eigenstates of the collision operators.

As the result, the eigenstate of L_H is given by

$$|F_\alpha^{(\nu)}\rangle\!\rangle = N_\alpha^{(\nu)1/2}[P^{(\nu)} + \mathscr{C}^{(\nu)}(Z_\alpha^{(\nu)})]\,|u_\alpha^{(\nu)}\rangle\!\rangle \tag{5.19}$$

where $N_\alpha^{(\nu)}$ is the normalization constant given by Eq. (A.22). The eigenfunction $|F_\alpha^{(\nu)}\rangle\!\rangle$ contains two parts, the privileged part (5.17) and its complement associated to the creation operator. We may give a compact form to Eq. (5.19). Indeed, assuming bicompleteness in the space $P^{(\nu)}$, we may always construct a set of states $\{\langle\!\langle \tilde{u}_\alpha^{(\nu)}|\}$ biorthogonal to $\{|u_\alpha^{(\nu)}\rangle\!\rangle\}$:

$$\langle\!\langle \tilde{u}_\alpha^{(\nu)}|u_\beta^{(\mu)}\rangle\!\rangle = \delta_{\nu,\mu}\delta_{\alpha,\beta}, \qquad \sum_\alpha |u_\alpha^{(\nu)}\rangle\!\rangle\langle\!\langle \tilde{u}_\alpha^{(\nu)}| = P^{(\nu)} \tag{5.20}$$

We have

$$L_0|u_\alpha^{(\nu)}\rangle\!\rangle = w_\nu|u_\alpha^{(\nu)}\rangle\!\rangle, \qquad \langle\!\langle \tilde{u}_\alpha^{(\nu)}|L_0 = \langle\!\langle \tilde{u}_\alpha^{(\nu)}|w_\nu \tag{5.21}$$

We can then introduce the "global" collision operators:

$$\theta_C^{(\nu)} \equiv \sum_\alpha \psi^{(\nu)}(Z_\alpha^{(\nu)})|u_\alpha^{(\nu)}\rangle\!\rangle\langle\!\langle \tilde{u}_\alpha^{(\nu)}| = \sum_\alpha |u_\alpha^{(\nu)}\rangle\!\rangle Z_\alpha^{(\nu)}\langle\!\langle \tilde{u}_\alpha^{(\nu)}| \tag{5.22}$$

We have

$$\theta_C^{(\nu)}|u_\alpha^{(\nu)}\rangle\!\rangle = Z_\alpha^{(\nu)}|u_\alpha^{(\nu)}\rangle\!\rangle, \qquad \langle\!\langle \tilde{u}_\alpha^{(\nu)}|\theta_C^{(\nu)} = \langle\!\langle \tilde{u}_\alpha^{(\nu)}|Z_\alpha^{(\nu)} \tag{5.23}$$

Moreover, we have

$$\theta_C^{(\nu)} = L_0 P^{(\nu)} + \lambda P^{(\nu)} L_V \mathbf{C}^{(\nu)} P^{(\nu)} \tag{5.24}$$

where we have introduced the "global" creation operators defined by

$$\mathbf{C}^{(\nu)} \equiv \sum_\alpha \mathscr{C}^{(\nu)}(Z_\alpha^{(\nu)})|u_\alpha^{(\nu)}\rangle\!\rangle\langle\!\langle \tilde{u}_\alpha^{(\nu)}| \tag{5.25}$$

They are off-diagonal operators:

$$\mathbf{C}^{(\nu)} = Q^{(\nu)}\mathbf{C}^{(\nu)}P^{(\nu)} \tag{5.26}$$

Operating with them on the privileged components we have [cf. Eq. (5.11)]

$$Q^{(\nu)} | F_\alpha^{(\nu)} \rangle\!\rangle = Q^{(\nu)} \mathbf{C}^{(\nu)} P^{(\nu)} | F_\alpha^{(\nu)} \rangle\!\rangle \tag{5.27}$$

The eigenstates of the Liouvillian are therefore

$$| F_\alpha^{(\nu)} \rangle\!\rangle = N_\alpha^{(\nu)1/2} (P^{(\nu)} + \mathbf{C}^{(\nu)}) | u_\alpha^{(\nu)} \rangle\!\rangle \tag{5.28}$$

This is the form we shall use repeatedly. Similarly, we have the left eigenstates of the Liouvillian:

$$\langle\!\langle \tilde{F}_\alpha^{(\nu)} | = \langle\!\langle \tilde{v}_\alpha^{(\nu)} | [P^{(\nu)} + \mathscr{D}^{(\nu)}(Z_\alpha^{(\nu)})] N_\alpha^{(\nu)1/2} = \langle\!\langle \tilde{v}_\alpha^{(\nu)} | (P^{(\nu)} + \mathbf{D}^{(\nu)}) N_\alpha^{(\nu)1/2} \tag{5.29}$$

Here $\langle\!\langle \tilde{v}_\alpha^{(\nu)} |$ are the privileged components of $\langle\!\langle F_\alpha^{(\nu)} |$,

$$\langle\!\langle \tilde{v}_\alpha^{(\nu)} | \equiv \langle\!\langle \tilde{F}_\alpha^{(\nu)} | P^{(\nu)} \tag{5.30}$$

and are left eigenstates of the collision operators

$$\langle\!\langle \tilde{v}_\alpha^{(\nu)} | \psi^{(\nu)}(Z_\alpha^{(\nu)}) = \langle\!\langle \tilde{v}_\alpha^{(\nu)} | Z_\alpha^{(\nu)} \tag{5.31}$$

In general,[7] $\langle\!\langle \tilde{v}_\alpha^{(\nu)} | \neq \langle\!\langle \tilde{u}_\alpha^{(\nu)} |$. In Eq. (5.29) $\mathscr{D}^{(\nu)}(z)$ is the "destruction-of-correlations operator," (or "destruction operator") and its explicit form is given in Appendix A. The destruction operator also describes irreducible off-diagonal transitions:

$$\mathscr{D}^{(\nu)}(z) = P^{(\nu)} \mathscr{D}^{(\nu)}(z) Q^{(\nu)} \tag{5.32}$$

For example, we may have "destruction (of correlation) processes" such as those represented graphically in Fig. 3.

This operator is related to the collision operator by [cf. Eq. (5.16)]

$$\psi^{(\nu)}(z) = P^{(\nu)} L_0 + \lambda P^{(\nu)} \mathscr{D}^{(\nu)}(z) L_V P^{(\nu)} \tag{5.33}$$

[7] The proof is as follows: For $Z_\alpha^{(\nu)} \neq Z_\beta^{(\nu)}$, we have

$$Z_\alpha^{(\nu)} \langle\!\langle \tilde{v}_\alpha^{(\nu)} | u_\beta^{(\nu)} \rangle\!\rangle = \langle\!\langle \tilde{v}_\alpha^{(\nu)} | \psi^{(\nu)}(Z_\alpha^{(\nu)}) | u_\beta^{(\nu)} \rangle\!\rangle \neq \langle\!\langle \tilde{v}_\alpha^{(\nu)} | \psi^{(\nu)}(Z_\beta^{(\nu)}) | u_\beta^{(\nu)} \rangle\!\rangle = Z_\beta^{(\nu)} \langle\!\langle \tilde{v}_\alpha^{(\nu)} | u_\beta^{(\nu)} \rangle\!\rangle$$

Therefore, $\langle\!\langle \tilde{v}_\alpha^{(\nu)} | u_\beta^{(\nu)} \rangle\!\rangle \neq 0$. This implies that $\langle\!\langle \tilde{v}_\alpha^{(\nu)} | \neq \langle\!\langle \tilde{u}_\alpha^{(\nu)} |$, because $\langle\!\langle \tilde{u}_\alpha^{(\nu)} |$ is orthogonal to $| u_\beta^{(\nu)} \rangle\!\rangle$.

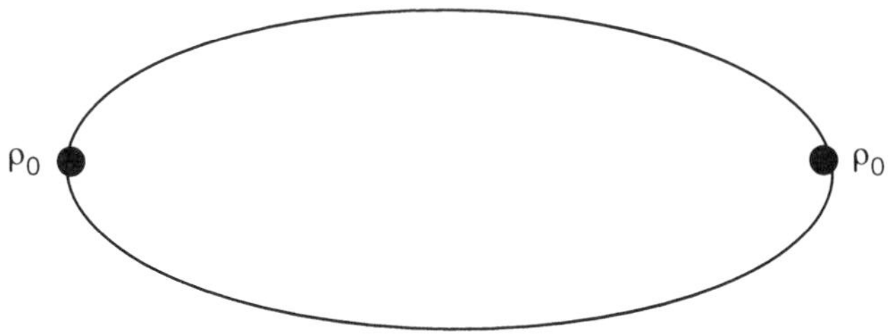

Figure 3. Graphical representation of the destruction-of-correlations operator $\mathscr{D}^{(\nu)}(z)$ for $\nu = 0$. The lowest-order contribution in the series expansion by λ (i.e., λ^2 order).

Assuming again a bicompleteness relation in the space $P^{(\nu)}$, we may construct a set of states $\{|v_\alpha^{(\nu)}\rangle\!\rangle\}$ biorthogonal to $\{\langle\!\langle\tilde{v}_\alpha^{(\nu)}|\}$ [see Eq. (A.17)]. Then we can introduce the "global" destruction operators $\mathbf{D}^{(\nu)}$

$$\mathbf{D}^{(\nu)} \equiv \sum_\alpha |v_\alpha^{(\nu)}\rangle\!\rangle\langle\!\langle\tilde{v}_\alpha^{(\nu)}|\mathscr{D}^{(\nu)}(Z_\alpha^{(\nu)}) \tag{5.34}$$

They are off-diagonal operators:

$$\mathbf{D}^{(\nu)} = P^{(\nu)}\mathbf{D}^{(\nu)}Q^{(\nu)} \tag{5.35}$$

We can also introduce the "global" collision operator $\theta_D^{(\nu)}$ associated with the destruction operator [cf. Eq. (5.24)]:

$$\theta_D^{(\nu)} = L_0 P^{(\nu)} + \lambda P^{(\nu)}\mathbf{D}^{(\nu)}L_V P^{(\nu)} \tag{5.36}$$

Generally,

$$\theta_D^{(\nu)} \neq \theta_C^{(\nu)} \tag{5.37}$$

But both operators share the same eigenvalues $Z_\alpha^{(\nu)}$, which are the same as the eigenvalues of the Liouvillian [see Eq. (A.20)]. Moreover, they satisfy the intertwining relation [33] [see Eq. (A.25)]:

$$\theta_C^{(\nu)}A^{(\nu)} = A^{(\nu)}\theta_D^{(\nu)} \tag{5.38}$$

where $A^{(\nu)}$ is related to the normalization constant $N_\alpha^{(\nu)}$ through the relation[8] [see Eq. (A.22)]:

$$A^{(\nu)} \equiv P^{(\nu)}(1 + \mathbf{D}^{(\nu)}\mathbf{C}^{(\nu)})^{-1} = \sum_\alpha |u_\alpha^{(\nu)}\rangle\!\rangle N_\alpha^{(\nu)} \langle\!\langle \tilde{v}_\alpha^{(\nu)}| \tag{5.39}$$

Of special interest are $\theta_C^{(0)}$ and $\theta_D^{(0)}$, which are the collision operators corresponding to the vacuum of correlations $\nu = 0$, because they lead to well-known kinetic equations for the momentum distribution function in the thermodynamic limit. For example, let us consider a weakly coupled system. Since $\psi^{(0)}(z)$ of the order λ^2, the eigenvalues $Z_\alpha^{(0)}$ are also of the same order λ^2 [see Eq. (5.15)],

$$Z_\alpha^{(0)} = \lambda^2 Z_{\alpha 2}^{(0)} + \lambda^3 \cdots \tag{5.40}$$

This implies that the lowest-order contribution of $\psi^{(0)}(Z_\alpha^{(0)})$ is evaluated at the point $Z_\alpha^{(0)} = 0$ in Taylor-series expansion. At this point, there are two possible analytic continuations. Let us denote them $z = \pm i\epsilon$ where ϵ is a positive infinitesimal $\epsilon \to 0+$. Then we have (to second order of λ)

$$\theta_C^{(0)\pm} = \lambda^2 \theta_2^{(0)\pm} + \lambda^3 \cdots \tag{5.41}$$

with

$$\begin{aligned}\lambda^2 \theta_2^{(0)\pm} &= \lambda^2 \psi_2^{(0)}(\pm i\epsilon) = \lambda^2 P^{(0)} L_V Q^{(0)} \frac{1}{\pm i\epsilon - L_0} Q^{(0)} L_V P^{(0)} \\ &= \lambda^2 \sum_{k\mathcal{N}} \sum_{P\mathcal{N}} P^{(0)} L_V Q^{(0)} |k, P\rangle\!\rangle \frac{1}{\pm i\epsilon - (k \cdot v)} \langle\!\langle k, P| Q^{(0)} L_V P^{(0)}\end{aligned} \tag{5.42}$$

In the continuous spectrum limit of k, the singularities of the denominator are located above or below the contour integral over k, depending on the direction of the analytic continuation. Then we have

$$\begin{aligned}\lambda^2 \langle\!\langle 0, P| \theta_2^{(0)\pm} |\rho_0\rangle\!\rangle &= \pm \frac{\lambda^2}{\Omega^2} \sum_{j>n}^{\mathcal{N}} \sum_{\mathbf{k}} |V_{|\mathbf{k}|}|^2 \partial_{jn}^{\mathbf{k}/2} \pi i \delta(\mathbf{k} \cdot \mathbf{g}_{jn}) \partial_{jn}^{\mathbf{k}/2} \rho_0(P) \\ &= \pm \frac{\lambda^2}{\Omega^2} \sum_{j>n}^{\mathcal{N}} \sum_{\mathbf{k}} 2\pi i |V_{|\mathbf{k}|}|^2 \delta(E_{\mathbf{P}_j+\mathbf{k}} + E_{\mathbf{P}_n-\mathbf{k}} - E_{\mathbf{P}_j} - E_{\mathbf{P}_n}) \\ &\quad \times [\rho_0(\mathbf{P}_j + \mathbf{k}, \mathbf{P}_n - \mathbf{k}, \{P\}^{\mathcal{N}-2}) - \rho_0(P)]\end{aligned} \tag{5.43}$$

[8] In general, $f(A^{(\nu)}) \neq \sum_\alpha |u_\alpha^{(\nu)}\rangle\!\rangle f(N_\alpha^{(\nu)}) \langle\!\langle \tilde{v}_\alpha^{(\nu)}|$. Therefore, this is not the spectral decomposition of $A^{(\nu)}$.

where

$$\mathbf{g}_{jn} \equiv \mathbf{v}_j - \mathbf{v}_n = \mathbf{P}_j/m_j - \mathbf{P}_n/m_n \tag{5.44}$$

and we have used the formula (for $\epsilon \to 0+$)

$$\frac{\epsilon}{w^2 + \epsilon^2} \to \pi\delta(w) \tag{5.45}$$

Equation (5.43) are the Pauli collision operators. They are antihermitian operators. By the choice $+i\epsilon$ of analytic continuation, the collision operator has nonvanishing negative imaginary eigenvalues associated with *diffusive* processes in momentum space (see the examples in the Appendices and also [4,29]). In contrast $-i\epsilon$ leads to *antidiffusive* processes:

$$\lambda^2 Z_{\alpha 2}^{(0)\pm} = \mp i\lambda^2 \gamma_{\alpha 2} \tag{5.46}$$

with $\gamma_{\alpha 2} \geq 0$ real. To be consistent with the semigroup oriented toward the future (i.e., with Im $Z_\alpha^{(0)} \leq 0$), we have to choose the branch $+i\epsilon$. If we choose the branch $-i\epsilon$, we obtain a semigroup oriented toward the past. Hereafter, we shall always choose the branch $+i\epsilon$ and abbreviate the superscript "+."

As mentioned, we generally have $\theta_C^{(0)} \neq \theta_D^{(0)}$, but in the weak coupling limit they reduce to the same Pauli collision operator:

$$\theta_D^{(0)} \approx \theta_C^{(0)} \approx \lambda^2 \theta_2^{(0)} \tag{5.47}$$

Also, in the low concentration limit, $\theta^{(0)}$ reduces to the quantum Boltzman collision operator [4].

The example discussed just above shows that the analytic continuation in the collision operator is not trivial. Indeed, $+i\epsilon$ gives the eigenvalues that have opposite sign, $-i\lambda^2\gamma_{\alpha 2}$. To evaluate higher-order corrections of the collision operator, we should keep the location of the singularity at $k \cdot v = -i\lambda^2\gamma_{\alpha 2}$ relative to the contour the same as the location at $k \cdot v = +i\epsilon$ in Eq. (5.42). For example, to evaluate the next-order correction, we *first* evaluate the contour integration in Eq. (5.42) for $\epsilon > 0$, then continue analytically to $\epsilon \to -\lambda^2\gamma_{\alpha 2}$. This leads to the concept of "delayed" analytic continuation [40] or "complex distributions" [see Eqs. (A.7)–(A.9) for more detail].

Starting with the analytic continuation for the eigenstates $|F_\alpha^{(0)}\rangle\!\rangle$ corresponding to the vacuum of correlations $\nu = 0$, and imposing the bi-orthogonality condition of the eigenstates in Eq. (5.5), one can determine the analytic continuation for arbitrary degrees of correlations ν. We present the result in Appendix A.

Needless to say, analytic continuation has meaning only with continuous variables. However, it is convenient to use summation signs for discrete momenta and for wave vectors when we discuss the thermodynamic limit. Hereafter, we often use conventional notations of the summation sign of the discrete variables, but with the parameter ϵ of the analytic continuation, such as $(w_\nu \pm i\epsilon)^{-1}$ in Eq. (5.42). For the continuous spectrum limit $\Omega \to \infty$, the propagator becomes the distribution

$$\frac{1}{w_\nu \pm i\epsilon} \to \mathscr{P}\frac{1}{w_\nu} \mp i\pi\delta(w_\nu) \tag{5.48}$$

where $\mathscr{P}$ stands for the principal part. The use of the δ function $\delta(w_\nu)$ is possible only because we consider the wave vector $\mathbf{k}$ as a continuous variable. For finite ϵ, the delta function $\pi\delta(w_\nu)$ may be aproximated by the Lorentzian distribution $\epsilon/(w_\nu^2 + \epsilon^2)$. To obtain a consistent evaluation for the delta function in terms of the box normalization formalism, there should be enough discrete states around the peak of the Lorentzian. Therefore, our expressions have to be understood in the continuous limit $\Delta k = 2\pi/L \to 0$ and $\epsilon \to 0+$ with the condition

$$\frac{|dw_\nu/dk|\,\Delta k}{\epsilon} \to 0 \tag{5.49}$$

The reader should understand notations such as Eq. (3.5) by keeping in mind Eq. (5.49).

Eq. (5.42) shows that the global collision operators coincide to the irreducible collision operator in the lowest-order approximation of λ. Because $Z_\alpha^{(0)}$ starts with order λ^2, we have, up to the second order [see Eqs. (5.13) and (A.2)],

$$\mathbf{C}^{(0)} = \mathscr{C}^{(0)}(+i\epsilon) + O(\lambda^3), \qquad \mathbf{D}^{(0)} = \mathscr{D}^{(0)}(+i\epsilon) + O(\lambda^3) \tag{5.50}$$

The difference $\mathbf{C}^{(0)}$ from $\mathscr{C}^{(0)}(+i\epsilon)$ and $\mathbf{D}^{(0)}$ from $\mathscr{D}^{(0)}(+i\epsilon)$ begins with order λ^3. As a result, the difference of the global collision operators $\theta^{(0)}$ from the irreducible collision operator $\psi^{(0)}(+i\epsilon)$ begins with order λ^4.

We note that the contribution of the Pauli operator in Eq. (5.43) occurs only when the integration over wave vectors satisfies Poincaré's resonance condition $\mathbf{k} \cdot \mathbf{g}_{jn} = 0$, which is equivalent to $E_{\mathbf{p}_j+\mathbf{k}} + E_{\mathbf{p}_j-\mathbf{k}} = E_{\mathbf{p}_j} + E_{\mathbf{p}_n}$. This means that the dissipation has a dynamical origin associated with the "nonintegrability" of LPS due to Poincaré's resonances. The Pauli operator leads to "Brownian motion." Instead of separate dynamical events described by each interaction λL_V, we have events "coupled" by the resonance condition $\delta(\mathbf{k} \cdot \mathbf{g}_{jn})$ (see Fig. 2). The diffusion process is "irreducible" to any wave-function dynamics. Each eigenvalue of the diffusion operator cannot be represented by a difference of two numbers, so the Rydberg–Ritz principle does not hold for LPS. This corresponds to a striking difference between nonintegrable and integrable systems, where each eigenvalue of the Liouvillian is simply a difference of two eigenvalues of the Hamiltonian in the latter. We have "non-Schrödinger" processes due to the Poincaré resonances.

In the correlation subspace $P^{(\nu)}$, the collision operator $\theta_C^{(\nu)}$ leads to a natural generalization of the kinetic theory. We then obtain (e.g., to the lowest order) the contribution of the collision operators for $\nu \neq 0$ (for the semigroup oriented toward the future):

$$\theta_C^{(\nu)} \approx \theta_D^{(\nu)} \approx L_0 P^{(\nu)} + \lambda^2 \theta_2^{(\nu)}$$
$$= L_0 P^{(\nu)} + \lambda^2 \sum_{\mu} P^{(\nu)} L_V P^{(\mu)} \frac{-1}{w_\mu - w_\nu + i\epsilon_{\mu\nu}} P^{(\mu)} L_V P^{(\nu)} \quad (5.51)$$

where the parameter of the analytical continuation $\epsilon_{\mu\nu}$ is defined in Eq. (A.14). For a given order of correlation ν, the corrrelation in the intermediate state μ could be higher or lower than ν. However, to achieve a diagonal transition, wave-vector transfer cannot be arbitrary. This restricts the wave vectors of the intermediate states and leads to extra volume factors Ω^{-1} throughout the interaction [see Eq. (3.50)]. This implies that $\lambda^2 \theta_2^{(\nu)}$ in Eq. (5.51) gives nonvanishing contributions in the thermodynamic limit only when the intermediate states μ involve more particles than the states in $P^{(\nu)}$, because the summation over the extra particle leads to extra factor $\mathcal{N}$, which then compensates for the factor Ω^{-1}. In other words, the intermediate states should have a higher degree of correlations than $P^{(\nu)}$. This is a special case of Henin's theorem [35] which states that:

> Nonvanishing contributions to an *irreducible* diagonal transition between $P^{(\nu)}$ come only from processes which consist of intermediate states with higher order of correlations than $P^{(\nu)}$.

This fixes the direction of the analytic continuation of the denominator in Eq. (5.51) as $\epsilon_{\mu\nu} = -\epsilon$ (within the semigroup oriented toward the future). Then we obtain [see Eq. (4.32)]

$$\lambda^2 \theta_2^{(\nu)} = \lambda^2 P^{(\nu)} L_V Q^{(\nu)} \frac{1}{w_\nu + i\epsilon - L_0} Q^{(\nu)} L_V P^{(\nu)} = \lambda^2 \psi_2^{(\nu)}(w_\nu + i\epsilon) \quad (5.52)$$

In Appendix I we present an example of the eigenvalue problem for the collision operators (5.47) and (5.51), as well as of L_H, for a simple $\mathcal{N}$-body system of the so-called "perfect Lorentz gas" [29]. There we show that there is no equilibrium mode with $Z_\alpha^{(\nu)} = 0$ for $\nu \neq 0$. The equilibriun state lies in the subspace associated with the vacuum of correlations [i.e., in the subspace associated with $\Pi^{(0)}$, as defined in Eq. (6.16)].[9]

In summary we have obtained the explicit form of the "complex spectral representation" of L_H [see Eq. (5.6)] and therefore of the evolution operator $\mathscr{U}(t)$:

$$\langle\langle M | \mathscr{U}(t) | \rho(0) \rangle\rangle = \sum_{\nu} \sum_{\alpha} \langle\langle M | F_{\alpha}^{(\nu)} \rangle\rangle e^{-iZ_{\alpha}^{(\nu)}t} \langle\langle \tilde{F}_{\alpha}^{(\nu)} | \rho(0) \rangle\rangle \tag{5.53}$$

This spectral decomposition involves the spectral decomposition of the dissipative collision operators.

However, the existence of the collision operator is only a *necessary* condition to observe irreversibility. In addition, we have to discuss the class of distribution functions ρ on which our complex spectral decomposition acts. In subsequent sections we apply our spectral representation to various situations. In simple cases (finite number of particles and density matrices with finite Hilbert norm), we recover the usual results of wave-function dynamics without any dissipation even though we deal with LPS. Still, there are many situations where our new "non-Schrödinger" effects can be observed (see Sections X–XIII). Our basic result is that we obtain new solutions of the quantum Liouville–von Neumann equation which lie outside ordinary quantum mechanics.

VI. NONUNITARY TRANSFORMATIONS AND SUBDYNAMICS

In the previous section we showed that L_H shares the same eigenvalues $Z_{\alpha}^{(\nu)}$ with the dissipative operators $\theta_C^{(\nu)}$ and $\theta_D^{(\nu)}$ [see Eqs. (5.23) and (A.21)]. This implies that nonunitary transformation operators exist which lead to *similitude relations* between the Liouvillian and the collision operators [14,22] (hereafter the index B also stands for C or D according to the collision operators (5.22) or (5.36) we consider). The similitude relation is

$$\Lambda_B L_H \Lambda_B^{-1} = \Theta_B \tag{6.1}$$

where

$$\Theta_B \equiv \sum_{\nu} \theta_B^{(\nu)} \tag{6.2}$$

[9] This is a general property of the systems in thermodynamic limit, as one can prove $\psi^{(0)}(+i\epsilon)f(H_0) = 0$ [that leads to $L_H F^{eq}(H) = 0$] in every order of the coupling constant λ for arbitrary analytic functions f [4].

As can be easily verified, the nonunitary transformations Λ and their inverses are given by [see Eqs. (5.28) and (5.29)]

$$\Lambda_C = \sum_\nu \sum_\alpha |u_\alpha^{(\nu)}\rangle\!\rangle\langle\!\langle \tilde{F}_\alpha^{(\nu)}| N_\alpha^{(\nu)1/2} = \sum_\nu A^{(\nu)}\hat{\Phi}_\nu^D \tag{6.3a}$$

$$\Lambda_C^{-1} = \sum_\nu \sum_\alpha |F_\alpha^{(\nu)}\rangle\!\rangle\langle\!\langle \tilde{u}_\alpha^{(\nu)}| N_\alpha^{(\nu)-1/2} = \sum_\nu \hat{\Phi}_\nu^C \tag{6.3b}$$

and

$$\Lambda_D = \sum_\nu \sum_\alpha |v_\alpha^{(\nu)}\rangle\!\rangle\langle\!\langle \tilde{F}_\alpha^{(\nu)}| N_\alpha^{(\nu)-1/2} = \sum_\nu \hat{\Phi}_\nu^D \tag{6.3c}$$

$$\Lambda_D^{-1} = \sum_\nu \sum_\alpha |F_\alpha^{(\nu)}\rangle\!\rangle\langle\!\langle \tilde{v}_\alpha^{(\nu)}| N_\alpha^{(\nu)1/2} = \sum_\nu \hat{\Phi}_\nu^C A^{(\nu)} \tag{6.3d}$$

where the operators $\hat{\Phi}_\nu^B$ are defined by

$$\hat{\Phi}_\nu^C \equiv P^{(\nu)} + \mathbf{C}^{(\nu)} \quad \text{and} \quad \hat{\Phi}_\nu^D \equiv P^{(\nu)} + \mathbf{D}^{(\nu)} \tag{6.4}$$

Because L_H shares the same eigenvalues with $\theta_B^{(\nu)}$, we have the "intertwining relations" [14]

$$L_H \hat{\Phi}_\nu^C = \hat{\Phi}_\nu^C \theta_C^{(\nu)} \quad \text{and} \quad \hat{\Phi}_\nu^D L_H = \theta_D^{(\nu)} \hat{\Phi}_\nu^D \tag{6.5}$$

One can easily verify these relations by acting on the eigenstates of the collision operators. We note that the similitude relation (6.1) leads to the intertwining relations (6.5), and vice versa. These relations were obtained previously [14,22]. The existence of the two different transformations Λ_C and Λ_D suggests the possibility of another transformation operator Λ associated with a more symmetrical form of the collision operator. This was shown in previous papers [22] and in *I*. We present the symmetrical form in Appendix B.

As is well known, unitary transformations U exist for integrable systems which lead to[10]

$$U L_H U^\dagger = L_0 \tag{6.6}$$

We expect that when dissipative effects are negligible, the relations (6.1) would reduce to

$$\Lambda_B L_H \Lambda_B^{-1} = L_0 \tag{6.7}$$

[10] In general, the diagonalization of the Liouvillian L_H by unitary transformations leads to a renormalized Liouvillian L_0' (instead of L_0) which gives frequency shifts associated to diagonal transitions. However, for the case where there is no bound state, the renormalization effects are negligible because for these interactions they lead to terms of order Ω^{-1}.

We shall verify this expectation later. However, because the complex spectral representation uses both analytic continuations (see Appendix A), Eq. (6.7) is not a unitary transformation even for the integrable case. As a result, we see that integrable LPS are diagonalized through both a nonunitary and a unitary transformation. We come back to this problem in Section VIII.

Using Λ, we may introduce the transformed distribution function ρ_B and the transformed observables M_B:

$$|\rho_B(t)\rangle\rangle \equiv \Lambda_B|\rho(t)\rangle\rangle, \qquad \langle\langle M_B(t)| \equiv \langle\langle M(t)|\Lambda_B^{-1} \tag{6.8}$$

The new states ρ_B obey [see Eq. (6.2)]

$$i\frac{\partial}{\partial t}|\rho_B(t)\rangle\rangle = \Theta_B|\rho_B(t)\rangle\rangle \tag{6.9}$$

Since $\theta_B^{(\nu)}$ are operations acting on $P^{(\nu)}$ subspace, Eq. (6.9) actually represents "kinetic equations" for $P^{(\nu)}|\rho_B(t)\rangle\rangle$ in each correlation subspace:

$$i\frac{\partial}{\partial t}P^{(\nu)}|\rho_B(t)\rangle\rangle = \theta_B^{(\nu)}P^{(\nu)}|\rho_B(t)\rangle\rangle \tag{6.10}$$

This represents a set of Pauli type kinetic equations [4]; each function $P^{(\nu)}|\rho_B(t)\rangle\rangle$ evolves independently. This represents the general formulation of "kinetic theory." We shall consider examples later. We note that

$$\int d^{\mathcal{N}}P\rho_0^B(|P,t) = \sum_{P^{\mathcal{N}}}\langle\langle 0, P|\rho_B(t)\rangle\rangle = \sum_{P^{\mathcal{N}}}\langle\langle 0, P|\rho(t)\rangle\rangle = 1 \tag{6.11}$$

This is a direct consequence of the relation (3.60) as well as of the definition of Λ_B in Eq. (6.3).

Similarly, the new observables $M_B(t)$ obey

$$i\frac{\partial}{\partial t}\langle\langle M_B(t)| = \langle\langle M_B(t)|\Theta_B \tag{6.12}$$

which leads again to a set of equations

$$i\frac{\partial}{\partial t}\langle\langle M_B(t)|P^{(\nu)} = \langle\langle M_B(t)|P^{(\nu)}\theta_B^{(\nu)} \tag{6.13}$$

The transformation (6.8) preserves the expectation value of M:

$$\langle M \rangle_t = \langle\langle M(0) | \rho(t) \rangle\rangle = \langle\langle M_B(0) | \rho_B(t) \rangle\rangle \tag{6.14}$$

Using the solution of the eigenvalue problem of the collision operator $\theta_C^{(\nu)}$, for example, the expectation value is

$$\langle M \rangle_t = \sum_\nu \sum_\alpha \langle\langle M_C(0) | u_\alpha^{(\nu)} \rangle\rangle e^{-iZ_\alpha^{(\nu)} t} \langle\langle \tilde{u}_\alpha^{(\nu)} | \rho_C(0) \rangle\rangle \tag{6.15}$$

The nonunitary transformations Λ preserve the reality of the states $\langle\langle 0, P | \rho \rangle\rangle = \langle P | \rho | P \rangle$ (see Appendix B for the proof). But the transformed states $\langle p | \rho_B | p \rangle$ cannot be considered as probability distribution functions, because Λ does not preserve positivity. This is a direct consequence of the causal evolution of dynamics combined with the analytic continuations (A.5) and (A.6) (see Appendix C). However, these states play an important role because they lead to block diagonal equations and permit us to introduce "Lyapounov functions" for dynamical systems (see the next section).

In our earlier work, we repeatedly introduced the concept of "subdynamics" [33–45]. To see the relation of subdynamics to the complex spectral representation, let us introduce projection operators $\Pi^{(\nu)}$ [see Eqs. (5.39) and (6.4)]:

$$\Pi^{(\nu)} = \Lambda_B^{-1} P^{(\nu)} \Lambda_B = \sum_\alpha | F_\alpha^{(\nu)} \rangle\rangle \langle\langle \tilde{F}_\alpha^{(\nu)} | \tag{6.16}$$

We note that

$$\Pi^{(\nu)\dagger} \neq \Pi^{(\nu)} \tag{6.17}$$

Equation (6.16) leads to the familiar form [33–45]

$$\Pi^{(\nu)} = \hat{\Phi}_\nu^C A^{(\nu)} \hat{\Phi}_\nu^D = (P^{(\nu)} + \mathbf{C}^{(\nu)}) A^{(\nu)} (P^{(\nu)} + \mathbf{D}^{(\nu)}) \tag{6.18}$$

These operators satisfy the orthogonality and completeness relations

$$\Pi^{(\nu)} \Pi^{(\mu)} = \Pi^{(\nu)} \delta_{\nu,\mu} \quad \text{and} \quad \sum_\nu \Pi^{(\nu)} = 1 \tag{6.19}$$

as well as the commutation relation with L_H,

$$L_H \Pi^{(\nu)} = \Pi^{(\nu)} L_H \tag{6.20}$$

$\Pi^{(\nu)}$ is an extension of $P^{(\nu)}$ to the total Liouvillian L_H.

Because these projection operators commute with the Liouvillian, each component $\mathcal{U}(t)\Pi^{(\nu)}$ satisfies separate equations of motion:

$$
\begin{aligned}
\langle\langle M | \mathcal{U}(t)\Pi^{(\nu)} | \rho(0)\rangle\rangle &= \langle\langle M | \Pi^{(\nu)}\mathcal{U}(t) | \rho(0)\rangle\rangle \\
&= \langle\langle M | \hat{\Phi}_{\nu}^{C} e^{-i\theta_{C^{(\nu)}}t} A^{(\nu)} \hat{\Phi}_{\nu}^{D} | \rho(0)\rangle\rangle \\
&= \langle\langle M | \hat{\Phi}_{\nu}^{C} A^{(\nu)} e^{-i\theta_{D^{(\nu)}}t} \hat{\Phi}_{\nu}^{D} | \rho(0)\rangle\rangle \qquad (6.21)
\end{aligned}
$$

For this reason, the projection operators $\Pi^{(\nu)}$ are associated with *subdynamics.*

As an illustration, let us consider the evolution of a state which is initially in the vacuum of correlations:

$$
| \rho(0)\rangle\rangle = P^{(0)} | \rho(0)\rangle\rangle \qquad (6.22)
$$

We now show that the time evolution leads to the correlations that satisfy the volume dependence given in Eq. (4.18). From Eqs. (6.19)–(6.21), we have

$$
\begin{aligned}
| \rho(t)\rangle\rangle &= (P^{(0)} + \mathbf{C}^{(0)}) e^{-i\theta_{C^{(0)}}t} A^{(0)} P^{(0)} | \rho(0)\rangle\rangle \\
&\quad + \sum_{\nu(\neq 0)} \mathbf{C}^{(\nu)} e^{-i\theta_{C^{(\nu)}}t} A^{(\nu)} \mathbf{D}^{(\nu)} P^{(0)} | \rho(0)\rangle\rangle \\
&= (P^{(0)} + \lambda \mathbf{C}_1^{(0)} + \lambda^2 \cdots) e^{-i\lambda^2 \theta_{2^{(0)}}t} (1 + \lambda^2 A_2^{(0)} + \lambda^3 \cdots) | \rho(0)\rangle\rangle \\
&\quad + \lambda^2 \mathbf{C}_1^{(2)} e^{-i(L_0 P^{(2)} + \lambda^2 \theta_{2^{(2)}})t} \mathbf{D}_1^{(2)} P^{(0)} | \rho(0)\rangle\rangle + \lambda^3 \cdots \qquad (6.23)
\end{aligned}
$$

where the subscripts n in the operators represent their λ^n-order contributions, and the superscript (ν) corresponds to νth-order correlations. By calculations similar to Eq. (4.20) for each term in Eq. (6.23), one can easily verify that the volume dependence for all correlation components are in agreement with Eq. (4.18). The reader can find the detailed estimation of the volume dependence in our earlier articles [4,29]. This shows that the class of singular distribution functions (4.18) is not only form invariant in time, but it acts even as an *attractor.*

VII. LYAPOUNOV FUNCTIONS–$\mathcal{H}$ THEOREMS

The nonunitary transformations have led to the similitude relation (6.1) between the total Liouvillian L_H and the collision operators. As a consequence, we may introduce transformed states and observables (6.8) whose time evolutions are described only by the $P^{(\nu)}$ components in each correlation subspace. This permits us to introduce "Lyapounov functions" which are dynamical analogues of Boltzmann's $\mathscr{H}$-function (i.e., "entropy") for dynamical systems [14,15]. The existence of entropy is the consequence of the complex, irreducible spectral representation of the Liouvillian.

To illustrate this statement, let us consider first the *generic*, reduced, single-particle momentum distribution function defined by

$$\varphi_1(\mathbf{P}_j, t) = \langle\!\langle \hat{\varphi}_{\mathbf{P}_j} | \rho(t) \rangle\!\rangle = \int d^{\mathscr{N}} P' \delta(\mathbf{P}'_j - \mathbf{P}_j) \rho_0(|P', t) \tag{7.1}$$

with

$$\begin{aligned} \langle\!\langle \hat{\varphi}_{\mathbf{P}_j} | &\equiv \int d^{\mathscr{N}} Q \int d^{\mathscr{N}} P' \delta(\mathbf{P}'_j - \mathbf{P}_j) \langle\!\langle Q, P' | \\ &= L^{3\mathscr{N}/2} \int d^{\mathscr{N}} P' \delta(\mathbf{P}'_j - \mathbf{P}_j) \langle\!\langle 0, P' | P^{(0)} \end{aligned} \tag{7.2}$$

where $\langle\!\langle 0, P' |$ in the right-hand side of Eq. (7.2) is written in the wave-number representation with $k = 0$. We have, for example [see Eq. (4.25)],

$$\langle\!\langle \hat{\mathbf{P}}_j | = \int d\mathbf{P}_j \mathbf{P}_j \langle\!\langle \hat{\varphi}_{\mathbf{P}_j} | \tag{7.3}$$

We note that

$$\langle\!\langle 0, P' | \hat{\varphi}_{\mathbf{P}_j} \rangle\!\rangle \langle\!\langle \hat{\varphi}_{\mathbf{P}_j} | \rho \rangle\!\rangle = L^{3\mathscr{N}/2} \delta(\mathbf{P}'_j - \mathbf{P}_j) \varphi_1(\mathbf{P}_j) \tag{7.4}$$

Hence, the hermitian operator $| \hat{\varphi}_{\mathbf{P}_j} \rangle\!\rangle \langle\!\langle \hat{\varphi}_{\mathbf{P}_j} |$ preserves positivity. The reduction to the single particle distribution function does not change the sign of the distribution function ρ.

We now consider the transformed single particle distribution function [see Eq. (6.8)], for example, for $j = 1$,

$$\varphi_1^B(\mathbf{P}_1, t) = \langle\!\langle \hat{\varphi}_{\mathbf{P}_1} | P^{(0)} | \rho_B(t) \rangle\!\rangle \tag{7.5}$$

We have [see Eq. (6.11)]

$$\int d\mathbf{P}_1 \varphi_1^B(\mathbf{P}_1, t) = 1 \tag{7.6}$$

Then, a Lyapounov function associated with this distribution function may be defined by

$$\bar{\mathscr{H}}^B_{\varphi_1}(t) = \int d\mathbf{P}_1 [\varphi_1^B(\mathbf{P}_1, t)]^2 \tag{7.7}$$

where

$$[\varphi_1^B(\mathbf{P}_1, t)]^2 = \langle\langle \rho(t) | \Lambda_B^\dagger | \hat{\varphi}_{\mathbf{P}_1} \rangle\rangle \langle\langle \hat{\varphi}_{\mathbf{P}_1} | \Lambda_B | \rho(t) \rangle\rangle \tag{7.8}$$

Note that our Lyapounov function is defined in the $P^{(0)}$ space that does not depend on wave vectors. All integrations over wave vectors which appear in the intermediate states are already performed. The Lyapounov function (7.7) is thus a well-defined function (and not a *distribution*), and (7.8) is a positive quantity.[11]

We have, from Eq. (6.10) (e.g., for $B = C$),

$$[\varphi_1^B(\mathbf{P}_1, t)]^2 = \sum_{\alpha, \beta} e^{-i(Z_\alpha(0) - Z_\beta(0)c.c.)t} \langle\langle \rho_C(0) | \tilde{u}_\beta^{(0)} \rangle\rangle \\ \times \langle\langle u_\beta^{(0)} | \hat{\varphi}_{\mathbf{P}_1} \rangle\rangle \langle\langle \hat{\varphi}_{\mathbf{P}_1} | u_\alpha^{(0)} \rangle\rangle \langle\langle \tilde{u}_\alpha^{(0)} | \rho_C(0) \rangle\rangle \tag{7.9}$$

All decay modes are damped for $t > 0$. Moreover, we now show that the damping is monotonic. Taking the time derivative of Eq. (7.8), we obtain

$$\frac{\partial}{\partial t} [\varphi_1^B(\mathbf{P}_1, t)]^2 = -\langle\langle \rho_B(t) | \mathscr{K}_B^{(0)}(\mathbf{P}_1) | \rho_B(t) \rangle\rangle \tag{7.10}$$

where $\mathscr{K}_B^{(0)}$ is defined by

$$\mathscr{K}_B^{(0)}(\mathbf{P}_1) \equiv | \hat{\varphi}_{\mathbf{P}_1} \rangle\rangle \langle\langle \hat{\varphi}_{\mathbf{P}_1} | i\theta_B^{(0)} + (i\theta_B^{(0)})^\dagger | \hat{\varphi}_{\mathbf{P}_1} \rangle\rangle \langle\langle \hat{\varphi}_{\mathbf{P}_1} | \tag{7.11}$$

where $\mathscr{K}_B^{(0)}$ is a hermitian operator. Eigenstates of $\mathscr{K}_B^{(0)}$ are normalizable in terms of the Hilbert norm. This is in contrast to the eigenstates of the Liouvillian L_H in the extended class of density matrices.[12] Thus, eigenvalues of $\mathscr{K}_B^{(0)}$ are real and the left eigenstates are the hermitian conjugate of the right eigenstates.

[11] This is not necessarily true for $\nu \neq 0$ because $\langle\langle \nu | \rho_B(t) \rangle\rangle$ are "complex" distributions [see Eq. (A.9)]. We thank M. de Haan for help in the clarification of the definition of the Lyapounov function.

[12] The eigenstates of $\mathscr{K}_B^{(0)}$ consist only of the privileged components, which are normalizable. In contrast, the eigenstates of L_H include the nonprivileged components associated with the creation operators, by which they become nonnormalizable.

Let us assume that the spectral decomposition of $\mathscr{K}_B^{(0)}$ is known:

$$\mathscr{K}_B^{(0)}(\mathbf{P}_1) = \sum_\beta \gamma_\beta(\mathbf{P}_1)|w_\beta(\mathbf{P}_1)\rangle\!\rangle\langle\!\langle w_\beta(\mathbf{P}_1)| \tag{7.12}$$

where γ_β are real numbers and

$$\begin{aligned} \langle\!\langle w_\beta(\mathbf{P}_1)|w_{\beta'}(\mathbf{P}_1)\rangle\!\rangle &= \delta_{\beta,\beta'}, \\ \sum_\beta |w_\beta(\mathbf{P}_1)\rangle\!\rangle\langle\!\langle w_\beta(\mathbf{P}_1)| &= \int d^{\mathcal{N}}P'\delta(\mathbf{P}'_1 - \mathbf{P}_1)|0, P'\rangle\!\rangle\langle\!\langle 0, P'| \end{aligned} \tag{7.13}$$

As mentioned, the operator $|\hat{\varphi}_{\mathbf{P}_1}\rangle\!\rangle\langle\!\langle\hat{\varphi}_{\mathbf{P}}|$ preserves the positivity. The reduction does not change the sign of the collision operator. Therefore, $\mathscr{K}_B^{(0)}$ is a nonnegative operator:

$$\gamma_\beta(\mathbf{P}_1) \geq 0 \tag{7.14}$$

Then we have

$$\frac{\partial}{\partial t}[\varphi_1^B(\mathbf{P}_1, t)]^2 = -\sum_\beta \gamma_\beta(\mathbf{P}_1)|\langle\!\langle w_\beta(\mathbf{P}_1)|\rho_B(t)\rangle\!\rangle|^2 \leq 0 \tag{7.15}$$

The evolution of $[\varphi_1^B(\mathbf{P}_1, t)]^2$ is therefore monotonic. As a consequence [see Eq. (7.7)],

$$\frac{d}{dt}\bar{\mathscr{H}}_{\varphi_1}^B(t) \leq 0 \tag{7.16}$$

Hence, the $\mathscr{H}$ theorem holds. For the nonintegrable case, we have the even stronger condition $\gamma_\beta > 0$ for some components. Then, $\bar{\mathscr{H}}_{\varphi_1}^B(t)$ monotonously decreases for $t > 0$, until all decay modes disappear and the system approaches equilibrium. Contrary to Boltzmann's $\mathscr{H}$ theorem, our $\mathscr{H}$ theorem is valid for all λ (or concentrations) for which the spectral decomposition of $\mathscr{K}_B^{(0)}$ can be determined.

Instead of the Lyapounov function (7.7), we can introduce the more familiar form of the $\mathscr{H}$-function such as

$$\mathscr{H}_{\varphi_1}^B(t) = \int d\mathbf{P}_1\varphi_1^B(\mathbf{P}_1, t)\log\varphi_1^B(\mathbf{P}_1, t) \tag{7.17}$$

Taking the time derivative, we obtain

$$\frac{d}{dt}\mathcal{H}^B_{\varphi_1}(t)=\frac{1}{2}\int d\mathbf{P}_1\,\frac{1}{\varphi^B_1(\mathbf{P}_1,t)}(\log\varphi^B_1(\mathbf{P}_1,t)+1)\frac{\partial}{\partial t}[\varphi^B_1(\mathbf{P}_1,t)]^2\le 0 \quad (7.18)$$

Again we recover the $\mathcal{H}$ theorem. In the lowest order of λ (or of the concentration), the transformation (7.5) is not necessary (i.e., $\Lambda_B \approx 1$) and Boltzmann's formulation is recovered.

One can also introduce an extensive $\mathcal{H}$-function which is proportional to the number of degrees of freedom. Let us consider the generic r-particle reduced momentum distribution functions φ_r given by

$$\varphi_r(\{P\}^r,t)=\langle\!\langle\hat{\varphi}_{Pr}|\rho(t)\rangle\!\rangle=\int d^{\mathcal{N}}P'\delta(\mathbf{P}'_1-\mathbf{P}_1)\cdots\delta(\mathbf{P}'_r-\mathbf{P}_r)\rho_0(|P',t) \quad (7.19)$$

with [cf. Eq. (7.2)]

$$\langle\!\langle\hat{\varphi}_{Pr}|=L^{3\mathcal{N}/2}\int d^{\mathcal{N}}P'\delta(\mathbf{P}'_1-\mathbf{P}_1)\cdots\delta(\mathbf{P}'_r-\mathbf{P}_r)\langle\!\langle 0,P'|P^{(0)} \quad (7.20)$$

The transformed distribution function is then given by

$$\varphi^B_r(\{P\}^r,t)=\langle\!\langle\hat{\varphi}_{Pr}|P^{(0)}|\rho_B(t)\rangle\!\rangle \quad (7.21)$$

It is well known that by assuming the reduced momentum distribution, functions are factorized at a certain time (e.g., at $t=0$), their factorization property propagates in time by the Liouville–von Neumann evolution for the class of the density matrices in Eq. (4.18) in the thermodynamic limit (see [4.29] for the proof):

$$\varphi_r(\{P\}^r,t)=\varphi_1(\mathbf{P}_1,t)\cdots\varphi_1(\mathbf{P}_r,t) \quad (7.22)$$

All contributions that destroy the factorizability are higher order in $1/\Omega$, which vanishes in the thermodynamic limit. A straightforward extension of the proof leads to the factorization property of the transformed distribution functions:

$$\varphi^B_r(\{P\}^r,t)=\varphi^B_1(\mathbf{P}_1,t)\cdots\varphi^B_1(\mathbf{P}_r,t) \quad (7.23)$$

Now, we can define a $\mathcal{H}$-function associated with r particles through the transformed distribution functions

$$\mathcal{H}^B_r(t)=\int d^rP\varphi^B_r(\{P\}^r,t)\log\varphi^B_r(\{P\}^r,t)=r\mathcal{H}^B_{\varphi_1}(t) \quad (7.24)$$

where we have used Eq. (7.6). This is indeed proportional to the number of degrees of freedom.

Let us recall that we have an infinite set of the kinetic equations (6.10) for each subspace $P^{(\nu)}$. This allows us to generalize the Lyapounov functions for more general cases associated with each subspace. However, since the generalization involves a careful manipulation of the distributions of the wave vectors, we shall present the generalization in a separate paper [47].

VIII. INTEGRABILITY CONDITIONS AND LONG-RANGE CORRELATIONS

We shall now show that Poincaré divergences lead to long-range correlations in nonequilibrium conditions. To introduce this problem, it is convenient to define the observables:

$$\langle\langle \tilde{M}^B(t)| \equiv \langle\langle M(0)| \Lambda_B \mathscr{U}(t) \tag{8.1}$$

where we have introduced the notation with a tilde as well as the superscript B, instead of the subscript, to distinguish this quantity from $\langle\langle M_B(t)|$ introduced in Eq. (6.8). As mentioned [see Eq. (6.12)], the time evolution of $\langle\langle M_B(t)|$ is generated by the collision operator Θ_B. In contrast, the time evolution of the new transformed observables $\langle\langle \tilde{M}^B(t)|$ is generated by the Liouvillian L_H.

The important property of $\tilde{M}^B$ is that when M is in a single correlation subspace $P^{(\nu)}$, then $\tilde{M}^B$ is in the $\Pi^{(\nu)}$ subspace. For example, let us assume that

$$\langle\langle M(0)| = \langle\langle M(0)| P^{(\nu)} \tag{8.2}$$

Then we have [see Eq. (6.3)]

$$\begin{aligned}\langle\langle \tilde{M}^D(0)| \Pi^{(\nu)} &= \sum_\alpha \langle\langle M(0)| v_\alpha^{(\nu)}\rangle\rangle\langle\langle \tilde{F}_\alpha^{(\nu)}| N_\alpha^{(\nu)-1/2}\Pi^{(\nu)} \\ &= \sum_\alpha \langle\langle M(0)| v_\alpha^{(\nu)}\rangle\rangle\langle\langle \tilde{F}_\alpha^{(\nu)}| N_\alpha^{(\nu)-1/2} = \langle\langle \tilde{M}^D(0)|\end{aligned} \tag{8.3}$$

For this case, we have [see Eq. (6.21)]

$$\langle\langle \tilde{M}^D(t)| = \sum_\alpha \langle\langle M(0)| v_a^{(\nu)}\rangle\rangle e^{-iZ_\alpha^{(\nu)}t}\langle\langle \tilde{v}_\alpha^{(\nu)}| \hat{\Phi}_\nu^D = \langle\langle M(0)| e^{-i\theta_D^{(\nu)}t}\hat{\Phi}_\nu^D \tag{8.4}$$

Of special interest is the case where $\nu = 0$, because this leads to "invariants" of motion for integrable systems. To see this, let us consider the transformed momenta which are in the $P^{(0)}$ subspace [see Eq. (4.26)]:

$$\langle\langle \tilde{\mathbf{P}}_i^D(t)| = \langle\langle \hat{\mathbf{P}}_i | \Lambda_D \mathscr{U}(t) \tag{8.5}$$

We have

$$\langle \tilde{\mathbf{P}}_i^D(t) \rangle_\rho \equiv \langle\langle \tilde{\mathbf{P}}_i^D(t) | \rho(0) \rangle\rangle = \langle\langle \hat{\mathbf{P}}_i | e^{-i\theta_{D^{(0)}} t} \hat{\Phi}_0^D | \rho(0) \rangle\rangle \tag{8.6}$$

In the limit of vanishing coupling constant $\lambda \to 0$, they reduce to the unperturbed momenta evaluated on the density matrix $\rho(0)$.

Suppose the diagonal transitions are negligible, that is, $\theta_D^{(0)} = 0$. Then the transformed momenta reduce to the invariants of motion [see Eq. (9.14)]:

$$\langle \tilde{\mathbf{P}}_i^D(t) \rangle_\rho = \langle\langle \hat{\mathbf{P}}_i | \hat{\Phi}_0^D | \rho(0) \rangle\rangle = \langle \tilde{\mathbf{P}}_i^D(0) \rangle_\rho \tag{8.7}$$

Therefore, when the conditions (a) diagonal transitions are negligible, and (b) the right-hand side of Eq. (8.7) exists, the relations (8.7) define a set of $3\mathcal{N}$ "new" momenta $\tilde{\mathbf{P}}_i^D$ which are invariants of motion. The system is *integrable* in the sense of Poincaré. We shall call the conditions (a) and (b) the "integrability conditions."

Let us discuss in detail these integrability conditions. We first consider the case where the number of particles $\mathcal{N}$ is finite, and the density matrices are regular, as given by Eq. (4.1), with no delta function singularity in their momentum representation. We assume that the coupling is small, that is, $\lambda \ll 1$. Expanding Eq. (8.6) in powers of λ, we have

$$\begin{aligned} \langle \tilde{\mathbf{P}}_i^D(t) \rangle_\rho = \sum_{k\mathcal{N}} \int d^{\mathcal{N}}P \int d^{\mathcal{N}}P' \mathbf{P}_i \langle\langle 0, P | e^{-i\lambda^2 \theta_{2^{(0)}} t} \\ \times (1 + \lambda \mathbf{D}_1^{(0)} + \lambda^2 \mathbf{D}_2^{(0)}) | k, P' \rangle\rangle \tilde{\rho}_k(P', 0) + \lambda^3 \cdots \end{aligned} \tag{8.8}$$

Here $\lambda^n \theta_n^{(0)}$ are $\lambda^n \mathbf{D}_n^{(0)}$ are the nth-order approximations of the corresponding operators [see Eq. (5.42) and (5.50) and the expansions in Eq. (A.2)]. To the second order in λ, we obtain

$$\begin{aligned} \langle \tilde{\mathbf{P}}_i^D(t) \rangle_\rho = \sum_{k\mathcal{N}} \int d^{\mathcal{N}}P \int d^{\mathcal{N}}P' \mathbf{P}_i \\ \times \langle\langle 0, P | [1 + \lambda \mathbf{D}_1^{(0)} + \lambda^2 (\mathbf{D}_2^{(0)} - i\theta_2^{(0)} t) + \lambda^3 \cdots] | k, P' \rangle\rangle \tilde{\rho}_k(P', 0) \end{aligned} \tag{8.9}$$

We shall in succession study the effect of the destruction operator **D** and the collision operator θ. The contribution from $\mathbf{D}_n^{(0)}$ corresponds to off-diagonal transitions, while $\theta_2^{(0)}$ corresponds to diagonal transitions in the space $P^{(0)}$.

An example of a regular distribution function is

$$|\rho(0)\rangle\rangle = |Q^0, P^0\rangle\rangle \tag{8.10}$$

The Fourier coefficients in Eq. (4.1) are given by

$$\tilde{\rho}_k(P, t) = \delta_\Omega(P - P^0)e^{-ik\cdot Q^0} \tag{8.11}$$

To first order in λ, we have (e.g., for $i = 1$)

$$\begin{aligned}
\tilde{\mathbf{P}}_1^D(Q^0, P^0, t) &\equiv \langle\langle \tilde{\mathbf{P}}_1^D(t)|Q^0, P^0\rangle\rangle \\
&= \mathbf{P}_1^0 - \lambda\frac{1}{\Omega}\sum_{n>j}^{\mathcal{N}}\sum_{\mathbf{k}}\int d^{\mathcal{N}}P\mathbf{P}_1\partial_{jn}^{\mathbf{k}/2}\frac{V_{|\mathbf{k}|}}{\mathbf{k}\cdot\mathbf{g}_{jn} - i\epsilon} \\
&\quad\times e^{-i\mathbf{k}\cdot(\mathbf{Q}_j{}^0 - \mathbf{Q}_n{}^0)}\delta(P - P^0) + O(\lambda^2) \\
&= \mathbf{P}_1^0 + \lambda\frac{1}{\Omega}\sum_{n=2}^{\mathcal{N}}\sum_{\mathbf{k}}\frac{V_{|\mathbf{k}|}}{k\cdot\mathbf{g}_{1n}^0 - i\epsilon}\mathbf{k}e^{-i\mathbf{k}\cdot(\mathbf{Q}_1{}^0 - \mathbf{Q}_n{}^0)} + O(\lambda^2)
\end{aligned} \tag{8.12}$$

To obtain the last line in Eq. (8.12) we have used Eq. (3.58) and again retained the "connected" contribution to the labeled particle 1.

Similarly, the second-order contribution $\mathbf{D}_2$ is given by

$$\begin{aligned}
\overline{[\tilde{\mathbf{P}}_1^D(Q^0, P^0, t)]}_{\lambda^2\mathbf{D}_2} &= -\lambda^2\frac{1}{\Omega^2}\sum_{n=2}^{\mathcal{N}}\sum_{\mathbf{k}}\sum_{\mathbf{k}'}{}'\int d^{\mathcal{N}}P\mathbf{P}_1\partial_{jn}^{\mathbf{k}/2}\frac{V_{|\mathbf{k}|}}{\mathbf{k}\cdot\mathbf{g}_{1n} - i\epsilon} \\
&\quad\times\partial_{1n}^{(\mathbf{k}-\mathbf{k}')/2}\frac{V_{|\mathbf{k}-\mathbf{k}'|}}{\mathbf{k}'\cdot\mathbf{g}_{1n} - i\epsilon}e^{-i\mathbf{k}\cdot(\mathbf{Q}_1{}^0 - \mathbf{Q}_n{}^0)}\delta(P - P^0)
\end{aligned} \tag{8.13}$$

where the bar denotes the particular term we are looking at. This term comes from binary correlations. To this order, we must also retain the effect of ternary correlations, which we do not write here. The prime on the summation sign over $\mathbf{k}'$ in Eq. (8.13) denotes that we exclude $\mathbf{k}' = 0$. This restriction results because $\mathbf{D}^{(0)}$ is the off-diagonal transition [expressed by $Q^{(0)}$ in Eq. (5.35)]. This restriction is necessary to obtain finite contributions for nonintegrable systems. However, the restriction can be removed for the integrable case which we consider now. Indeed, the term corresponding to

$\mathbf{k}' = 0$ in Eq. (8.13) is of order $(\epsilon\Omega)^{-1}$ and can be neglected by the condition (5.49).

For the diagonal transition we have [see Eq. (5.43)]

$$[\tilde{\mathbf{P}}_1^D(Q^0, P^0, t)]_{\lambda^2\theta_2}$$

$$= i\lambda^2 t \frac{1}{\Omega^2} \sum_{n=2}^{\mathcal{N}} \sum_{\mathbf{k}} \int d^{\mathcal{N}} P \mathbf{P}_1 \partial_{1n}^{\mathbf{k}/2} \frac{|V_{|\mathbf{k}|}|^2}{\mathbf{k} \cdot \mathbf{g}_{1n} - i\epsilon} \partial_{1n}^{\mathbf{k}/2} \delta(P - P^0)$$

$$= \frac{\lambda^2 t}{\Omega^2} \sum_{n=2}^{\mathcal{N}} \sum_{\mathbf{k}} \int d^{\mathcal{N}} P \mathbf{P}_1 2\pi |V_{|\mathbf{k}|}|^2 \delta(E_{\mathbf{P}_1+\mathbf{k}} + E_{\mathbf{P}_n-\mathbf{k}} - E_{\mathbf{P}_1} - E_{\mathbf{P}_n})$$

$$\times (\eta_1^{\mathbf{k}} \eta_n^{-\mathbf{k}} - 1)\delta(P - P^0) \tag{8.14}$$

For any finite $\mathcal{N}$, the diagonal transition (8.14) is negligible, because this term is proportional to Ω^{-1} [see Eq. (3.5) for the summation over $\mathbf{k}$]. This shows that for the regular-density matrices, the diagonal transition is indeed negligible. On the contrary, if density matrices are singular, the diagonal transition is not negligible. We come back to the singular case in the following sections.

We can perform the integration of Eq. (8.12) over the wave vector $\mathbf{k}$ explicitly for a short-range Gaussian repulsive interaction given by

$$V(|\mathbf{Q}|) = V_0 e^{-Q^2/4a^2} = \frac{1}{\Omega} \sum_{\mathbf{l}} B e^{-a^2 l^2} e^{i\mathbf{l} \cdot \mathbf{Q}} \tag{8.15}$$

where $B \equiv V_0 a^3/\pi^{3/2}$. Detailed calculations are presented in the previous paper *I*. Here, we present only the result. We put

$$\mathbf{r}_n \equiv \mathbf{Q}_n^0 - \mathbf{Q}_1^0 \tag{8.16}$$

Let us denote the unit vectors of the polar coordinates of $\mathbf{g}_{1n}^0$ in an arbitrary reference system by $(\hat{v}_n, \hat{\theta}_n, \hat{\phi}_n)$, where $\hat{v}_n$ is the unit vector in the longitudinal direction of $\mathbf{g}_{1n}^0$, $\hat{\theta}_n$ in the transversal direction of $\mathbf{g}_{1n}^0$ parallel to the direction of the angle θ_n, and $\hat{\phi}_n$ in the transversal direction of $\mathbf{g}_{1n}^0$ parallel to the angle ϕ_n. Let us also introduce the notations

$$r_{n1} = (\mathbf{r}_n \cdot \hat{v}_n), \qquad r_{n2} = (\mathbf{r}_n \cdot \hat{\theta}_n), \qquad r_{n3} = (\mathbf{r}_n \cdot \hat{\phi}_n) \tag{8.17}$$

Then, to first order in λ, Eq. (8.12) leads to

$$\tilde{\mathbf{P}}_1^D(Q^0, P^0, t) = \mathbf{P}_1 + \frac{\pi^2 \lambda B}{a^3} \sum_{n=2}^{\mathcal{N}} \frac{1}{g_{1n}^0} e^{-(1/4a^2)(r_{n2}^2 + r_{n3}^2)}$$

$$\times \left\{ \frac{\hat{v}_n}{\sqrt{\pi}} e^{-(r_{n1}^2/4a^2)} - \frac{r_{n2}\,\hat{\theta}_n + r_{n3}\,\hat{\phi}_n}{2a} \left[1 + \operatorname{erf}\left(\frac{r_{n1}}{2a} \right) \right] \right\} \quad (8.18)$$

where $g_{1n}^0 = |\mathbf{g}_{1n}^0|$ and the error function is defined by

$$\operatorname{erf}(x) = \frac{2}{\sqrt{\pi}} \int_0^x e^{-t^2}\, dt \quad (8.19)$$

We note that in spite of the short-range interaction, the effect of the interaction in the transversal direction does not disappear for $r_{n1} \to +\infty$. After this limit is taken in Eq. (8.18), we have

$$\tilde{\mathbf{P}}_1^D(Q^0, P^0, t) \to \mathbf{P}_1^0 - \frac{\pi^2 \lambda B}{a^4} \sum_{n=2}^{\mathcal{N}} \frac{1}{g_{1n}^0} e^{-(1/4a^2)(r_{n2}^2 + r_{n3}^2)} (r_{n2}\,\hat{\theta}_n + r_{n3}\,\hat{\phi}_n) \quad (8.20)$$

where $g_{1n}^0 = |\mathbf{g}_{1n}^0|$. There is a long-range correlation between the particle 1 and n whatever their distance. This results from the resonance singularity at $\mathbf{k} \cdot \mathbf{g}_{1n}^0 = 0$ in Eq. (8.12) (see *I* for more detail). The order of Eq. (8.20) is $O(\lambda \mathcal{N})$. Similarly, one can show that the order of Eq. (8.13) from the binary correlations is $\lambda^2 \mathcal{N}$, and from the ternary correlations in λ^2 the contribution is $O(\lambda^2 \mathcal{N}^2)$, and so on. As a result, if the number of particles $\mathcal{N} \to \infty$, each term in Eq. (8.12) generally diverges. For Eq. (8.12) to be an invariant of motion, $\mathcal{N}$ should be finite. Even if $\mathcal{N}$ is finite, but too large, the series expansion of Eq. (8.12) in λ may not converge.

This has far reaching consequences. In the thermodynamic limit $\mathcal{N} \to \infty$ we have divergences. As we see, to avoid these divergences, we have to introduce a random element in the initial conditions. This changes radically the form of dynamics in the thermodynamic limit.

Let us now present our results in a form closer to the usual S-matrix theory in quantum mechanics.

IX. LINEAR AND NONLINEAR LIPPMANN–SCHWINGER EQUATIONS

In this section we reconsider the problem of integrability from a slightly different aspect, which is related to the Lippmann–Schwinger equations in the Liouville space.

In Section V we derived the complex spectral representation of L_H by solving the eigenvalue problem of the collision operator $\psi^{(\nu)}$ in Eq. (5.15). Note that Eq. (5.15) is a "nonlinear eigenvalue problem," because the collision operator $\psi^{(\nu)}$ itself depends on $Z_\alpha^{(\nu)}$. The unknown eigenvalues $Z_\alpha^{(\nu)}$ appear in the propagator inside the collision operator.

Alternatively, if we first determine the operators $\mathbf{C}^{(\nu)}$ and $\mathbf{D}^{(\nu)}$, we can construct the global collision operators $\theta_B^{(\nu)}$, which do not explicitly depend on the eigenvalues $Z_\alpha^{(\nu)}$ [see Eq. (5.24)]. Then, using the solutions of the "linear" eigenvalue problem for $\theta_B^{(\nu)}$ in Eq. (5.23), we can construct the solutions of the eigenvalue problem of L_H through the intertwining relations of Eq. (6.5) [14,22]. Using this approach, the nonlinearity of the problem appears in the equations for $\mathbf{C}^{(\nu)}$ and $\mathbf{D}^{(\nu)}$. Indeed, the intertwining relations, Eqs. (6.5) and (5.24), lead to nonlinear equations for $\hat{\Phi}_\nu^C$ and $\hat{\Phi}_\nu^D$, through which we can determine $\mathbf{C}^{(\nu)}$ and $\mathbf{D}^{(\nu)}$ [22,36,37]:

$$L_0 \hat{\Phi}_\nu^C - \hat{\Phi}_\nu^C L_0 = -L_V \hat{\Phi}_\nu^C + \hat{\Phi}_\nu^C L_V \hat{\Phi}_\nu^C \tag{9.1a}$$

$$L_0 \hat{\Phi}_\nu^D - \hat{\Phi}_\nu^D L_0 = \hat{\Phi}_\nu^D L_V - \hat{\Phi}_\nu^D L_V \hat{\Phi}_\nu^D \tag{9.1b}$$

Let us operate $\hat{\Phi}_\nu^C$ and $\hat{\Phi}_\nu^D$ on the eigenstates of the unperturbed Liouvillian L_0. We put

$$|\Phi_\nu^C\rangle\!\rangle \equiv \hat{\Phi}_\nu^C |\nu\rangle\!\rangle \quad \text{and} \quad \langle\!\langle \Phi_\nu^D| \equiv \langle\!\langle \nu | \hat{\Phi}_\nu^D \tag{9.2}$$

In general, these states are *not* the eigenstates of L_H.

From Eq. (9.1) we derive the nonlinear equations

$$\begin{aligned} |\Phi_\nu^C\rangle\!\rangle = |\nu\rangle\!\rangle &+ \sum_\mu \frac{-1}{w_\mu - w_\nu + i\epsilon_{\mu\nu}} P^{(\mu)} Q^{(\nu)} \lambda L_V |\Phi_\nu^C\rangle\!\rangle \\ &+ \sum_\mu \frac{1}{w_\mu - w_\nu + i\epsilon_{\mu\nu}} P^{(\mu)} Q^{(\nu)} |\Phi_\nu^C\rangle\!\rangle\langle\!\langle \nu | \lambda L_V |\Phi_\nu^C\rangle\!\rangle \end{aligned} \tag{9.3a}$$

$$\begin{aligned} \langle\!\langle \Phi_\nu^D| = \langle\!\langle \nu| &+ \sum_\mu \langle\!\langle \Phi_\nu^D | \lambda L_V Q^{(\nu)} P^{(\mu)} \frac{1}{w_\nu - w_\mu + i\epsilon_{\nu\mu}} \\ &+ \sum_\mu \langle\!\langle \Phi_\nu^D | \lambda L_V | \nu\rangle\!\rangle\langle\!\langle \Phi_\nu^D | Q^{(\nu)} P^{(\mu)} \frac{-1}{w_\nu - w_\mu + i\epsilon_{\nu\mu}} \end{aligned} \tag{9.3b}$$

where the parameter of the analytic continuation $\epsilon_{\nu\mu}$ is defined in Eq. (A.14), and we have implied the limit $\Omega \to \infty$, as usual. We have imposed the boundary conditions

$$|\Phi_\nu^C\rangle\!\rangle = |\nu\rangle\!\rangle \quad \text{and} \quad \langle\!\langle \Phi_\nu^D| = \langle\!\langle \nu| \qquad (\text{for } \lambda = 0) \tag{9.4}$$

By iterating Eq. (9.3) we can construct the explicit form of $\mathbf{C}^{(\nu)}$ and $\mathbf{D}^{(\nu)}$ in powers of λ. We can then construct $A^{(\nu)}$ through Eq. (5.39), and thus $\Pi^{(\nu)}$ as well as $\theta_B^{(\nu)}$ in powers of λ. We shall call Eqs. (9.3) the "nonlinear Lippmann–Schwinger equations," and shall show that they correspond to a nonlinear extension of the Lippmann–Schwinger equations in the Liouville space.

The nonlinear terms in Eqs. (9.3) involve the contribution from the diagonal transitions associated with the collision operators [see Eq. (5.23)]:

$$\langle\!\langle \nu|\lambda L_V|\Phi_\nu^C\rangle\!\rangle = \langle\!\langle \nu|\theta_C^{(\nu)}|\nu\rangle\!\rangle - w_\nu, \qquad \langle\!\langle \Phi_\nu^D|\lambda L_V|\nu\rangle\!\rangle = \langle\!\langle \nu|\theta_D^{(\nu)}|\nu\rangle\!\rangle - w_\nu \tag{9.5}$$

Consider the case in which the contribution from the diagonal transitions in the left-hand side of these expressions are negligible. Then we have

$$\theta_B^{(\nu)} = L_0 P^{(\nu)} \tag{9.6}$$

This implies that the eigenstates of $\theta_B^{(\nu)}$ are the unperturbed states $|\nu\rangle\!\rangle$, and the eigenvalues of L_H are w_ν—the same as for L_0. The diffusive contributions vanish; that is, the dynamic evolution becomes time symmetric. Also, we have [again neglecting the diagonal transitions; see Eq. (A.22)]

$$N_\alpha^{(\nu)} = 1 \tag{9.7}$$

as well as

$$A^{(\nu)} = P^{(\nu)} \tag{9.8}$$

Combining these equations with Eq. (6.5), we have

$$L_H|\Phi_\nu^C\rangle\!\rangle = w_\nu|\Phi_\nu^C\rangle\!\rangle, \qquad \langle\!\langle \Phi_\nu^D|L_H = \langle\!\langle \Phi_\nu^D|w_\nu \tag{9.9}$$

that is, the states Φ_ν^C and Φ_ν^D are eigenstates of L_H with the same real eigenvalues w_ν as L_0.

Then Eqs. (9.3) reduce to the "linear" equations

$$
\begin{aligned}
|\Phi_\nu^C\rangle\!\rangle &= |\nu\rangle\!\rangle + \sum_\mu \frac{-1}{w_\mu - w_\nu + i\epsilon_{\mu\nu}} P^{(\mu)} Q^{(\nu)} \lambda L_V |\Phi_\nu^C\rangle\!\rangle \\
\langle\!\langle \Phi_\nu^D| &= \langle\!\langle \nu| + \sum_\mu \langle\!\langle \Phi_\nu^D| \lambda L_V Q^{(\nu)} P^{(\mu)} \frac{1}{w_\nu - w_\mu + i\epsilon_{\nu\mu}}
\end{aligned}
\tag{9.10}
$$

In the limit of $\Omega \to \infty$, we can neglect the restriction associated with $Q^{(\nu)}$.

These are Lippmann–Schwinger type of equations in the Liouville space. For this case, we have

$$\langle\!\langle \Phi_\nu^D | \Phi_\mu^C \rangle\!\rangle = \delta_{\nu,\mu}, \qquad \sum_\nu |\Phi_\nu^C\rangle\!\rangle\langle\!\langle \Phi_\nu^D| = 1 \tag{9.11}$$

and the spectral decomposition of the evolution operator becomes

$$e^{-iL_H t} = \sum_\nu |\Phi_\nu^C\rangle\!\rangle e^{-iw_\nu t} \langle\!\langle \Phi_\nu^D| \tag{9.12}$$

Moreover, the transformations Λ_B in Eq. (6.3) reduce to

$$\Lambda_B = \Lambda_I \tag{9.13}$$

where

$$\Lambda_I = \sum_\nu |\nu\rangle\!\rangle\langle\!\langle \Phi_\nu^D|, \qquad \Lambda_I^{-1} = \sum_\nu |\Phi_\nu^C\rangle\!\rangle\langle\!\langle \nu| \tag{9.14}$$

and

$$\Lambda_B L_H \Lambda_B^{-1} = \Lambda_I L_H \Lambda_I^{-1} = L_0 \tag{9.15}$$

We have inserted the index I to emphasize that Λ_I is associated with integrable systems, because Eq. (9.15) holds only for this case. In a previous paper [11], we presented the explicit form of the solutions for Eq. (9.10) for potential scattering (see also [1]).

Let us consider the case in which interaction among the particles is "transient." The number of particles is finite. For this situation, asymptotic states exist before and after scattering. This is the situation to which the S-matrix theory applies. In analogy with the S-matrix theory in wave-function space, we can introduce the asymptotic states in the Liouville space, which are the "Möller scattering states" $\Phi_\nu^\pm$ and are defined as the solution to the equation

$$|\Phi_\nu^\pm\rangle\rangle = |\nu\rangle\rangle + \frac{1}{w_\nu - L_0 \pm i\epsilon} Q^{(\nu)}\lambda L_V |\Phi_\nu^\pm\rangle\rangle \tag{9.16}$$

Again, we may neglect the restriction to $Q^{(\nu)}$ in the limit of $\Omega \to \infty$. The relation of the Möller states in the Liouville space to the ones in the wave-function space is presented in detail in reference 11.

The formal solutions of Eq. (9.16) are given by

$$|\Phi_\nu^\pm\rangle\rangle = |\nu\rangle\rangle + \frac{1}{w_\nu - L_0 \pm i\epsilon} Q^{\nu}\mathscr{T}(w_\nu \pm i\epsilon)|\nu\rangle\rangle \tag{9.17}$$

where the $\mathscr{T}$-matrix is the solution of the integral equation [cf. Eq. (A.4)].

$$\mathscr{T}(z) = \lambda L_V + \mathscr{T}(z)\frac{1}{z - L_0}\lambda L_V \tag{9.18}$$

They also satisfy

$$L_H|\Phi_\nu^\pm\rangle\rangle = w_\nu|\Phi_\nu^\pm\rangle\rangle \tag{9.19}$$

as well as (for localized wave packets)

$$\langle\langle\Phi_\nu^+|\Phi_\mu^+\rangle\rangle = \delta_{\nu,\mu}, \qquad \sum_\nu |\Phi_\nu^+\rangle\rangle\langle\langle\Phi_\nu^+| = 1 \tag{9.20}$$

and

$$e^{-iL_Ht} = \sum_\nu |\Phi_\nu^+\rangle\rangle e^{-iw_\nu t}\langle\langle\Phi_\nu^+| \tag{9.21}$$

and similar relations for Φ_ν^-. The states Φ_ν^+ correspond to the "retarded" solutions of the scattering, and Φ_ν^- to the "advanced" solutions.

Equation (9.21) is the unitary spectral decomposition of the evolution operator. Moreover, we can introduce the unitary transformations (for the case in which there are no bounded states as for repulsive forces).

$$U_\pm = \sum_\nu |\nu\rangle\rangle\langle\langle\Phi_\nu^\pm|, \qquad U_\pm^\dagger = \sum_\nu |\Phi_\nu^\pm\rangle\rangle\langle\langle\nu| \tag{9.22}$$

which lead to

$$U_+ L_H U_+^\dagger = L_0 \tag{9.23}$$

and a similar relation for U_-.

The structure of Λ_I is quite similar to that of $U_\pm$. However, owing to the difference in the analytic continuations between Eqs. (9.10) and (9.16), these transformations are different. For example, the eigenstates corresponding to the vacuum of correlations (i.e., the states with zero eigenvalue $w_0 = 0$) are given for Eq. (9.10) by

$$|\Phi_0^C\rangle\rangle = |\Phi_0^+\rangle\rangle, \qquad \langle\langle\Phi_0^D| = \langle\langle\Phi_0^-| \tag{9.24}$$

Because the complex spectral representation uses both analytic continuations in Eq. (9.14), Λ_I is a nonunitary transformation even for the case in which dissipation is neglected. Nevertheless, because of the bicompleteness relation in Eq. (9.11), the spectral representations in Eqs. (9.12) and (9.21) lead to the same evolution of the distribution function $|\rho(t)\rangle\rangle$.

It is remarkable that integrable LPS admit both the nonunitary transformation, Eq. (9.14), as well as to the unitary ones, Eq. (9.22)[13]. However, there is a significant difference between the two. As shown in Appendix D, the singularity $\sim O(\mathcal{N}/\Omega\epsilon)$ appears, owing to the Poincaré resonances in the inner product $\langle\langle\Phi_{0,p'}^-|\Phi_{0,p''}^-\rangle\rangle$ for the unitary transformations. For $\mathcal{N}$ finite, this singularity is harmless, because we must first take the limit $\Omega \to \infty$ before taking the limit $\epsilon \to 0+$ [see Eq. (5.49)]. However, the unitary transformations cannot be extended to nonintegrable systems in the thermodynamic limit, because Ω^{-1} is compensated for by $\mathcal{N}$ in this limit. In contrast, the nonunitary transformation regularizes the Poincaré divergence as $\langle\langle\Phi_{0,p'}^D|\Phi_{0,p''}^C\rangle\rangle \sim O(\mathcal{N}/\Omega)$ (see Appendix D). Hence, Eq. (9.14) has a natural extension for the nonintegrable systems where the time symmetry is broken (see also *I*).

Let us now apply these results to construct the exact invariants of motion for all order of λ for systems where the diagonal transitions are negligible. As discussed in the previous section, the systems described by the regular-density matrices corresponds this case. Corresponding to Eq. (8.12) we have, for Eq. (9.14),

$$\tilde{\mathbf{P}}_i^D(Q^0, P^0) = L^{3N/2} \int d^{\mathcal{N}}P' \mathbf{P}'_i \langle\langle\Phi_{0,P'}^D|Q^0, P^0\rangle\rangle \tag{9.25}$$

[13] The nonuniqueness of the spectral decomposition including a nonunitary spectral decomposition has also been observed for the Friedrichs model in quantum mechanics [15].

Substituting Eq. (9.17) for (9.24), we obtain (for integrable systems)

$$\tilde{\mathbf{P}}_i^D(Q^0, P^0) = \mathbf{P}_i + \lim_{\Omega \to \infty} \int d^{\mathcal{N}} P' \mathbf{P}'_i$$

$$\times \sum_{k^{\mathcal{N}}} \langle\!\langle 0, P' | \mathscr{T}(+i\epsilon) | k, P \rangle\!\rangle \frac{1}{+i\epsilon - k \cdot v^0} e^{-ik \cdot Q^0} \qquad (9.26)$$

Hence, the existence of the $\mathscr{T}$-matrix corresponds to the integrability condition (b) mentioned at Eq. (8.7). As mentioned in Section VIII, for short-range repulsive interactions and not too large number of particles $\mathcal{N}$, this condition is satisfied.[14] The system is then integrable in the sense of Poincaré, in spite of the divergence of the small denominator at the resonance point $k \cdot v^0 = 0$ in the integrand of Eq. (9.26). This is the most striking difference of LPS from systems with discrete spectrum. For LPS, the effect of the Poincaré resonances can be treated as a well-defined object with a distribution $i\pi\delta(k \cdot v^0)$; in the discrete case, the resonances lead to a meaningless expression with divergence.

Even when there is no analytic solution described by the Born series in λ of the $\mathscr{T}$-matrix, there may exist nonanalytic solutions of Eq. (9.18), for example, they may occur for attractive forces. We hope to discuss this situation elsewhere.

Equation (9.26) presents examples of "singular invariants" (because the Fourier components of the invariants are singular at the resonance $k \cdot v = 0$) first introduced by one of the authors [4,48] (see also [49,50]). For classical systems, the invariants of Eq. (9.26) corresponds to the Hamilton–Jacobi invariants of motion (see *I*).

In summary, systems containing a small number of particles described by regular-density matrices are expected to be *integrable*. On the contrary, if the distribution functions are singular, or the number of particles approaches infinity, the system is no longer integrable. Then, one can observe dissipative effects in LPS. In the following sections we discuss these nonintegrable situations, which cannot be described within the framework of the usual wave-function theory.

X. PERSISTENT POTENTIAL SCATTERING

In the previous sections, we constructed the invariants of motion in Eq. (9.26) for systems with a finite number of particles and described by regular-

[14] For more than two-body systems, we need a careful discussion of the analyticity of the $\mathscr{T}$-matrix, as performed by Faddeev for the three-body collision [51]. We shall not discuss this problem here.

density matrices. We now show that the new momenta defined in Eq. (8.12) are no longer invariants of motion when they are associated with singular-density matrices.

Let us integrate Eq. (8.12) over Q:

$$I(P, t) = \int d^{\mathcal{N}}Q\tilde{\mathbf{P}}_1^D(Q, P, t) \tag{10.1}$$

The diagonal transition in Eq. (8.14) now gives a finite contribution, whereas the off-diagonal transitions in Eqs. (8.12) and (8.13) vanish because of the restriction by $Q^{(0)}$ in the $\mathbf{D}^{(0)}$ operator. For example, let us consider the simplest case of $\mathcal{N} = 1$ for potential scattering, Eq. (3.62). We have discussed this simple example in detail in our previous paper [11]. From the expression corresponding to Eq. (8.14) for potential scattering we obtain [11]

$$\frac{d}{dt} I(\mathbf{P}_1, t) = \lambda^2 \int d\mathbf{k} \int d\mathbf{P}'_1 2\pi |V_{|\mathbf{k}|}|^2 \delta(E_{\mathbf{P}_1'+\mathbf{k}} - E_{\mathbf{P}_1'})(\eta_1^{\mathbf{k}} - 1)\delta(\mathbf{P}'_1 - \mathbf{P}_1) \tag{10.2}$$

We see that $I(\mathbf{P}_1, t)$ evolves in time.

One can understand this result as follows: The integration corresponds to the introduction of a nonlocal ensemble which has a delta-function singularity in its momentum representation:

$$\rho_{\mathbf{k}}(\mathbf{P}_1, 0) = \rho_0^d(\mathbf{P}_1, 0)\delta_\Omega(\mathbf{k}) + \rho'_{\mathbf{k}}(\mathbf{P}_1, 0) \tag{10.3}$$

where we assume that ρ_0^d and $\rho'_{\mathbf{k}}$ do not depend on Ω in the limit of large volumes. We have the boundary condition (for $|\mathbf{Q}_1| \to \infty$)

$$\rho^W(\mathbf{Q}_1, \mathbf{P}_1) \to \rho_0^d(\mathbf{P}_1) \tag{10.4}$$

Because of this singularity, the effect of the diagonal transitions are amplified Ω times. As a result, $I(\mathbf{P}_1, t)$ evolves in time.[15] Physically, this corresponds to a situation where we continuously send "test" particles toward a

[15] For this case nonnegligible diagonal transitions appear only in the vacuum of correlation. Hence, the analytic continuation for the diagonal operators are also uniquely determined with the complex distribution, as in the case of the thermodynamic limit shown in Eq. (5.52).

single potential.[16] We assume that the interaction between the test particles is negligible as compared with their interaction with the potential. Moreover, we assume that the test particles are distributed with a finite concentration in space. Therefore, the interaction between the particles with the potential is "persistent." There are no asymptotic states for this scattering process. This situation goes beyond the usual S-matrix theory.

Corresponding to Eq. (10.2), we obtain, to the lowest order of λ for the ensemble of Eq. (10.3),

$$\frac{d}{dt}\langle\langle \tilde{\mathbf{P}}_1^D(t)\,|\,\rho(0)\rangle\rangle = \lambda^2 \int d\mathbf{k} \int d\mathbf{P}_1 \mathbf{P}_1 2\pi\, |V_{|\mathbf{k}|}|^2 \delta(E_{\mathbf{P}_1+\mathbf{k}} - E_{\mathbf{P}_1}) \times [\rho_0^d(\mathbf{P}_1 + \mathbf{k}, 0) - \rho_0^d(\mathbf{P}_1, 0)] \quad (10.5)$$

Therefore $\langle\langle \tilde{\mathbf{P}}_1^D(t)\,|\,\rho(0)\rangle\rangle$ evolves in time when associated with the singular distribution function, Eq. (10.3). In the right-hand side of Eq. (10.6) we recognize the Pauli operator [see Eq. (5.43)]. Dissipative processes are enhanced by the delta-function singularity in Eq. (10.3). The system is *non-integrable* for the persistent interaction described by the singular-density matrices.

In the evolution of $\langle\langle \mathbf{P}_1^{\nu}(t)\,|\,\rho(0)\rangle\rangle$, generally higher-order contributions in time appear, as $(-i\theta_a^{(0)}t)^n$ with $n \geq 2$ [see Eq. (6.21)]. However, as one can easily see, a repetition of diagonal transitions always leads to the extra volume factor Ω^{-1} for the singular case we consider in this section. All higher-order contributions t^n in time with $n \geq 2$ are negligible in the large-volume limit [11]. The evolution of $\tilde{\mathbf{P}}_1^D(t)$ is strictly linear in time. In previous papers [11,12], we have performed numerical simulations for the singular situation discussed here. The agreement between the theoretical prediction and the numerical results is excellent.

Because of the linear time dependence of $\langle\langle \tilde{\mathbf{P}}_1^D(t)\,|\,\rho(0)\rangle\rangle$ in Eq. (10.5), however, the system cannot approach equilibrium in a finite time. This is in contrast with the systems studied in Section XI, where we investigate the evolution of dynamical systems which are described in the thermodynamic limit by singular with L_1-normalizable distributions such as Eq. (4.18).

We note that the evolution of the transformed momentum $\langle\langle \tilde{\mathbf{P}}_1^D(t)\,|\,\rho(0)\rangle\rangle$ gives a finite contribution in Eq. (10.5) even though it is evaluated over singular ρ that have no L_1 norm. In this simple example, the divergence in $\langle\langle \hat{\mathbf{P}}_1^D(t)\,|\,\rho(0)\rangle\rangle$ appears only in a trivial component which is associated with the *free* motion. All effects coming from the interaction converge. Physi-

[16] In the usual S-matrix description $\mathcal{N}/\Omega \to 0$, while we consider here situations where $\mathcal{N}/\Omega =$ finite. This corresponds to the nonlocal distribution (10.3). The L_1 norm of Eq. (10.3) represents the number of particles.

cally, the divergence can easily be understood, because we are continuously sending test particles toward the potential. A detector behind the potential registers this incident flow of test particles. Simply by putting the detector in a direction that is not parallel to the flow, one avoids this diverging contribution.

XI. NONEQUILIBRIUM STATISTICAL MECHANICS AND FLOW OF CORRELATIONS

We now consider the singular-density matrices of Eq. (4.18). As mentioned in Section IV, canonical equilibrium belongs to this class. The main difference from the one considered in the previous section is that although the distribution functions are singular in the Fourier representation, they have a well-defined L_1 norm. The time evolution of this class of ensembles is the main subject of nonequilibrium statistical mechanics (NESM). Its time dependence has been investigated in our earlier work [4,29,32]. All results obtained from NESM can be recovered by our complex spectral representation. This includes the derivation of the Pauli master equation, the Boltzmann equation, and, more generally, non-Markovian master equations. Because this class of ensembles leads to non-Schrödinger contributions, we concluded that these contributions result from approximations introduced in the solution of the Liouville–von Neumann equation. We see now that these results are exact consequences of the solution of the eigenvalue problem of the Liouvillian for singular-distribution functions outside the Hilbert space. It has also been shown that this class of density matrices approaches equilibrium at $t \to \infty$ [4]. This is confirmed by our formulation of the $\mathscr{H}$ theorem in Section VII.

By definition this class of density matrices belongs to the domain of the nonunitary transformation Λ. For example, let us evaluate the transformed momentum $\tilde{\mathbf{P}}_1^D(0)$ associated with the ensemble in Eq. (4.18). As in Eq. (8.12), we have, to first order in λ,

$$\langle\langle \tilde{\mathbf{P}}_1^D(0) \,|\, \rho(0) \rangle\rangle = \int d^{\mathcal{N}}P \mathbf{P}_1 \rho_0(|P, 0) + \frac{\lambda}{\Omega^2} \sum_{n=2}^{\mathcal{N}} \sum_{\mathbf{k}} \int d^{\mathcal{N}}P$$

$$\times\, \mathbf{k} \frac{V_k}{\mathbf{k}\cdot \mathbf{g}_{1n} - i\epsilon} \rho_{\mathbf{k},-\mathbf{k}}(\mathbf{P}_1, \mathbf{P}_n \,|\, P^{\mathcal{N}-2}, 0) + O(\lambda^2) \quad (11.1)$$

Note the difference in the volume dependence between Eqs. (11.1) and (8.12). There is no more divergence. The transformed momentum has a well-defined value of order c in the thermodynamic limit [see Eq. (2.2)]. One can easily verify that Eq. (11.1) is well defined to an arbitrary order of

λ (see I). In contrast to the regular distributions discussed in Section VIII, ensembles described by the distribution function in Eq. (4.18) are in the domain of the nonunitary transformations Λ in the thermodynamic limit.

Let us now evaluate its time evolution. As mentioned before, $\tilde{\mathbf{P}}_1^D(t)$ is in the $\Pi^{(0)}$ subspace. Hence, we can apply the formula given in Eq. (6.21). Then, we obtain

$$\frac{d}{dt}\langle\langle\tilde{\mathbf{P}}_1^D(t)\,|\,\rho(0)\rangle\rangle = \int d^{\mathcal{N}}P\mathbf{P}_1\,\frac{\partial}{\partial t}\,\rho_0(P,t) + O(\lambda^3) \tag{11.2}$$

where $\rho_0(P, t)$ satisfies [under the integration over the momentum in Eq. (11.2)] the Pauli master equation:

$$\frac{\partial}{\partial t}\,\rho_0(P,t) = \frac{2\pi\lambda^2}{\Omega}\sum_{n=2}^{\mathcal{N}}\int d\mathbf{k}\,|\,V_{|\mathbf{k}|}\,|^2\delta(E_{\mathbf{P}_1+\mathbf{k}} + E_{\mathbf{P}_n-\mathbf{k}} - E_{\mathbf{P}_1} - E_{\mathbf{P}_n})$$
$$\times\,[\rho_0(\mathbf{P}_1+\mathbf{k},\,\mathbf{P}_n-\mathbf{k},\,\{P\}^{\mathcal{N}-2},\,t) - \rho_0(P,t)] + O(\lambda^3) \tag{11.3}$$

In the thermodynamic limit, the right-hand side of Eq. (11.3) gives a finite contribution of order c.

The result, Eq. (11.3), is quite similar to Eq. (10.5), but there is an interesting difference. The right-hand side of Eq. (11.3) depends on time, while it does not in Eq. (10.5). In the situation considered here, summations appear over new particles owing to repeated collisions $(-i\theta_B^{(0)}t)^n$. As such summation over particles leads to a factor $\mathcal{N}$, we can no longer neglect the higher contribution of t^n with $n \geq 2$ [see the discussion in Eq. (10.5)]. Because of this nonlinear contribution in time, the system approaches equilibrium in a finite time scale $t_r \sim (\lambda^2 c)^{-1}$ (see also the example in Appendix I).

The complex spectral representation as well as the $\mathcal{H}$ theorem permit us to understand the mechanism of irreversibility, which is associated with the "flow of correlations" [4,16].

Let us now consider the evolution of binary correlations g_2:

$$g_2(\mathbf{Q}_1, \mathbf{Q}_2, \mathbf{P}_1, \mathbf{P}_2, t) = \int d^{\mathcal{N}-2}Q\int d^{\mathcal{N}-2}P$$
$$\times\,\frac{1}{\Omega}\sum_{\mathbf{k}} e^{i\mathbf{k}\cdot(\mathbf{Q}_1-\mathbf{Q}_2)}\rho_{\mathbf{k},\,-\mathbf{k}}(\mathbf{P}_1, \mathbf{P}_2\,|\,P^{\mathcal{N}-2}, t)$$
$$= L^{3\mathcal{N}/2}\int d^{\mathcal{N}-2}Q\int d^{\mathcal{N}-2}P$$
$$\times\sum_{\mathbf{k}} e^{i\mathbf{k}\cdot(\mathbf{Q}_1-\mathbf{Q}_2)}\langle\langle\mathbf{k},\,-\mathbf{k},\,\{0\}^{\mathcal{N}-2},\,P\,|\,\rho(t)\rangle\rangle \tag{11.4}$$

where

$$
\begin{aligned}
|\rho(t)\rangle\!\rangle &= \sum_{\nu} \sum_{\alpha} e^{-iZ_{\alpha}^{(\nu)}t} |F_{\alpha}^{(\nu)}\rangle\!\rangle\langle\!\langle \tilde{F}_{\alpha}^{(\nu)} | \rho(0)\rangle\!\rangle \\
&= \sum_{\nu} (P^{(\nu)} + \mathbf{C}^{(\nu)}) e^{-i\theta_C^{(\nu)}t} A^{(\nu)} (P^{(\nu)} + \mathbf{D}^{(\nu)}) | \rho(0)\rangle\!\rangle
\end{aligned}
\tag{11.5}
$$

The time evolution of ρ results from the superposition of all paths corresponding to the various subdynamics. Let us assume that the system is initially in the vacuum of correlation:

$$
|\rho(0)\rangle\!\rangle = P^{(0)} |\rho(0)\rangle\!\rangle \tag{11.6}
$$

We assume that the coupling is weak. The traditional approximation for weakly coupled system in kinetic theory is the $\lambda^2 t$-approximation corresponding to subdynamics $\Pi^{(0)}$ [4,52]. Then one only retains contributions of order $(\lambda^2 t)^n$ with $n \geq 0$. For the evolution of correlations, however, this leads to the trivial result that all correlations vanish in the weakly coupled limit $\lambda \to 0$. Hence, we have to go beyond this approximation. We shall keep terms to order $\lambda(\lambda^2 t)^n$ in Eq. (11.5). We call this the "$\lambda(\lambda^2 t)^n$ approximation." It should be emphasized that the $\lambda^2 t$ approximation describes only the asymptotic evolution in time. As the result, one cannot discuss causality in this approximation. In contrast, our approximation is applicable to *all time scales*. It is easy to extend this procedure to higher-order approximations.

Thus, we have the contribution from $\Pi^{(0)}$ associated with the vacuum of correlation, and from $\Pi^{(k,-k)}$ associated with binary correlations (all other contributions are negligible in this approximation):

$$
\begin{aligned}
\langle\!\langle \mathbf{k}, -\mathbf{k}, \{0\}^{\mathcal{N}-2}, P | \rho(t)\rangle\!\rangle &= \lambda \langle\!\langle \mathbf{k}, -\mathbf{k}, \{0\}^{\mathcal{N}-2}, P | \\
&\quad \times \left\{ \mathbf{C}_1^{(0)} e^{-i\lambda^2 \theta_2^{(0)} t} + e^{-i(\mathbf{k}\cdot\mathbf{g}_{12} + \lambda^2 \theta_2^{(k,-k)})t} \mathbf{D}_1^{(k,-k)} \right\} P^{(0)} | \rho(0)\rangle\!\rangle
\end{aligned}
\tag{11.7}
$$

where

$$
\mathbf{C}_1^{(0)} + \mathbf{D}_1^{(k,-k)} = 0 \tag{11.8}
$$

and

$$\int d^{\mathcal{N}}P' \langle\!\langle \mathbf{k}, -\mathbf{k}, \{0\}^{\mathcal{N}-2}, P \,|\, \lambda \mathbf{C}_1^{(0)} \,|\, 0, P' \rangle\!\rangle \rho_0(P')$$

$$= -\lambda \frac{V_{|\mathbf{k}|}}{\mathbf{k}\cdot\mathbf{g}_{12} - i\epsilon} \partial_{12}^{\mathbf{k}/2} \rho_0(P, 0)$$

$$= -\frac{\lambda V_{|\mathbf{k}|}}{E_{\mathbf{P}_1+\mathbf{k}/2} + E_{\mathbf{P}_2-\mathbf{k}/2} - E_{\mathbf{P}_1-\mathbf{k}/2} - E_{\mathbf{P}_2+\mathbf{k}/2} - i\epsilon}$$

$$\times [\rho_0(|\mathbf{P}_1 + \tfrac{1}{2}\mathbf{k}, \mathbf{P}_2 - \tfrac{1}{2}\mathbf{k}, \{P\}^{\mathcal{N}-2}, 0)$$

$$- \rho_0(|\mathbf{P}_1 - \tfrac{1}{2}\mathbf{k}, \mathbf{P}_2 + \tfrac{1}{2}\mathbf{k}, \{P\}^{\mathcal{N}-2}, 0)] \qquad (11.9)$$

To evaluate Eq. (11.4) explicitly, we have to specify the interaction and solve the eigenvalue problem for the collision operators $\lambda^2\theta_2^{(0)}$ and $\mathbf{k}\cdot\mathbf{g}_{12} + \lambda^2\theta_2^{(k,-k)}$. For a simple system such as the "perfect Lorentz gas" (classical and quantum), we can solve the eigenvalue problem explicitly (see Appendix I). We shall present the detailed calculations for the causal evolution of the correlations in separate papers [57]. Here, we offer a sketch of the results.

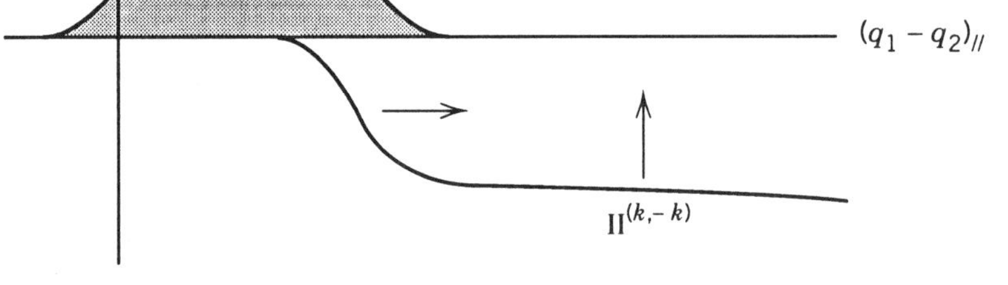

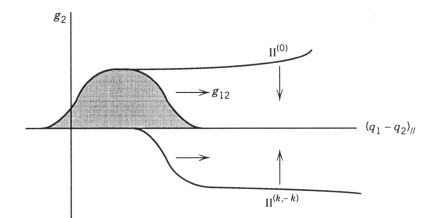

Figure 4. Causal propagation of binary correlations g_2 in the parallel direction $(\mathbf{q}_1 - \mathbf{q}_2)_{\parallel}$ to the relative velocity $\mathbf{g}_{12}$. The contribution from the $\Pi^{(0)}$ diverges exponentially as a function of $(\mathbf{q}_1 - \mathbf{q}_2)_{\parallel}$ (see the discussion in Appendix C). The contribution in the space $\Pi^{(0)}$ damps in time, while the contribution in the space $\Pi^{(k,-k)}$ shifts with the relative velocity $\mathbf{g}_{12}$ and damps. Outside the noncausal region $(\mathbf{q}_1 - \mathbf{q}_2)_{\parallel} > |\mathbf{g}_{12}|t$, the contribution in the space $\Pi^{(0)}$ is canceled by the contribution in $\Pi^{(k,-k)}$. As time goes on, a long-range correlation is built up by the resonance.

We first note that the effects of the two subspaces $\Pi^{(0)}$ and $\Pi^{(k,-k)}$ cancel at $t = 0$ [see Eq. (11.8)]. For short time scales, effects of dissipation coming from the collision operators are negligible. The evaluation of the integral over the wave vector $\mathbf{k}$ in Eq. (11.4) with Eq. (11.8) is quite similar to Eq. (8.8) with Eq. (8.18) (see *I* for detail calculations). Then, similarly to Eq. (8.20), the result shows that the binary correlation in $\Pi^{(0)}$ remains finite for $|\mathbf{Q}_1 - \mathbf{Q}_2| \to \infty$. This results from the resonance singularity at $\mathbf{k} \cdot \mathbf{g}_{12} - 0$ in Eq. (11.8). In the $\Pi^{(0)}$ subspace, the resonance effect leads to the long-range correlations between particles 1 and 2, whatever their distance. Here the long-range correlations due to the Poincaré resonances do not lead to any divergences. Indeed, the effect in subspace $\Pi^{(k,-k)}$ is simply a shift in space with a sign opposite to the contribution in $\Pi^{(0)}$ for short time scale, because $\mathbf{Q}_1 - \mathbf{Q}_2$ is replaced by $\mathbf{Q}_1 - \mathbf{Q}_2 - \mathbf{g}_{12}t$. Hence, following the causal evolution of the system, the total contribution of the binary correlation develops in space. Owing to the Poincaré resonances the long-range correlations are built up as time goes on (see also Appendix C (Fig. 4).

The long-range correlations are associated only with nonequilibrium modes; for equilibrium mode, we have $\rho_0(P) = f(H_0)$ in Eq. (11.8), which leads to

$$\partial_{12}^{\mathbf{k}/2} f(H_0) = f\left(E_{\mathbf{P}_1+\mathbf{k}/2} + E_{\mathbf{P}_2-\mathbf{k}/2} + \sum_{i=3}^{\mathcal{N}} E_{\mathbf{P}_i}\right) - f\left(E_{\mathbf{P}_1-\mathbf{k}/2} + E_{\mathbf{P}_2+\mathbf{k}/2} + \sum_{i=3}^{\mathcal{N}} E_{\mathbf{P}_i}\right) \quad (11.10)$$

Hence, there is no resonance singularity at $E_{\mathbf{P}_1+\mathbf{k}/2} + E_{\mathbf{P}_2-\mathbf{k}/2} = E_{\mathbf{P}_1-\mathbf{k}/2} + E_{\mathbf{P}_2+\mathbf{k}/2}$ for the equilibrium mode. This is quite satisfactory, as it shows that nonequilibrium situations are associated properties.

For long time scales of order $t \sim \lambda^{-2}$, the effect of the dissipation coming from the collision operators is no longer negligible. Effects from nonequilibrium modes both in $\Pi^{(0)}$ and in $\Pi^{(k,-k)}$ subspaces vanish for long time scales owing to the repeated collisions with particles in the medium. In Eq. (11.4), only equilibrium short-range binary correlations remain around each particle. However, during this process, ternary nonequilibrium correlations are built up, then they decay to equilibrium correlations, then to fourth-order correlations, and so on. As time goes on, nonequilibrium correlations propagate over a larger distance and transfer the correlations among more particles. We then have a directed "flow of correlations." This flow finally disappear in the "sea" of highly multiple, incoherent correlations [4]. The meaning of irreversibility acquires an intuitive sense in terms of the "dynamics of correlations" based on the complex spectral representation. The $\mathscr{H}$ property and the flow of correlations are direct consequences of the existence of the semigroup that results from the complex spectral representation of the Liouvillian. We come back to this problem briefly in Section XIV (and in more detail in Appendix E).

We shall not try to summarize the results obtained when starting with singular-density matrices, Eq. (4.18), and applying complex spectral decomposition. This would involve a summary of NESM [4]. We want only to emphasize that here we have a striking example of the emergence of non-Schrödinger contributions.

XII. THE THERMODYNAMIC LIMIT

In Section XI we considered the singular-density matrices of Eq. (4.18). As mentioned, regular-density matrices do not belong to the domain of the nonunitary transformations Λ because they lead to divergence of $\tilde{\mathbf{P}}_1^D$ in Eq. (8.12) in the thermodynamic limit. As compared with Eq. (11.1), the initial correlations in Eq. (8.12) are too large. Still, we can consider an interesting class of regular-density matrices which belong to the domain of Λ even in the thermodynamic limit.

To illustrate this situation, let us consider a system that consists of wave packets associated with each particle. We assume that the wave packets are well separated in space from each other at $t = 0$. Then, $\mathbf{Q}_n^0$ in Eq. (8.12) can be considered as a center of each wave packet. Let us suppose that $\mathbf{Q}_n^0$ are chosen *randomly*. Here, random means that the algorithm to write the sequence $\mathbf{Q}_1^0, \mathbf{Q}_2^0, \mathbf{Q}_3^0, \ldots$ is "incompressible" [53]. Then, in the thermodynamic limit, the summation over n and $\mathbf{k}$ in Eq. (8.12) gives a contribution of order

$$\frac{1}{\Omega}\sum_{\mathbf{k}}\sum_{n}^{\mathcal{N}} f_{\mathbf{k}}\, e^{-i\mathbf{k}\cdot(\mathbf{Q}_1{}^0-\mathbf{Q}_n{}^0)} \sim \frac{\sqrt{\mathcal{N}L^3}}{L^3} \tag{12.1}$$

As a consequence, the right-hand side of Eq. (8.12) gives a finite contribution of order $\sqrt{c}$ in this limit. One can verify this estimate by taking the average of the square of the absolute value of Eq. (12.1) over $\mathbf{Q}_n^0$, assuming a uniform distribution of $\mathbf{Q}_n^0$ in space. In this estimate, we have to take the thermodynamic limit *after* taking the average. This shows that the square is of order c in the thermodynamic limit.

This estimate of the concentration dependence is valid only for the ensemble average over the random distribution of the initial positions. For each given sequence $\mathbf{Q}_1^0, \mathbf{Q}_2^0, \mathbf{Q}_3^0, \ldots$, the value of Eq. (12.1) may change significantly. However, this estimate guarantees a finite value of Eq. (12.1) in the thermodynamic limit for *almost all single* sequences of the initial values

$\mathbf{Q}_1^0$, $\mathbf{Q}_2^0$, $\mathbf{Q}_3^0$, ..., because the average of the square of absolute value is finite. The random numbers are generic points in configuration space [53].

Note that if we first replace the summation over the wave vector by the integral in each individual term in the summation over n in Eq. (8.12) [as in Eq. (8.20)], then take the limit $\mathcal{N} \to \infty$, assuming a random distribution of the particles, we would obtain a diverging contribution of order $\sqrt{\mathcal{N}}$. This shows that we have to perform the summation over $\mathcal{N}$ and $\mathbf{k}$ simultaneously in the thermodynamic limit for regular-density matrices. Recognizing the difference between these two limiting procedures is essential to understanding the origin of dissipative processes. We have tested the two different estimates for a simple example by numerical simulations in *I*. The results confirmed our expectation and are in agreement with Eq. (12.1).

As a result of random initial conditions, the destruction operator in Eq. (8.12), as well as in Eq. (9.13), gives a finite contribution even for initial conditions with regular Fourier coefficients, Eq. (4.1), in the thermodynamic limit. Moreover, the collision operator θ_2 (which corresponds to a diagonal transition) in Eq. (8.14) also gives a finite contribution in the thermodynamic limit, which is of order c regardless of the random or coherent choice of the initial values $\mathbf{Q}_n^0$, because θ_2 does not depend on $\mathbf{Q}_n^0$. One can easily see that for every order in λ, the destruction operator gives a finite contribution. Hence, this class of random initial conditions belongs to the domain of Λ. As mentioned in Section V, the system approaches equilibrium, which is in $\Pi^{(0)}$. In the $\Pi^{(0)}$ subspace, the correlation is generated from the vacuum of correlation $P^{(0)}$ through the creation operator $\mathbf{C}^{(0)}$ [see Eq. (5.27)]. As illustrated in Eq. (4.20), the interaction λL_V (hence the creation operator) introduces an extra volume factor Ω^{-1} as compared with the states in the vacuum of correlation. This is a general property of the $\Pi^{(0)}$ subspace, and one can easily verify that the states in the $\Pi^{(0)}$ subspace satisfy the delta-function singularity in Eq. (4.18). Therefore, the delta-function singularity in Fourier space emerges as time goes on, even if we start from a nonsingular distribution function. The class of singular distribution functions given in Eq. (4.18) acts as an attracter.

In conclusion, the maintenance of regular Fourier coefficients and the existence of a thermodynamic limit are incompatible. Whenever the thermodynamic limit exists, the regular distribution function approaches the class of singular distribution functions of Eq. (4.18).

XIII. ILLUSTRATIONS

Applications of the complex spectral representation abound. Essentially, all problems in nonequilibrium statistical physics lead to this representation.

In addition, many other interesting phenomena can be treated using our methods. In the appendices, we present several examples.

In Appendix F, we present the simplest example of persistent scattering for potential scattering. We consider the scattering for the nonlocal distribution of the particles in space. The solution of the eigenvalue problem of the Liouvillian leads to microcanonical equilibrium. This corresponds to a mixture, and cannot be written as a single product of wave functions. We then illustrate our method, examining three-body persistent scattering in Appendix G. We consider scattering for large wave packets in space. We then observe the process during the collision for times which are less than the duration of collision t_c. For large wave packets, we may consider situations where t is large with respect to characteristic frequencies, but smaller than t_c. We show that this leads to a dramatic effect, because classes of secular effects appear only for $t \ll t_c$ and disappear in the frame of the S-matrix theory for $t \gg t_c$. Our results have been confirmed by numerical simulations. As a result of this effect, the three-body scattering cross section is different in each situation. In the frame of our asymptotic theory for $t \ll t_c$, the system still does not "know" if we started with a large, but finite, wave packet or if we started with a delocalized density matrix with the delta-function singularity in momentum space.

In Appendix H we consider a simple example of matter–field interacting systems (or of quantum optics), the Friedrichs model. This is a well-known model for the spontaneous emission of the field by an unstable excited state. We consider our complex spectral representations for two different situations: in the first, the field is localized in space; in the second, the space-energy density of the field is finite. There is no delta-function singularity in the field in the first case, while there is in the second. As we shall see, the evolution of the density matrices is reducible to wave functions for the regular case; it is irreducible for the singular cases.

As an example of many-body systems, we present the perfect Lorentz gas in Appendix I. In the thermodynamic limit, the system approaches equilibrium in a finite time scale. Using this model, we discuss the collapse of the wave function as a dynamical process. The transition from "potentially" to "actuality" in quantum mechanics is realized through the transition of a wave function to a mixture described by the density matrix.

The eigenvalue problems of the Hamiltonian for some one-dimensional $\mathcal{N}$-particle systems are known to be solvable [54]. In Appendix K, we show that these solutions are applicable only to regular-density matrices. For the thermo-dynamic limit, these solutions lead to divergence. In this limit, the system approaches equilibrium and there is no solution to the eigenvalue problem of the Hamiltonian. We have to solve the eigenvalue problem of the Liouvillian for the class of singular-density matrices given by Eq. (4.18).

Finally, we illustrate anharmonic lattice for both quantum and classical mechanics in Appendix L. In the quantum case, this gives an example of field–field interacting systems. Since normal modes and angle-action variables are collective variables, the interactions are *automatically persistent.* The situation is therefore simpler than for the interacting particles considered in the previous sections. We then show that the anharmonic force destroys the Hilbert space structure in the thermodynamic limit. Therefore, we have here a simple example where our extended formulation of dynamics applies.

The presentation of our illustrations in the appendices are short. The reader is invited to peruse the original papers for more details.

XIV. CONCLUDING REMARKS

We believe that our approach answers several important questions. We see now how irreversibility emerges in the fundamental dynamic description. This is in accordance with the fact that on all levels of observation nature appears as asymmetric in time. Our formulation also avoids the paradoxes associated with the traditional formulation of quantum mechanics. In agreement with the well-known statements of Bohr and Rosenfeld, measurement is associated with irreversibility, but irreversibility implies a semigroup description. The quantum superposition principle is then broken and we derive a purely dynamical description of measurement.

Our approach, therefore, involves a radical change of perspective. In the traditional view, the universe was associated with a dynamical group; irreversibility was assumed to arise from our approximations. We now invert the perspective [55]: the group description corresponds only to simple situations when we may consider a number of particle isolated from the rest of the universe.

It should be emphasized that irreversibility is an "emergent" property. We need LPS and the thermodynamic limit. This limit does not imply that the number of particles $\mathcal{N}$ is infinite, but that we can use asymptotic expansions neglecting terms such as $1/\mathcal{N}$. The system has to be large enough so that the distinction between *intensive* and *extensive* variables makes sense.

A dynamical group can be split into two semigroups; one oriented toward our future, the other toward our past. It has often been asked why all systems share the same time direction. An essential element of the answer lies in cosmology. It is likely that all matter now present in our universe had been formed at an early stage (probably associated with the so-called Planck time, $\sim 10^{-44}$ sec). In this view, the universe forms a single indecomposable dynamical system. The question of the nature of the "initial" conditions is still a subject of controversy, but it seems likely that the early universe was a thermodynamic LPS, because at a very early stage we can introduce concepts such as temperature and pressure. We then come back to the semigroup description.

Let us introduce the notation $\Lambda^{\pm}$ for the nonunitary transformations where Λ^+ is associated with the complex eigenvalues of L_H with Im $Z_\alpha^{(\nu)} \leq 0$, and the semigroup described by Λ^- is associated with the eigenvalues that are complex conjugates of $Z_\alpha^{(\nu)}$. We may distinguish two situations. There are initial conditions which belong to the domain of both Λ^+ and Λ^-. For example, the vacuum of correlations is in the domain of both. Let us follow the evolution in a semigroup, say in Λ^+. This would correspond by definition to an evolution into the "future." The main point is that there are *no dynamical processes* which would transform Λ^+ to Λ^-. Moreover, the Λ inversion leads to divergences. In short, the Λ inversion involves contributions of the form $P^{(\nu)}\mathbf{D}^{(\nu)-}\mathbf{C}^{(\nu)+}P^{(\nu)}$ (where the destruction operator $\mathbf{D}^{(\nu)-}$ refers to Λ^- and the creation operator $\mathbf{C}^{(\nu)+}$ to Λ^+). To the lowest order (see Appendix E for more detail),

$$\mathbf{C}^{(\nu)+} \sim \frac{1}{w_\nu - i\epsilon}, \qquad \mathbf{D}^{(\nu)-} \sim \frac{1}{w_\nu + i\epsilon} \tag{14.1}$$

where $\epsilon \to 0+$ is a positive infinitesimal. Therefore, the product

$$\mathbf{D}^{(\nu)-}\mathbf{C}^{(\nu)+} \sim \frac{1}{|w_\nu - i\epsilon|^2} \sim \frac{1}{\epsilon}\,\delta(w_\nu) \to \infty \tag{14.2}$$

We can also consider this problem from a different point of view—the "time inversion." The Λ inversion is a complex conjugation which corresponds in quantum mechanics to a time inversion. Again, time inversion is not a spontaneous process included in a semigroup. The simplest way to discuss its meaning is to consider the behavior of the $\mathscr{H}$-functions, as introduced in Section VII. (An example is discussed in detail in a recent paper [15].) Starting with a certain value of $\mathscr{H}$ at $t = 0$, this function decreases monotonically inside Λ^+ until t_0, when we proceed to a time (or velocity) inversion. At this point, the $\mathscr{H}$-function has a discontinuity (see Fig. 5)—it jumps to a higher value. This jump is easy to understand because velocity inversion leads to additional "long-range correlations" (particles that have collided in the past are bound to recollide; see also the discussion in Section XI). The value of this jump in $\mathscr{H}$ increases rapidly when t_0 increases. For large times, we again have a divergence. This corresponds to what we may call an "entropy barrier." Time inversion corresponds to an *entropy flow* coming from *outside*, and not to spontaneous processes described by a decrease in $\mathscr{H}$ or positive entropy production *inside* the system.

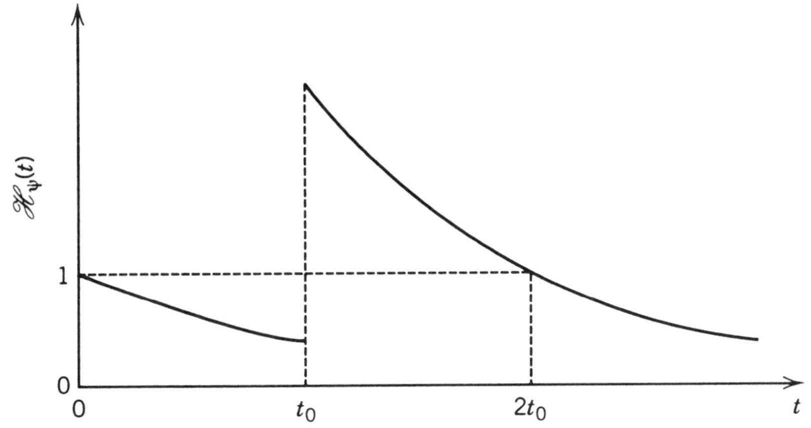

Figure 5. Behavior of the $\mathscr{H}$-function by a time (or velocity) inversion. We perform the inversion at $t = t_0$. After the jump, it still decreases monotonically.

Obviously, it is meaningless to speak of an entropy flow for the universe in the framework of our Hamiltonian description. (It is already difficult to imagine how we could proceeed to our own time inversion!) However, this does not prohibit time inversion for a limited ensemble of particles, as realized for example in spin-echo experiments.

Finally, let us just mention that in a time-asymmetrical universe, other symmetry breaking associated with the appearance of "dissipative structures" in far from equilibrium conditions becomes the rule (e.g., see [56]). To study these problems, we need to specify well defined constraints which prevent the system to reach equilibrium. We hope to investigate these problems in a separate paper.

APPENDIX A. COMPLEX SPECTRAL REPRESENTATION

In this appendix, we display some formulae used in the complex spectral representations. For the derivations of the formulae, see the previous paper *I*.

As in Eqs. (5.12) and (5.13), we have, for the "destruction-of-correlations" operator,

$$\mathscr{D}^{(\nu)}(z) = P^{(\nu)}\lambda L_V Q^{(\nu)} \frac{1}{z - Q^{(\nu)} L_H Q^{(\nu)}} \tag{A.1}$$

and

$$\mathscr{D}^{(\nu)}(z) = \sum_{n=0}^{\infty} P^{(\nu)} \lambda L_V Q^{(\nu)} \frac{1}{z - L_0} \left(Q^{(\nu)} \lambda L_V Q^{(\nu)} \frac{1}{z - L_0} \right)^n \tag{A.2}$$

These expansions correspond to a sequence of "irreducible transitions," because the intermediate states are orthogonal to the initial state in the space $P^{(\nu)}$ [4].

The substitution of z by $Z_\alpha^{(\nu)}$ in Eq. (A.1) [and in Eq. (5.12)] leads to the destruction operator $\mathscr{D}^{(\nu)}(Z_\alpha^{(\nu)})$ in Eq. (5.29), and also to a similar expression for the creation operator $\mathscr{C}^{(\nu)}(Z_\alpha^{(\nu)})$ in Eq. (5.28). However, we have to be careful when analyzing the continuation of $(z - L_0)^{-1}$ to avoid divergences associated with the Poincaré resonances [6]. This is achieved by using the analytic continuation in Eqs. (A.5) and (A.6) defined below. This leads to

$$P^{(\mu)} \mathscr{C}^{(\nu)}(Z_\alpha^{(\nu)}) P^{(\nu)} = P^{(\mu)} \frac{-1}{(w_\mu - Z_\alpha^{(\nu)})_{C\mu\nu}} \mathscr{T}_C^{(\nu)}(Z_\alpha^{(\nu)}) P^{(\nu)} \tag{A.3a}$$

$$P^{(\nu)} \mathscr{D}^{(\nu)}(Z_\alpha^{(\nu)}) P^{(\mu)} = P^{(\nu)} \mathscr{T}_D^{(\nu)}(Z_\alpha^{(\nu)}) \frac{1}{(Z_\alpha^{(\nu)} - w_\mu)_{D\mu\nu}} P^{(\mu)} \tag{A.3b}$$

where we have introduced the "$\mathscr{T}$-matrices" that satisfy the recurrence relations

$$\mathscr{T}_C^{(\nu)}(z) = \lambda Q^{(\nu)} L_V + \sum_\mu \lambda Q^{(\nu)} L_V P^{(\mu)} \frac{-1}{(w_\mu - z)_{C\mu\nu}} \mathscr{T}_C^{(\nu)}(z) \tag{A.4a}$$

$$\mathscr{T}_D^{(\nu)}(z) = \lambda L_V Q^{(\nu)} + \sum_\mu \mathscr{T}_D^{(\nu)}(z) \frac{1}{(z - w_\mu)_{D\mu\nu}} P^{(\mu)} \lambda L_V Q^{(\nu)} \tag{A.4b}$$

The analytic continuations of the denominators in Eqs. (A.3) and (A.4) are deduced from the biorthogonality condition of the eigenstates [11]. The result is as follows: To Specify the analytic continuations in Eq. (A.3), we define the index d_ν of the "degree of correlations" of the unperturbed state $|\nu\rangle\rangle$ as the integer which is the minimum number of interactions λL_V required to raise the state $|\nu\rangle\rangle$ from the state $|0, p\rangle\rangle$, the "vacuum of correlations." The degree of the correlation for $|0, p\rangle\rangle$ is $d_0 = 0$. The second term of the Fourier component in Eq. (5.7) corresponds to $d_\nu = 1$, the third term to $d_\nu = 2$, and so on. Note that the sign of the wave vectors is irrelevant to the definition of the degree of correlation, that is, $d_{\nu_-} = d_{\nu_+}$, where $\nu_\pm \equiv (\pm k, P)$.

Then the analytic continuations in Eq. (A.3) are given by

$$P^{(\mu)} \frac{-1}{(w_\mu - Z_\alpha^{(\nu)})_{C\mu\nu}} \equiv P^{(\mu)} \frac{-1}{[w_\mu - z]^+_{Z_\alpha^{(\nu)}}} \quad \text{for } d_\mu \geq d_\nu \tag{A.5a}$$

$$P^{(\mu)} \frac{-1}{(w_\mu - Z_\alpha^{(\nu)})_{C\mu\nu}} \equiv P^{(\mu)} \frac{-1}{w_\mu - Z_\alpha^{(\nu)}} \quad \text{for } d_\mu < d_\nu \tag{A.5b}$$

and

$$\frac{1}{(Z_\alpha^{(\nu)} - w_\mu)_{D\nu\mu}} P^{(\mu)} \equiv \frac{1}{[z - w_\mu]^+_{Z_\alpha^{(\nu)}}} P^{(\mu)} \quad \text{for } d_\nu < d_\mu \tag{A.6a}$$

$$\frac{1}{(Z_\alpha^{(\nu)} - w_\mu)_{D\nu\mu}} P^{(\mu)} \equiv \frac{1}{Z_\alpha^{(\nu)} - w_\mu} P^{(\mu)} \quad \text{for } d_\nu \geq d_\mu \tag{A.6b}$$

Here,

$$\int_{\mathbf{R}} dw \frac{f(w)}{[w - z]^+_{Z_\alpha^{(\nu)}}} = \lim_{\epsilon \to 0+} \sum_{n=0}^{\infty} \int_{\mathbf{R}} dw \frac{(-i\gamma)^n}{(w - w' - i\epsilon)^{n+1}} f(w) \tag{A.7}$$

and

$$\int_{\mathbf{R}} dw \frac{f(w)}{w - Z_\alpha^{(\nu)}} = \lim_{\epsilon \to 0+} \sum_{n=0}^{\infty} \int_{\mathbf{R}} dw \frac{(-i\gamma)^n}{(w - w' + i\epsilon)^{n+1}} f(w) \tag{A.8}$$

where $Z_\alpha^{(\nu)} = w' - i\gamma$ with w' and $\gamma \geq 0$ real, and the integrations are performed with a suitable test function $f(w)$ on the real axis $\mathbf{R}$. We can perform the summation of the geometrical series in Eq. (A.7) by introducing the "complex distribution" defined by [14]:

$$\int_{\mathbf{R}} dw \frac{f(w)}{[w - z]^+_{Z_\alpha^{(\nu)}}} \equiv \lim_{z \downarrow Z_\alpha^{(\nu)}} \int_{\mathbf{R}} dw \frac{f(w)}{w - z} \tag{A.9}$$

Here, $z \downarrow Z_\alpha^{(\nu)}$ means the "delayed" analytic continuations, that is, we first evaluate the integration in the upper-half plane of z (i.e., Im $z > 0$) and then take the limit $z \to Z_\alpha^{(\nu)}$ in the lower-half plane [40].

We can also represent Eq. (A.9) in terms of a "complex delta function" defined by [with a suitable test function $f(z)$]

$$\int_{\mathbf{R}} dw\, f(w) \delta_c(w - z) = f(z) \tag{A.10}$$

Then Eq. (A.9) can be written as (for $Z_\alpha^{(\nu)} = w' - i\gamma$ with w' and γ real with $\gamma \geq 0$)

$$\int_{\mathbf{R}} dw \frac{f(w)}{[w - z]^{+}_{Z_\alpha^{(\nu)}}} = \int_{\mathbf{R}} dw \left[\frac{1}{w - Z_\alpha^{(\nu)}} + 2\pi i \delta_c(w - Z_\alpha^{(\nu)}) \right] f(w) \quad \text{(A.11)}$$

There is another branch of the analytic continuation in Eq. (A.3), which corresponds to the complex conjugate of Eqs. (A.5) and (A.6). But we shall not consider this branch, because it leads to the eigenstates belonging to another semigroup oriented toward our past.

For the lowest-order contribution of Eqs. (A.5) and (A.6) in perturbation expansion, the analytic continuation above reduces to the relation

$$\left[P^{(\mu)} \frac{-1}{L_0 - z} \right]_{z = w_\nu - i\epsilon_{\mu\nu}} \quad \text{(A.12)}$$

for the creation operators $C^{(\nu)}(Z_\alpha^{(\nu)})$, and

$$\left[\frac{1}{z - L_0} P^{(\nu)} \right]_{z = w_\mu + i\epsilon_{\mu\nu}} \quad \text{(A.13)}$$

for the creation operators $D^{(\mu)}(Z_\alpha^{(\mu)})$. Here, $\epsilon_{\mu\nu}$ is defined by

$$i\epsilon_{\mu\nu} = \begin{cases} -i\epsilon & \text{for } d_\mu \geq d_\nu \\ +i\epsilon & \text{for } d_\mu < d_\nu \end{cases} \quad \text{(A.14)}$$

and ϵ is a positive infinitesimal $\epsilon \to 0+$. Reading from the right to the left, we can unify Eqs. (A.12) and (A.13) in a single form (see [45] for arbitrary orders in the perturbation expansion):

$$\frac{1}{w_\mu - w_\nu + i\epsilon_{\mu\nu}} \quad \text{(A.15)}$$

This relation was first proposed to obtain irreversible kinetic equations and was called the $i\epsilon$-rule [39,45]. As mentioned, we can deduce this relation from the biorthogonality condition of the eigenstates in the complex spectral representation (see [11] and *I*).

The global destruction operator $\mathbf{D}^{(\nu)}$ in Eq. (5.28) is then defined by

$$\mathbf{D}^{(\nu)} \equiv \sum_\alpha |v_\alpha^{(\nu)}\rangle\!\rangle\langle\!\langle \tilde{v}_\alpha^{(\nu)}| \mathscr{D}^{(\nu)}(Z_\alpha^{(\nu)}) \quad \text{(A.16)}$$

where we have introduced a complete set of states $\{|v_\alpha^{(\nu)}\rangle\!\rangle\}$ biorthogonal to $\{\langle\!\langle\tilde{v}_\alpha^{(\nu)}|\}$ [cf. Eq. (5.20)],

$$\langle\!\langle\tilde{v}_\alpha^{(\nu)}|v_\beta^{(\mu)}\rangle\!\rangle = \delta_{\nu,\mu}\delta_{\alpha,\beta}, \qquad \sum_\alpha |v_\alpha^{(\nu)}\rangle\!\rangle\langle\!\langle\tilde{v}_\alpha^{(\nu)}| = P^{(\nu)} \tag{A.17}$$

which satisfy

$$L_0|v_\alpha^{(\nu)}\rangle\!\rangle = w_\nu|v_\alpha^{(\nu)}\rangle\!\rangle, \qquad \langle\!\langle\tilde{v}_\alpha^{(\nu)}|L_0 = \langle\!\langle\tilde{v}_\alpha^{(\nu)}|w_\nu \tag{A.18}$$

We have [cf. Eqs. (5.11) and (5.27)]

$$\langle\!\langle\tilde{F}_\alpha^{(\nu)}|Q^{(\nu)} = \langle\!\langle\tilde{F}_\alpha^{(\nu)}|P^{(\nu)}\mathscr{D}^{(\nu)}(Z_\alpha^{(\nu)}) = \langle\!\langle\tilde{F}_\alpha^{(\nu)}|P^{(\nu)}\mathbf{D}^{(\nu)} \tag{A.19}$$

The global collision operator $\theta_D^{(\nu)}$ associated with $\mathbf{D}^{(\nu)}$ is given by [cf. Eq. (5.25)]

$$\theta_D^{(\nu)} \equiv \sum_\alpha |v_\alpha^{(\nu)}\rangle\!\rangle\langle\!\langle\tilde{v}_\alpha^{(\nu)}|\psi^{(\nu)}(Z_\alpha^{(\nu)}) = \sum_\alpha |v_\alpha^{(\nu)}\rangle\!\rangle Z_\alpha^{(\nu)}\langle\!\langle\tilde{v}_\alpha^{(\nu)}| \tag{A.20}$$

We then have

$$\theta_D^{(\nu)}|v_\alpha^{(\nu)}\rangle\!\rangle = Z_\alpha^{(\nu)}|v_\alpha^{(\nu)}\rangle\!\rangle, \qquad \langle\!\langle\tilde{v}_\alpha^{(\nu)}|\theta_D^{(\nu)} = \langle\!\langle\tilde{v}_\alpha^{(\nu)}|Z_\alpha^{(\nu)} \tag{A.21}$$

The normalization constant $N_\alpha^{(\nu)}$ in Eqs. (5.28) and (5.29) is given by

$$N_\alpha^{(\nu)} = [\langle\!\langle\tilde{v}_\alpha^{(\nu)}|(P^{(\nu)} + \mathbf{D}^{(\nu)}\mathbf{C}^{(\nu)})|u_\alpha^{(\nu)}\rangle\!\rangle]^{-1} \tag{A.22}$$

Putting

$$(A^{(\nu)})^{-1} \equiv P^{(\nu)} + \mathbf{D}^{(\nu)}\mathbf{C}^{(\nu)} \tag{A.23}$$

we obtain

$$(A^{(\nu)})^{-1} = \sum_\alpha |v_\alpha^{(\nu)}\rangle\!\rangle(N_\alpha^{(\nu)})^{-1}\langle\!\langle\tilde{u}_\alpha^{(\nu)}| \tag{A.24}$$

and its inverse operator in $P^{(\nu)}$ subspace is given by Eq. (5.39). Hence, we have

$$\langle\!\langle\tilde{u}_\alpha^{(\nu)}|\theta_C^{(\nu)}A^{(\nu)}|v_\beta^{(\nu)}\rangle\!\rangle = Z_\alpha^{(\nu)}N_\alpha^{(\nu)}\delta_{\alpha,\beta} = \langle\!\langle\tilde{u}_\alpha^{(\nu)}|A^{(\nu)}\theta_D^{(\nu)}|v_\beta^{(\nu)}\rangle\!\rangle \tag{A.25}$$

This leads to the intertwining relationship of $A^{(\nu)}$ with the collision operators in Eq. (5.39).

APPENDIX B. STAR-UNITARY TRANSFORMATIONS AND HERMITICITY PRESERVATION

In this appendix we display a symmetry property, called "star conjugation," of the projection operators $\Pi^{(\nu)}$ and the nonunitary transformation Λ_B [22,33,38]. We also discuss the preservation of the reality of the Wigner distribution functions $\langle\langle Q, P | \rho \rangle\rangle$ by the nonunitary transformations.

The projection operators $\Pi^{(\nu)}$ are not hermitian [see Eq. (6.17)]. They satisfy more general symmetry relations. To see this, let us first observe the relation between the "Schrödinger picture" of the evolution of states in Eq. (2.3) and the "Heisenberg picture" of observables in Eq. (3.48). By interchanging the role of the distribution functions and observables $|\rho(t)\rangle\rangle \leftrightarrow |M(t)\rangle\rangle$, as well as changing the sign of $L_H \to -L_H$, one picture leads to the other. To obtain a consistent description in which each subdynamics generate the evolution approach to equilibrium toward our future in both states and in observables, we have to use the spectral decomposition for the density matrices:

$$|\rho(t)\rangle\rangle = \sum_\nu \sum_\alpha |F_\alpha^{(\nu)}\rangle\rangle\langle\langle \tilde{F}_\alpha^{(\nu)}| e^{-iL_H t} |\rho(0)\rangle\rangle \tag{B.1}$$

while for the observables:

$$|M(t)\rangle\rangle = \sum_\nu \sum_\alpha |\tilde{F}_\alpha^{(\nu)}\rangle\rangle\langle\langle F_\alpha^{(\nu)}| e^{+iL_H t} |M(0)\rangle\rangle \tag{B.2}$$

Let us denote the "Heisenberg–Schrödinger interchanging operation" by the notation " ′ ", and call this the "prime operation" [14,38]. Comparing Eqs. (B.1) and (B.2), we have[17]

$$(|F_\alpha^{(\nu)}\rangle\rangle)' = |\tilde{F}_\alpha^{(\nu)}\rangle\rangle, \qquad (\langle\langle \tilde{F}_\alpha^{(\nu)}|)' = \langle\langle F_\alpha^{(\nu)}|, \qquad L_H' = -L_H \tag{B.3}$$

The $P^{(\nu)}$ component of the first relation in Eq. (B.3) leads to [see Eqs. (5.28) and (5.29)]

$$(|u_\alpha^{(\nu)}\rangle\rangle)' = |\tilde{v}_\alpha^{(\nu)}\rangle\rangle \tag{B.4}$$

and so on. We note that

$$\Pi^{(\nu)\prime} = \Pi^{(\nu)\dagger} \neq \Pi^{(\nu)} \tag{B.5}$$

[17] By definition, we have $(W')' = W$ and also $(W^*)^* = W$ for the star conjugation introduced below. The prime operation inverts bra's and ket's.

A combination of the prime operation with hermitian conjugation is called the "star–hermitian conjugation" and is denoted by "*" [39]. Then we have

$$(e^{-iL_Ht}|F_\alpha^{(\nu)}\rangle\!\rangle\langle\!\langle\tilde{F}_\alpha^{(\nu)}|)^* = [e^{+iL_Ht}(|F_\alpha^{(\nu)}\rangle\!\rangle\langle\!\langle\tilde{F}_\alpha^{(\nu)}|)']^\dagger$$
$$= (e^{+iL_Ht}|\tilde{F}_\alpha^{(\nu)}\rangle\!\rangle\langle\!\langle F_\alpha^{(\nu)}|)^\dagger = |F_\alpha^{(\nu)}\rangle\!\rangle\langle\!\langle\tilde{F}_\alpha^{(\nu)}|e^{-iL_Ht} \quad \text{(B.6)}$$

as well as

$$(e^{-iL_Ht}|F_\alpha^{(\nu)}\rangle\!\rangle\langle\!\langle\tilde{F}_\alpha^{(\nu)}|)^* = (|\tilde{F}_\alpha^{(\nu)}\rangle\!\rangle\langle\!\langle F_\alpha^{(\nu)}|e^{+iL_Ht})' = |F_\alpha^{(\nu)}\rangle\!\rangle\langle\!\langle\tilde{F}_\alpha^{(\nu)}|e^{-iL_Ht} \quad \text{(B.7)}$$

Hence we obtain

$$(e^{-iL_Ht}|F_\alpha^{(\nu)}\rangle\!\rangle\langle\!\langle\tilde{F}_\alpha^{(\nu)}|)^* = e^{-iL_Ht}|F_\alpha^{(\nu)}\rangle\!\rangle\langle\!\langle\tilde{F}_\alpha^{(\nu)}| \quad \text{(B.8)}$$

and

$$\Pi^{(\nu)*} = \Pi^{(\nu)}, \qquad L_H^* = -L_H \quad \text{(B.9)}$$

$\Pi^{(\nu)}$ is a star–hermitian operator, while L_H is an anti-star–hermitian operator. Star-hermiticity is a natural generalization of the usual conjugates of hermiticity. Equations (B.9) lead to

$$A^{(\nu)*} = A^{(\nu)}, \qquad \mathbf{C}^{(\nu)*} = \mathbf{D}^{(\nu)} \quad \text{(B.10)}$$

as well as (see also *I*)

$$(|u_\alpha^{(\nu)}\rangle\!\rangle Z_\alpha^{(\nu)}\langle\!\langle\tilde{u}_\alpha^{(\nu)}|)^* = -|v_\alpha^{(\nu)}\rangle\!\rangle Z_\alpha^{(\nu)}\langle\!\langle\tilde{v}_\alpha^{(\nu)}| \quad \text{(B.11)}$$

that is, $\theta_C^{(\nu)*} = -\theta_D^{(\nu)}$, and these collision operators are antiskew star symmetric. Applying these operations to Λ_B, we have

$$\Lambda_C^* = \Lambda_D^{-1}, \qquad \Lambda_D^* = \Lambda_C^{-1} \quad \text{(B.12)}$$

that is (for $B' \neq B$ with $B, B' = C, D$),

$$\Lambda_B\Lambda_{B'}^* = \Lambda_{B'}^*\Lambda_B = 1 \quad \text{(B.13)}$$

We can introduce even more symmetric transformation operators [22,33,38]. Let us define

$$\Lambda = \sum_\nu A^{(\nu)1/2}(P^{(\nu)} + \mathbf{D}^{(\nu)}), \qquad \Lambda^{-1} = \sum_\nu (P^{(\nu)} + \mathbf{C}^{(\nu)})A^{(\nu)1/2} \quad \text{(B.14)}$$

This is a "star-unitary" operator:

$$\Lambda\Lambda^* = \Lambda^*\Lambda = 1 \tag{B.15}$$

Then we have

$$\langle\langle \hat{M} | \mathscr{U}(t) | \rho(0) \rangle\rangle = \langle\langle \hat{M} | \Lambda^* \exp(-i\Theta t)\Lambda | \rho(0) \rangle\rangle \tag{B.16}$$

where Θ is a new collision operator defined by

$$\Theta \equiv \Lambda L_{\rm H} \Lambda^* \tag{B.17}$$

This satisfies

$$\Theta = \sum_{\nu} \theta^{(\nu)} \tag{B.18}$$

with

$$\theta^{(\nu)} = (A^{(\nu)})^{-1/2} \theta_C^{(\nu)} (A^{(\nu)})^{1/2} = (A^{(\nu)})^{1/2} \theta_D^{(\nu)} (A^{(\nu)})^{-1/2} \tag{B.19}$$

as well as

$$\theta^{(\nu)*} = -\theta^{(\nu)} \tag{B.20}$$

that is, $\Theta^* = -\Theta$, which is an anti-star–hermitian operator. Moreover, the new transformation operator leads to the same subdynamics $\Pi^{(\nu)}$ as Eq. (6.16):

$$\Pi^{(\nu)} = \Lambda^* P^{(\nu)} \Lambda \tag{B.21}$$

Note that the nonunitary transformations preserve the hermiticity of the density matrices (see below). But the transformed states $\langle\langle 0, P | \Lambda_B | \rho \rangle\rangle = \langle P | (\Lambda_B \rho) | P \rangle$ and $\langle\langle 0, P | \Lambda | \rho \rangle\rangle$ cannot be considered as a probability distribution functions because they do not preserve the positivity of the distribution function. This is a direct consequence of the causal evolution of the dynamics combined with the analytic continuation, Eqs. (A.5) and (A.6), involving complex distributions (see Appendix C). However, these states play an essential role because they permit us to introduce $\mathscr{H}$-functions for dynamics, as shown in Section VII.

Next we prove that the nonunitary transformations, Eq. (6.3), preserve the hermiticity of the density matrices. We verify the relations

$$(\mathbf{C}^{(k)})^a = \mathbf{C}^{(k)}, \qquad (\mathbf{D}^{(k)})^a = \mathbf{D}^{(k)}, \qquad (A^{(k)})^a = A^{(k)} \tag{B.22}$$

as well as

$$(\theta_B^{(k)})^a = -\theta_B^{(k)} \tag{B.23}$$

where we have specified the correlation ν by its value of the wave vector k and abbreviated the momentum arguments P. Here, the *associated* superoperator $\mathcal{Q}^a$ is defined through the relation [38]

$$(\mathcal{Q}\,|\,A^+\rangle\!\rangle)^+ = \mathcal{Q}^a\,|\,A\rangle\!\rangle \tag{B.24}$$

where "+" is the adjunction in the wave-functions space [cf. Eqs. (3.11) and (3.14)]. Equation (B.24) leads to

$$\langle\!\langle B\,|\,\mathcal{Q}^a\,|\,A\rangle\!\rangle = \langle\!\langle B^+\,|\,\mathcal{Q}\,|\,A^+\rangle\!\rangle^{c.c.} \tag{B.25}$$

In the momentum representation, we have

$$\langle\!\langle p;\, p'\,|\,\mathcal{Q}^a\,|\,p'';\, p'''\rangle\!\rangle = \langle\!\langle p';\, p\,|\,\mathcal{Q}\,|\,p''';\, p''\rangle\!\rangle^{c.c.} \tag{B.26}$$

which is equivalent to

$$\langle\!\langle k,\, P\,|\,\mathcal{Q}^a\,|\,k',\, P'\rangle\!\rangle = \langle\!\langle -k,\, P\,|\,\mathcal{Q}\,|\,-k',\, P'\rangle\!\rangle^{c.c.} \tag{B.27}$$

in the Wigner representation. For example, we have

$$(iL_H)^a = iL_H, \qquad (L_H)^a = -L_H \tag{B.28}$$

Equation (B.23) and the last equality of Eq. (B.22) are the consequence of the first two equalities of Eqs. (B.22) and (B.28).

Equation (B.22) shows that these operators are adjoint-symmetrical superoperators. The adjoint-symmetrical superoperators preserve the self-adjoint character of the supervector on which they act, that is,

$$\text{if} \quad |\,A\rangle\!\rangle = |\,A^+\rangle\!\rangle \quad \text{and} \quad \mathcal{Q} = \mathcal{Q}^a, \qquad \text{then } (\mathcal{Q}\,|\,A\rangle\!\rangle)^+ = \mathcal{Q}\,|\,A\rangle\!\rangle \tag{B.29}$$

Equation (B.22) leads to

$$\Lambda_B = \Lambda_B^a \tag{B.30}$$

Hence, the transformed density matrices $\Lambda_B|\rho\rangle\rangle$ are hermitian operators in the wave-function space. Here we prove only the first equality of Eq. (B.22); the proof of the second is essentially the same.

We begin with the nonlinear Lippmann–Schwinger equation (9.3a). In the Wigner representation, we have

$$\langle\langle l|\Phi_k^C\rangle\rangle = \langle\langle l|k\rangle\rangle + \sum_{k'} \frac{-1}{w_l - w_k + i\epsilon_{lk}} \langle\langle l|\lambda L_V|k'\rangle\rangle\langle\langle k'|\Phi_k^C\rangle\rangle$$
$$+ \sum_{k',k''} \frac{1}{w_l - w_k + i\epsilon_{lk}} \langle\langle l|\Phi_k^C\rangle\rangle\langle\langle k|\lambda L_V|k''\rangle\rangle\langle\langle k''|\Phi_k^C\rangle\rangle \quad \text{(B.31)}$$

where we have abbreviated the momentum arguments, for example, $\langle\langle l|$ for $\langle\langle l, P|$. Using Eq. (3.50) and changing the sign of l and k as well as of the dummy variables k' and k'', we obtain, by a straightforward calculation,

$$\langle\langle -l|\Phi_{-k}^C\rangle\rangle^{c.c.} = \langle\langle l|k\rangle\rangle + \sum_{k'} \frac{-1}{w_l - w_k + i\epsilon_{lk}} \langle\langle l|\lambda L_V|k'\rangle\rangle\langle\langle -k'|\Phi_{-k}^C\rangle\rangle^{c.c.}$$
$$+ \sum_{k',k''} \frac{1}{w_l - w_k + i\epsilon_{lk}} \langle\langle -l|\Phi_{-k}^C\rangle\rangle^{c.c.}\langle\langle k|\lambda L_V|k''\rangle\rangle\langle\langle -k''|\Phi_{-k}^C\rangle\rangle^{c.c.} \quad \text{(B.32)}$$

where we have used Eq. (B.28) and the relations $w_{-k} = -(k \cdot v) = -w_k$. We have also used $\epsilon_{-l,-k} = \epsilon_{l,k}$, because the degree of correlation does not depend on the sign of nonvanishing wave vectors, that is $d_{-k} = d_k$ [see the definition below Eq. (A.4b)].

This shows that $\langle\langle -l|\Phi_{-k}^C\rangle\rangle^{c.c.}$ satisfies Eq. (B.31), as does $\langle\langle l|\Phi_k^C\rangle\rangle$. Moreover we have the same boundary condition:

$$\langle\langle -l|\Phi_{-k}^C\rangle\rangle^{c.c.} = \langle\langle l|\Phi_k^C\rangle\rangle = \langle\langle l|k\rangle\rangle \qquad \text{for } \lambda = 0 \quad \text{(B.33)}$$

Hence we obtain, for any λ,

$$\langle\langle -l|\Phi_{-k}^C\rangle\rangle^{c.c.} = \langle\langle l|\Phi_k^C\rangle\rangle \quad \text{(B.34)}$$

Because of the relation [see Eqs. (6.4) and (9.2)]

$$\langle\langle l|\Phi_k^C\rangle\rangle = \langle\langle l|(P^{(k)} + \mathbf{C}^{(k)})|k\rangle\rangle \quad \text{(B.35)}$$

we obtain

$$\langle\langle -l|\mathbf{C}^{(-k)}|-k\rangle\rangle^{c.c.} = \langle\langle l|\mathbf{C}^{(k)}|k\rangle\rangle \quad \text{(B.36)}$$

which gives us the first equality of Eq. (B.22). Similarly, one can also prove the second equality of Eq. (B.22).

APPENDIX C. EXPONENTIALLY GROWING CONTRIBUTIONS AND CAUSAL EVOLUTION

The most striking consequence of the analytic continuations of Eq. (A.5) is that the transformed state $|\rho_B\rangle\!\rangle$ in Eq. (6.8) does not preserve the positivity of the probability. This is a result of the complex distribution in Eq. (A.5a), which leads to an exponentially growing contribution in space (the so-called "exponential catastrophe" [9]). Let us consider a one-dimensional integration over l with a suitable test function $f(l)$ and with $v > 0$ and $Q > 0$:

$$I(Q) = \int_{-\infty}^{+\infty} \frac{f(l)}{[lv - z]^{+}_{Z_\alpha(v)}} e^{ilQ}\, dl \tag{C.1}$$

Then, the residue Res[I] of this integration at the pole $l = Z_\alpha^{(v)}/v$ is given by

$$\text{Res}[I] = \frac{2\pi i}{v} f(Z_\alpha^{(v)}/v) e^{iZ_\alpha(v)Q/v} \tag{C.2}$$

This gives the exponentially growing contribution in Q for decay modes with Im $Z_\alpha^{(v)} < 0$.

This type of contribution in space is necessary to ensure the causal evolution of the decay modes, because in time the damping factor $\exp(-iZ_\alpha^{(v)}t)$ requires a space dependence given by $\exp[iZ_\alpha^{(v)}(Q - vt)/v]$. For $v > 0$ with finite time t, this diverges in the limit of $Q \to +\infty$. However, the bicompleteness relation of the eigenstates ensures that the contribution from $Q > vt$ in the distribution function vanishes (see Fig. 4). As a result, some components of the transformed states should have negative values for $Q > vt$. The causal evolution in terms of the complex spectral representation has been fully analyzed for the Friedrichs model, which is a model for the spontaneous emission of photons by an exited atom [57].

APPENDIX D. SINGULARITY IN UNITARY TRANSFORMATION

In this appendix we show that the inner product $\langle\!\langle \Phi_{0,P}^{-} | \Phi_{0,P'}^{-} \rangle\!\rangle$ for the unitary transformation leads to a singularity at $\epsilon = 0$, while $\langle\!\langle \Phi_{0,P}^{D} | \Phi_{0,P'}^{C} \rangle\!\rangle$ for the nonunitary transformation regularizes the singularity.

Because the inner products are distributions, we evaluate this term in conjunction with an integration over P as (e.g., with the momentum $\mathbf{P}_1$)

$$\begin{aligned}\int d^{\mathcal{N}}P\mathbf{P}_1\langle\langle\Phi_{0,P}^{-}|\Phi_{0,P'}^{-}\rangle\rangle &= \mathbf{P}_1 - \frac{\lambda^2}{\Omega^2}\sum_{n>j}^{\mathcal{N}}\sum_{\mathbf{k}}\int d^{\mathcal{N}}P\mathbf{P}_1\partial_{jn}^{\hbar\mathbf{k}/2}\frac{|V_k|^2}{|\mathbf{k}\cdot\mathbf{g}_{jn}+i\epsilon|^2}\partial_{jn}^{\hbar\mathbf{k}/2}\delta(P-P') + O(\lambda^3)\\ &= \mathbf{P}_1 - \frac{\lambda^2}{\hbar\Omega^2}\sum_{n=2}^{\mathcal{N}}\sum_{\mathbf{k}}\left(\partial_{1n}^{\hbar\mathbf{k}/2}\frac{|V_k|^2}{|\mathbf{k}\cdot\mathbf{g}'_{1n}+i\epsilon|^2}\mathbf{k}\right)_{\substack{\mathbf{P}_{1'}=\mathbf{P}_1\\ \mathbf{P}_{n'}=\mathbf{P}_n}} + O(\lambda^3)\\ &= \mathbf{P}_1 + O(\mathcal{N}/\epsilon\Omega)\end{aligned} \tag{D.1}$$

To obtain the second equality in Eq. (D.1), we performed an integration by parts over the momenta [see the notes at Eqs. (3.61) and (3.58)].

Owing to Poincaré resonances, a singularity, $\sim\epsilon^{-1}$, appears in Eq. (D.1). For $\mathcal{N}$ finite, this singularity is harmless, because we must first take the limit $\Omega\to\infty$ before taking the limit $\epsilon\to 0+$ [see Eq. (5.49)]. We see that the unitary transformations cannot be extended to nonintegrable systems in the thermodynamic limit. In contrast, the nonunitary transformation regularizes the Poincaré divergence as

$$\int d^{\mathcal{N}}P\mathbf{P}_1\langle\langle\Phi_{0,P}^{D}|\Phi_{0,P'}^{C}\rangle\rangle \sim \mathbf{P}_1 + \frac{\mathcal{N}}{\Omega^2}\sum_{\mathbf{k}}\left[\frac{1}{(\mathbf{k}\cdot\mathbf{v}+i\epsilon)^2} + c.c.\right] \sim \mathbf{P}_1 + O(\mathcal{N}/\Omega) \tag{D.2}$$

In contrast, the nonunitary transformation, Eq. (9.14), has a natural extension for the nonintegrable systems where the time symmetry is broken.

APPENDIX E. DYNAMICAL GROUPS AS APPROXIMATIONS OF DYNAMICAL SEMIGROUPS

In this appendix we show that once the universe as a whole is in a semigroup oriented toward a well-defined direction of time, it is impossible to prepare a system that belongs to the other semigroup–oriented toward the opposite direction of time.

Let us introduce the new notations $\Lambda_B^{\pm}$ for the nonunitary transformations, where Λ_B^{+} is identical to Λ_B in Eq. (6.3), which is associated with the complex eigenvalues of L_H with Im $Z_\alpha^{(\nu)} \leq 0$. In the semigroup described by Λ_B^{+}, equilibrium states are achieved in the direction of our future at $t > 0$.

In contrast, the semigroup described by Λ_B^- is associated with eigenvalues that are complex conjugate to $Z_\alpha^{(\nu)}$, and equilibrium states are achieved in the direction of our past at $t < 0$.

Infinitely many initial conditions belong to both semigroups. For example, any states in the vacuum of correlations are in the domain of Λ_B^+ and Λ_B^-.

$$P^{(0)}\rho \in \Lambda_B^\pm \tag{E.1}$$

There is an ambiguity in the direction of time. Let us suppose that this ambiguous situation occurred early in the development of the universe. We put $t = 0$ at that moment. There are two exclusive possibilities in the evolution of the universe; one is the semigroup Λ^+ and the other is in Λ^-. Let us then follow the evolution in one semigroup, say in Λ^+ (the choice of semigroup is not important). For $t > 0$, we have

$$|\rho^+(t)\rangle\!\rangle = \sum_\nu \sum_\alpha e^{-iZ_\alpha^{(\nu)}t} |F_\alpha^{(\nu)+}\rangle\!\rangle\langle\!\langle \tilde{F}_\alpha^{(\nu)+} | P^{(0)} | \rho(0)\rangle\!\rangle \tag{E.2}$$

where we have also introduced the superscript + to indicate that these objects belong to the semigroup Λ_B^+. The interaction leads to correlations among the particles. For a long time, the resonance builds up the "long range correlations" (see Section XI). Let us then consider a contribution to the μth-order correlation coming from the $\Pi^{(\nu)}$ subspace at $t = t_0$ for $t_0 > 0$:

$$P^{(\mu)}\Pi^{(\nu)+} |\rho^+(t_0)\rangle\!\rangle = \sum_\alpha e^{-iZ_\alpha^{(\nu)}t_0} P^{(\mu)}\mathscr{C}^{(\nu)+}(Z_\alpha^{(\nu)}) | u_\alpha^{(\nu)+}\rangle\!\rangle N_\alpha^{(\nu)+}$$
$$\times \langle\!\langle \tilde{v}_\alpha^{(\nu)+} | \mathscr{D}^{(\nu)+}(Z_\alpha^{(\nu)}) P^{(0)} | \rho(0)\rangle\!\rangle \tag{E.3}$$

We now ask if this component also belongs to the domain of Λ_B^-. To determine this, we evaluate $\Lambda_B^- |\rho^+(t)\rangle\!\rangle$. Then $\Lambda_B^- |\rho^+(t_0)\rangle\!\rangle$ involves the following contribution coming from a diagonal transition through the intermediate states $d_\mu > d_\nu$ [see Eq. (A.3) and Henin's theorem presented at Eq. (5.52)]:

$$\Lambda_B^- \Pi^{(\nu)+} |\rho^+(t_0)\rangle\!\rangle \sim \sum_\mu P^{(\nu)}\mathscr{D}^{(\nu)-} P^{(\mu)}\mathscr{C}^{(\nu)+} P^{(\nu)}$$
$$= \sum_\mu P^{(\nu)}\mathscr{T}_D^{(\nu)c.c.} P^{(\mu)} \frac{1}{|[z - w_\mu]_{Z_\alpha}^+|^2} P^{(\mu)}\mathscr{T}_C^{(\nu)} P^{(\nu)} \tag{E.4}$$

Recall that in the thermodynamic limit, this expression has a well-defined meaning when the inner products with observables (4.22) that depend on a finite number of degrees of freedom are taken [see also Eq. (5.8)]. Then, the inner products become a well-defined function of the concentration $c = \mathcal{N}/L^3$.

In general, there are diffusion modes in space $\Pi^{(\nu)}$ with $\nu \neq 0$. We have illustrated this fact for the perfect Lorentz gas in Appendix I. The diffusion modes are associated with small wave vectors **k**, which correspond to microscopic scales, such as the hydrodynamic scale in space [29]. The characteristic feature of the diffusion modes is that their eigenvalues are purely imaginary and are proportional to $-i|\mathbf{k}|^2$ (see Appendix I). As mentioned, resonances lead to long-range correlations.

As we wait, more long-range correlations build up, continually involving more particles. This implies that smaller and smaller wave vectors **k** contribute to Eq. (E.4). Hence, for sufficiently large t_0 and a finite concentration of the system c, Eq. (E.4) has a contribution from the diffusion mode which diverges as

$$\langle\langle M|\Lambda_B^-|\rho^+(t_0)\rangle\rangle \sim \int dw_\mu \frac{1}{|[z - w_\mu]^+_{-i|\mathbf{k}|^2}|^2} \sim \frac{1}{|\mathbf{k}|^2} \to \infty \qquad \text{for } |\mathbf{k}| \to 0 \tag{E.5}$$

and $|\rho^+(t_0)\rangle\rangle$ is no longer in the domain of Λ_B^-. We note that this divergence appears only in the *diagonal transitions*. Therefore, we could neglect the divergence if the system would be integrable (see Sections VIII and IX). If the universe would be integrable, $|\rho^+(t_0)\rangle\rangle$ would still be in the domain of Λ_B^-.

Let us now ask which operation on the system leads to the class of initial conditions belonging to the other semigroup. From the discussion above, it is clear that if one can prepare the states in which all components in each subspaces $\Pi^{(\nu)}$ are complex conjugates of $|\rho^+(t_0)\rangle\rangle$, then the states are in the domain of Λ_B^-. Let us denote the operator corresponding to this operation by $\hat{I}_{\mathcal{N}}$. We have

$$\hat{I}_{\mathcal{N}} P^{(\mu)}\Pi^{(\nu)}|\rho^+(t_0)\rangle\rangle \sim P^{(\mu)}[\mathscr{C}^{(\nu)}(Z_\alpha^{(\nu)})]^{c.c.} P^{(\nu)} \tag{E.6}$$

Therefore,

$$\hat{I}_{\mathcal{N}}|\rho^+(t_0)\rangle\rangle = |\rho^-(t_0)\rangle\rangle \in \Lambda_B^- \tag{E.7}$$

In the Wigner representation, one can realize this operation by changing the sign of the "momentum" (or "velocity") variable P of all particles while keeping the "coordinate" Q (or wave vector k) unchanged. Indeed, this operation changed the sign of the Liouvillian $\hat{I}_{\mathcal{N}} L_H = -L_H$ [see Eqs. (3.38) and (3.50)] and leads the analytic continuation in the opposite direction; for example, for the creation operator $\lambda C_1^{(0)}$,

$$\hat{I}_{\mathcal{N}} \frac{-1}{L_0 - i\epsilon} Q^{(0)} L_V P^{(0)} = \frac{-1}{L_0 + i\epsilon} Q^{(0)} L_V P^{(0)} \tag{E.8}$$

In quantum mechanics, it is well known that the complex conjugation corresponds to the "time (or velocity) inversion" (e.g., see [60]). Hence, if we could invert the direction of the velocity of *all particles* in the universe, we could obtain an initial condition belonging to the other semigroup. However, we can only operate a *part* of the universe which consists of n particles where $n \ll \mathcal{N}$.

Let us isolate this subsystem from the rest of universe and consider its time evolution. We label the particles in the subsystem by the indices j where $1 \leq j \leq n$. As we focus our attention only on this isolated subsystem, we have to evaluate the expectation value of observables [e.g., M in Eq. (4.22)] over *specific* distribution functions associated with the subsystem. If n is sufficiently small, we can invert the velocity of the particles in the subsystem (such as in the spin-echo experiment). However, elements of the subsystem will eventually encounter the elements in the rest of the universe, so that this isolation can be achieved only for a finite time scale, say t_i. Let us then denote the "velocity inversion" on the subsystem $\hat{I}_n$. We compare the following three expectation values over the specific distribution functions:

$$M(t) = \langle\langle M | e^{-iL_H(t-t_0)} | \rho^+(t_0) \rangle\rangle \tag{E.9a}$$

$$M_{n-}(t) = \langle\langle M | e^{-iL_H(t-t_0)} \hat{I}_n | \rho^+(t_0) \rangle\rangle \tag{E.9b}$$

$$M_{\mathcal{N}-}(t) = \langle\langle M | e^{-iL_H(t-t_0)} \hat{I}_{\mathcal{N}} | \rho^+(t_0) \rangle\rangle \tag{E.9c}$$

For example, we have

$$\hat{I}_n \frac{1}{k \cdot v - i\epsilon} = \frac{1}{-\sum_{j=1}^{n} \mathbf{k}_j \cdot \mathbf{v}_j + \sum_{r=n+1}^{\mathcal{N}} \mathbf{k}_r \cdot \mathbf{v}_r - i\epsilon} \tag{E.10}$$

while

$$\hat{I}_{\mathcal{N}} \frac{1}{k \cdot v - i\epsilon} = \frac{1}{-k \cdot v - i\epsilon} \tag{E.11}$$

As an example, consider the evolution of the momentum of particle 1 [see Eq. (4.25)]. The first-order contribution in λ is given by

$$\langle\langle \hat{\mathbf{P}}_1 | \rho(t) \rangle\rangle|_\lambda = \lambda \sum_{r=2}^{\mathcal{N}} \sum_{\mathbf{k}_r} \sum_{P^{\mathcal{N}}, P'^{\mathcal{N}}} \mathbf{P}'_1$$
$$\times \langle\langle 0, P' | (\mathbf{D}_1^{(0)} + \mathbf{C}_1^{(k,-k)} e^{-i\mathbf{k}\cdot\mathbf{g}_{1r}t}) | \mathbf{k}_r, -\mathbf{k}_r, \{0\}^{\mathcal{N}-2}, P \rangle\rangle$$
$$\times \langle\langle \mathbf{k}_r, -\mathbf{k}_r, \{0\}^{\mathcal{N}-2}, P | \rho(0) \rangle\rangle \tag{E.12}$$

where

$$\mathbf{D}_1^{(0)} + \mathbf{C}_1^{(k,-k)} = 0 \tag{E.13}$$

and

$$\langle\langle 0, P' | \lambda \mathbf{D}_1^{(0)} | \mathbf{k}, -\mathbf{k}, \{0\}^{\mathcal{N}-2}, P \rangle\rangle = \frac{\lambda}{\Omega} V_{|\mathbf{k}|} \partial_{1r}^{\mathbf{k}/2} \frac{1}{\mathbf{k} \cdot i\epsilon - \mathbf{g}_{1r}} \delta^{kr}(P' - P) \tag{E.14}$$

Similarly to Eq. (11.7), we can show the causal evolution of Eq. (E.12) (see *I*). Hence, before the subsystem encounters the surrounding system, that is, during the time scale $|t - t_0| \leq t_i$, the nonvanishing contribution in Eq. (E.12) comes from the particles $2 \leq r \leq n$. The effects of all other particles outside the subsystem are negligible.

Because both operations $\hat{I}_n$ and $\hat{I}_{\mathcal{N}}$ change the sign of velocities of the particles with the indices $2 \leq r \leq n$ in the denominators of Eq. (E.14), the values of $M_{n-}(t)$ and of $M_{\mathcal{N}-}(t)$ are different from $M(t)$. However, because of the causality, the value of $M_{n-}(t)$ is the same as $M_{\mathcal{N}-}(t)$ for $|t - t_0| \leq t_i$. During this time scale, the values of the observables are the same as those in semigroup Λ_B^+, and we recover group properties. This is the situation with which the S-matrix theory deals.

For a sufficiently large time scale $|t - t_0| \gg t_i$, the subsystem encounters surrounding particles. After many collisions with N_s surrounding particles (with $N_s \gg n$), we cannot distinguish $M_{n-}(t)$ from $M(t)$ (their difference is proportional to n/N_s). Then the observer notices that the subsystem is still embedded in the same semigroup as the one before the velocity inversion.

In contrast with the traditional view, we come to the following conclusion: *dynamical groups of isolated systems are deduced through approximations from a semigroup oriented toward a well-defined direction of time.*

Note that the above discussion is closely related to the appearance of the "entropy barrier" in the process of "velocity inversion."

Starting with a certain value of Lyapounov function at $t = 0$, its value monotonically decreases by the evolution (E.2) until the moment $t = t_0$ before the velocity inversion. Now let us estimate the value of the Lyapounov function coming from Eq. (E.6) after the velocity inversion. To estimate this, we have to evaluate the value of the transformed state $\Lambda_B^+ \hat{I}_{\mathcal{N}} | \rho^+(t_0) \rangle\!\rangle$. By a similar manipulation of Eq. (E.4) (for $d_\mu > d_\nu$),

$$\begin{aligned} \Lambda_B^+ \hat{I}_{\mathcal{N}} \Pi^{(\nu)} | \rho^+(t_0) \rangle\!\rangle &\sim \sum_\mu P^{(\nu)} \mathscr{D}^{(\nu)+} P^{(\mu)} [\mathscr{C}^{(\nu)+}]^{c.c.}\, P^{(\nu)} \\ &= \sum_\mu P^{(\nu)} \mathscr{T}_D^{(\nu)} P^{(\mu)} \frac{1}{|[z - w_\mu]^+_{Z_\alpha^{(\nu)}}|^2} P^{(\mu)} \mathscr{T}_C^{(\nu)c.c.}\, P^{(\nu)} \end{aligned} \tag{E.15}$$

This leads to a similar divergence to Eq. (E.5) in the diffusion modes:

$$\Lambda_B^+ \hat{I}_{\mathcal{N}} | \rho^+(t) \rangle\!\rangle \sim \frac{1}{|\mathbf{k}|^2} \to \infty \qquad \text{for } |\mathbf{k}| \to 0 \tag{E.16}$$

Therefore, the value of the Lyapounov functions defined as the square of Eq. (E.16) becomes larger for larger t_0. Let us then define the "entropy barrier" at t_0 as an absolute value of the difference of the Lyapounov functions after and before the velocity inversion. This barrier increases as a function of t_0, and diverges for large t_0. A quantitative description of the growth of the entropy barrier has been presented in detail for the Friedrichs model in a previous paper [15].

APPENDIX F. PERSISTENT POTENTIAL SCATTERING AND COLLAPSE OF WAVE FUNCTIONS

In the standard theory of scattering (e.g., see [58–61]), one considers asymptotically free "in" and "out" states interacting during a finite time t_c. This leads to the S-matrix description of the collision process, which is valid for times of observation t larger than t_c. For large wave packets, we may consider different situations where t, while large with respect to characteristic frequencies, is smaller than t_c. This leads to an asymptotic theory that is different from the usual S-matrix approach. "Intermediate" states may now appear, which result from *secular effects* for $t \ll t_c$ but vanish for $t \gg t_c$. The consideration of such intermediate time scales is the normal procedure in many-body situations, for example, chemical reactions.

For a typical momentum k (with mass $m = 1$), the duration of the collision is given by $t_c \sim (|k|\eta)^{-1}$ where $1/\eta$ is the size of the wave packet in configuration space. Then one can summarize the two different asymptotic

TABLE I
Two Approaches for Determining Asymptotic Limit

S-Matrix	Statistical Mechanics
$\|k\|\eta\rangle\!\rangle\epsilon$ First $t \to \infty$ then $\eta \to 0$	$\|k\|\eta\langle\!\langle\epsilon$ First $\eta \to 0$ then $t \to \infty$

limits in Table I, where $\epsilon \sim 1/t$. It is interesting that the extension of scattering theory to finite times, corresponding to the statistical mechanics approach, can only be performed on the level of density matrices and *not* on the level of wave functions. The reason is that the transition probability can no longer be factored into a product of wave functions.

The simplest example is a persistent potential scattering in one-dimensional space as associated to delocalized density matrices. We first take the asymptotic limit $\eta \to 0$. The wave function then reduces to a plane wave, and the density matrix has a delta singularity in the momentum representation, Eq. (10.3) [11]. As mentioned in Section X, this corresponds to situations where we continuously send test particles toward the scattering center. There are no asymptotic free states. This situation is therefore not covered by the usual S-matrix theory. We can solve the eigenvalue problem of the Liouvillian for the class of singular functions belonging to Eq. (5.7) in any order of λ without approximations. We have presented the proofs in detail in a previous paper [11]; here we give only the solutions.

Of special interest is the vacuum of correlations ($\nu = 0$), which contains the equilibrium solutions. Here we present the solutions in $\Pi^{(0)}$ (see [11] for the complete solutions of the eigenvalue problem).

As mentioned in Section V, an essential step is to obtain the eigenfunctions of the collision operator $\theta^{(0)}$ which reduces here to the usual Pauli operator ($\Lambda_C = \Lambda_D$, moreover the left and right eigenfunctions are identical in this example). In one-dimensional quantum scattering, we obtain:

$$\theta^{(0)}P^{(0)}|F_{\kappa\pm}^{(0)}\rangle\!\rangle = Z_{\kappa,+}^{(0)}P^{(0)}|F_{\kappa\pm}^{(0)}\rangle\!\rangle \tag{F.1}$$

whose Fourier representations with $k = 0$ are

$$\langle\!\langle 0, l|P^{(0)}|F_{\kappa+}^{(0)}\rangle\!\rangle = \frac{1}{\sqrt{2}}[\delta(l-\kappa) + \delta(l+\kappa)] \tag{F.2a}$$

$$\langle\!\langle 0, l|P^{(0)}|F_{\kappa-}^{(0)}\rangle\!\rangle = \frac{1}{\sqrt{2}}[\delta(l-\kappa) - \delta(l+\kappa)] \tag{F.2b}$$

and whose eigenvalues are, respectively,

$$Z^{(0)}_{\kappa,+} = 0, \qquad Z^{(0)}_{\kappa,-} = \frac{-2i\sigma_\kappa}{\Omega} \tag{F.3}$$

where σ_κ is the "scattering cross section" for the potential scattering.

The eigenfunctions are superpositions of distributions and cannot be written as products of amplitudes. The symmetrical mode shown in Eq. (F.2a) corresponds to equilibriun, while the antisymmetrical mode in Eq. (F.2b) corresponds to a decay process. In this simplified model, the relaxation time is proportional to $1/\Omega$. To obtain a finite relaxation time, we have to consider the succession of scattering (as in a medium with finite concentration). We present the example that leads to a finite relaxation time in Appendix I.

The equilibrium mode of the Liouvillian is simply the microcanonical equilibrium

$$|F^{(0)}_{\kappa+}\rangle\rangle = \frac{1}{\sqrt{2}}\,\delta(H-\omega_\kappa) \tag{F.4}$$

where ω_κ is determined by the initial condition. The equilibrium state is a *mixture* that cannot be reduced to a product of wave functions. We see that our approach unifies both equilibrium and nonequilibrium processes.

APPENDIX G. THREE-BODY SCATTERING

Our next example is a three-body scattering. For simplicity, we consider scattering in one dimension and assume repulsive delta-function interactions (see also Appendix K).

In a previous paper [27] we calculated the various contributions leading to secular effects starting from large wave packets. We showed that the new asymptotic theory leads to a dramatic effect, because classes of secular effects appear only for $t \ll t_c$ and disappear in the frame of the S-matrix theory for $t \gg t_c$ (see Table I). Our results have been confirmed by numerical simulations [27]. Because of this effect, the scattering three-body cross section is different in each situation. In the framework of our asymptotic theory for $t \ll t_c$, the system still does not "know" if we started with a large, but finite size of wave packet or with a delocalized density matrix with the delta-function singularity in momentum space. In this case, we obtained complex, irreducible spectral representations of the Liouvillian. The results of our new asymptotic approach remain valid for all times.

We assume that the three particles have the same mass $m = 1/2$ and the Hamiltonian is given by

$$H = -\sum_{a=1}^{3} \frac{\partial^2}{\partial R_a^2} + \sqrt{2}\lambda \sum_{a<b}^{3} v_{ab}\, \delta(R_a - R_b) \tag{G.1}$$

where R_a is the position of particle a, and v_{ab} is a constant that characterizes the interaction between particles a and b. We also use the standard notation $v_c \equiv v_{ab}$ where (a, b, c) are chosen cyclically from (1, 2, 3), such as $v_3 = v_{12}$. We introduce the factor $\sqrt{2}$ in order to have a more compact expression for V when we write this in terms of the Jacobi coordinates. We assume $v_c > 0$, so that there are no bounded states. Moreover, we assume the interaction is weak, $\lambda \ll 1$. Hence, we can evaluate the transition probability of the scattering by the lower-order approximation in the perturbation expansion in λ. For the repulsive interaction which we consider in this paper, the extension to arbitrary order on λ is straightforward, using the t-matrix formulation for the Hamiltonian [51].

In the center-of-mass system we have

$$K_1 + K_2 + K_3 = 0 \tag{G.2}$$

where K_i are the unperturbed momenta. Because the three particles have the same mass, the sum $R_1 + R_2 + R_3$ is proportional to the position of the center of mass. We put $R_1 + R_2 + R_3 = 0$. We now introduce Jacobi's coordinates, defined by

$$k_{xa} = \frac{K_b - K_c}{\sqrt{2}}, \qquad k_{ya} = \sqrt{\tfrac{3}{2}}\, K_a \tag{G.3}$$

and the same transformation from R_a to r_a. The unperturbed energy can be written, for any $a = 1, 2,$ or 3, as

$$\omega_k \equiv \sum_{i=1}^{3} K_i^2 = k_{xa}^2 + k_{ya}^2 \tag{G.4}$$

The three sets of Jacobi's variables are related by a rotation with the angle $\phi_0 \equiv 2\pi/3$:

$$\begin{pmatrix} k_{xa} \\ k_{ya} \end{pmatrix} = \begin{pmatrix} \cos\phi_0 & \sin\phi_0 \\ -\sin\phi_0 & \cos\phi_0 \end{pmatrix} \begin{pmatrix} k_{xb} \\ k_{yb} \end{pmatrix} \tag{G.5}$$

and the same transformation between (r_{xa}, r_{ya}) and (r_{xb}, r_{yb}). Therefore, the variables (r_{xa}, r_{ya}) and (k_{xa}, k_{ya}) are the components of vector $\mathbf{r}$ and $\mathbf{k}$ in the

ath coordinate system, respectively. The magnitude of these vectors is invariant under the change of coordinate systems.

After eliminating the total momentum, the Hamiltonian (G.1) can be written in terms of Jacobi's coordinates as

$$H = H_0 + \lambda V = -\frac{\partial^2}{\partial r_{xa}^2} - \frac{\partial^2}{\partial r_{ya}^2} + \lambda \sum_{c=1}^{3} v_c\, \delta(r_{xc}) \tag{G.6}$$

We can interpret Eq. (G.6) as the Hamiltonian that represents the evolution of a single "particle" in two-dimensional space, interacting with the three walls. The three walls are located at $r_{xc} = 0$ (i.e., the r_{yc} axis) for $c = 1$, 2, and 3, respectively. The angles between the walls are $\phi_0 = 2\pi/3$.

Let us consider a specific scattering where the collision between particles 1 and 2 is followed by a collision between particles 2 and 3. Our complex spectral representation is applicable for finite time scales, and predicts several secular effects during the collision. The strongest secular effect is proportional to t^2. In this process [Fig. 6(a)], the unperturbed energy is preserved for each two-body collision. This corresponds to the "rescattering

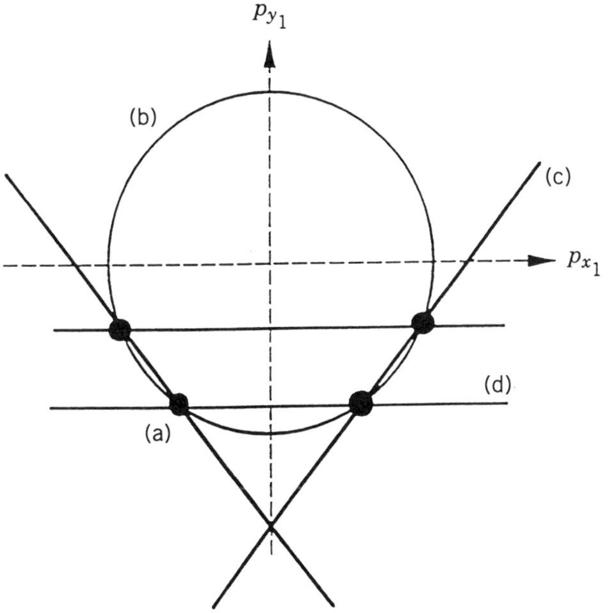

Figure 6. Theoretical prediction of the distribution of the momentum during the collision by the complex spectral representation.

process." For a given initial momentum, the final momentum is distributed among the four points indicated in Fig. 6. There are three other secular effects which are proportional to t: (b) in which the energy is preserved between the initial and final state, but not in the intermediate states; (c) between the final and intermediate states, but not between the intermediate and initial states; and (d) between the intermediate and initial states, but not between the final and intermediate states.

In contrast, the usual S-matrix theory predicts the distribution of the final momentum after the collision, as shown in Fig. 7. The difference comes from process (d), which is transient and disappears after the collision.

We display the results of a numerical simulation performed for an initial condition with a large wave packet. Figure 8 shows the process during the collision, and Fig. 9 shows after the collision. These results verify our prediction.

In summary, our extension of quantum mechanics outside the Hilbert space, based on the density matrix, allows us to predict new phenomena which lie outside quantum mechanics, based on wave functions in the Hilbert space. The effect of these phenomena is essential, because the order

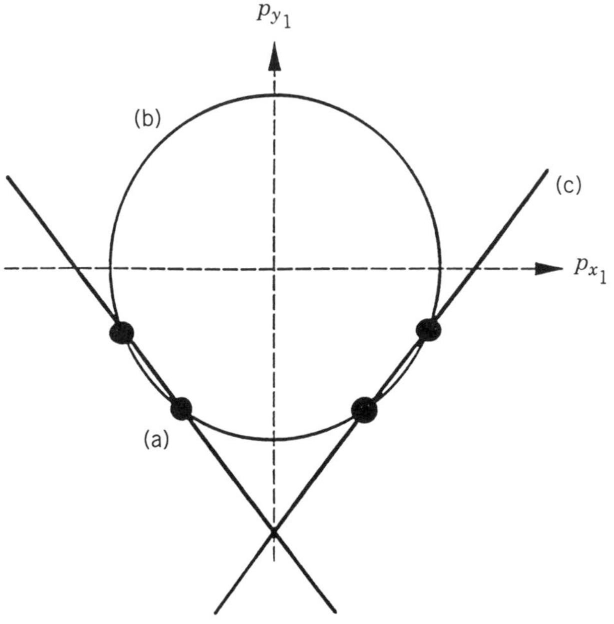

Figure 7. Theoretical prediction of the distribution of the momentum after the collision by the traditional S-matrix theory.

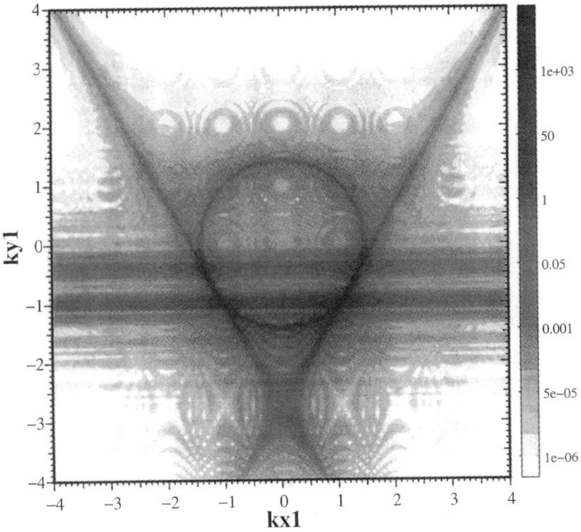

Figure 8. Numerical result of the distribution of the momentum during the collision.

of magnitude of the effect is the same as the effects predicted by the S-matrix theory.

For a large but finite wave packet, these effects are transient. We should distinguish these effects from transient effects in resonance scattering. The

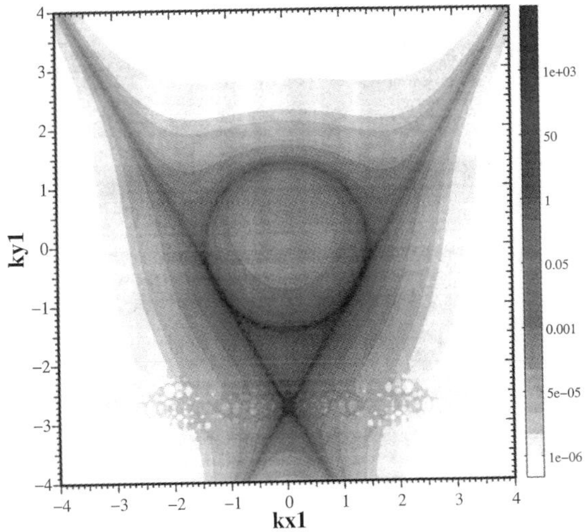

Figure 9. Numerical result of the distribution of the momentum after the collision.

effect of the resonance scattering appears in the S-matrix theory as a broadening of the line shape after the collision, while our new effect, associated with the nonlinear terms in "nonlinear" Lippmann–Schwinger equations (9.3), disappears after the collision.

We also note that a typical situation of chemical kinetics corresponds to the thermodynamic situation. The interactions are persistent. Therefore, the reaction rate of chemical reactions can in general not be evaluated by the traditional quantum mechanics based on wave functions in the Hilbert space.

APPENDIX H. THE FRIEDRICHS MODEL

A simple example of matter–field interacting systems is the Friedrichs model, in which a discrete state $|1\rangle$ corresponding to a bare unstable particle is coupled to a continuous state $|k\rangle$ corresponding to field modes. The Hamiltonian is given by (with the unit $\hbar = 1$)

$$H = H_0 + \lambda V = \omega_1 |1\rangle\langle 1| + \sum_k \omega_k |k\rangle\langle k| + \lambda \sum_k V_k(|k\rangle\langle 1| + |1\rangle\langle k|) \tag{H.1}$$

For simplicity, we consider a one-dimensional system. We assume that the dispersion relation is $\omega_k = |k|$. As in the text, we use a normalization of the states by Kronecker's delta (with L as the size of the box, and $\Omega = L/2\pi$). In the limit of large volume, we should read a summation sign of the spectrum as an appropriate integration sign [e.g., see Eq. (3.5)]. The state $|\alpha\rangle = \{|1\rangle, |k\rangle\}$ forms a complete orthonormal set, that is,

$$\sum_\alpha p^{(\alpha)} = 1, \qquad p^{(\alpha)} p^{(\beta)} = p^{(\alpha)} \delta_{\alpha\beta} \tag{H.2}$$

with

$$H_0 = \sum_\alpha \omega_\alpha p^{(\alpha)} \tag{H.3}$$

where we have introduced the projection operator in the *wave-function space* (hereafter, $\alpha = 1$ or k)

$$p^{(\alpha)} \equiv |\alpha\rangle\langle\alpha| \tag{H.4}$$

We assume that the continuous spectrum spreads from 0 to $+\infty$, and that the discrete spectrum of H_0 is embedded in the continuous one (i.e.,

$\omega_k \geq 0$ and $\omega_1 > 0$). We assume that the interaction $V_{|k|}$ is real and that the volume dependence of the interaction is given by

$$V_k \sim \Omega^{-1/2} \tag{H.5}$$

We further assume that the coupling constant is small and that the total Hamiltonian has no discrete spectrum associated with a real eigenvalue.

We shall illustrate the complex spectral representations for two different situations: in the first, the field is localized in space and its space density over the whole space is infinitesimal; in the second, the space density is finite. For the first case, there is no delta-function singularity in the field; there is a singularity in the second case. To some extent, the first case corresponds to "zero temperature" of the field, and the second to "finite temperature." As we shall see, the evolution of the density matrices is reducible to wave functions for the regular cases; it is irreducible for the singular cases.

1. *Regular Case.* Let us assume that the field is localized in space. For this case, the solution of the eigenvalue problem in the *Hilbert space* is known as Friedrichs' solution [62] and is given by

$$H\,|\,\phi_k^F\rangle = \omega_k\,|\,\phi_k^F\rangle \tag{H.6}$$

Here

$$|\,\phi_k^F\rangle = |\,k\rangle + \frac{\lambda V_k}{\eta^+(\omega_k)}\left(|\,1\rangle + \sum_{k'} \frac{\lambda V_{k'}}{\omega_k - \omega_{k'} + i\epsilon}\,|\,k'\rangle\right) \tag{H.7}$$

and

$$\eta(z) \equiv z - \omega_1 - \sum_k \frac{\lambda^2 V_k}{z - \omega_k} \tag{H.8}$$

and the plus or minus superscript on $\eta^{\pm}(\omega_k)$ denotes the function defined by analytic continuation of the argument from the upper half (or lower half) of the plane; for example, $\eta^{\pm}(\omega) = \eta(\omega \pm i\epsilon)$ with the Eq. (5.49).

The locality of the field implies that

$$\rho_{11} \sim \sum_k \rho_{kk} \to \Omega \int dk\, \rho_{kk} \qquad \text{for } \Omega \to \infty \tag{H.9}$$

Hence, $\rho_{11} \sim \Omega \rho_{kk}$ in the large-volume limit. To some extent, the discrete spectrum plays the role of a singular component of the density matrix.

Indeed, on the level of wave functions, the discrete components act as singular components: $\langle 1 | \psi \rangle \sim \sqrt{\Omega} \langle k | \psi \rangle$. The projection operator $p^{(1)}$ in Eq. (H.4) singles out the singular component of the wave functions. A direct consequence is that complex spectral representations of the Hamiltonian H already exist in the wave-function space (in addition to the Friedrichs solution given above). However, in this case, the complex spectral representations for L_H in the Liouville space can be reduced to a product of the complex spectral representations of H. This is a striking difference from the situation discussed in the text and also that discussed later in this appendix.

We constructed the complex spectral representations of H in a previous paper [15]. Here, we include part of the results (without proofs) in a form parallel to that used in Section V.

We have the right and left eigenstates of H:

$$H | \varphi_\alpha \rangle = z_\alpha | \varphi_\alpha \rangle, \qquad \langle \hat{\varphi}_\alpha | H = z_\alpha \langle \tilde{\varphi}_\alpha | \tag{H.10}$$

where

$$z_1 = \tilde{\omega}_1 - i\gamma, \qquad z_k = \omega_k \tag{H.11}$$

and $\tilde{\omega}_1$ is the renormalized energy and γ (>0) is the damping rate satisfying $\eta^+(\tilde{\omega}_1 - i\gamma) = 0$.

We have (with $q^{(\nu)} = 1 - p^{(\nu)}$)

$$| \varphi_\alpha \rangle = N_\alpha^{1/2} (p^{(\alpha)} + c^{(\alpha)}) | \alpha \rangle, \qquad \langle \tilde{\varphi}_\alpha | = \langle \alpha | (p^{(\alpha)} + d^{(\alpha)}) N_\alpha^{1/2} \tag{H.12}$$

For example, for $\alpha = 1$ we have [15]

$$c^{(1)} | 1 \rangle = \sum_k \frac{\lambda V_k}{[z - \omega_k]_{z_1}^+} | k \rangle, \qquad \langle 1 | d^{(1)} = \sum_k \frac{\lambda V_k}{[z - \omega_k]_{z_1}^+} \langle k | \tag{H.13}$$

and $N_1 = 1/\eta^{+\prime}(z_1)$, where the prime denotes the derivative with respect to the argument. Similarly, the eigenstates for $\alpha = k$ are written in terms of the complex distributions [15]. The states $| \varphi_1 \rangle$ and $\langle \hat{\varphi}_1 |$ are examples of the "Gamow states" [8].

We have a complex spectral representation for H,

$$H = \sum_\alpha z_\alpha \pi^{(\alpha)} \tag{H.14}$$

where the projection operators

$$\pi^{(\alpha)} \equiv | \varphi_\alpha \rangle \langle \tilde{\varphi}_\alpha | \tag{H.15}$$

form a complete and orthogonal set in the rigged Hilbert space [7,10]

$$\sum_{\alpha} \pi^{(\alpha)} = 1, \qquad \pi^{(\alpha)}\pi^{(\beta)} = \pi^{(\alpha)}\delta_{\alpha\beta} \tag{H.16}$$

Again, the projection operators are nonhermitian:

$$(\pi^{(\alpha)})^{\dagger} \neq \pi^{(\alpha)} \tag{H.17}$$

The nonunitary transformation operators in the wave-function space are defined by [cf. Eq. (6.3)]

$$\Lambda = \sum_{\alpha} |\alpha\rangle\langle\tilde{\varphi}_{\alpha}|, \qquad \Lambda^{-1} = \sum_{\alpha} |\varphi_{\alpha}\rangle\langle\alpha| \tag{H.18}$$

This leads to the "collision operator" in the wave function space [cf. Eq. (5.22)]:

$$\theta = \Lambda H \Lambda^{-1} = |1\rangle(\tilde{\omega}_1 - i\gamma)\langle 1| + \sum_{k} |k\rangle\omega_k\langle k| \tag{H.19}$$

The Lyapounov function associated with the wave function $|\psi(t)\rangle$ is given by $\mathscr{H}_{\psi}(t) = \langle\psi(t)|\Lambda^{+}\Lambda|\psi(t)\rangle$, which monotonically decreases in time for $t > 0$. The behavior of the Lyapounov function under the "velocity inversion," as well as the "entropy barrier" generated by the velocity inversion, have been investigated in detail for this model (see Fig. 5) [15].

We note that the decay state $|\varphi_1\rangle$ introduced cannot be a "physical" state, because it is a complex distribution. As mentioned in Appendix C, this leads to exponential growth of the field distribution over large distances. Still, in a suitable space, these eigenstates are complete (H.16), and we can use them as an expansion basis of a given initial condition to investigate its time evolution [15].

2. *Singular Case.* Next let us consider the case when the density matrix is given by

$$\rho_{kk'} = \rho_k^d \delta_{\Omega}(k - k') + \rho'_{kk'}, \qquad \rho_{11} = \rho_1^d \tag{H.20}$$

where

$$\rho_k^d \sim \rho'_{kk'} \sim \rho_1^d \tag{H.21}$$

The field component has a delta singularity. In this situation, we can also construct the exact form of the complex spectral representation which is irreducible to wave functions; however, there is no solution of the eigenvalue problem in the Hilbert space.

Physically, Eq. (H.20) corresponds to the situation in which a widely spread wave packet of the field is overlapping with the atom. The wave packet is so wide that the atom cannot distinguish the packet from a plane wave during the duration of interaction t_c, as in the case of persistent scattering we consider in Appendices F and G. This situation is somewhat similar to the system with a "finite temperature." In the plane wave limit, (i.e., $t_c \to \infty$), each component in Eq. (H.21) is of order $1/\Omega$. In this limit the atom has an infinitesimal energy, while the total energy of the field is finite. As a result, the atom cannot influence the field, whereas the field can influence the atom.

We can solve the eigenvalue problem of L_H in the extended-function space. We present the calculations in detail elsewhere [57]. Restricting ourselves to the vacuum of correlations, the unperturbed projection operators are given by

$$P^{(0)} = |1;\,1\rangle\!\rangle\langle\!\langle 1;\,1| + \sum_k |k;\,k\rangle\!\rangle\langle\!\langle k;\,k|, \qquad P^{(0)} = 1 - Q^{(0)} \tag{H.22}$$

We solve the eigenvalue problem

$$L_H |F_\alpha^{(0)}\rangle\!\rangle = Z_\alpha^{(0)} |F_\alpha^{(0)}\rangle\!\rangle \tag{H.23}$$

in the form

$$\langle\!\langle k;\,k'|F_\alpha^{(0)}\rangle\!\rangle = F_k^d(\alpha)\delta_\Omega(k-k') + F'_{kk'}(\alpha), \qquad \langle\!\langle 1;\,1|F_\alpha^{(0)}\rangle\!\rangle = F_1^d(\alpha) \tag{H.24}$$

with the conditions

$$F_k^d \sim F'_{kk'} \sim F_1^d, \qquad \xi_\alpha \sim 1 \tag{H.25}$$

where

$$\xi_\alpha \equiv \Omega Z_\alpha^{(0)} \tag{H.26}$$

Combining Eqs. (5.23) and (H.25) leads to

$$\sum_k \theta_{11;\,kk}^{(0)} F_k^d(\alpha) + \theta_{11;\,11}^{(0)} F_1^d(\alpha) = 0 \tag{H.27a}$$

$$\theta_{kk;\,11}^{(0)} F_1^d(\alpha) + \sum_l \theta_{kk;\,ll}^{(0)} F_l^d(\alpha) = \xi_\alpha F_k^d(\alpha) \tag{H.27b}$$

where we have neglected terms, such as $\xi_\alpha F'_{kk}(\alpha)/\Omega$, which are $1/\Omega$ times smaller than the terms written in Eq. (H.27).

The explicit form of the collision operator, as well as of the creation and destruction operators, has been derived by de Haan and Henin [40], and we shall not write them here. We give only the result of the eigenstates of L_H [57]:

For equilibrium states with zero eigenvalue $Z_{\alpha+}^{(0)} = 0$,

$$|F_{\alpha+}^{(0)}\rangle\!\rangle = |\varphi_\alpha^F; \varphi_\alpha^F\rangle\!\rangle + |\varphi_{-\alpha}^F; \varphi_{-\alpha}^F\rangle\!\rangle \tag{H.28}$$

which is a mixture, where φ_α^F are Friedrichs' solution (H.7) of the eigenstates of the Hamiltonian. This is microcanonical equilibrium and can be written with a suitable normalization constant C (in the limit $\Omega \to \infty$):

$$|F_{\alpha+}^{(0)}\rangle\!\rangle = C\delta(H - \omega_\alpha) \tag{H.29}$$

For nonequilibrium states,

$$|F_{\alpha-}^{(0)}\rangle\!\rangle = |\alpha; \alpha\rangle\!\rangle - |-\alpha; -\alpha\rangle\!\rangle \tag{H.30}$$

with imaginary eigenvalues [cf. Eq. (F.3)]

$$Z_{\alpha-}^{(0)} = -4\pi i \frac{\lambda^2 V_\alpha^2}{|\eta^+(\omega_\alpha)|^2} \sim O\left(\frac{1}{\Omega}\right) \tag{H.31}$$

This shows that there are indeed complex eigenvalues of L_H which are *not* the difference of the complex eigenvalue of the Hamiltonian.

For a given energy ω_α of the field, the probability of finding the atom in the excited state in the equilibrium state is proportional to

$$\langle\!\langle 1; 1 | F_{\alpha+}^{(0)}\rangle\!\rangle = \frac{2\lambda^2 V_\alpha^2}{|\eta^+(\omega_\alpha)|^2} \tag{H.32}$$

Our equilibrium solution, Eq. (H.29), offers an interesting model of unstable particles. Because of the nonintegrability of unstable states, the definition of the unstable particles are still controversial. Indeed, by definition, Hamiltonians of nonintegrable systems cannot be diagonalized in the Hilbert space. Therefore, we cannot associate unstable particles as eigenstates of the Hamiltonian, which we do for stable particles. As mentioned above, the complex eigenstate $|\varphi_1\rangle$ is not a physical state because of the exponential catastrophe. The unstable state $|1\rangle$ is also not the unstable particle, because it cannot behave purely exponentially (there are significant deviations from the exponential decay in short and long time scale) [63,64]. These deviations destroy the nondistinguishability of the unstable particles by their differing ages. We need a suitable cloud of the field surrounding the exited particle, which we can see in the equilibrium states (H.28). When the background field is switched off, the exited state starts to decay. The unstable particle is defined only in the Liouville space, because the cloud of the field is a mixture and is irreducible to a single wave function. This problem is considered in a separate paper.

APPENDIX I. QUANTUM LORENTZ GAS

As an example of many-body systems, we consider here the quantum perfect Lorentz gas in one-dimensional space. The explicit form of the complex spectral representation of this model was derived in reference 65. We present a summary of the results without their proofs. This model corresponds to the motion of a light test particle of mass m_1 scattered by a large ensemble of heavy particles of mass m_n. The Hamiltonian is given by Eq. (2.1) with a short-range repulsive interaction $V(|Q_j - Q_n|)$. Moreover, we impose the condition

$$V(|Q_j - Q_n|) = \begin{cases} V(|Q_1 - Q_n|) & \text{for } j = 1 \\ 0 & \text{otherwise} \end{cases} \tag{I.1}$$

The perfect Loretz gas is defined as follows:

1. We assume $m_1/m_n \ll 1$ for $n \geq 2$. Hence, there is a nonnegligible effect of the interaction only over the light particle, and we can neglect the effect of the interaction for the heavy particles.

2. We assume that the heavy particles are uniformly distributed in space at $t = 0$ with a finite concentration $c = (\mathcal{N} - 1)/L$. The density matrix associated with the heavy particles $\rho_h(0)$ is given by a plane wave,

$$\rho_h(0) = |P_2^0, \ldots, P_{\mathcal{N}}^0; P_2^0, \ldots, P_{\mathcal{N}}^0\rangle\!\rangle \tag{I.2}$$

which belongs to the class given by Eq. (4.18). The density matrix for the total system at $t = 0$ is therefore

$$\rho(0) = \rho_1(0) \otimes \rho_h(0) \tag{I.3}$$

where $\rho_1(0)$ is the density matrix associated with the test particle.

We consider a weakly coupled system $\lambda \ll 1$, and we put $m_1 = 1$ for the test particle. Then, the results we obtain can be summarized as follows:

1. $\Pi^{(0)}$ *Subspace.* In the λ^2 order approximation, the collision operators $\theta_C^{(0)}$ and $\theta_D^{(0)}$ are the same [see Eq. (5.52)]. Let us denote them by $\lambda^2\theta_2^{(0)}$. Then, the matrix element of $\lambda^2\theta_2^{(0)}$ is given by

$$\lambda^2 \langle\!\langle 0, P \,|\, \theta_2^{(0)} \,|\, 0, P' \rangle\!\rangle = \lambda^2 \langle\!\langle 0, p_1 \,|\, \theta_0 \,|\, 0, p_1' \rangle\!\rangle \prod_{n=2}^{\mathcal{N}} \delta^{kr}(P_n - P_n') \tag{I.4}$$

where

$$\lambda^2 \langle\!\langle 0, p_1 \,|\, \theta_0 \,|\, 0, p_1' \rangle\!\rangle = i(2\pi)^2 \lambda^2 c \int dl \,|\, V_l \,|^2 (\eta_1^{l/2} - \eta_1^{-l/2}) \times \delta(l \cdot p_1)(\eta_1^{l/2} - \eta_1^{-l/2}) \delta^{kr}(p_1 - p_1') \tag{I.5}$$

Hereafter, we use a lowercase letter for the variables of the test particle, such as p_1. The collision operator θ_0 is a linear (antihermitian) operator acting on the momentum of the test particle. This linearity is a characteristic feature of the perfect Lorentz model, and thanks to it, we can solve explicitly the eigenvalue problem for the collision operator. The solutions are given by (for $w > 0$ and with $\mu = 0$ or 1)

$$\theta_0 \,|\, g_{w,\mu} \rangle\!\rangle = Z_{w,\mu}^{(0)} \,|\, g_{w,\mu} \rangle\!\rangle \tag{I.6}$$

where

$$Z_{w,0}^{(0)} = 0, \qquad Z_{w,1}^{(0)} = -i\xi_w \equiv -2i \,\frac{(4\pi)^2 \lambda^2 c}{w} \,|\, V_{2w} \,|^2 \tag{I.7}$$

Writing $f(p_1) = \langle\!\langle p_1 \,|\, f \rangle\!\rangle$ for the function of the momentum of the test particle, the eigenstates are given by

$$\langle\!\langle p_1 \,|\, g_{w,\mu} \rangle\!\rangle = \frac{1}{\sqrt{2}} [\delta(p_1 - w) + (-1)^{\mu} \delta(p_1 + w)] \tag{I.8}$$

where $\mu = 0$ (1) corresponds to equilibrium (nonequilibrium) modes, which form complete orthogonal bases for functions of momentum p_1. The relaxation times of the nonequilibrium modes are given by

$$t_w \equiv 1/\xi_w \tag{I.9}$$

2. $\Pi^{(k)}$ *Subspace with* $k \neq 0$. In the same approximation, the collision operators with nonvanishing wave vector $k \neq 0$ for all correlation subspaces have the same form, given by [see Eq. (5.52)]

$$\langle\!\langle k, P \,|\, \theta_C^{(k)} \,|\, k', P' \rangle\!\rangle \approx \langle\!\langle k, P \,|\, \theta_D^{(\nu)} \,|\, k', P' \rangle\!\rangle$$

$$\approx \langle\!\langle k_1, p_1 \,|\, \theta_k \,|\, k'_1, p'_1 \rangle\!\rangle \prod_{n=2}^{\mathcal{N}} \delta^{kr}(P_n - P'_n) \tag{I.10}$$

where

$$\langle\!\langle k_1, p_1 \,|\, \theta_k \,|\, k'_1, p'_1 \rangle\!\rangle = (k_1 \cdot p_1)\delta^{kr}(k_1 - k'_1)\delta^{kr}(p_1 - p'_1) + \lambda^2 \langle\!\langle 0, p_1 \,|\, \theta_0 \,|\, 0, p'_1 \rangle\!\rangle \tag{I.11}$$

The solution of the eigenvalue problem is given by

$$\theta_k \,|\, g_{w,\mu}^{(k)} \rangle\!\rangle = Z_{w,\mu}^{(k)} \,|\, g_{w,\mu}^{(k)} \rangle\!\rangle, \qquad \langle\!\langle \tilde{g}_{w,\mu}^{(k)} \,|\, \theta_k = Z_{w,\mu}^{(k)} \langle\!\langle \tilde{g}_{w,\mu}^{(k)} \,| \tag{I.12}$$

where

$$Z_{w,\mu}^{(k)} = -\frac{i}{2}\,\xi_w + (-1)^\mu w \sqrt{k_1^2 - k_w^2} \tag{I.13}$$

and

$$k_w \equiv \frac{\xi_w}{2w} \tag{I.14}$$

is the order of the inverse of the "mean free path." Equation (I.13) shows that *all* eigenstates in $\Pi^{(k)}$ subspace with $k \neq 0$ are decay modes. The eigenstates $g_{w,\mu}^{(k)}$ and $\tilde{g}_{w,\mu}^{(k)}$ are written by a superposition of the eigenstates of θ_0 with k-dependent coefficients (see [65] for the explicit forms).

There is a critical value k_w. For small k_1, that is, $|k_1| < k_w$ (which corresponds to large scale in space), we have

$$Z_{w,0}^{(k)} = -i\,\frac{\xi_w}{2}\left(\frac{k_1}{k_w}\right)^2 + O\left[\left(\frac{k_1}{k_w}\right)^4\right] \tag{I.15a}$$

$$Z_{w,1}^{(k)} = -i\xi_w + O\left[\left(\frac{k_1}{k_w}\right)^2\right] \tag{I.15b}$$

For this case, the eigenvalues are purely imaginary numbers, and are proportional to $|k_1|^2$. They correspond to the diffusion mode in space.

For large k_1, that is, $|k_1| \geq k_w$, the eigenvalues are complex numbers and their imaginary part is given by

$$\text{Im } Z^{(k)}_{w,\mu} = -\frac{\xi_w}{2} \tag{I.16}$$

and the real part of the eigenvalue approaches (for $|k_1| \gg k_w$)

$$\text{Re } Z^{(k)}_{w,\mu} \to (-1)^{\mu} |k_1| \cdot w \tag{I.17}$$

The real part generates a shift in space of the Wigner function. Because of the condition $|k_1| \gg k_w$, this shift appears only for "far from equilibrium" modes.

The diffusion mode with small nonvanishing k is the slowest damped mode among all damped modes. All other modes decay with a relaxation time of order $t_r \sim (\lambda^2)^{-1}c$. The time scale t_d of damping for the diffusion mode depends as usual on the wave vector as $t_d \sim (\lambda^2 c |k_1|^2)^{-1}$. The larger the scale is in space, the slower the damping. Hence, in any finite time scale, the effect of the diffusion mode does not vanish for sufficiently small $|k|$.

We can construct the eigenstates of the Liouvillian L_H by acting $\mathbf{C}^{(\nu)}$ and $\mathbf{D}^{(\nu)}$ to the eigenstates of the collision operators presented above. Of special interest is the equilibrium eigenstate; all other eigenstates vanish as time goes on. Applying $\lambda \mathbf{C}_1^{(0)}$ to $|g_{w,0}\rangle\rangle$ in Eq. (I.8), we obtain the equilibrium mode of L_H to the first order of λ [with Eq. (I.2)]:

$$|F^{eq}(w)\rangle\rangle = \delta[H_1(q_1, p_1) - \tfrac{1}{2}w^2] \otimes \rho_h(0) \tag{I.18}$$

where

$$H_1(q_1, p_1) = \tfrac{1}{2}p_1^2 + \lambda V(|q_1|) \tag{I.19}$$

This corresponds to the microcanonical equilibrium for the test particle.

APPENDIX J. COLLAPSE OF WAVE FUNCTIONS

In this appendix, we show that the collapse of wave functions is a dynamical process which occurs outside the Hilbert space for LPS. The system we want to measure may be an integrable system and may be in a pure state at a given time. In contrast, the measurement devices are nonintegrable LPS. A coupling of the system with a measurement device leads to a nonintegrable LPS, which as a whole is outside the Hilbert space. The coupled system evolves toward a mixed state. The transition from the "potentiality" to the "actuality" is realized through this irreversible process. We illustrate this process using the perfect Lorentz gas discussed in Appendix I. In this example, the system is the test particle. The heavy particles play the role of LPS.

We assume that the initial condition is in the pure state given by Eqs. (I.3) with (I.2) and that

$$|\rho_1(0)\rangle\!\rangle = |\psi;\psi\rangle\!\rangle \tag{J.1}$$

where

$$|\psi\rangle = N_c^{1/2}(c_1|\phi_a\rangle + c_2|\phi_b\rangle) \tag{J.2}$$

with

$$|c_1|^2 + |c_2|^2 = 1 \tag{J.3}$$

and N_c is a normalization constant:

$$N_c = (1 + 2\,\mathrm{Re}[c_1^* c_2 \langle\phi_a|\phi_b\rangle])^{-1} \tag{J.4}$$

In this appendix, we use the notation c_1^* (instead of $c_1^{c.c.}$) for the complex conjugate of c_1.

We have $\langle\psi|\psi\rangle = 1$. Here ϕ_c (hereafter $c = a$ and b) is a Gaussian wave packet for the test particle defined by

$$\langle p|\phi_c\rangle = \frac{1}{\sqrt{\eta\pi^{1/2}\Omega}}\, e^{-(p-p_c)^2/2\eta^2} e^{-ipx_c} \tag{J.5}$$

Here, $\eta > 0$ is the width of the wave packet in momentum space. We have $\langle\phi_c|\phi_c\rangle = 1$. We are interested in the case with large wave packet in space, especially in the limit $\eta \to 0$.To save notations, we drop the index 1 from the variables associated with test particle 1 in this appendix. The wave packet has a peak at $p = p_c$ in momentum space and at $q = x_c$ in configuration space. We assume that

$$|p_c| \gg \eta \quad \text{and} \quad |p_a - p_b| \gg \eta \tag{J.6}$$

so that each wave packet has a well-defined momentum p_c. Then we have

$$1 \gg \langle \phi_a | \phi_b \rangle \xrightarrow[\eta \to 0]{} 0 \tag{J.7}$$

We have the Fourier expansion

$$\langle\!\langle q, p | \rho_1(0) \rangle\!\rangle = \frac{1}{L} \sum_k \tilde{\rho}_k(p) e^{ikq} \tag{J.8}$$

where

$$\tilde{\rho}_k(p) = \frac{N_c}{\eta\sqrt{\pi}} [(|c_1|^2 e^{-(p-p_a)^2/\eta^2} e^{-ikx_a} + |c_2|^2 e^{-(p-p_b)^2/\eta^2} e^{-ikx_b}) e^{-k^2/4\eta^2}$$
$$+ (c_1 c_2^* e^{-ip(x_a - x_b)} + c.c.) e^{-(p-p_{ab})^2/\eta^2} e^{-ikx_{ab}} e^{-(k-k_{ab})^2/4\eta^2}] \tag{J.9}$$

and

$$p_{ab} = \tfrac{1}{2}(p_a + p_b), \qquad x_{ab} = \tfrac{1}{2}(x_a + x_b), \qquad k_{ab} = p_a - p_b \tag{J.10}$$

The Fourier component $\tilde{\rho}_k(p)$ has two sharp peaks, one at $k = 0$ and another at $k = k_{ab}$. The appearance of the latter manifests the existence of nonnegligible "quantum correlations," which is proportional to $c_1 c_2^*$. In the limit $\eta \to 0$, $\tilde{\rho}_k(p)$ has delta-function singularities in k. In the Wigner representation the quantum correlations $c_1 c_2^*$ are associated with the "inhomogeneous" components with nonvanishing wave vector k_{ab}, while the probabilities $|c_1|^2$ and $|c_2|^2$ are associated with the homogeneous components with $k = 0$. Because of the quantum correlations, the momentum value of the test particle is *indefinite* at $t = 0$ [2].

We now show that coupling this system with the LPS in Eq. (I.2) leads to the collapse of the wave function. Because of the two sharp peaks at $k = 0$ and at $k = k_{ab}$ in the Fourier component $\tilde{\rho}_k(p)$, the contributions from the two subspaces $P^{(0)}$ and $P^{(k_{ab})}$ dominate in the initial condition for the large wave packet limit $\eta \to 0$. As mentioned in Section IV, the inhomogeneous components evolve independently from the homogeneous components. Hence, we can discuss the evolution of the total density matrix associated with the initial condition in $P^{(0)}$ separately from the others. Let us first consider the evolution associated with this initial condition:

$$|\rho(t)\rangle\!\rangle_0 \equiv e^{-iL_H t} P^{(0)} |\rho(0)\rangle\!\rangle \tag{J.11}$$

The complex spectral representation gives

$$|\rho(t)\rangle\rangle_0 = (P^{(0)} + \mathbf{C}^{(0)})e^{-i\theta_{C^{(0)}}t}A^{(0)}P^{(0)}|\rho(0)\rangle\rangle + \sum_{\nu\neq 0}(P^{(\nu)} + \mathbf{C}^{(\nu)})\,e^{-i\theta_{C^{(\nu)}}t}A^{(\nu)}\mathbf{D}^{(\nu)}P^{(0)}|\rho(0)\rangle\rangle \quad \text{(J.12)}$$

In the spectral decomposition of the collision operators $\theta_C^{(0)}$ with Eq. (I.6) and $\theta_C^{(k)}$ with Eq. (I.12), the components with $w = p_c$ give the dominating contributions. The contribution from the $\Pi^{(0)}$ subspace approaches microcanonical equilibrium with the arguments $w = p_a$ and $w = p_b$ in time scale t_{p_c} in Eq. (I.9). Moreover, in the contribution from the $\Pi^{(\nu)}$ subspace, the correlations created from the vacuums of correlation have a dominant contribution around the wave vector k, whose order is the inverse of a, which is the size of the interaction range and which we assume is much less than the mean distance of the heavy particles. With $w = p_c$ and with small λ and low concentration of the heavy particles, we have $|k| \gg k_w$. Therefore, all correlations propagated by $\theta_C^{(\nu)}$ in Eq. (J.12) vanish with the same time scale as t_{p_c} [see Eq. (I.16)]. In summary, for $t > \max(t_{p_a}, t_{p_b})$, we obtain

$$|\rho(t)\rangle\rangle_0 \to |\rho^{eq}\rangle\rangle = \{|c_1|^2\delta[H_1(q,p) - \tfrac{1}{2}p_a^2] + |c_2|^2\delta[H_1(q,p) - \tfrac{1}{2}p_b^2]\} \otimes \rho_h(0) \quad \text{(J.13)}$$

Next we consider the contribution from the inhomogeneous components related to the initial quantum correlation at $k = k_{ab}$. We denote the density matrix associated with these components by $|\rho(t)\rangle\rangle_1$. In terms of the complex spectral representation, the components with $w = p_{ab}$ now give predominant contributions for the evolution of $|\rho(t)\rangle\rangle_1$ [see Eq. (J.9)]. There are two situations for the initial quantum correlation: $|k_{ab}| > k_{p_{ab}}$, corresponding to Eq. (I.16), and $|k_{ab}| < k_{p_{ab}}$, corresponding to Eq. (I.15).

In the first case, the initial quantum correlation shifts in space, damps in time, and then vanishes in the time scale $t_{p_{ab}}$. In the second case, the initial quantum correlation does not shift in space, but still vanishes in time with a slower time scale of order $t_d \equiv t_{p_{ab}}(k_{p_{ab}}/|k_{ab}|)^2$ associated with the diffusion mode [see Eq. (I.15)].

Combining these results, we obtain, for the evolution of the whole density matrix (for $t > \max(t_{p_a}, t_{p_b}, t_{p_{ab}}, t_d)$),

$$|\rho(t)\rangle\rangle = |\rho(t)\rangle\rangle_0 + |\rho(t)\rangle\rangle_1 \to |\rho^{eq}\rangle\rangle \quad \text{(J.14)}$$

Starting with a pure state, Eqs. (I.3), (I.2), and (J.12), the whole system approaches in finite time scale a mixed state with two peaks weighted by $|c_1|^2$ and $|c_2|^2$. During this time scale, the quantum correlation is destroyed by irreversible processes. The transition from potentiality to the actuality has been achieved through the dynamical collapse of the wave functions as a result of the Poincaré resonances.

APPENDIX K. ON "EXACTLY SOLVABLE" ONE-DIMENSIONAL SYSTEMS

The eigenvalue problems of the Hamiltonian for some one-dimensional $\mathcal{N}$-particle systems are known to be solvable [54]. An example is a system interacting via a repulsive delta-function potential. This problem was solved by Lieb and Liniger [66] for interacting bosons and generalized by Yang [68] for distinguishable particles. Their solutions have been used to study some properties of the ground state and the spectrum of excitation [67,69]. In this appendix, we show that their solutions are applicable only to regular-density matrices. For the thermodynamic limit, their solutions lead to divergences. In this limit the system approaches equilibrium. There is no solution for the eigenvalue problem of the Hamiltonian, because its existence is incompatible with the approach to equilibrium. We have to solve the eigenvalue problem of the Liouvillian for the class of singular density matrices given by Eq. (4.18).

The Hamiltonian of the "solvable" system is given by

$$H = H_0 + \lambda V = -\sum_{i=1}^{\mathcal{N}} \frac{\partial^2}{\partial x_i^2} + 2\lambda \sum_{j>i}^{\mathcal{N}} \delta(|x_i - x_j|) \tag{K.1}$$

For a system of distinguishable particles contained in a "box" of size L, with periodic boundary conditions, the eigenstates of the Hamiltonian can be found from Bethe's hypothesis [68,71]:

$$H|\phi_k^L\rangle = \omega_k |\phi_k^L\rangle \tag{K.2}$$

where $k = (k_1, \ldots, k_{\mathcal{N}})$, as usual, and

$$\omega_k = \sum_{i=1}^{\mathcal{N}} k_i^2 \tag{K.3}$$

and in the coordinate representation,

$$\langle x|\phi_k^L\rangle = \sum_S \theta(x_{S_1})\theta(x_{S_2} - x_{S_1}) \cdots \theta(x_{S_{\mathcal{N}}} - x_{S_{\mathcal{N}-1}})\theta(L - x_{S_{\mathcal{N}}}) \times \sum_R A_{R,S}(\lambda, k) \exp\left(i \sum_j^{\mathcal{N}} x_{S_j} k_{R_j}\right) \tag{K.4}$$

Here, $\theta(x)$ is the step function; $\theta(x) = 0$ for $x < 0$, $\theta(x) = 1$ for $x \geq 0$, $R \equiv (R_1, \ldots, R_{\mathcal{N}})$ and $S \equiv (S_1, \ldots, S_{\mathcal{N}})$ are permutations of $(1, 2, \ldots, \mathcal{N})$, and $A_{R,S}(\lambda, k)$ are constants to be determined. The explicit form of $A_{R,S}(\lambda, k)$ is presented in reference 68, but is not necessary for our discussion.

In the limit $L \to \infty$, we have a continuous spectrum for momentum k, and the eigenstates are expressed by the Möller scattering states:

$$\lim_{L\to\infty} |\phi_k^L\rangle = |\phi_k\rangle = |k\rangle + \frac{1}{\omega_k + i\epsilon - H_0} t^+(\omega_k)|k\rangle \tag{K.5}$$

where $t^{\pm}(\omega_k) \equiv t(\omega_k \pm i\epsilon)$ is the t-matrix in the wave-function space and $\epsilon \to 0+$ as usual. The t-matrix is obtained by taking the Fourier transform of Eq. (K.4) (a similar expression for interacting bosons can be found from the results of Thacker [70]):

$$\langle l|t^+(\omega_k)|k\rangle = \delta^{kr}(l^T - k^T)(\omega_k - \omega_l + i\epsilon)\sum_{R,S}[A_{R,S}(\lambda, k) - 1] \times \prod_{m-1}^{\mathcal{N}-1} \frac{i}{\sum_{n=1}^{m}(k_{R_n} - l_{S_n}) + i\epsilon} \tag{K.6}$$

where l^T is the total momentum:

$$l^T \equiv \sum_{i=1}^{\mathcal{N}} l_i \tag{K.7}$$

It should be emphasized that Eq. (K.6) is obtained by taking the continuous spectrum limit $L \to \infty$ by keeping $\mathcal{N}$ finite, that is, $\mathcal{N}/L \to 0$. Therefore, even if we consider the case $\mathcal{N} \to \infty$ in Eq. (K.6), this does not correspond to the thermodynamic limit (2.2).

From Eq. (K.5) we have

$$\langle\!\langle \rho|L_H|\phi_k; \phi_k\rangle\!\rangle = \sum_{l^{\mathcal{N}}, l'^{\mathcal{N}}} \langle\!\langle \rho|l; l'\rangle\!\rangle \times \left\{\frac{i\epsilon}{w_{l'l} + i\epsilon}[t_{lk}^+(\omega_k)\delta^{kr}(l' - k) - \delta^{kr}(l - k)t_{kl'}^-(\omega_k)] + \frac{2i\epsilon}{(w_{kl} + i\epsilon)(w_{kl'} - i\epsilon)} t_{lk}^+(\omega_k)t_{kl'}^-(\omega_k)\right\} \tag{K.8}$$

where $w_{kl} = \omega_k - \omega_l$, as usual. For regular density matrices ρ, we obtain

$$\lim_{\epsilon\to 0+} \langle\!\langle \rho|L_H|\phi_k; \phi_k\rangle\!\rangle = 0 \tag{K.9}$$

Hence $|\phi_k; \phi_k\rangle\!\rangle = |\phi_k\rangle\langle\phi_k|$ is an invariant of motion, and this is consistent with the solution of the eigenvalue problem, Eq. (K.2).

However, if ρ belongs to the class of singular density matrices given by Eq. (4.18), such as

$$\langle\!\langle l; l' | \rho \rangle\!\rangle \sim \rho_0^d(l) \prod_{i=1}^{\mathcal{N}} \delta(l_i - l_i') \tag{K.10}$$

then we have

$$\langle\!\langle \rho | L_H | \phi_k; \phi_k \rangle\!\rangle \sim \sum_{l,\mathcal{N}} \rho_0^d(l)\{\delta^{kr}(l-k)[t_{lk}^+(\omega_k) - t_{kl}^-(\omega_k)] + 2\pi i \delta(\omega_k - \omega_l)|t_{lk}^+(\omega_k)|^2\} \tag{K.11}$$

where we have used Eq. (5.45) in the last term of Eq. (K.8). The right-hand side of Eq. (K.11) is not only nonvanishing, but diverges in the limit of the thermodynamic limit.[18] Indeed, Eq. (K.6) leads to an ill-defined expression for $\epsilon \to 0+$ even in the sense of distributions, that is [see a similar divergence in Eq. (D.1)],

$$|t_{lk}^+(\omega_k)|^2 \sim \frac{1}{|\sum_{n=1}^m (k_{R_n} - l_{S_n}) + i\epsilon|^2} \to \infty \tag{K.12}$$

We note that Eq. (K.11) diverges even in conjunction with reduced observables, for example,

$$\sum_k^{\mathcal{N}} \langle\!\langle \rho | L_H | \phi_k; \phi_k \rangle\!\rangle k_1 \to \infty \tag{K.13}$$

Therefore, $|\phi_k\rangle\langle\phi_k|$ is no longer an invariant of motion. This shows that $|\phi_k\rangle$ is *not* the eigenstate of the Hamiltonian in conjunction with the singular density matrices.

The Hamiltonian (K.1) is simply a special case of the Hamiltonian (2.1). As discussed in the text, the $\mathcal{N}$-particle systems described by the Hamiltonian (2.1) are nonintegrable in the thermodynamic limit. Therefore, a consistent description of system (K.1) in the thermodynamic limit is not given by Yang's solution, Eq. (K.2), but requires the complex spectral representations presented in this article.

[18] Strictly speaking, it is meaningless to apply the thermodynamic limit to Eq. (K.8), because we have already emphasized at Eq. (K.6) that Eq. (K.5) is not the solution in this limit. The following discussion is simply a reconfirmation of this fact.

APPENDIX L. ANHARMONIC LATTICES

Our last example is the problems associated with anharmonic lattices, which is of great interest because it brings us close to nonlinear field theory. We consider the thermodynamic limit for which the number of particles $\mathcal{N} \to \infty$. Moreover, we require the distinction between intensive and extensive variables as maintained in this limit. For example, the displacement of a single particle has to remain finite as well as the density of energy $H/\mathcal{N}$. Anharmonic lattices have long been studied by researchers who observed these conditions [4,72]. From our point of view, they are interesting because they are indeed LPS, and their interactions are *automatically persistent* (remember that normal modes and angle-action variables are collective variables). The situation is therefore simpler than that for the interacting particles considered in the text. We show that the anharmonic force destroys the Hilbert space structure in the thermodynamic limit. Therefore, we have here a simple example where our extended formulation of dynamics applies.

We first consider a classical lattice and use angle-action variables α_k and J_k to describe a normal mode with wave vector k and frequency ω_k. For simplicity, we consider one-dimensional lattices. We assume that $\mathcal{N}$ atoms with mass m are equally spaced with a distant a in the equilibrium position. We follow closely the notation in [4] and [73] and give only minimum details.

If we restrict ourselves to three-phonon processes, the Hamiltonian is given by

$$H = H_0 + \lambda V = \sum_k \omega_k J_k + \lambda \sum_{k, k', k''}' \sum_{\sigma, \sigma', \sigma'' = \pm 1} V_{\sigma k, \sigma' k', \sigma'' k''} \times \left(\frac{J_k J_{k'} J_{k''}}{\omega_k \omega_{k'} \omega_{k''}} \right)^{1/2} e^{i(\sigma\alpha_k + \sigma\alpha_{k'} + \sigma\alpha_{k''})} \tag{L.1}$$

where $\omega_{-k} = \omega_k$, and the prime on the summation sign over k, k', k'' denotes that it is restricted to vectors on the reciprocal lattice. Here (with integer $1 \le n \le \mathcal{N}$)

$$k = k_n = \frac{2\pi n}{\Omega_a} \qquad \text{with } \Omega_a \equiv \mathcal{N} a \tag{L.2}$$

and the summation over k is a short-hand notation of the summation over n from $n = 1$ to $n = \mathcal{N}$. Hereafter, we abbreviate the argument n in k_n.

Note that, as can easily be verified (see [4]),

$$V_{k,k',k''} \sim \frac{1}{\sqrt{\Omega_a}} \tag{L.3}$$

As usual in the limit $\Omega_a \to \infty$ with a finite a (i.e., $\mathcal{N} \to \infty$), the summation over the wave vectors is replaced by the integration $(1/\Omega_a)\sum_k \to \int dk$.

The unperturbed Hamiltonian H_0 in Eq. (L.1) corresponds to the unperturbed Liouvillian:

$$L_0 = -i \sum_k \omega_k \frac{\partial}{\partial \alpha_k} \tag{L.4}$$

We have the eigenstate [for $\{v\} \equiv (v_{k_1}, v_{k_2}, \ldots, v_{k_N})$ and so on]:

$$L_0 | \{v\} \rangle\!\rangle = \sum_k v_k \omega_k | \{v\} \rangle\!\rangle \tag{L.5}$$

where

$$\langle\!\langle \{\alpha\}, | \{v\} \rangle\!\rangle = \frac{1}{(2\pi)^{\mathcal{N}/2}} e^{i \sum v_k \alpha_k} \tag{L.6}$$

We could include the action $\{J\}$ in Eq. (L.6), similarly to the momentum P in Eq. (3.41), but to simplify the notation we treat J as a parameter (see [4]).

For the interaction, the matrix element is given by

$$\langle\!\langle \{v\} | L_V | \{v'\} \rangle\!\rangle = \frac{1}{(2\pi)^{\mathcal{N}}} \int_0^{2\pi} \cdots \int_0^{2\pi} d\alpha_{k_1} \cdots d\alpha_{k_{\mathcal{N}}}\, e^{-i \sum v_k \alpha_k^n} L_V e^{i \sum v_{k'} \alpha_k} \tag{L.7}$$

For the only nonvanishing matrix elements [4], we obtain

$$\begin{aligned} &\langle\!\langle v_k, v_{k'}, v_{k''}, \{v\}^{\mathcal{N}-3} | L_V | v_k \sigma, v_{k'} \sigma', v_{k''} \sigma'', \{v\}^{\mathcal{N}-3} \rangle\!\rangle \\ &\quad = V_{\sigma k, \sigma k', \sigma k''} \left(\left[\frac{v_k}{2J_k} \right] + \left[\sigma \frac{\partial}{\partial J_k} \right] \right) \left(\frac{J_k J_{k'} J_{k''}}{\omega_k \omega_{k'} \omega_{k''}} \right)^{1/2} \end{aligned} \tag{L.8}$$

where

$$\left[\sigma \frac{\partial}{\partial J_k} \right] \equiv \sigma \frac{\partial}{\partial J_k} + \sigma' \frac{\partial}{\partial J_{k'}} + \sigma'' \frac{\partial}{\partial J_{k''}} \tag{L.9}$$

and so on.

All this is easily extended to the quantum case. [73] Using creation a_k^+ and annihilation operators a_k, the Hamiltonian operator is

$$H = \hbar \sum_k (\omega_k + \tfrac{1}{2}) b_k^{+1} b_k^{-1} + \lambda \hbar^{3/2} \sum_{\sigma, k} \frac{V_{\sigma k, \sigma k', \sigma k''}}{(\omega_k \omega_{k'} \omega_{k''})^{1/2}} b_k^{\sigma} b_{k'}^{\sigma'} b_{k''}^{\sigma''} \quad \text{(L.10)}$$

where

$$b_k^{+1} \equiv a_k^+, \qquad b_k^{-1} \equiv a_k \quad \text{(L.11)}$$

It is easy to write the matrix elements of the three-phonon interaction using the usual occupation number representation $\langle \{n\} | V | \{n'\} \rangle$.

Similar to the interacting particle case, it is convenient to introduce the "Wigner representation" for the occupation number representation by introducing the two sets of variables:

$$N_k = \tfrac{1}{2}(n_k + n'_k), \qquad \nu_k = n_k - n'_k \quad \text{(L.12)}$$

The variables N_k plays the same role in quantum mechanics as the action variable in classical mechanics, while ν_k plays the same role as the variable ν_k we introduced in Eq. (L.6) (for more details, see [4,31]).

We also use the notation

$$\langle n | A | n' \rangle = A_{n-n'}\left(\frac{n+n'}{2}\right) = A_\nu(N) = \langle\langle \nu, N | A \rangle\rangle \quad \text{(L.13)}$$

The transition from the quantum description to the classical one can be performed easily by putting

$$J_k = \hbar N_k \quad \text{(L.14)}$$

and then taking the limit $\hbar \to 0$ while keeping J_k finite.

We may write the Liouville–von Neumann equation for the quantum mechanical density matrix in terms of the Wigner representation, but we do not do so here. The reader can find the details in [73].

Let us consider closely the limit $\mathcal{N} \to \infty$ for the harmonic lattice with $\lambda = 0$. We first examine the classical case. The displacement u_n of the nth atom from the equilibrium position is given by [4]:

$$u_n = \sqrt{\frac{2}{\mathcal{N} m}} \sum_k e^{ikna} \sqrt{\frac{J_k}{\omega_k}} e^{i\alpha_k} \quad \text{(L.15)}$$

The condition

$$u_n \to \text{finite} \qquad \text{for } \Omega_a \to \infty \tag{L.16}$$

requires that

$$\sum_k e^{i\alpha_k} \to \sqrt{\mathcal{N}} \qquad \text{for } \Omega_a \to \infty \tag{L.17}$$

The angle variables must, therefore, behave as "stochastic variables" to which we can apply the law of large numbers. Not all initial conditions are compatible with Eq. (L.16). If Eq. (L.17) is not satisfied, we have to leave the model of anharmonic lattices. Note that this condition means that the sequence $\alpha_{k_1}, \alpha_{k_2}, \ldots$ with $k_j = (2\pi j/\mathcal{N}a)$ is "incompressible" [53] and has therefore, a larger probability (cf. the discussion in Section XIII). It corresponds to a stochastic sequence among the real number sequences for $0 \leq \alpha_{k_j} < 2\pi$.

For quantum case, the corresponding operator for the displacement is given with the annihilation operator a_k by

$$u_n = \sum_k \sqrt{\frac{2\hbar}{\mathcal{N} m\omega_k}}\, e^{ikna} a_k \tag{L.18}$$

The eigenstates of the annihilation operator are known as the "coherent states" and are given with c-numbers N_k and α_k by [74]

$$a_k \,|\, a_k\rangle = \sqrt{N_k}\, e^{i\alpha_k} \,|\, a_k\rangle \tag{L.19}$$

One can discuss these states in terms quite parallel to the classical case.

We now turn to the statistical description. As could be expected, this description is equivalent to the individual descriptions of a harmonic lattice. Now let us improve Hilbert space structure on the statistical description. The Hilbert norm of the distribution function is given by

$$\begin{aligned}\langle\!\langle \rho \,|\, \rho \rangle\!\rangle &= \sum_{\{\nu\}} \langle\!\langle \rho \,|\, P^{(\nu)} \,|\, \rho \rangle\!\rangle = \int d\{J\} \sum_{\{\nu\}} |\, \rho_{\{\nu\}}(\{J\}) \,|^2 \\ &= \int d\{J\} \Big[\, |\, \rho_0(\{J\}) \,|^2 + \sum_k |\, \rho_{1_k}(\{J\}) \,|^2 \\ &\qquad + \sum_k |\, \rho_{2_k}(\{J\}) \,|^2 + \sum_{kk'} |\, \rho_{1_k, 1_{k'}}(\{J\}) \,|^2 + \cdots \Big]\end{aligned} \tag{L.20}$$

This norm is preserved in time. Here $\rho_{\{\nu\}} \equiv P^{(\nu)}\rho$ and we introduce the projection operators defined by

$$P^{(0)} \equiv |\{0\}\rangle\!\rangle\langle\!\langle\{0\}|, \qquad P^{(\nu_k)} = |\nu_k, \{0\}^{N-1}\rangle\!\rangle\langle\!\langle \nu_k, \{0\}^{N-1}|, \ldots \tag{L.21}$$

They satisfy

$$P^{(\nu)}P^{(\mu)} = P^{(\nu)}\delta^{kr}(\{\nu\} - \{\mu\}), \qquad \sum_{\{\nu\}} P^{(\nu)} = 1 \tag{L.22}$$

where we have abbreviated $P^{(\{\nu\})}$ as $P^{(\nu)}$. The "vacuum of correlations" is again defined as the states in $P^{(0)}$ space. We note that the variables ν_k remain integers, while k becomes a continuous variable in the limit Ω_k. Hence, there is a well-defined $P^{(0)}$ even when we begin with the continuous variable [cf. the note at Eq. (4.23)]. In this sense, anharmonic lattices are simpler than interacting particles.

To obtain a finite Hilbert norm for $\Omega_a \to \infty$, well-defined conditions have to be satisfied. Indeed, the norm (L.20) must converge for $\Omega_a \to \infty$. This implies that

$$\rho_0 \sim 0(1), \qquad \rho_{1_k} \sim 1/\sqrt{\Omega_a}, \qquad \rho_{1_k, 1_{k'}, 1_{k''}} \sim 1/\Omega_a, \ldots \tag{L.23}$$

(recall that the summation $\sum'_{kk'k''}$ is over $k + k' + k'' = 0$ or a vector of the reciprocal lattice). To guarantee that Eq. (L.20) is finite we could, of course, also have, for example, $\rho_{1_k, 1_{k'}, 1_{k''}} \sim \Omega_a^{-3/2}$. But if $\rho_{1_k 1_{k'} 1_{k''}} \sim \Omega_a^{-1/2}$ the norm diverges and the limit $\Omega_a \to \infty$ leads outside the Hilbert space. This is the situation we meet with anharmonic lattices.

The Hilbert space structure is equivalent to the individual description, including the randomness condition of Eq. (L.17). Indeed, using Eqs. (L.15) [or Eq. (L.18) for the quantum case] and (L.23) we have

$$\langle u_n \rangle \sim \frac{1}{\sqrt{\Omega_a}} \sum_k \int d\{J\} \sqrt{J_k}\, \rho_{-1_k} \sim 0(1) \tag{L.24}$$

We may calculate in the same way other averages such as $\langle u_n u_{n'} \rangle$ or $\langle u_n u_{n'} u_{n''} \rangle$.

Note that using Eqs. (L.15) and (L.23),

$$\langle u_n u_{n'} u_{n''} \rangle \sim \frac{1}{\Omega_a^{3/2}} \sum_{kk'k''}' \rho_{-1_k, -1_{k'}, -1_{k''}} \sim \frac{1}{\Omega_a^{3/2}} \Omega_a^2 \frac{1}{\Omega_a} \sim \frac{1}{\sqrt{\Omega_a}} \to 0 \tag{L.25}$$

There appear only "even" correlations for harmonic lattices.

We now come to anharmonic lattices in which both the individual description and the Hilbert space structure are incompatible with the thermodynamic limit $\mathcal{N} \to \infty$. In thermodynamic equilibrium (equipartition theorem),

$$\langle H_0 \rangle \sim \mathcal{N} \sim \Omega_a \tag{L.26}$$

as well as

$$\langle V \rangle \sim \Omega_a \tag{L.27}$$

In contrast, using Eq. (L.23), corresponding to the Hilbert space structure, we obtain at most

$$\langle V \rangle \sim \sum_{kk'k''}' V_{kk'k''} \rho_{1_k, 1_{k'}, 1_{k''}} \sim \frac{1}{\sqrt{\Omega_a}} \Omega_a^2 \frac{1}{\Omega_a} \sim \sqrt{\Omega_a} \tag{L.28}$$

This shows that the thermodynamic equilibrium (L.27) lies outside the Hilbert space (see also [4] and its Appendix III). To obtain Eq. (L.27) we need stronger "correlations," such as

$$\rho_{1_k, 1_{k'}, 1_{k''}} \sim 1/\sqrt{\Omega_a} \tag{L.29}$$

but then the Hilbert space norm diverges.

This is a strong indication that the approach to equilibrium requires us to give up the Hilbert space description, as with interacting particles. There is, however, an interesting difference. In the case of interacting particles, the Hilbert space norm vanishes in the limit $\mathcal{N} \to \infty$ [see Eq. (4.21)], whereas here it diverges.

For the anharmonic lattice, the interaction couples among each correlation subspace $P^{(\nu)}$. Using Eq. (L.3) we see immediately that Eq. (L.8) leads precisely to Eq. (L.29), that is,

$$\rho_{1_k, 1_{k'}, 1_{k''}}(t) \sim \langle\langle 1_k, 1_{k'}, 1_{k''}, \{0\}^{\mathcal{N}-3} | L_V | \{0\} \rangle\rangle \rho_0(0) \sim V_{k,k',k''} \rho_0(0) \sim \frac{1}{\sqrt{\Omega_a}} \tag{L.30}$$

Correlations are *amplified* by anharmonic effects and bring us outside the Hilbert space.

The individual description in terms of wave function (or trajectory) is destroyed as well. Indeed, because of Poincaré resonances, we have Pauli (or Fokker–Planck) collision operators for the anharmonic lattice, as with interacting particles [4,73]. This then leads us to complex spectral representations of the Liouvillian.

The situation studied here is closely related to nonlinear field theory, in which we also expect to go outside the Hilbert space. Therefore, it becomes interesting to reconsider some of the basic problems of modern physics, such as renormalization, from this point of view. We hope to investigate this problem in a separate paper.

ACKNOWLEDGMENTS

This chapter incorporates much previous work performed mainly by Professor C. George, Professor F. Henin, Professor A. Grecos, Dr. I. Antoniou, and Dr. S. Tasaki. We have also benefited from the constructive criticism and suggestions of Dr. I. Antoniou and Dr. B. Misra. We are grateful to Professor C. George, Dr. M. de Haan, and Professor E. C. G. Sudarshan for their helpful comments. We wish to thank G. Ordóñez and T. Miyasaka for performing computer simulations. We also acknowledge the U. S. Department of Energy Grant No. DE-FG03-94ER14465, the Robert A. Welch Foundation Grant No. F-0365, and the European Community Contract No. PSS*0823 for support of this work.

REFERENCES

1. T. Petrosky and I. Prigogine, *Chaos, Solitons & Fractals*, **7**, 441 (1996). In this chapter we refer to this paper as *I*.
2. A. Shimony, "Conceptual foundations of quantum mechanics," in *The New Physics*, (ed.), P. Davies (Cambridge University Press, New York, 1993).
3. N. Bohr and L. Rosenfeld, *Dan. Vid. Selsk. mat.-fys. Medd.*, **12**(8) (1933); *Phys. Rev.*, **78**, 794 (1950).
4. I. Prigogine, *Non-Equilibrium Statistical Mechanics* (Wiley, New York, 1962).
5. I. Prigogine, *From Being to Becoming* (W. H. Freeman, New York, 1980).
6. T. Petrosky and I. Prigogine, *Physica* **147A**, 439 (1988).
7. A. Böhm and M. Gadella, *The Rigged Hilbert Space and Quantum Mechanics*, Springer Lecture Notes on Physics, Vol. 78 (Springer, New York, 1978).
8. A. Böhm and M. Gadella, *Dirac Kets, Gamow Vectors and Gelfand Triplets*, Springer Lecture Notes on Physics, Vol. 348 (Springer, New York, 1989).
9. A. Böhm, M. Gadella, and G. B. Mainland, *Am. J. Phys.* **57**, 1103 (1989).
10. I. Antoniou and I. Prigogine, *Physica* **192A**, 443 (1993).
11. T. Petrosky and I. Prigogine, *Chaos, Solitons & Fractals* **4**, 311 (1994).
12. T. Petrosky and I. Prigogine, *Proc. Natl. Acad. Sci. USA* **90**, 9393 (1993).
13. T. Petrosky and I. Prigogine, *Phys. Lett. A* **182**, 5 (1993).
14. T. Petrosky and I. Prigogine, *Physica* **175A**, 146 (1991).
15. T. Petrosky, I. Prigogine, and S. Tasaki, *Physica* **173A**, 175 (1991).
16. T. Petrosky and I. Prigogine, *Can. J. Phys.* **68**, 670 (1990).
17. H. H. Hasegawa and W. C. Saphir, *Phys. Rev. A* **46**, 7401 (1992).
18. H. H. Hasegawa and D. J. Driebe, *Phys. Rev.* **E50**, 1781 (1994).

19. D. J. Driebe and H. H. Hasegawa, in *Instabilities and Nonequilibrium Structures V*, E. Tirapegui and W. Zeller (eds.) (Kluwer, The Netherlands, 1995).
20. I. Antoniou and S. Tasaki, *Physica* **190A**, 303 (1992).
21. I. Antoniou and S. Tasaki, *J. Phys. A: Meth. Gen.* **26**, 73 (1993).
22. I. Antoniou and S. Tasaki, *Int. J. Quant. Chem.* **46**, 425 (1993).
23. J. Kumičák and E. Brändas, *Int. J. Quant. Chem.* **32**, 669 (1987).
24. E. C. G. E. Sudarshan, C. Chiu, and V. Gorini, *Phys. Rev.* **D18**, 2914 (1978).
25. L. Schwartz, *Theorie des Distributions I, II* (Hermann, Paris, 1957, 1959).
26. I. Gelfand and N. Vilenkin, *Generalized Functions*, Vol. 4 (Academic, New York, 1964).
27. T. Petrosky, G. Ordonez, and T. Miyasaka, *Phys. Rev. A* **53**, 4075 (1996).
28. See, for example, *Time's Arrows Today*, S. F. Savit (ed.) (Cambridge University Press, New York, 1993).
29. R. Balescu, *Statistical Mechanics of Charged Particles* (Wiley, New York, 1963).
30. N. Mishima, T. Petrosky, and M. Yamazaki, *J. Stat. Phys.* **14**, 359 (1976).
31. R. Résibois, *Physica* **27**, 541 (1961).
32. R. Résibois, in *Physics of Many-body Systems*, E. Meeron (ed.) (Gordon and Breach, New York, 1966).
33. C. George, *Physica* **37**, 182 (1967).
34. C. George, *Bull. Classe Sci., Acad. Roy. Belg.* **53**, 623 (1967).
35. F. Henin, *Physica* **54**, 385 (1970).
36. I. Prigogine, C. George, and F. Henin, *Physica* **45**, 418 (1969).
37. C. George, I. Prigogine, and L. Rosenfeld, *Det Kong. Dan. Vid. Selskab Matematisk-fys. Med.* **38**, 1 (1972).
38. I. Prigogine, C. George, F. Henin, and L. Rosenfeld, *Chem. Scripta*, **4**, 5 (1973).
39. C. George, *Physica* **65**, 277 (1973).
40. M. de Haan and F. Henin, *Physica* **67**, 197 (1973).
41. A. Grecos, T. Guo, and W. Guo, *Physica* **80A**, 421 (1975).
42. A. Grecos and M. Theodosopulu, *Physica* **50A**, 749 (1976).
43. M. de Haan, C. George, and F. Mayné, *Physica* **92A**, 584 (1978).
44. C. George, F. Mayné, and I. Prigogine, *Adv. Chem. Phys.* **61**, 223 (1985).
45. T. Petrosky and H. Hasegawa, *Physica* **160A**, 351 (1989).
46. I. Antoniou, Z. Suchaneki, R. Laura, and S. Tasaki, "Quantum systems with diagonal singularity," in *Advances in Chemical Physics*, Vol. 99, I. Prigogine and S. Rice (eds.) (Wiley, New York, 1997), pp. 299–332.
47. T. Petrosky and I. Prigogine, unpublished.
48. P. Résibois and I. Prigogine, *Bull. Classe Sci., Acad. Roy. Belg.* **46**, 53 (1960).
49. I. Prigogine, A. Grecos, and C. George, *Celestial Mech.* **16**, 489 (1977).
50. T. Petrosky, *J. Stat. Phys.* **48**, 1363 (1987).
51. L. D. Faddeev and S. P. Merkuriev, *Quantum Scattering Theory for Several Particle Systems* (Kluwer, Dordrecht, 1993).
52. L. Van Hove, *Physica* **23**, 441 (1957).
53. G. J. Chaitin, *Information, Randomness and Incompleteness* (World Scientific, Singapore, 1987).

54. D. C. Mattis (ed.), *The Many-body Problem* (World Scientific, Singapore, 1993).
55. A. Rae, *Quantum, Illusion or Reality* (Cambridge University Press, New York, 1986).
56. N. Nicolis and I. Prigogine, *Exploring Complexity* (W. H. Freeman, New York, 1989).
57. T. Petrosky, unpublished.
58. M. Ross (ed.), *Quantum Scattering Theory. Selected Papers* (Indiana University Press, Bloomington, IN, 1963).
59. C. C. Grosjean, *Formal Theory of Scattering Phenomena* (Institut Interuniversitaire des Sciences Nucleaires, Monographie No 7, Bruxelles, 1960).
60. M. L. Goldberger and K. M. Watson, *Collision Theory* (Wiley, New York, 1965).
61. R. G. Newton, *Scattering Theory or Waves and Particles* (McGraw-Hill, New York, 1966).
62. K. Friedrichs, *Commun. Pure Appl. Math.* **1**, 361 (1948).
63. L. A. Khalfin, *JETP* **6**, 1053 (1948).
64. B. Misra and E. C. G. Sudarshan, *J. Math. Phys.* **18**, 756 (1977).
65. Z. L. Zhang, *Irreversibility and Extended Formulation of Classical and Quantum Non-integrable Dynamics, Doctoral dissertation*, The University of Texas at Austin, 1995.
66. E. H. Lieb and W. Linger, *Phys. Rev.* **130**, 1616 (1963).
67. E. H. Lieb, *Phys. Rev.* **130**, 1605 (1963).
68. C. N. Yang, *Phys. Rev. Lett.* **19**, 1312 (1967).
69. H. B. Thacker, *Rev. Mod. Phys.* **53**, 253 (1981).
70. H. B. Thacker, *Phys. Rev.* **D14**, 3508 (1976).
71. H. A. Bethe, *Z. Phys.* **71**, 205 (1931).
72. I. Prigogine and F. Henin, *J. Math. Phys.* **1**, 349 (1960).
73. F. Henin, I. Prigogine, C. Geroge, and F. Mayné, *Physica* **32**, 1828 (1966).
74. W. H. Louisell, *Quantum Statistical Properties of Radiation* (Wiley, New York, 1973).

UNSTABLE SYSTEMS IN GENERALIZED QUANTUM THEORY

E. C. G. SUDARSHAN, CHARLES B. CHIU, and G. BHAMATHI

Department of Physics and Center for Particle Physics, University of Texas, Austin, Texas

CONTENTS

Advances in Chemical Physics, Volume XCIX, Edited by I. Prigogine and Stuart A. Rice.
ISBN 0-471-16526-3

ABSTRACT

Phenomenological treatments of unstable states in quantum theory have been known for six decades and have been extended to more complex phenomena. But the twin requirement of causality ruling out a physical state with complex energy and the apparent decay of unstable states necessitates generalizing quantum mechanics beyond the standard Dirac formulation. Analytically continued dense sets of states and their duals provide the natural framework for a consistent and conceptually satisfying formulation and solution. Several solvable examples are used to illustrate the general formalism, and the differences from traditional phenomenological treatment (and its modern revivals) are noted. The unreliability of the singularities of the S-matrix as a criterion for determining the spectrum of states in the generalized theory is also brought out. The time evolution of unstable systems is characterized by three domains. Results in the decay of the neutral Kaon and its counterpart in higher-flavor-generations provide physically relevant and interesting unstable systems.

I. INTRODUCTION

The study of the decay of a metastable quantum system began with Gamow's theory [1] of alpha decay of atomic nuclei and Dirac's theory [2] of spontaneous emission of radiation by excited atoms. A general treatment of decaying systems was given by Weisskopf and Wigner [3], and by Breit and Wigner [4] (see for examples Bohm [5–8], Fonda, Ghirardi, and collaborators [9,10] and Yamaguchi and collaborators [11], all these gave a strictly exponential decay. Fermi [12] gave a simple derivation of the rate of transition following the work of Dirac; and this has come to be known as the Golden Rule. The close relationship between resonances and metastable decaying states had been noted in nuclear reactions by Bohr [13], Kapur and Peierls [14], and Peierls [15]; see also Matthews and Salam [16,17].

Siegert [18] was the first to associate the complex poles in the S-matrix of Wheeler [19] to quantum theory resonances. Peierls [20] seems to have been the first to seriously investigate the problem that the Breit–Wigner resonance model has complex energy states on the "physical sheet" in violation of the notion of causality in quantum mechanics [21]; he emphasized the need to relegate any such complex poles to an unphysical sheet in the analytic continuation of the scattering amplitude in the complex energy plane.

The exact solution of a model of decay going beyond the Breit–Wigner approximation of the Dirac model for metastable atoms was studied by Glaser and Källen [22], Höhler [23], and Nakanishi [24] following the field theoretical formulations of Lee [25] and an early work of Friedrichs [26]. Other models of a metastable system were studied by Moshinsky [27], Winter [28], Frey and Thiele [29], Levy [30], Williams [31], and Fleming [32].

Khalfin [33,34] had shown that, on general principles, if the Hamiltonian was bounded from below, the decay could not be strictly exponential. He used the Paley–Wiener theorem [35] to demonstrate the result. He also showed that there should be deviations from the exponential in the very large and very small time domains. Misra and Sudarshan [36] showed that for a wide class of systems, tests of nondecay repeated at arbitrarily small times prevent the decay of a metastable state—the so-called Zeno effect.

The question of irreversibility and the treatment of unstable states has been systematically pursued by Prigogine and his collaborators. Our interest in the conceptual questions has been stimulated by Prigogine's work and his important observation that an unstable particle, if it is autonomous, must obey the same decay law at all times. They must then be distinct from

Khalfin's [33] unstable states, which must age. Because the work of Prigogine and collaborators is presented elsewhere in this volume, we content ourselves with reference to their latest papers [37–39]. See also the point of view elaborated by Prigogine in *From Being to Becoming* [40].

In this article, we are concerned with a systematic and conceptually consistent development of the theory of metastable systems going beyond the Breit–Wigner model and its modern revivals. We shall follow several of our papers [36,41–46] over the past two decades.

A. Spectral Information of a Resonance

Quantum mechanics is defined in terms of vectors in Hilbert space with self-adjoint linear operators realizing dynamical variables. Self-adjoint operators have a real spectrum. For stationary states, we have point eigenvalues of the spectrum; scattering states are usually associated with the continuous part of the spectrum. What then about resonances and metastable resonances?

In standard quantum theory these also belong to the continuous spectrum bounded from below. The only signature of a resonance or a metastable state is a "spectral concentration" or a line shape. Because the line shape is affected by the background and by kinematical factors, we can usually extract only the center of the resonance peak and its width (full width at half maximum). It would be desirable to see these items emerge as spectral information: this is what the Breit–Wigner approximation does, but at a very high price—the violation of spectral boundedness. But the phenomena in which this situation obtains are many: deexcitation of atomic levels, alpha decay, formation of compound nuclei, and resonant scattering. Therefore, we need a more general formulation of quantum mechanics which has a richer spectral structure but does not violate physical principles.

B. Lorentz Line Shape and Breit–Wigner Approximation

The amplitude for a metastable state to overlap itself after evolution for a fixed time t is called the survival amplitude:

$$A(t) = \langle \psi | e^{-iHt} | \psi \rangle \tag{1.1}$$

Since, in general, its absolute value is less than 1, it is tempting to write

$$e^{-iHt} | \psi \rangle \rightarrow e^{-[iE_0 + (1/2)\gamma]t} | \psi \rangle \tag{1.2}$$

so that there is a complex eigenvalue. If we recognize that for negative time $|A(t)|$ is also less than 1, we may consider

$$e^{-iE_0t-(1/2)\gamma|t|}|\psi\rangle \tag{1.3}$$

as the evolute of the metastable state. Taking the Fourier transform of the exponential factor, there are contributions from both the negative and the positive time. We obtain, for $-\infty < \omega < \infty$,

$$\begin{aligned} f(\omega) &= \frac{1}{2\pi}\cdot\frac{-i}{(\omega - E_0) - (i\gamma/2)} + \frac{1}{2\pi}\cdot\frac{i}{(\omega - E_0) + (i\gamma/2)} \\ &= \frac{1}{\pi}\cdot\frac{(1/2)\gamma}{(\omega - E_0)^2 + (\gamma^2/4)} \end{aligned} \tag{1.4}$$

The last expression has the Lorentz line shape known from the response of a harmonically bound electron with a dissipative term. Note that the spectrum is unbounded from below. The spectral weight is an analytic function of ω with isolated poles at $\omega = E_0 \pm i\gamma/2$. Because $f(\omega)$ is nonzero along the entire real axis, there are two pieces of the piecewise analytic Fourier transform. One piece varies as $e^{-\gamma t/2}$ for positive t and zero for negative t; the other piece has $e^{\gamma t/2}$ for negative t and zero for positive t. Neither piece models an autonomous physical state because the state appears to be created or destroyed at $t = 0$ and has a purely exponential law. There are no states in the physical Hilbert space, that is, among states in the linear span of positive energy states which have such a property. (A "state" with such a time dependence can be synthesized only if one includes unphysical negative energy states along with the physical states.) If we require of these unphysical "states" the physical requirement of causality, that is, they vanish for negative times, we get the unique (though unphysical) choice

$$f(\omega) = \frac{1}{2\pi}\cdot\frac{i}{(\omega - E_0) + (i\gamma/2)} \tag{1.5}$$

$$|\psi(t)\rangle = \theta(t)e^{-iE_0t-(1/2)\gamma t}|\psi(0)\rangle \tag{1.6}$$

However, this state is not time-reversal invariant, and cannot be made time-reversal invariant without giving up the causality requirement. If we give up causality, we get back Eqs. (1.3) and (1.4).

The decomposition in Eq. (1.4) into two terms is the split of an unphysical state with a spectrum $-\infty < \omega < \infty$ into *two* unphysical states which are analytic in the lower and upper half planes and are therefore, respectively, causal and anticausal. This is a special case of a general decomposition which can be carried out for physical states and for unphysical states into the sum of functions analytic in half planes. This is discussed in detail, see Section IV.F.

Let us return to the Lorentz line shape (1.4). A classical physical context in which such a line shape arises is in the correlation function of a harmonically driven damped harmonic oscillator. Here the time-dependent amplitude $x(t)$ and the two-point correlation function are respectively described by

$$\frac{d^2x}{dt^2} + R\frac{dx}{dt} + \omega_0^2 x = ae^{-i\omega t}, \qquad x(t) = \frac{ae^{-i\omega t}}{\omega_0^2 - \omega^2 - i\omega R} \tag{1.7}$$

$$\begin{aligned}\langle \tilde{x}(0)x(t)\rangle &\equiv \frac{1}{2\pi}\int_{-\infty}^{\infty} \frac{a^*}{\omega_0^2 - \omega^2 + i\omega R}\cdot\frac{ae^{-i\omega t}}{\omega_0^2 - \omega^2 - i\omega R}\,d\omega \\ &\approx \frac{|a|^2}{4R\omega_0^2}\,e^{-(1/2)R|t|}\cos\omega_0 t \end{aligned}\tag{1.8}$$

In the last step, the approximation $R \ll \omega_0$ is assumed. Here the temporal behavior for the two-point correlation function, which is analogous to the survival amplitude, is exponentially damped for both positive and negative time. The frequency dependence has been used in models of dispersion relations for the refractive index of a dielectric in the Sellmeier formula [47] and in more detailed theories of the refractive index [48].

C. Lorentz Transformation on State with Complex Eigenvalues

Although Lorentz transformation is not the main concern of this chapter, it is instructive to digress here to see how a resonance state with a complex eigenvalue would transform under Lorentz transformation. For definiteness, consider the real spectrum of Eq. (1.5) with the corresponding time dependence given in Eq. (1.6). The real spectrum here consists of all energies, so when we make a fixed Lorentz transformation, we get all possible momenta—some positive and some negative—with a concentration around the value expected for energy m. This range of momenta may be expressed by a complex momentum suitably defined. We could work with real momenta but any fixed Lorentz transformation with the boost parameter η would produce not a unique momentum $m \sin h\eta$ but all momenta from $-\infty$ to $+\infty$.

To perform this analysis for the correct real spectrum $0 \leq \omega < \infty$ is not difficult; there is no state with complex energy $m - i(\Gamma/2)$ by itself, it must be accompanied by a complex background. Such a state will transform itself into complex momenta, but that is mostly the alternative expansion for a superposition of all momenta (in the same direction!). So there is no inconsistency.

In the narrow width approximation $\Gamma \ll m$, we can get a simple derivation of the behavior of the lifetime:

$$\left(m - \frac{i}{2}\Gamma, 0\right) = \to \left(E - \frac{i}{2}\Gamma', p\right)^2 - p^2 \tag{1.9}$$

with

$$\left(m - \frac{i}{2}\Gamma, 0\right) \to \left(E - \frac{i}{2}\Gamma', \vec{p}\right) \tag{1.10}$$

Equating the imaginary part on the two sides of the equation leads to

$$E\Gamma' = m\Gamma \tag{1.11}$$

So

$$\frac{\Gamma'}{\Gamma} = \frac{m}{E} = \sqrt{1 - \frac{v^2}{c^2}} \tag{1.12}$$

Thus while

$$\mathrm{Re}\left(m - \frac{i\Gamma}{2}\right) \to \gamma m = E, \qquad \mathrm{Im}\left(m - \frac{i\Gamma}{2}\right) \to \frac{\Gamma}{\gamma} = \Gamma' \tag{1.13}$$

Thus the width is reduced and the lifetime is increased!

D. Violation of the Second Law of Thermodynamics

The Breit–Wigner model (and its modern revivals) violate the spectral condition to obtain a strict exponential decay; or, more generally, the linear sum of a finite number of exponentials. This violation would, were it actually to occur, also violate the second law of thermodynamics [49]. Because states with arbitrarily large negative energy are admitted here, we can devise suitable interactions that take away arbitrarily large amounts of energy from the system. The first law of thermodynamics can be satisfied and yet the available energy from the system is arbitrarily large. Since this must not be possible, the unbounded spectrum by itself should not occur.

In the formalism presented in this article, there is a complex energy discrete state always accompanied by a complex background such that the real energy spectrum is always bounded from below. The restriction we have to impose to obtain this resolution is to have only such states as are derived from analytic continuations of physical states; then an isolated, discrete, complex-energy state must always be accompanied by a complex background with a real threshold.

E. Organization of This Chapter

Our discussion below is divided into two main topics: one concerns the characteristic region in the temporal evolution of unstable quantum systems and the other concerns the formulation of a consistent theory for an unstable quantum state.

1. Temporal Evolution of an Unstable Quantum System

When the energy spectrum of the unstable particle system is semibounded, one expects a deviation from pure exponential decay. This deviation occurs [33,34] in both the small and the large t-regions. In Section II, the three characteristic time regions [36,41], in the time evolution of a one-level unstable quantum system are discussed. In the small t-region, the time evolution of the system is sensitive to repeated measurements. When the expectation value of the energy of the system is finite, one expects the Zeno paradox, that is, frequent measurement of the unstable system leads to nondecay. However, when the expectation value of the energy is infinite, repeated measurement of the system would lead to a rapid decay of the system. In the large t-region, the survival probability has a power-law fall off, with the rate of the fall off governed by the threshold behavior of the semibounded spectra. There may be intricate interference phenomena [28] at the transition from the exponential decay to the power-law region.

The solution [50] to the multilevel unstable quantum system is presented in Section III. The most common example is the neutral Kaon system, which is a two-level system. Here the Lee-Oehme–Yang [51] model is the Breit–Wigner approximation for the two-level unstable quantum system. Within this approximation the unstable Kaons K_L and K_S can be written as superpositions of K^0 and $\bar{K}^0$, where K_L and K_S decay independently. When one takes into account that they are poles on the second sheet, there are cut contributions for the survival amplitudes in addition to the pole contributions. The cut contributions are particularly important in the very small and very large t-regions. Regeneration effects [52,53], that is, the transitions between K_L and K_S, are expected to be nonnegligible in these regions [50]. These and related issues for the neutral Kaon system are also discussed in Section III.

Thus far our attention has mainly been on the features of the time development of unstable quantum systems, which show the departure from pure exponential decay of the Breit–Wigner approximation. This deviation arises when one takes the continuum spectrum into account. Here resonance is a discrete pole contribution in the survival amplitude or, more generally, the transition amplitude on the unphysical sheet. This is in contrast to the Breit–Wigner approximation, where the resonance pole(s) are on the physical sheet. The "physical sheet" and the "unphysical sheet" designations used here have important distinctions. From the requirement of causality, it can be shown that transition amplitudes are analytic on the physical sheet. The presence of complex poles on the physical sheet, therefore, implies the violation of causality. Since we want to work with a causal theory, resonance poles must be identified with the second-sheet poles and deviation from exponential behavior in the time evolution is expected.

From the study of solvable models, it is likely that departure from the exponential decay law at presently accessible experimental time scales is numerically insignificant. Nevertheless, it is important to insist on having a consistent framework for the description of unstable states, which gives predictions coinciding with the Breit–Wigner approximation in the bulk of the middle region and at the same time allows extension to the very small and very large time regions. We proceed now to the generalized quantum system where the resonance pole will be identified as a generalized quantum state.

2. *A Theory for Unstable Quantum Systems*

As we explain in detail later, a consistent framework for the unstable state is achieved through the use of a generalized vector space of quantum states. Consider the integral representation defined by the scalar product between an arbitrary vector in the dense subset of analytic vectors in the physical-state space $\mathscr{H}$ and its dual vector: the integration is along the real axis. Keeping the scalar product fixed, the analytic vectors may be continued through the deformation of the integration contour. The deformed contour defines the generalized spectrum of the operator in the continued theory, which typically consists of a deformed contour in the fourth quadrant and the exposed singularities, if any, between the real axis and the deformed contour. We identify an unstable particle pole as a bona fide discrete state in the generalized space with a complex eigenvalue. Here the continuum states are defined along some complex contour γ, which is deformed in such a manner as to expose the unstable particle pole. The inner product and transition amplitudes are defined between states in $\mathscr{G}$ and its dual state in the corresponding dual space $\tilde{\mathscr{G}}$.

In Section IV, we discuss this analytic continuation approach. Several models are studied, and special attention is given to the unfolding of the generalized spectrum. We demonstrate how the analytic continuation is done for the Friedrichs–Lee model in the lowest sector and for the Yamaguchi [54] potential model. We show that the generalized spectrum obtained leads to the correct extended unitarity relation for the scattering amplitude. In this Section we demonstrate the possibility of having mismatches between poles in the S-matrix and the discrete states in the Hamiltonian, which may arise when $\mathscr{H}$ obtains also in the generalized $\mathscr{G}$ space. Finally, we consider the analytic continuation of the probability function and the operation of time-reversal invariance.

In Section V, we study the analytic continuation as applied to the multilevel system [55] and its application to the Bell–Steinberger relation [56] for the neutral Kaon system.

In Section VI, we extend our consideration to the three-body system. In particular, we consider a solvable model involving a three-body system, that is, the cascade model [44], which contains A, B, and C together with two species of quanta. The interactions are given by $A \to B\theta$, $B \to C\phi$. Here the second-sheet singularities are the resonance pole A^* and the branch cut $B^*\theta$. The analytic continuation [43] of this model is also discussed. The extended unitarity relation here can conveniently be displayed in terms of the generalized discontinuity relations. Section VII gives a summary and our conclusions.

II. TIME EVOLUTION OF AN UNSTABLE QUANTUM SYSTEM

In this section we study the time evolution of the so-called unstable particle system. By definition, an unstable particle is a nonstationary state which undergoes substantial changes in a time scale much larger than the natural time scales associated with the energy of the system. In this case, the "natural" evolution in time and the "decay transition" may be viewed as two separate kinds of time development. It would be profitable to think of the natural evolution as if it were accounted for by an unperturbed Hamiltonian and the decay transition as being brought about by an additional perturbation. Conversely, given a Hamiltonian with a point spectrum and a continuous spectrum, we may introduce perturbations which lead to "decay" of the states which belonged to the point spectrum and which were, therefore, stationary. In this way we can determine the precise time development of the system.

Many studies have been devoted to questions relating to deviations from the exponential decay law of particle decay processes. The time-reversal invariance requires that the slope of the survival probability at $t = 0$ be continuous, which admits two possibilities—it may be either 0 or ∞.

When the expectation value of the energy of the system is finite, this slope is zero, which leads to Zeno's paradox. The theorem on Zeno's paradox of Misra and Sudarshan [36], proves that nondecay results generally. Earlier work by Degasperes et al. [57] and Rau [58] showed that the limit of infinitely frequent interactions leads to nondecay. These are special cases of Zeno's paradox theorem. Some subsequent investigations of Zeno effect were performed by Chiu, Sudarshan, and Misra [41], Ghirardi et al. [10], Peres [59], Fleming [60], and Valanju [61,62]. The quantum Zeno effect has been verified by Itano et al. [63] using metastable atoms "interrogated" by microwaves.

On the other hand, for a quantum system where the energy expectation value is ∞, the slope of the survival probability at $t = 0$ is ∞. For this case, the repeated measurement of the system leads to a rapid decay of the system [41].

In the large t region, the survival probability has a power-law fall off in t, with the rate of the fall off governed by the threshold behavior of the semi-bounded spectra. Winter [28] studied a simple barrier-penetration problem to elucidate the time development of quasi-stationary states in the small-, intermediate-, and large-time regions. Some interference phenomena were observed. Our discussions below are based mainly on the paper by Chiu, Sudarshan, and Misra [41].

A. Deviation from Exponential-Decay Law at Small Time

We start with a brief recapitulation of the quantum-theoretical formalism for describing unstable states. Let $\mathscr{H}$ denote the Hilbert space formed by the unstable (undecayed) states of the system as well as the states of the decay products. The time evolution of this total system is then described by the unitary group $U_t = e^{-iHt}$, where H denotes the self-adjoint Hamiltonian operator of the system. For simplicity, we assume that there is exactly one unstable state represented by the vector $|M\rangle$ in $\mathscr{H}$, which must be orthogonal to all bound stationary states of the Hamiltonian H. Hence $|M\rangle$ is associated with the continuous spectrum. (In contrast to this simplified situation in quantum mechanics, the spectrum of the Liouville operator of a classical dynamical system, which is weakly mixing or nonmixing, must have a singular continuous part.) Thus, if F_λ denotes the spectral projections of the Hamiltonian,

$$H = \int \lambda \, dF_\lambda \equiv \int \lambda |\lambda\rangle\langle\lambda| \, d\lambda \tag{2.1}$$

The function $\langle M | F_\lambda | M\rangle$ is absolutely continuous, and its derivative

$$\rho(\lambda) = \frac{d}{d\lambda} \langle M | F_\lambda | M\rangle = \langle M | \lambda\rangle\langle\lambda | M\rangle \tag{2.2}$$

can be interpreted as the energy-distribution function of the state $|M\rangle$; that is, the quantity

$$\int_{E}^{E+dE} \rho(\lambda)\, d\lambda \tag{2.3}$$

is the probability that the energy of the state $|M\rangle$ lies in the interval $[E, E + dE]$.

The distribution function $\rho(\lambda)$ has the following general properties:

1. $\rho(\lambda) \geq 0$
2. $\int \rho(\lambda)\, d\lambda = 1$, corresponding to the normalization condition $\langle M | M \rangle = 1$
3. $\rho(\lambda) = 0$ for λ outside the spectrum of H

It may be noted that, in defining the energy-distribution function $\rho(\lambda)$ as we have done above, we have absorbed the customary density of states factor or the phase space factor $\sigma(\lambda)$ in $\rho(\lambda)$.

The conditions mentioned above are quite general and hold for any state orthogonal to the bound states of H. To identify it as an unstable particle state with a characteristic lifetime, its energy distribution function should satisfy certain additional conditions, which are discussed in Section II.C. In this section, we use only properties 1–3 of the energy-distribution function.

The nondecay probability $Q(t)$ (or the probability for survival) at the instant t for the unstable state $|M\rangle$ is given by

$$Q(t) = |\langle M | e^{-iHt} | M \rangle|^2 \tag{2.4}$$

Accordingly, the decay probability $P(t)$ at t is $1 - Q(t)$. The nondecay amplitude $a(t) = \langle M | e^{-iHt} | M \rangle$ is easily seen to be the Fourier transform of the energy-distribution function $\rho(\lambda)$:

$$a(t) = \langle M | e^{-iHt} | M \rangle = \int e^{-i\lambda t} d\langle M | F_\lambda | M \rangle \tag{2.5}$$

$$= \int e^{-i\lambda t} \rho(\lambda)\, d\lambda \tag{2.6}$$

The celebrated Paley–Wiener theorem [35] then shows that if the spectrum of H is bounded below, so that $\rho(\lambda) = 0$ for $\lambda < 0$, then $|a(t)|$ and hence $Q(t) = |a(t)|^2$ decreases to 0 as $t \to \infty$ less rapidly than any exponential function $e^{-\Gamma t}$. This is essentially Khalfin's argument proving the necessity of deviation from the exponential decay law at large time.

The following proposition shows that $Q(t)$ must deviate from the exponential decay at sufficiently small time as well. Let the spectrum of H be bounded below; assume further that the energy expectation value for the state $|M\rangle$ is finite:

$$\int \lambda\rho(\lambda)\, d\lambda < \infty \tag{2.7}$$

Then $Q(t) > e^{-\Gamma t}$ for sufficiently small t. We shall assume, without loss of generality, that the spectrum of H is confined to the positive semiaxis $[0, \infty]$.

To prove the proposition, it is sufficient to show that $Q(t)$ is differentiable and

$$\dot{Q}(0) \equiv \frac{d}{dt} Q(t)\bigg|_{t=0} > -\Gamma, \qquad (\Gamma > 0) \tag{2.8}$$

We shall in fact show that

$$\dot{Q}(0) = 0 \tag{2.9}$$

In view of the positivity of the operator H, the energy distribution function $\rho(\lambda) = 0$ for $\lambda < 0$. Thus, Eq. (2.7), together with the semiboundedness of the spectrum, implies that the function $\lambda\rho(\lambda)$ is absolutely integrable:

$$\int |\lambda|\, \rho(\lambda)\, d\lambda < \infty \tag{2.10}$$

The survival amplitude is defined by

$$a(t) = \int e^{-i\lambda t} \rho(\lambda)\, d\lambda \qquad \text{with } a(0) = 1 \tag{2.11}$$

The condition of Eq. (2.10) implies that $a(t)$ is differentiable for all t, since

$$|\dot{a}(t)| = \left| \int e^{-i\lambda t} \lambda\rho(\lambda)\, d\lambda \right| \leq \int |\lambda|\, \rho(\lambda)\, d\lambda < \infty \tag{2.12}$$

Thus, the derivative here is continuous. Now

$$a^*(t) = a(-t) \tag{2.13}$$

so that

$$\frac{d}{dt} a^*(t)\bigg|_{t=-s} = -\frac{d}{dt} a(t)\bigg|_{t=-s} = -\dot{a}(-s) \tag{2.14}$$

Since $Q(t) = a(t)a^*(t)$,

$$\frac{d}{dt} Q(t)\bigg|_{t=s} = a(-s)\dot{a}(s) - a(s)\dot{a}(-s) \tag{2.15}$$

In particular,

$$\dot{Q}(0) = \dot{a}(0_+) - \dot{a}(0_-) = 0 \tag{2.16}$$

since $a(0) = 1$ and $\dot{a}(t)$ is continuous so that $\dot{a}(0_+) = \dot{a}(0_-)$. We emphasize that the semiboundedness of H, which ensures the continuity of the derivative, is an essential ingredient in the proof. Otherwise, consider the usual Breit–Wigner weight function $\rho(\lambda) = 1/(1 + \lambda^2)$, for which

$$a(t) = \frac{1}{\pi}\int_{-\infty}^{\infty} \frac{e^{-i\lambda t}\, dx}{1 + \lambda^2} = e^{-|t|} \tag{2.17}$$

The magnitude of the corresponding derivative at $t = 0$ is

$$|\dot{a}(0)| = \frac{1}{\pi}\int_{-\infty}^{\infty} \frac{\lambda\, d\lambda}{1 + \lambda^2} \tag{2.18}$$

Notice that this integral diverges at both the lower and the upper limit; hence it is indefinite. This is manifested by the discontinuity at $t = 0$:

$$\dot{a}(0_+) = -1 \quad \text{and} \quad \dot{a}(0_-) = 1$$

The preceding proposition shows that at sufficiently small time, the nondecay probability $Q(t)$ falls off less rapidly than would be expected on the basis of the exponential decay law. Thus, if the unstable system is monitored for its existence at sufficiently small intervals of time, it would appear to be longer lived than if it were monitored at intermediate intervals, where the decay law is exponential. The quantum Zeno's paradox states that in the limit of continuous monitoring, the particle does not decay at all. In the present case of a one-dimensional subspace of undecayed (unstable) states, this conclusion follows as an immediate corollary to the preceding proposition. It can easily be seen that if the system prepared initially in the unstable state $|M\rangle$ is (selectively) monitored on its survival at the instants $0, t/n, \ldots, (n-1)t/n, t$, the probability for its survival is given by

$$Q\left(\frac{t}{n}\right)^n$$

Since $Q(t)$ is continuously differentiable and $\dot{Q}(0) = 0$, it can easily be shown that

$$\lim_{n\to\infty} Q\left(\frac{t}{n}\right)^n = 1 \tag{2.19}$$

independent of t. It is evident that the survival probability under discrete but frequent monitoring will be close to 1 provided that t/n is sufficiently small, so that the departure from the exponential decay law remains significant. It is thus important to estimate the time scale for which the small-time deviation from the exponential decay law is prominent.

B. Resonance Models for Decay Amplitudes

To estimate the parameters T_1 and T_2 which separate the intermediate-time domain, where the exponential decay law holds, from small- and large-time domains where deviations are prominent, we need to make a more specific assumption about the energy-distribution function $\rho(\lambda)$ of the unstable state $|M\rangle$. In fact, so far we have assumed only very general properties of $\rho(\lambda)$ that are not sufficient to warrant the identification that $|M\rangle$ represents an unstable state which behaves as a more or less autonomous entity with a characteristic lifetime.

To formulate this resonance requirement, we rewrite the nondecay amplitude as a contour integral. To this end, we consider the resolvent $R(z) \equiv (H - zI)^{-1}$ of the Hamiltonian H. This forms a (bounded) operator-valued analytic function of z on the whole of the complex plane except for the cut along the spectrum of H, which we take to be the real half axis $[0, \infty]$. Under mild restrictions on the state $|M\rangle$, for instance, when $|M\rangle$ lies in the domain of H^2, we have

$$e^{-iHt}|M\rangle = \frac{1}{2\pi i}\int_C e^{-izt}R(z)|M\rangle\, dz \tag{2.20}$$

where C is the contour shown in Fig. 1. The nondecay probability is then

$$a(t) = \langle M|e^{-iHt}|M\rangle = \frac{1}{2\pi i}\int_C e^{-izt}\beta(z)\, dz \tag{2.21}$$

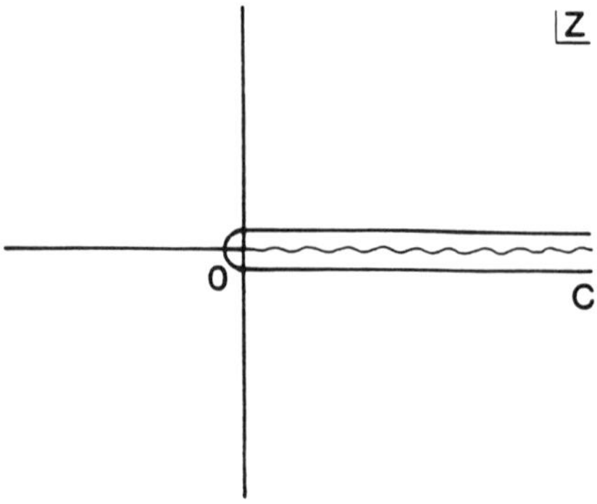

Figure 1. The contour C in the complex λ plane.

where

$$\beta(z) = \langle M | R(z) | M \rangle \tag{2.22}$$

The function $\beta(z)$ is uniquely determined by the energy-distribution function $\rho(\lambda)$ of $| M\rangle$ through the formula

$$\beta(z) = \int_0^\infty \frac{\rho(\lambda)}{\lambda - z}\, d\lambda \tag{2.23}$$

and in turn determines the distribution function $\rho(\lambda)$ through the formula

$$\rho(\lambda) = \lim_{\epsilon \to 0+} \frac{1}{2\pi i} [\beta(\lambda + i\epsilon) - \beta(\lambda - i\epsilon)] \tag{2.24}$$

The function $\beta(z)$ is analytic in the cut plane and is free of zeros there. We may thus introduce

$$\gamma(z) = \frac{1}{\beta(z)} \tag{2.25}$$

which is analytic and free of zeros in the cut plane. The nondecay probability is then given by

$$a(t) = \frac{i}{2\pi} \int_C \frac{e^{-izt}}{\gamma(z)}\, dz \tag{2.26}$$

where the contour C is illustrated in Fig. 1. This representation of $a(t)$ is quite general and does not yet incorporate the important resonance condition alluded to earlier. The resonance condition may be formulated as the requirement that the analytic continuation of $\gamma(z)$ in the second sheet possess a zero at $z = E_0 - \frac{1}{2}i\Gamma$ with $E_0 \gg \Gamma > 0$. Under this condition, the representation for $a(t)$ shows that it will have a dominant contribution $e^{-iEt}e^{-\Gamma t/2}$ from the zero of $\gamma(z)$ in the second sheet and certain correction terms to the exponential decay law arising from a "background" integral. An investigation of the corrections to the exponential decay law then amounts to an investigation of the background integral in Eq. (2.26). This approach of considering the deviation from the exponential decay law has been discussed in the past (see for example [23]). Here we investigate the detailed properties of the background integral by making a specific choice for $\gamma(z)$.

To facilitate the choice and to relate our results to investigations on the Lee model and the related Friedrichs model, we note that one can write (suitable subtracted) dispersion relations for $\gamma(z)$.

For instance, if $\gamma(z)$ has the asymptotic behavior

$$|\gamma(z) - z| \xrightarrow[|z|\to\infty]{} z^n \tag{2.27}$$

with $n \leq 0$, then

$$\gamma(z) = z - \lambda_0 + \frac{1}{\pi}\int_0^\infty \frac{|f(\lambda)|^2}{\lambda - z}\, d\lambda \tag{2.28}$$

with

$$|f(\lambda)|^2 = \frac{1}{2i} \lim_{\epsilon\to 0+} [\gamma(\lambda + i\epsilon) - \gamma(\lambda - i\epsilon)] \tag{2.29}$$

On the other hand, if $\gamma(z)$ satisfies Eq. (2.27) with $0 < n < 1$, then $\gamma(z)$ satisfies the once-subtracted dispersion relation. With the subtraction at $z = E_s$,

$$\gamma(z) = z - E_s + \gamma(E_s) + \frac{z - E_s}{\pi}\int_0^\infty \frac{|f(\lambda)|^2}{(\lambda - z)(\lambda - E_s)}\, d\lambda \tag{2.30}$$

It may be noted that the form (2.28) for $\gamma(z)$ is the one obtained in various model-theoretic descriptions of unstable states. All such descriptions picture the unstable state $|M\rangle$ as a normalized stationary state of an unperturbed Hamiltonian H_0 associated with a point spectrum of H_0

embedded in the continuous spectrum. The decay transition is caused solely by a perturbation H_I under suitable assumptions about H_I, for instance, that the transition amplitude of H_I between the states associated with the continuous spectrum of H_0 may be neglected in the evaluation of $a(t)$. The nondecay amplitude is given by Eqs. (2.26) and (2.28) or Eq. (2.30), where

$$|f(\lambda)|^2 = |\langle\lambda|H_I|M\rangle|^2 \tag{2.31}$$

with $|\lambda\rangle$ being the continuum eigenkets of H_0.

Next, define

$$k = z^{1/2}e^{i\pi/4} \tag{2.32}$$

and write

$$\gamma(z) = \tilde{\gamma}(k) = e^{i\pi/2}(k - k_+)(k - k_-)\xi(k) \tag{2.33}$$

with resonance poles as stated earlier at

$$z = E_0 - \frac{1}{2}\,i\Gamma \quad \text{and} \quad z = e^{2\pi i}E_0 + \frac{1}{2}\,i\Gamma \tag{2.34}$$

In the k plane they are at

$$k_\pm \simeq \pm k_0 + \delta \tag{2.35}$$

where $k_0 = E_0^{1/2}e^{i\pi/4}$ and $\delta = \Delta^{1/2}e^{i\pi/4}$ with $\Delta^{1/2} = \Gamma/4E_0^{1/2}$ (see Fig. 2). Substituting Eq. (2.33) into Eq. (2.26) and deforming the contour, we may write

$$\begin{aligned} a(t) &= \frac{i}{2\pi}\int_C \frac{e^{-k^2t}2k\,dk}{(k - k_+)(k - k_-)\xi(k)} \\ &= a_+(t) + a_1(t) + a_2(t) \end{aligned} \tag{2.36}$$

with

$$a_+(t) = \frac{1}{\xi(k_+)}\,\frac{k_+}{k_0}\,e^{-iE_0t}e^{-\Gamma t/2} \tag{2.37}$$

$$\begin{aligned} a_1(t) &= \frac{i}{2\pi}\int_{-\infty}^{\infty} \frac{e^{-k^2t}2k\,dk}{(k - k_+)(k - k_-)\xi(k)} \\ &\simeq \frac{i}{2\pi}\int_{-\infty}^{\infty} \frac{e^{-k^2t}k\,dk}{(k^2 - k_0^2)\xi(k)}\left(1 + \frac{2\,\delta k}{k^2 - k_0^2}\right) \end{aligned} \tag{2.38}$$

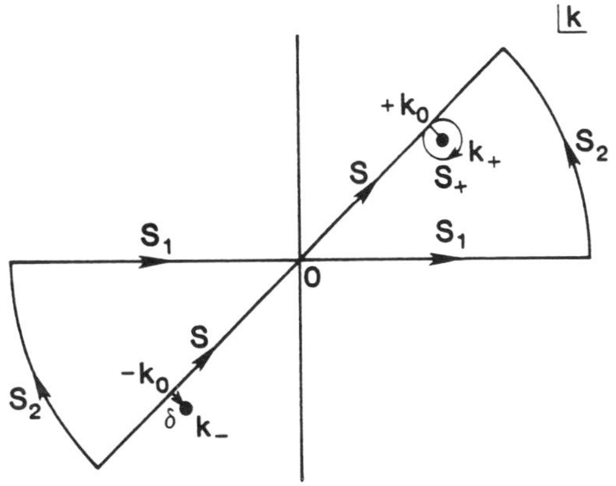

Figure 2. The contours defining the integrals shown in Eq. (2.36).

and $a_2(t)$ is a contribution that can be dropped owing to a suitable cancellation. These three parts are associated with the deformed contour

$$C \to S = S_+ + S_1 + S_2$$

illustrated in Fig. 2. Note that we do not have to include any contribution from k_-.

To proceed further, one has to make specific choices for $\xi(k)$. We may now restate our problem in the following fashion. Given an amplitude of the form (2.36) with a suitable choice for ξ, how does the decay probability behave as a function of time? What are the characteristic times T_1 and T_2 for the system? How sensitive are these conclusions in relation to the specific forms assumed for ξ? In the following section, we attempt to answer these questions.

C. Specific Decay Models and a Resolution of Zeno's Paradox

In reference 41, two specific choices for ξ are considered. For model I,

$$\xi(z) = 1 \tag{2.39}$$

This leads to a dispersion relation of the form of Eq. (2.30). For model II,

$$\xi(z) = \frac{\sqrt{-z} - B^{1/2}}{\sqrt{-z} - (B^{1/2} + 2\Delta^{1/2})} \tag{2.40}$$

This leads to the dispersion relation of the form of Eq. (2.28). We look at several aspects of these solutions.

1. *The Large-t Power Law and Its Geometric Interpretation*

The large-t behavior of the survival amplitude for both models is given by

$$|a(t)| \sim \text{const} \times \frac{1}{t^{3/2}} \tag{2.41}$$

A slower than exponential decay, as mentioned in Section II.B is expected from the general argument of Khalfin, though it could be like $\exp(-t^{1-\epsilon})$. On the other hand, the specific $t^{-3/2}$ law is not only a particular property of these special models, but a reflection of the kinematics of the decay process. We show this as follows. We write $|f(E)|^2 \equiv |\bar{f}(E)|^2\sigma(E)$, where $\sigma(E)$ is the phase-space weight factor. Then from Eqs. (2.26) and (2.29),

$$a(t) = \frac{1}{\pi}\int_0^\infty dE\, \frac{|\bar{f}(E)|^2}{|\gamma(E+i\epsilon)|^2}\,\sigma(E)e^{-iEt} \tag{2.42}$$

$$\simeq \frac{1}{\pi}\int_0^{1/t} dE\, \frac{|\bar{f}(E)|^2}{|\gamma(E+i\epsilon)|^2}\,\sigma(E)e^{-iEt}$$

$$\simeq \frac{1}{\pi}\int_0^{1/t} dE\, \sigma(E)e^{-iEt}$$

$$\simeq \frac{1}{\pi}\int_0^{\infty} dE\, \sigma(E)e^{-iEt} \tag{2.43}$$

for very large times, because of the rapid variation of the phase factor, provided the functions $\bar{f}(E)$ and $\gamma(E+i\epsilon)$ behave gently near zero. The phase-space factor $\sigma(E)$ has a power-law behavior in the neighborhood of the origin. For a nonrelativistic system $E = k^2/2m$,

$$\sigma(E) = 4\pi k^2 \frac{dk}{dE} \sim \sqrt{E} \tag{2.44}$$

whereas for a relativistic system $E = (k^2 + m^2)^{1/2} - m$, as $E \to 0$,

$$\sigma(E) = 4\pi k(E+m) \sim \sqrt{E} \tag{2.45}$$

Hence, in both cases Eq. (2.43) behaves as

$$\int_0^\infty dE\sqrt{E}\,e^{-iEt} = t^{-3/2}\int_0^\infty du\sqrt{u}\,e_{-iu} \tag{2.46}$$

Thus the inverse-cube dependence of the probability of nondecay $Q(t)$ may be related to the structure of the phase-space factor, provided the form factor $\tilde{f}(E)$ is gently varying.

This power-law dependence has a simple geometrical meaning: The "unstable particle" as such is not a new state, but a certain superposition of the decay products. These latter states have a continuum of energy eigenvalues. The precise manner in which the superposition is constituted depends on our definition of the unstable particle, and the development of this wave packet as a function of time depends on the dynamics of the system. Eventually the packet spreads so that the decay products separate sufficiently far to be outside each other's influence. Once this state is reduced, further expansion is purely kinematic, and the amplitude decreases inversely as the square root of the cube of time. Consequently, the overlap amplitude $a(t)$ also behaves in the same manner. The requirement of gentle variation of the form factor is precisely that the corresponding interaction becomes negligible beyond some large but finite distance.

In view of this geometric interpretation, we expect that any unstable system with well-behaved interactions would exhibit such a power law rather than an exponential law.

2. *Two Types of t Dependence Near t = 0*

The short-time behaviors of the probability $Q(t)$ given by two models are very different, which correlates with corresponding differences in the spectral moments. We recall that in the small-t region, the survival amplitude can formally be expanded on the same terms as the spectral moments, that is,

$$a(t) = \int e^{-i\lambda t}\rho(t)\, d\lambda \equiv 1 - i\langle\lambda\rangle t - \frac{\langle\lambda^2\rangle}{2} t^2 + \cdots \qquad (2.47)$$

However, in the small-t region where $E_0 t \ll 1$, both models allow the expansion in power of $\sqrt{E_0 t}$. For model I, we obtain

$$a(t) \to 1 - \text{const} + e^{i\pi/4} t^{1/2} \qquad (2.48)$$

which is compatible with the fact that without the form factor, from inspection of Eq. (2.36), the first spectral moment is infinite. Equation (2.48) leads to the decay rate, as $t \to 0$,

$$\dot{Q}(t) \propto \frac{1}{\sqrt{-t}} \to \infty \qquad (2.49)$$

For model II, we obtain

$$a(t) \to 1 - i \text{ const} + e^{-i\pi/4} t^{3/2} \tag{2.50}$$

which is compatible with $\langle \lambda \rangle$ being finite and $\langle \lambda^2 \rangle$ infinite. Equation (2.50) leads to

$$\dot{Q} \propto -t^{1/2} \to 0 \tag{2.51}$$

Model II is an example of the proposition considered in Section II.A, where the energy expectation value for the resonance state $\langle M | H | M \rangle$ is finite. From general arguments, we already concluded that as $t \to 0$, the decay rate should approach 0. Equation (2.51) is in agreement with this conclusion. If $\langle M | H | M \rangle$ does not exist, such as in model I, as $t \to 0$, the rate of decay is infinite. So the exponential law again does not hold. We see that in no case could the exponential law hold to arbitrarily small values of t. The conclusion that we have arrived at only depends on the basic notions of quantum mechanics; it is therefore quite general.

3. *Repeated Measurements in Short- and Long-Time Limits*

From the discussions above, we are led to two possibilities regarding the leading-term behavior of $Q(t)$ as $t \to 0$:

$$Q(t) \to 1 - \frac{\alpha}{\beta} t^{\beta} \quad \text{and} \quad \dot{Q}(t) \to -\alpha t^{\beta - 1}, \qquad \beta \neq 1 \tag{2.52}$$

Because $0 \leq Q(t) \leq Q(0)$, $\alpha > 0$ and $\beta > 0$ [we are not considering non-polynomial dependences such as $t^{\beta}(\log t)^{\gamma}$], the ranges $\beta < 1$ and $\beta > 1$ behave quite differently. In one case, the rate is becoming larger as $t \to 0$, and in other case, it is vanishing.

Now consider, as in Section II, the n measurements at times t/n, $2t/n, \ldots,$ t. In the limit of $n \to \infty$, the time interval t/n tends to zero. Hence, for arbitrarily small t as $n \to \infty$,

$$Q_n(t) \to \left[1 - \frac{\alpha}{\beta} \left(\frac{t}{n} \right)^{\beta} \right]^n \to \begin{cases} 1 & \beta > 1 \\ 0 & \beta < 1 \end{cases} \tag{2.53}$$

The first case corresponds to Zeno's paradox in quantum theory. In the second case, the limit as $n \to \infty$ is 0. Thus continuous observation would lead to a zero lifetime. The lesson is that quantum mechanics prevents us from determining the lifetime of an unstable particle with "infinite precision." There is a built-in tolerance of $\Delta t = T_1 \sim 1/E_0$, where E_0 is the distance in the energy plane of the resonance pole from the first nearby singularity. The latter is usually the threshold of the closest decay channel.

With the time interval $0 < t < T_1$, the time evolution is not governed by the exponential decay law of the unstable particle. Depending on the dynamics of the system, the apparent lifetime could be substantially lengthened or shortened.

It is also interesting to determine what happens in the long-time limit. We have seen that with reasonable dynamics, the asymptotic form is purely kinematic. What happens with repeated measurement? The wave packet has expanded beyond the range of interaction in accordance with the $t^{-3/2}$ amplitude law: The measurement collapses this packet to the size of the original packet we call the unstable particle, and the time evolution begins again. For $t/n > T_2$, we then have the behavior $(t/n)^{-3n/2}$. We attenuate the unstable-particle amplitude by repeated observation. Naturally there is now no question of continuous observation.

4. *Laboratory Observations on Unstable Particles and Possible Resolution of Zeno's Paradox*

In these discussions we have dealt with the uninterrupted time development of an unstable particle. What can we conclude about laboratory observations on unstable particles? Is it proper to apply these considerations to particles that cause a track in a bubble chamber?

The uninterrupted time evolution was, we saw above, characterized by three regions: (1) $0 < t < T_1$, the small-time region where $Q(t) \simeq 1 - (\alpha/\beta)t^{\beta}$, $\beta > 0$; (2) $T_1 < t < T_2$, the intermediate-time region where an exponential law holds; and (3) $t > T_2$, the large-time region where there is an inverse power-law behavior. Of these, the intermediate-time region alone satisfies the simple composition law

$$Q(t_1)Q(t_2) = Q(t_1 + t_2) \tag{2.54}$$

In this domain, therefore, a classical probability law operates, and the results for the two-step measurement are the same as for the one-step measurement.

If the particle is making a track or otherwise interacting with a surrounding medium and is thus an open system, the considerations we have made do not apply. Instead, we would have to account for the interpretation of the evolution by the interaction and a consequent reduction of the wave packet. The nondecay probability is now defined by the composition law

$$Q(t_1, t_2, \ldots, t_n) = Q(t_1)Q(t_2) \cdots Q(t_n) \tag{2.55}$$

Hence, if $t_1 = t_2 = \cdots = t_n - \tau$, we can write

$$Q(n\tau) = [Q(\tau)]^n \tag{2.56}$$

so that for times that are large compared with τ, the dependence is essentially exponential, independent of the law of quantum evolution $q(t)$. If the interruptions do not occur at equal intervals but are randomly distributed, the behavior is more complex, but this has been considered by Ekstein and Siegert [64] and Fonda et al. [9]. The pure exponential behavior is somewhat altered, but the power-law dependence of the long-time behavior of the uninterrupted time evolution is no longer obtained.

We wish to call particular attention to this result: This long-time behavior of the closed and open systems are essentially different. Classical probabilistic notions do not apply to the closed system. The reason is not difficult to discuss: Classical intuition is related to probabilities which are the directly "observed" quantities. But probabilities do not propagate. Propagation is for the amplitude. Despite this, it is difficult if not impossible to observe the differences between the two. To be able to see the difference we must reach the third domain $t > T_2$, but since T_2 is much larger than the mean lifetime, by the time this domain is reached, the survival probability is already many orders of magnitude smaller than unity. The variable T_2 may be estimated in following manner. For large t, in Eq. (2.38) the integrand peaks at $k^2 = 0$. Within the peak approximation, for the regular terms in the integrand set $k^2 = 0$ and set

$$\xi(k) \to \xi(0) = 1 \tag{2.57}$$

This leads to

$$a_1(t) \sim \frac{i}{2\pi} \cdot \frac{4\delta}{k_0^4} \int_0^\infty e^{-k^2 t} k^2 \, dk = \frac{2i\delta}{\pi k_0^4} \cdot \frac{1}{t^{3/2}} \int_0^\infty e^{-u^2} u^2 \, du$$

so the magnitude

$$|a(t)| \sim \text{const}\left(\frac{\Gamma}{E_0}\right)^{5/2} \frac{1}{\Gamma t^{3/2}}$$

T_2 is the time where the exponential pole term has the same magnitude as this term, solving for

$$a_1(T_2) = \text{const}\left(\frac{\Gamma}{E_0}\right)^{5/2} \frac{1}{\Gamma T_2^{3/2}} = \exp\left(\frac{-\Gamma}{2} T_2\right)$$

For $E_0 \gg \Gamma$, we obtain

$$T_2 \sim \frac{5}{\Gamma} \ln \frac{E_0}{\Gamma} + \frac{3}{\Gamma} \ln\left(5 \ln \frac{E_0}{\Gamma}\right) \tag{2.58}$$

Notice that our estimate here is not sensitive to the details of the form factor assumed as long as $\xi(0) = 1$, which is certainly more general than the models considered. Take the example of the decay of a charged pion, $\pi \rightarrow \mu\nu$

$$\Gamma = (3 \times 10^{-8} \text{ sec})^{-1}$$

This leads to $T_2 \sim 190/\Gamma$. So, by the time the power law is operative, $Q(t) < 10^{-80}$. Clearly this is outside of the realm of detection.

In the small-time domain we have other physical considerations that may prevent the conditions for Zeno's paradox from manifesting. This is ultimately to be traced to the atomic structure of matter and therefore to our inability to monitor the unstable system continuously. For example, in our model II, where Zeno's paradox is operative, in the Appendix of reference 41 one finds $T_1 \sim 10^{-14}\Gamma \sim 10^{21}$ sec for charged-pion decay. On the other hand, we have checkpoints at interatomic distances, a time of the order of $10^{-8}(3 \times 10^{10}) \simeq 3 \times 10^{-19}$ sec. We have no way of monitoring the natural evolution of a system for finer times. Within the present range of technology, according to the estimates, one is unable to observe the deviation of the exponential decay law [65].

This resolution of Zeno's paradox is quite satisfactory as resolutions go in modern physics, but it raises a more disturbing question: Is the continued existence of a quantum world unverifiable? Is the sum total of experience (of the quantum world) a sequence of still frames that we insist on endowing with a continuity? (See also [66].) Is this then the resolution of Zeno's paradox?

One special context which may point to the operation of the Zeno effect in high-energy physics is in hadron–nucleus collisions. The collisions with successive nucleons inside a complex nucleus by an incoming hadron are in times of the order of the Zeno time and we would therefore expect a partial quenching of particle production in such nuclear collisions. This effect has been systematically studied by Valanju [61,62]. The Zeno time in high-energy hadron–nucleus and nucleus–nucleus collisions has also been subsumed as the "formation time" or the "healing time," (for examples, see [67 and 68]).

III. MULTILEVEL UNSTABLE SYSTEMS AND THE KAON SYSTEM

In this section we study multilevel unstable quantum systems. The most common case in particle physics is the $K^0\bar{K}^0$ system, that is, the "strange and antistrange" meson system.

A. Introduction

About four decades ago, Gell-Mann and Pais [69] pointed out that K^0 and $\bar{K}^0$ communicated via the decay channels and, therefore, the decay contained two superpositions K_1 and K_2, which were the orthonormal combinations of K^0 and $\bar{K}^0$, which were, respectively, even and odd under charge conjugation. With the discovery of parity and charge conjugation violation and CP conservation, the terms K_1 and K_2 were redefined to correspond to, respectively, CP-even and -odd superpositions. With the discovery of the small CP violation, qualitatively new phenomena were obtained with nonorthonormal short- and long-lived neutral Kaons K_S and K_L. Lee, Oehme, and Yang [51] formulated the necessary generalization of the Weisskopf–Wigner formalism, which has since been used in the discussion of the empirical data. This phenomenological theory has the same shortcoming as the Weisskopf–Wigner theory and the Breit–Wigner formalism, as discussed earlier. For subsequent theoretical discussions on the LOY model, see, in particular, the papers by Sachs and by Kenny and Sachs [70,71].

Khalfin [52,53] has pointed out some of these theoretical deficiencies and gave estimates of the departure from the Lee–Oehme–Yang (LOY) theory to be expected in the neutral-Kaon system as well as in the $D^0\bar{D}^0$ and $B^0\bar{B}^0$ systems. He asserts that there are possibly measurable "new CP-violation effects." We have reexamined this question in detail, formulated a general solvable model, and studied the exact solution [50]. While bearing out the need to upgrade the LOY formalism to be in accordance with the boundedness from below the total Hamiltonian, our estimates of the corrections are more modest than Khalfin's. We review Khalfin's work to pose the problem and establish notation.

In the LOY formalism, the short- and long-lived particles are liner combinations of K^0 and $\bar{K}^0$:

$$\begin{pmatrix} |K_S\rangle \\ |K_L\rangle \end{pmatrix} = U \begin{pmatrix} |K^0\rangle \\ |\bar{K}^0\rangle \end{pmatrix} \qquad U = \begin{pmatrix} p & -q \\ p' & q' \end{pmatrix} \tag{3.1}$$

with $|p|^2 + |q|^2 = 1$ and $|p'|^2 + |q'|^2 = 1$. The parameters p, q, p', q' are complex; their phases may be altered by redefining the phases of $|K_S\rangle$ and

$|K_L\rangle$. Generally, the states are not orthogonal, but linearly independent:

$$\langle K_L | K_S \rangle = p'^* p - q'^* q \neq 0 \tag{3.2}$$

Let j denote K^0, $\bar{K}^0$ and α denote K_S, K_L. Equation (3.1) can be rewritten as

$$|\alpha\rangle = \sum_j |j\rangle |f\rangle |\alpha\rangle \equiv \sum_j |j\rangle R_{j\alpha} \tag{3.3}$$

where $R = U^T$. For a right eigenstate $|\alpha\rangle$, let the corresponding left eigenstate be $\langle\tilde{\alpha}|$. Then in terms of the oblique bases,

$$|j\rangle = \sum_\alpha |\alpha\rangle\langle\tilde{\alpha}|j\rangle = \sum_\alpha |\alpha\rangle R_{\alpha j}^{-1} \tag{3.4}$$

Let the "time-evolution matrix" of K^0 and $\bar{K}^0$ states be defined by

$$\begin{bmatrix} |K^0(t)\rangle \\ |\bar{K}^0(t)\rangle \end{bmatrix} = A(t) \begin{pmatrix} |K^0\rangle \\ |\bar{K}^0\rangle \end{pmatrix} \tag{3.5}$$

with $A_{jk}(t) = \langle j | e^{-iHt} | k \rangle$, and the corresponding matrix in the K_S and K_L bases by

$$\begin{bmatrix} |K_S(t)\rangle \\ |K_L(t)\rangle \end{bmatrix} = B(t) \begin{pmatrix} |K_S\rangle \\ |K_L\rangle \end{pmatrix} \tag{3.6}$$

with $B_{\alpha\beta}(t) = \langle \tilde{\alpha} | e^{-iHt} | \beta \rangle$. The matrices A and B can be related in the following way:

$$\begin{aligned} A_{kj} &= \sum_{\alpha,\beta} \langle k|\alpha\rangle\langle\tilde{\alpha}|e^{-iHt}|\beta\rangle\langle\tilde{\beta}|j\rangle \\ &= (RBR^{-1})_{kj} \end{aligned} \tag{3.7}$$

As in the LOY theory, for the time being, if we were to assume that K_L and K_S do not regenerate into each other, but otherwise have generic time evolutions:

$$B(t) = \begin{bmatrix} S(t) & 0 \\ 0 & L(t) \end{bmatrix} \tag{3.8}$$

Then

$$A(t) = RB(t)R^{-1}$$
$$= \frac{1}{pq' + p'q} \begin{bmatrix} pq'S + qp'L & -pp'(S-L) \\ -qq'(S-L) & qp'S + pq'L \end{bmatrix} \tag{3.9}$$

At this point, let us invoke CPT invariance, which implies $A_{11} = A_{22}$ or $pq'(S-L) = qp'(S-L)$. Since K_L and K_S are states with distinct masses and lifetimes, $S - L \neq 0$. In turn, $p/q = p'/q'$. The states $|K_S\rangle$ and $|K_L\rangle$ are defined only to within phases of our choice; we may therefore set $p' = p$ and $q' = q$. At this point we relax the normalization condition on p and q and write $|p|^2 + |q|^2 = \zeta^2$. The transformation matrix and its inverse are now given by

$$R = \frac{1}{\zeta}\begin{pmatrix} p & p \\ -q & q \end{pmatrix}, \qquad R^{-1} = \frac{\zeta}{2pq}\begin{pmatrix} q & -p \\ q & p \end{pmatrix} \tag{3.10}$$

We adhere to this convention in the rest of this paper. Equation (3.9) also implies that the ratio of the off-diagonal elements, that is, the ratio of the transition amplitude of $\bar{K}^0$ to K^0 to that of K^0 to $\bar{K}^0$, is given by

$$r(t) = \frac{A_{12}(t)}{A_{21}(t)} = \frac{p^2}{q^2} = \text{const} \tag{3.11}$$

To sum up, the assumptions that (1) K_S and K_L are definite superpositions of K^0 and $\bar{K}^0$ states, (2) there is no regeneration between K_S and K_L, and (3) *CPT* invariance holds, imply the constancy of $r(t)$. Khalfin's theorem [52,53] states that if the ratio $r(t)$ of Eq. (3.11) is constant, the magnitude of this ratio must be unity. His proof follows.

The matrix elements $A_{jk}(t)$ are given by the Fourier transform of the corresponding energy spectra, that is,

$$A_{jk}(t) = \int_0^\infty d\lambda \, e^{-i\lambda t} C_{jk}(\lambda) \tag{3.12}$$

where

$$C_{jk}(\lambda) = \sum_n \langle j | \lambda n \rangle \langle \lambda n | k \rangle \tag{3.13}$$

The summation is over different channels; λ is the energy variable. To be precise, it is the difference between the relevant energy and the threshold value. So $\lambda = 0$ is the lower bound of the spectrum. Using the sesquilinear property of the inner product, that is, $\langle A \mid B \rangle^* = \langle B \mid A \rangle$, Eq. (3.13) implies that

$$C_{jk}(\lambda) = C^*_{kj}(\lambda) \tag{3.14}$$

Now we explore the consequence when Eq. (3.11) holds. Denote $r(t)$ by the appropriate constant r; one may write

$$\begin{aligned} D(t) &= A_{12}(t) - rA_{21}(t) \\ &= \int_0^\infty d\lambda\, e^{-i\lambda t}[C_{12}(\lambda) - rC_{12}(\lambda)] \\ &= 0 \end{aligned} \tag{3.15}$$

Based on the integral representation, with λ being positive, $D(t)$ may now be extended as the function of the complex variable t. Since $e^{-i\lambda t} = e^{-i\lambda \operatorname{Re} t} \cdot e^{\lambda \operatorname{Im} t}$, the function $D(t)$ can now be defined in the entire lower half plane. By the Paley–Wiener theorem [35], $D(t)$ is also defined at the boundary of the function in the lower half plane. So

$$D(t) = 0 \qquad \text{for } -\infty < t < \infty \tag{3.16}$$

The inverse Fourier transform of $D(t)$ implies

$$C_{12}(\lambda) - rC_{21}(\lambda) = C_{12}(\lambda) - rC^*_{12}(\lambda) = 0 \qquad \text{or } |r| = 1 \tag{3.17}$$

This conclusion contradicts the expectation of the LOY theory. In particular, when there is CP violation, it is expected that

$$|r| = \left|\frac{p}{q}\right|^2 = \text{const} \neq 1 \tag{3.18}$$

We have investigated the situation in the framework of the Friedrichs–Lee model in the lowest section with the particle V_1 and its antiparticle V_2. They are coupled to an arbitrary number of continuum $N\theta$ channels. We express the time-evolution matrix in terms of pole contributions plus a background contribution. We show that because of the form-factor effect, both the correction to the pole contribution and the background contribution give rise to a tiny regeneration between K_L and K_S. This invalidates one of the original assumptions needed to deduce the constancy of the ratio $r(t)$. Therefore, in the generic Fredrichs–Lee model, the assumption that K_L and K_S are fixed superpositions of K^0 and $\bar{K}^0$ states is not valid. In the remainder of this section, we set up the dynamical system which involves multilevels and multichannels and investigate the generalization of Khalfin's theorem. We also look at the solution to the neutral Kaon problem beyond the Wigner–Weisskopf approximation. We show that in our solution the ratio $[A_{21}(t)/A_{12}(t)]$ does depend on time, which invalidates one of the assumptions of the Khalfin theorem, and predicts insignificant but nonzero departure from LOY model values in the region where the resonance pole contribution is dominating.

B. Multilevel Systems and Time-Evolution Matrix

1. *Eigenvalue Problem*

In the generalized Friedrichs–Lee model, the Hamiltonian is given by

$$\begin{aligned} H = \sum_{j,k} m_{jk} V_j^\dagger V_k + \sum_{n=1}^{N} \mu_n N_n^\dagger N_n + \int_0^\infty d\omega\, \omega \phi^*(\omega)\phi(\omega) \\ + \int_0^\infty d\omega \sum_{j,n} g_{jn}(\omega) V_j N_n^\dagger \phi^*(\omega) \\ + \int_0^\infty d\omega \sum_{j,n} g_{jn}^*(\omega) V_j^\dagger N_n \phi(\omega) \end{aligned} \tag{3.19}$$

Here the bare particles are V_1, V_2, $N_n (1 \le n \le N)$, and θ particles. The following number operators commute with the Hamiltonian:

$$\begin{aligned} Q_1 &= \sum_j V_j^\dagger V_j + \sum_n N_n^\dagger N_n \\ Q_2 &= \sum_n N_n^\dagger N_n + \int d\omega\, \phi^*(\omega)\phi(\omega) \end{aligned} \tag{3.20}$$

Denote the corresponding eigenvalues by q_1 and q_2. The Hilbert space of the Hamiltonian is divided into sectors, each with a different assignment of q_1 and q_2 values. We will only consider the eigenstates of the lowest nontrivial sector, where $q_1 = 1$ and $q_2 = 0$. Here the bare states are labeled by $|V_1\rangle$, $|V_2\rangle$, and $|n, \omega\rangle$, with $n = 1, 2, \ldots, N$. Since there are N independent continuum states, for each eigenvalue λ, there are N independent eigenstates which can be written as

$$|\lambda, n\rangle = \sum_j |V_j\rangle[a_\lambda]_{jn} + \int_0^\infty d\omega \sum_m |m, \omega\rangle[b_\lambda(\omega)]_{mn} \tag{3.21}$$

where

$$[a_\lambda]_{jn} = \langle V_j | \lambda, n\rangle, \qquad [b_\lambda(\omega)]_{mn} = \langle m, \omega | \lambda, n\rangle \tag{3.22}$$

In Eq. (3.21), the integration variable of the $|m, \omega\rangle$ state, ω, begins from 0. Therefore, it now stands for the difference between the energy of the state and the threshold energy.

Using the Einstein summation convention, the corresponding eigenvalue equation is given by

$$\begin{bmatrix} m_{ij} & g_{il}(\omega') \\ g^\dagger_{mj}(\omega) & \omega\delta(\omega - \omega')\delta_{ml} \end{bmatrix} \begin{Bmatrix} [a_\lambda]_{jn} \\ [b_\lambda(\omega')]_{ln} \end{Bmatrix} = \lambda \begin{Bmatrix} [a_\lambda]_{in} \\ [b_\lambda(\omega)]_{mn} \end{Bmatrix} \tag{3.23}$$

For brevity, hereafter we suppress the matrix indices. Equation (3.23) leads to

$$(\lambda I - m)a_\lambda = \langle g(\omega')b_\lambda(\omega')\rangle \tag{3.24}$$

$$(\lambda - \omega)b_\lambda(\omega) = g^\dagger(\omega)a_\lambda \tag{3.25}$$

$$\langle \cdots \rangle \equiv \int_0^\infty d\omega \cdots$$

We choose the boundary condition such that, in the uncoupled limit, b_λ is given by

$$[b_\lambda(\omega)]_{mn} = \delta(\lambda - \omega)\delta_{mn} \tag{3.26}$$

Such a solution is given by

$$b_\lambda(\omega) = \delta(\lambda - \omega)I + \frac{g^\dagger(\omega)a_\lambda}{\lambda - \omega + i\epsilon} \tag{3.27}$$

Substituting Eq. (3.27) into Eq. (3.24) leads to

$$(\lambda I - m)a_\lambda = g(\lambda) + \left\langle \frac{g(\omega')g^\dagger(\omega')}{\lambda - \omega' + i\epsilon} \right\rangle a_\lambda \tag{3.28}$$

or

$$a_\lambda = K^{-1} g \tag{3.29}$$

where

$$\begin{aligned} K &= \lambda I - m - G(\lambda) \\ &= \begin{bmatrix} \lambda - m_{11} - G_{11} & -m_{12} - G_{12} \\ m_{21} - G_{21} & \lambda - m_{22} - G_{22} \end{bmatrix} \end{aligned} \tag{3.30}$$

with

$$\begin{aligned} G(\lambda + i\epsilon) &= \left\langle \frac{g(\omega) g^\dagger(\omega)}{\lambda - \omega + i\epsilon} \right\rangle \\ &= \int_0^\infty d\omega \, \frac{g(\omega) g^\dagger(\omega)}{\lambda - \omega + i\epsilon} \end{aligned} \tag{3.31}$$

2. *Time-Evolution Matrix*

It follows from Eq. (3.31) that, for λ real, the 2×2 matrix G is

$$\begin{aligned} [G(\lambda + i\epsilon)]^\dagger &= \int_0^\infty d\omega \, \frac{g(\omega) g^\dagger(\omega)}{\lambda - \omega - i\epsilon} \\ &= G(\lambda - i\epsilon) \end{aligned} \tag{3.32}$$

This in turn implies the identity that, for real λ,

$$\begin{aligned} G(\lambda + i\epsilon) - G^\dagger(\lambda + i\epsilon) &= -2ig(\lambda) g^\dagger(\lambda) \\ &= K^\dagger(\lambda + i\epsilon) - K(\lambda + i\epsilon) \end{aligned} \tag{3.33}$$

The time-evolution matrix is easily evaluated:

$$\begin{aligned} A_{kj}(t) &= \langle k | e^{-iHt} | j \rangle \\ &= \int_0^\infty d\lambda \, e^{-i\lambda t} \sum_n \langle k | \lambda n \rangle \langle \lambda n | j \rangle \\ &= \int_0^\infty d\lambda \, e^{-i\lambda t} [a(\lambda) a^\dagger(\lambda)]_{kj} \end{aligned} \tag{3.34}$$

From Eqs. (3.29) and (3.33),

$$aa^\dagger = K^{-1}gg^\dagger(K^{-1})^\dagger + K^{-1}\left[\frac{K^\dagger - K}{-2\pi i}\right](K^{-1})^\dagger$$

$$= \frac{i}{2\pi}[K^{-1} - (K^{-1})^\dagger] \tag{3.35}$$

Substituting Eq. (3.35) into Eq. (3.34), we get

$$A_{kj}(t)I = \frac{i}{2\pi}\int_0^\infty d\lambda\, e^{-i\lambda t}\{K^{-1}(\lambda + i\epsilon) - [K^{-1}(\lambda + i\epsilon]^\dagger\}_{kj} \tag{3.36}$$

But

$$[K^{-1}(\lambda + i\epsilon)]^\dagger = \{[\lambda - m - G(\lambda + i\epsilon)]^\dagger\}^{-1}$$
$$= [K(\lambda - i\epsilon)]^{-1} \tag{3.37}$$

Based on Eq. (3.37), Eq. (3.36) can be written in a contour integral representation (see Fig. 3):

$$A_{kj}(t)I = \frac{i}{2\pi}\int_C d\lambda\, e^{-i\lambda t}[K^{-1}(\lambda)]_{kj}$$
$$= \frac{i}{2\pi}\int_C d\lambda\, e^{-i\lambda t}\frac{N_{kj}(\lambda)}{\Delta(\lambda)} \tag{3.38}$$

where

$$\Delta = \det K \tag{3.39}$$

and

$$N(\lambda) = \text{Cof } K = \text{Cof}(\lambda\delta_{kj} - m_{kj} - G_{kj}) \tag{3.40}$$

We recall that the cofactor of the element of A_{kj} of a square matrix A is equal to $(-)^{k+j}$ times the determinant of the matrix which A becomes when kth row and jth column are deleted.

Because $G(\lambda)$ is defined through the dispersion integral (3.31), the λ dependence of G, and in turn, the integrand of Eq. (3.38) may be extended to the entire cut plane of λ.

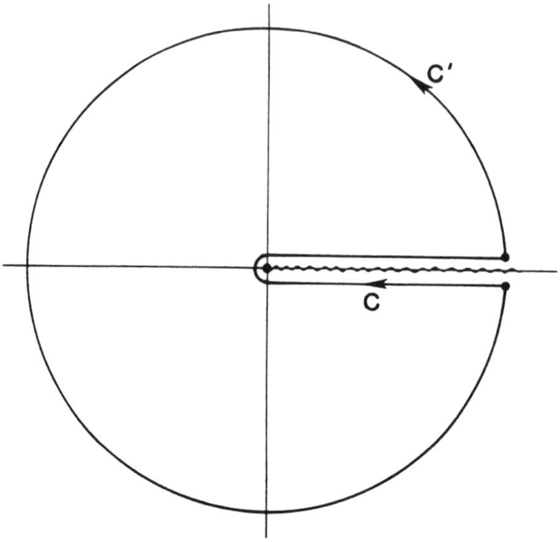

Figure 3. The contours C and C' in the complex λ plane.

3. *Completeness Relation*

At $t = 0$, from Eqs. (3.34) and (3.38),

$$\begin{aligned} A_{kj}(0) &= \int_0^\infty d\lambda \sum_n \langle k \mid \lambda n\rangle\langle \lambda n \mid j\rangle \\ &= \frac{i}{2\pi} \int_C d\lambda \frac{N_{kj}(\lambda)}{\Delta(\lambda)} \end{aligned} \tag{3.41}$$

From Eqs. (3.36) and (3.37) the asymptotic behaviors are

$$\begin{aligned} N_{kj}(\lambda) &\to \lambda^{n-1} \qquad \text{for } k = j \\ N_{kj}(\lambda) &\to \lambda^{n-2} \qquad \text{for } k \neq j \\ \Delta(\lambda) &\to \lambda^n \end{aligned} \tag{3.42}$$

Deform the contour as depicted in Fig. 3. Since the integrand is analytic, using Eq. (3.42),

$$A_{kj}(0) = -\frac{i}{2\pi} \int_{C'} d\lambda \frac{N_{kj}(\lambda)}{\Delta(\lambda)} = \delta_{kj} \tag{3.43}$$

or

$$\int_0^\infty d\lambda \sum_n \langle k \mid \lambda n\rangle\langle \lambda n \mid j\rangle = \delta_{kj} \tag{3.44}$$

which is the completeness relation.

C. Applications to Neutral Kaon System

1. *Formalism*

So far our treatment has been general. Now we want to specialize to the neutral-K system. We identify K^0 and $\bar{K}^0$ as V_1 and V_2. The K_S and K_L are the unstable particles which correspond to the second-sheet zeros of the determinant of the matrix K:

$$K = \begin{pmatrix} \lambda - m_{11} - G_{11} & m_{12} + G_{12} \\ m_{21} + G_{21} & \lambda - m_{11} - G_{11} \end{pmatrix} \tag{3.45}$$

where we have applied the CPT theorem and set $m_{22} + G_{22} + G_{11}$. The discontinuity of the G-matrix is given by

$$\begin{aligned} \frac{G_{kj}(\lambda + i\epsilon) - G^*_{kj}(\lambda + i\epsilon)}{2i} &= -[g(\lambda)g^\dagger(\lambda)]_{kj} \\ &= -\pi \sum_n \langle k \mid H \mid \lambda n\rangle\langle \lambda n \mid H \mid j\rangle \\ &\qquad \text{with } k, j = K^0, \bar{K}^0 \end{aligned} \tag{3.46}$$

In the Weisskopf–Wigner approximation, $G_{kj}(\lambda)$ is replaced by its imaginary part, evaluated at the resonance mass

$$G_{kj}(\lambda) = -i\,\frac{\Gamma_{kj}}{2} \tag{3.47}$$

This is the approximation of the LOY model, where the eigenvalue problem of the type

$$K\psi = \lambda\psi \quad \text{or} \quad \begin{pmatrix} AB \\ CA \end{pmatrix}\begin{pmatrix} r \\ s \end{pmatrix} = \lambda\begin{pmatrix} r \\ s \end{pmatrix} \tag{3.48}$$

is considered.

We digress a little to examine the solution of this eigenvalue problem in order to establish the relationship between r and s and the mass and width parameters. For the neutral K-system,

$$A = m_{11} - i\,\frac{\Gamma_{11}}{2}, \qquad B = m_{12} - i\,\frac{\Gamma_{12}}{2}, \qquad C = m_{21} - i\,\frac{\Gamma_{21}}{2}$$

The complex eigenvalues are

$$\lambda_L = m_L - i\,\frac{\Gamma_L}{2}, \qquad \lambda_S = m_S - i\,\frac{\Gamma_S}{2}$$

Substituting these quantities back in the eigenvalue equations, we obtain

$$\mathrm{Tr}\,K = 2A = \lambda_L + \lambda_S \quad \text{or} \quad A = \frac{1}{2}(\lambda_L + \lambda_S) = m_{11} - i\,\frac{(\Gamma_L + \Gamma_S)}{4} \tag{3.49}$$

In terms of the eigenvalues and the components of the corresponding eigenvectors,

$$B = \frac{r}{2s}(\lambda_L - \lambda_S)$$

$$C = \frac{s}{2r}(\lambda_L - \lambda_S)$$

or

$$\left(\frac{r}{s}\right)^2 = \frac{B}{C} \tag{3.50}$$

Making the correspondence between the definition of the K_L and K_S states defined earlier, for the K_L state we get

$$\frac{r}{s} = \sqrt{\frac{B}{C}} = \frac{p}{q} \quad \text{or} \quad \psi_L = N\begin{pmatrix} p \\ q \end{pmatrix} \tag{3.51}$$

and for the K_S state:

$$\frac{r}{s} = \sqrt{\frac{B}{C}} = \frac{p}{q} \quad \text{or} \quad \psi_S = N\begin{pmatrix} p \\ -q \end{pmatrix} \tag{3.52}$$

2. *The Ratio* $[A_{12}(t)/A_{21}(t)]$

We proceed to evaluate the ratio $A_{12}(t)/A_{21}(t)$ within the Weisskopf–Wigner approximation. Again we write

$$\Delta = (\lambda - \lambda_S)(\lambda - \lambda_L) \tag{3.53}$$

except that now λ_S and λ_L do depend on Λ. We are interested in the effect due to λ-dependence of G. For our purpose, we find it to be adequate to work with a common form factor and write

$$G_{kj}(\lambda) = -i\,\frac{\Gamma_{kj}}{2}\,F(\lambda) \tag{3.54}$$

where $\Gamma_{kj}/2$ is independent of λ. Then

$$\begin{aligned}
\lambda_S &= m_{11} - i\,\frac{\Gamma_{11}}{2}\,F(\lambda) + d(\lambda)\\
\lambda_L &= m_{11} - i\,\frac{\Gamma_{11}}{2}\,F(\lambda) - d(\lambda)\\
d(\lambda) &= \left\{\left[m_{12} - i\,\frac{\Gamma_{12}}{2}\,F(\lambda)\right]\left[m_{12}^* - i\,\frac{\Gamma_{12}^*}{2}\,F(\lambda)\right]\right\}^{1/2}
\end{aligned} \tag{3.55}$$

The transition amplitude

$$A_{12}(t) = \frac{i}{2\pi}\int_C d\lambda\, e^{-i\lambda t}\,\frac{m_{12} - i(\Gamma_{12}/2)F(\lambda)}{2d(\lambda)}\left[\frac{1}{\lambda - \lambda_S} - \frac{1}{\lambda - \lambda_L}\right] \tag{3.56}$$

The contour C here is illustrated in Fig. 3. It is to be deformed according to Fig. 4, such that the integral can be written as the sum of pole contribution and the background contribution. We further assume that $F(\lambda)$ varies in hadronic scale ($\sim$1 GeV), so that it is a smooth function in the neighborhood of $\lambda = m_S, m_L$. Expanding $F(\lambda)$ about $\lambda = m_{11}$, at $\lambda = m_S$ and $\lambda = m_L$ the corresponding form factors are respectively given by

$$F_S = 1 + F'd, \qquad F_L = 1 - F'd \qquad d = d(m_{11}) \tag{3.57}$$

Deforming the contour in the manner indicated in Fig. 4, the pole term gives

$$A_{12}(t)\Big|_{\text{pole}} \simeq \frac{p}{2q}\,[(1 - \delta_{12})e^{-i\lambda_S t} - (1 + \delta_{12})e^{-i\lambda_L t}]$$

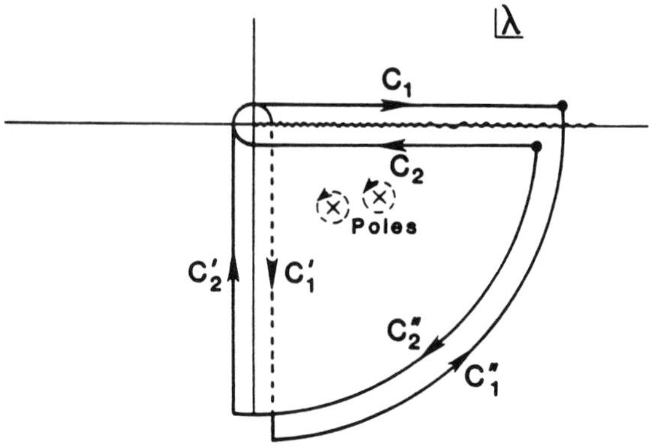

Figure 4. Illustration of the deformation of contours C_1 and C_2 into the pole contributions plus the background contribution.

with

$$\delta_{12} = i\,\frac{\Gamma_{12} F'}{2}\,\frac{p}{q} \sim O\left(\frac{\Gamma_{12}}{2m_{11}}\right) \tag{3.58}$$

$$A_{21}(t)\Big|_{\text{pole}} \simeq \frac{q}{2p}\,[(1-\delta_{21})e^{-i\lambda_S t} - (1+\delta_{21})e^{i\lambda_L t}]$$

with

$$\delta_{21} = i\,\frac{\Gamma^*_{12} F'}{2}\,\frac{q}{p} \sim O\left(\frac{\Gamma_{12}}{2m_{11}}\right) \tag{3.59}$$

So

$$\Gamma(t)\Big|_{\text{pole}} \simeq \frac{A_{12}(t)}{A_{21}(t)}\Big|_{\text{pole}} = \frac{p^2}{q^2}\,(1+\cdots) \tag{3.60}$$

where the "$\cdots$" term carries a time dependence wherever

$$\delta_{12} \neq \delta_{21} \quad \text{or} \quad \frac{\Gamma_{12}}{\Gamma^*_{12}} \cdot \frac{q^2}{p^2} \neq 1 \tag{3.61}$$

The amount of departure is bounded by the order of magnitude of δ_{12} which is $O(\Gamma_{12}/m_{11})$.

For neutral Kaons,

$$\frac{\Gamma_{12}}{2} \sim \frac{\Gamma_S}{2} \sim 5 \times 10^{10}\ \text{sec}^{-1}$$
$$m_{11} = m_K - 2m_\pi \sim 200\ \text{MeV} \sim 3 \times 10^{23}\ \text{sec}^{-1} \tag{3.62}$$

So

$$\delta \sim 0.2 \times 10^{-13}$$

The background term also contributes to the t dependence of the ratio $\Gamma(t)$. From general arguments it can be shown that

$$\left|\frac{A_{12}(t)}{A_{21}(t)}\right|_{bk} = 1 \tag{3.63}$$

In the very small t region and in the very large t regions, where the background-term contribution is significant and when $p/q \neq 1$, a further departure of the value of p^2/q^2 from the Weisskopf–Wigner approximation may be expected.

3. *Regeneration Effect*

Next, we demonstrate that there is a regeneration effect in the present solution, which invalidates one of the assumptions stated in Section III.A, leading to the conclusion of the constancy of the magnitude of the ratio r. The presence of the regeneration effect is inferred by the presence of the nondiagonal element in the time-evolution matrix B of Eq. (3.6). Based on Eqs. (3.9) and (3.38),

$$\begin{aligned} B(t) &= R^{-1}A(t)R \\ &= \frac{i}{2\pi}\int d\,\lambda e^{-i\lambda t}\,\frac{R^{-1}NR}{\Delta} \end{aligned} \tag{3.64}$$

with

$$R^{-1}NR = \frac{1}{2pq}\begin{bmatrix} 2pqN_{11} - (N_{12}q^2 + N_{21}p^2) & N_{12}q^2 - N_{21}p^2 \\ -N_{12}q^2 + N_{21}p^2 & 2pqN_{11} + (N_{12}q^2 N_{21}p^2) \end{bmatrix} \tag{3.65}$$

We focus our attention on the element B_{12}, which leads to the regeneration of K_S from K_L:

$$\begin{aligned} B_{12} &= N_{12}q^2 - N_{21}p^2 \\ &= \left(m_{12} - i\frac{\Gamma_{12}}{2}F\right)q^2 - (m_{12}^* F)p^2 \\ &= -i(F-1)\left(\frac{\Gamma_{12}}{2}q^2 - \frac{\Gamma_{12}^*}{2}p^2\right) \end{aligned} \tag{3.66}$$

In the last step, we used the relations $p^2 = m_{12} - i(\Gamma_{12}/2)$ and $q^2 = m_{12}^* - i(\Gamma_{12}^*/2)$. So

$$B_{12}(t) = v\frac{i}{2\pi}\int_C d\lambda e^{i\lambda t}\frac{F(\lambda)-1}{\Delta} \tag{3.67}$$

where $v = 2\,\mathrm{Im}[(\Gamma_{12}/2)m_{12}^*]$. So the regeneration correction occurs only when $v \neq 0$, that is, when there is CP violation. Deforming the contour, we get

$$B_{12}(t) = B_{12}(t)\Big|_{\text{poles}} + B_{12}(t)\Big|_{bk} \tag{3.68}$$

with

$$\begin{aligned} B_{12}(t)\Big|_{\text{poles}} &= \frac{v}{2d}\left[(F_S - 1)e^{-i\lambda_S t} - (F_L - 1)e^{-i\lambda_L t}\right] \\ &= \frac{vF'}{2}\left(e^{-i\lambda_S t} + e^{-i\lambda_L t}\right) \end{aligned} \tag{3.69}$$

and

$$B_{12}(t)\Big|_{bk} = -vJ(t) \tag{3.70}$$

Here $J(t)$ represents the background integral. It is complicated to evaluate $J(t)$ for general values of t. However, for both small- and large-t regions for the simple form of form factors, the background integral is manageable. In the small-t region, it can be shown that [50]

$$B_{12}(t) \propto \mathrm{Im}(\Gamma_{12}m_{12}^*)t \tag{3.71}$$

Here, the first power in t is the expected time dependence for the transition amplitude. Furthermore, there is always the Zeno region, in the sense that frequent observation would inhibit the transition from the "1" state to "2" state and also vice versa. For large t,

$$B_{12}(t)\bigg|_{bk} \propto \operatorname{Im}(\Gamma_{12} m_{12}^*) \frac{1}{t^{3/2}} \tag{3.72}$$

Once again the inverse power law associated with a geometric expansion picture is obtained.

In summary, we see from this analysis that a quantum system with two metastable states which communicate with each other exhibits interesting phenomena in its time evolution. For its short-time behavior, the quantum Zeno effect obtains; for very long-time behavior, there is a regeneration effect even in a vacuum, unless the long- and short-lived superpositions are strictly orthogonal. In the Kaon complex, the short-lived particle K_S has passed from the exponential regime to the inverse power regime before appreciable decay of the K_L or regeneration of the K_S takes place. The CPT invariance making the diagonal elements of the decay matrix in the K^0, $\bar{K}^0$ basis equal is crucial to the nature of the time evolution. In the study of communicating metastable states in atomic physics, such an additional constraint of CPT is not present: consequently, the decay exhibits richer features. We present the general study elsewhere. Suffice it to observe here that the asymptotic and Zeno-region time dependence are very much the same as with a single-metastable state decay: This is not surprising, because the generic arguments apply without restriction to the number of channels involved.

IV. GENERALIZED QUANTUM SYSTEM: ONE-LEVEL SYSTEM

Thus far our attention has mainly been focused on the features in the time development of unstable quantum systems, which show the departure from the pure exponential decay of the Breit–Wigner approximation. This deviation arises when one takes the continuum spectrum into account. Here resonance is a pole in the survival amplitude or, more generally, in a transition amplitude, on the second sheet. This is to be in contrast with the Breit–Wigner approximation, where the resonance poles are on the physical sheet. The "physical sheet" and the "second sheet" designations here have important distinctions. From the requirement of causality, it can be shown that transition amplitudes are analytic on the physical sheet. The presence of complex poles on the physical sheet implies the violation of causality. Since we want to work with a causal theory, resonance poles must be identified with the second-sheet poles and the deviation of the exponential behavior in the time evolution is expected. We then proceed to consider the generalized quantum system through the use of analytic continuation. Within this framework, the resonance pole may be identified as a generalized quantum state.

A. Introduction

As alluded to in Section I, orthodox quantum mechanics is formulated in a vector space over complex numbers with a sesquilinear inner product [72,73]. In most applications the vector space is a separable complete space and often taken to be a Hilbert space [5,74,75]. The vector space, except in cases of "spin" systems with a finite basis, is made up of L^2 functions of one or more variables or a vector of such functions. The dynamical variables are taken to be linear operators of finite norm. Among them the self-adjoint operators form a preferred class and the observables are usually identified with them.

But it is convenient to deal with unbounded operators like the canonical coordinate, momentum, or the Hamiltonian. Such operators do not have an action on the whole vector space because they could make the length of the image vector unbounded and thus not in the space; so we have to restrict the "domain" of the unbounded operator.

Even a further departure is often needed when we deal with an operator with a continuous spectrum: it is useful to introduce ideal vectors [72] with distribution-valued scalar products.

When the vector space is realized by functions of a certain class, it may be possible to consider analytic continuation of such function spaces with an associated bilinear form but with two analytic vector spaces being defined: the basic vector space and the space of linear functionals on this space. Of course, this generalization could have been considered without analytic continuation. If the base space topology becomes stronger, the dual space topology becomes weaker and vice versa. In a Hilbert space, the two topologies are the same (completeness of all Cauchy sequences!) with a reflexive antilinear transformation connecting the base space (ket) vectors and the dual space (bra) vectors [72]. In the context of density operators this has been emphasized by Segal [76]. In the context of vectors in a Hilbert space this formalism due to Gelfand [77], and amplified by Antoine [78] and Bohm [7], is called the Rigged Hilbert space. While such a generalization is by choice for Hilbert spaces, both in the Segal context and in the course of analytic continuation, the dichotomy between the base space and the dual enters automatically.

Dirac introduced the notion of analytic continuation of vector spaces in the context of the "extensor" representations of the Lorentz group in the 1940s, followed by Kuriyan, Mukunda, and Sudarshan [79], who obtained the master analytic representations of noncompact groups. Nakanishi [24] had employed the notion of an analytically continued set of "wave functions" in the context of a treatment of unstable particles in quantum mechanics. The first systematic generalization of the quantum vector space by analytic continuation was formulated by Sudarshan, Chiu, and Gorini [42]. Rigorous treatment of the problem with careful attention to functional analytic details have since been given [80].

The problem of decaying particles, scattering resonances, and generic metastable states in quantum physics continues to be of current interest. The long-time behavior departing from exponential decay exhibited by Khalfin [33], the short-time Zeno behavior [34,36], and the detailed transition behavior of quantum metastable excitations constitute a complex of rich phenomenology [41]. It has further been enriched by the multitude of features in the neutral Kaon decay and that of other such particles [50] and in the cascade decay phenomena. Recently, Yamaguchi [81] raised important questions about the behavior of decay amplitudes and the possibility that short- and long-lived Kaons are orthogonal whether CP is conserved or not. From a somewhat different point of view Tasaki, Petrosky, and Prigogine [37] have considered this question with special attention to the breaking of time symmetry in decay.

Apart from these questions, there has been some lack of precision concerning analytic continuation and scattering amplitude singularities: not enough attention has been paid to redundant zeros and discrete states buried in the continuum.

Complex variables, analytic functions, and topology are only aids to the mathematical discussion of physical phenomena; an essential part of the task is the proper identification and interpretation of the mathematical results. Not all quantum theories involving analytic continuations are alike, nor is their scope the same; several treatments are lacking in one aspect or the other. For example, many authors act as if poles in the analytic continuation are the only relevant singularities [8]. On the contrary, we show that the treatment of scattering amplitudes involving unstable particles requires complex branch points. We have therefore made a specific attempt to spell out in some detail the theory that we introduce. The use of solvable models enables us to illustrate many relevant features of the theory.

The most important point that we emphasize is that only suitable dense sets in the analytically continued spaces have a corresponding dense set of states in the space with which we begin the analytic continuation. Individual states in one space may or may not have analytical partners in the generalized spaces. The analytic continuation is therefore basis dependent and not every vector in the continuation may have direct physical interpretation. *The poles are examples of such objects.*

The outline of our presentation below is as follows. In Sections IV.B and C, the generalized vector space of quantum states is used to study the correspondence between the physical state space $\mathscr{H}$ and its continuation $\mathscr{G}$. We begin with the observation that the scalar product between an arbitrary vector *in the dense subset of analytic vectors* in $\mathscr{H}$ and its dual vector has an integral representation. While keeping the scalar product fixed, the analytic vectors may be "analytically continued" through the deformation of the integration contour. A typical analytically continued integral representation of present interest integrates along a deformed contour in the fourth quadrant of the complex energy plane and encircles those "exposed" singularities on the second sheet, if any (i.e., those between the real axis and the deformed contour). *The deformed contour, together with the exposed singularities, constitutes the generalized spectrum of the operator in the continued theory.*

In Sections IV.D and E, simple two-body models, the Friedrichs–Lee and the Yamaguchi, in the lowest sector are studied with special attention to the unfolding of the generalized spectrum. Here the "exposed" singularities, if present at all, are simple poles. We defer more complex situations involving multiresonance levels and an arbitrary number of two-body decay channels to Section V and a case with three-body decay channels to Section VI.

In Section IV.F, we observe that the predictions based on $\mathscr{H}$ and $\mathscr{G}$ are expected to be the same. Since a pure exponential time dependence is not possible for states in $\mathscr{H}$, it should not be possible for states in $\mathscr{G}$. On the other hand, the Breit–Wigner resonance does correspond to a pure exponential decay and it realizes the semigroup of time evolution. However, in such a case, one needs to give up the positivity of energy and define states with all possible values of energy from $-\infty$ to $+\infty$.

In Section IV.G, we recall the two possible disparities between poles in the S-matrix and the discrete states in the Hamiltonian. In particular, there can be a pole in the S-matrix without a corresponding state in the complete states of the Hamiltonian. Conversely, there may be a discrete state of the Hamiltonian, which does not have the corresponding pole in the S-matrix. We show that these disparities continue to be admissible in the generalized vector space. In Section IV.H we consider the analytic continuation of the probability function and the operation of time-reversal invariance.

Our concluding remarks are given in Section IV.I. Two distinct views on what constitutes an unstable particle are contrasted. One view is to identify an unstable particle as a physical state of the system which ceases to exist as a discrete eigenstate of the total Hamiltonian. The survival amplitude of the unstable particle *cannot ever be strictly exponential in time.* There is no autonomy in its time development. *It ages.* Therefore, the unstable particle does not furnish a representation of the time-translation group. The other view is to identify the unstable particle as a discrete state in the generalized space $\mathscr{G}$. It has a pure exponential time dependence. The time evolutions form a semigroup. Although the latter appears to be elegant, it is deduced at the expense of giving up the very starting premise of the lower boundedness of the energy spectrum.

B. Vector Spaces and Their Analytic Continuation

1. Vector Spaces $\mathscr{H}$ and $\mathscr{H}'$ in Conventional Formalism

Consider an infinite dimensional vector space $\mathscr{H}$ over the field of complex numbers [72] with vectors ψ, ϕ, ... Then, if a, b are complex numbers, $a\psi + b\phi$ is also a vector, and so are finite linear combinations. If $\{|e^{(r)}\rangle\}$ is a countable basis, then any vector ψ can be approximated to any desired limit by linear combinations of the form $\sum a_n^{(r)} |e^{(r)}\rangle = |\psi_n\rangle$, where the sequence $\{|\psi_n\rangle\}$ converges to ψ. A linear operator is a linear map from vectors in $\mathscr{H}$ to vectors in $\mathscr{H}$. The linear functional mapping each vector in $\mathscr{H}$ to a complex number constitute the dual vector space $\mathscr{H}'$ to $\mathscr{H}$. A basis $\{f^{(s)}\}$ in the dual vector space $\mathscr{H}'$ may be obtained by considering the linear functional

$$|e^{(r)}\rangle \xrightarrow{f^{(s)}} \delta_{rs} \quad \text{and the correspondence:} \quad |e^{(r)}\rangle \leftrightarrow \langle f^{(r)}| \tag{4.1}$$

Thus we can put the basis vectors into one-to-one correspondence, but the correspondence is antilinear:

$$a|e^{(r)}\rangle + b|e^{(s)}\rangle \leftrightarrow a^{\dagger}\langle f^{(r)}| + b^{\dagger}\langle f^{(s)}| \tag{4.2}$$

The linear functional can be thought of as the scalar product of vectors in $\mathscr{H}$, $\mathscr{H}'$ bilinear in them:

$$\phi \xrightarrow{\psi} (\psi, \phi) \equiv \langle \psi | \phi \rangle; \qquad \psi \in \mathscr{H}', \phi \in \mathscr{H} \tag{4.3}$$

or as a sesquilinear form in $\mathscr{H}$ by making use of the antilinear correspondence (4.2) between bra and ket vectors.

Given the basis vectors and the notion of scalar products, we can introduce the completeness identity. If we have a bra $\langle \psi |$ and a ket $|\phi\rangle$, we can define a linear operator by the vector valued linear functional:

$$|\chi\rangle \to \langle\psi|\chi\rangle|\phi\rangle \tag{4.4}$$

and identify it with the linear operator

$$A = |\phi\rangle\langle\psi| \tag{4.5}$$

In particular we can introduce the linear operator

$$\sum_r |e^{(r)}\rangle\langle e^{(r)}|$$

which acting on any vector $|\phi\rangle$ reproduces itself:

$$\begin{aligned}\sum_{r=1}^{\infty} |e^{(r)}\rangle\langle e^{(r)}|\phi\rangle &= \sum_{r,s} |e^{(r)}\rangle(e^{(r)}|a_s|e^{(s)}\rangle \\ &= \sum_{r,s} a_s\delta_{rs}|e^{(r)}\rangle = |\phi\rangle\end{aligned}$$

Hence it is the unit operator:

$$\sum_{r=1}^{\infty} |e^{(r)}\rangle\langle e^{(r)}| = \mathbb{1} \tag{4.6}$$

This is the *completeness identity* and provides a *resolution of the identity.* A linear operator V is *isometric* if for every vector ϕ,

$$\langle V\phi|V\phi\rangle = \langle\phi|\phi\rangle \tag{4.7}$$

Given an operator A, its *adjoint* operator $A^\dagger$ is defined by

$$\langle\phi|A\psi\rangle = \langle A^\dagger\phi|\psi\rangle \tag{4.8}$$

An isometric operator V satisfies the relation

$$V^\dagger V = \mathbb{1} \tag{4.9}$$

The adjoint is an antilinear operator valued function of operators. An operator whose adjoint coincides with itself is called *selfadjoint.*

$$A^\dagger = A \tag{4.10}$$

An isometric operator is *unitary* if in addition to (4.9) it satisfies

$$VV^{\dagger} = \mathbb{1} \tag{4.11}$$

If a linear operator C has the form

$$C = \sum_n c_n | e^{(n)} \rangle \langle e^{(n)} | \tag{4.12}$$

for some convergent sequence $\{c_n\}$ and some basis $\{| e^{(n)} \rangle\}$ it is said to be *completely continuous.* A completely continuous operator is the discrete (possibly infinite) sum of *projections*:

$$C = \sum_n c_n \Pi_n; \qquad \Pi_n = | e^{(n)} \rangle \langle e^{(n)} | \tag{4.13}$$

with

$$\Pi_n \Pi_m = \Pi_n \delta_{nm}; \qquad \sum_n \Pi_n = \mathbb{1} \tag{4.14}$$

Equations (4.12) and (4.13) also give the *spectral decomposition* of a completely continuous operator:

$$C | e^{(n)} \rangle = c_n | e^{(n)} \rangle \tag{4.15}$$

For any operator A we can consider the *resolvent* as the analytic operator-valued function

$$R(z; A) = (A - z\mathbb{1})^{-1} \tag{4.16}$$

$R(z)$ is regular acting on $\mathscr{H}$ everywhere except for the values

$$z = c_n$$

which constitute the spectrum of A. More generally, for any operator A, the set of points (discrete or continuous, finite or infinite) where the resolvent operator fails to be regular in $\mathscr{H}$ (i.e., the action of $R(z)$ considered as an analytic function of z is not regular for any vector in $\mathscr{H}$) is called the *spectrum* of A.

For a selfadjoint operator with a continuous spectrum, there may be no normalizable eigenvector in $\mathscr{H}$. In all the explicit examples we have considered, the continuous spectrum has no normalizable eigenvectors. One can either introduce ideal eigenvectors (of infinite length!) following Dirac, or consider a continuous family of spectral projections $\Pi(\lambda)$ for eigenvalues "less than" λ by introducing a notion of ordering in the continuous spectrum (when it is possible!) and writing a Steiltjes operator valued integral generalizing the spectral decomposition and completeness identity (4.12), (4.13), (4.14):

$$A = \int \lambda \, d\Pi(\lambda) \tag{4.17}$$

$$\int d\Pi(\lambda) = \mathbb{1}\,; \qquad \Pi(\lambda)\Pi(\mu) = \Pi(\lambda), \qquad \mu \geq \lambda \tag{4.18}$$

So far, we have considered the generic form A, the Hilbert space $\mathscr{H}$, and the vectors in $\mathscr{H}$. In the study of quantum systems the space $\mathscr{H}$ is realized in terms of the states of the system and the generic form of the state vectors is in terms of square integrable functions of one or more real variables. A dense subset of such L^2 functions is the class of analytic functions (restricted to real values of the arguments).

2. *Analytic Continuation of Vector Spaces*

This dense subset of $\mathscr{H}$ can be analytically continued. But there are many choices of analytic L^2 functions with varying domains of analyticity and correspondingly many choices of $\mathscr{G}$ and $\tilde{\mathscr{G}}$. The dense sets of analytic functions form a partially ordered set; and continuations using functions analytic in a domain that coincide with the analytic continuation using functions analytic in another domain, and will coincide within their common domain of convergence. Linear relationships are preserved; and we can define *analytic linear operators* to be those that, acting on an analytic function, produce another analytic function. Needless to say, the notion of analytic continuation is in terms of the specific L^2 function realization of the space $\mathscr{H}$ and the domain in which $\mathscr{G}$ is defined depends on the dense subset chosen. Because the correspondence between vectors in $\mathscr{H}$ and $\mathscr{H}'$ is antilinear, we must analytically continue these spaces separately to produce a family of generalized spaces $\mathscr{G}$ and $\tilde{\mathscr{G}}$.

The notion of resolvent and spectrum applies to the generalized family of spaces $\mathscr{G},\tilde{\mathscr{G}}$. The eigenvectors are now *right* eigenvectors in $\mathscr{G}$ and *left* eigenvectors in $\tilde{\mathscr{G}}$. For every vector in $\mathscr{H}$, we have its dual vector in $\mathscr{H}'$. The product of the analytic continuations of a dense set of vectors in $\mathscr{H}$ (and hence $\mathscr{H}'$) are in $\mathscr{G},\tilde{\mathscr{G}}$ and may be called the norm of the vector in $\mathscr{G}$. With respect to this norm, we can define Cauchy sequences.

Because the analytic continuation is for both $\mathscr{H}$ and $\mathscr{H}'$ to $\mathscr{G}$ and $\tilde{\mathscr{G}}$, scalar products and matrix elements of analytic linear operators are preserved. To this extent, the analytic vectors and operators can be thought of as having different representations in the family of spaces $\mathscr{G},\tilde{\mathscr{G}}$, which could correspond to the analytic vectors and linear operators in $\mathscr{H}$. However, the analytic continuation is not of the entire space $\mathscr{H}$ into the completion of $\mathscr{G}$, with the norm defined as the product of the vector in $\mathscr{G},\tilde{\mathscr{G}}$ associated with the vectors in $\mathscr{H},\mathscr{H}'$. In particular, there are vectors in $\mathscr{G}$ which may not have a counterpart in $\mathscr{H}$ and vice versa; for example, there are discrete states in $\mathscr{G}$ which have no counterpart in $\mathscr{H}$.

Finally, because the analytic continuation depends on the functional form of the state vectors as a function of its arguments, the relevant dynamical labels must be chosen. In the study of Hamiltonian systems, we often have a *total energy* label as well as the values of a *comparison Hamiltonian* energy. On writing the ideal eigenstates of the total Hamiltonian as a function of the comparison Hamiltonian energy, we look for analytic vectors; this can be done if the total Hamiltonian *represented* in terms of the functions of comparison Hamiltonian energies is *analytic*. The existence of the comparison ("free") Hamiltonian and its essential role in scattering theory where the "in" and "out" states are defined has been known for some time [82]. Formal scattering theory does make use of this representation to go "slightly off" the real axis as far as the *scattering amplitude* is concerned. The analytic continuation of scattering amplitude was extended to its various sheets by many authors [14,83,84]. However, except for the work of Nakanishi [24] and Sudarshan, Chiu, and Gorini [42] (see also [6] and [37]), there was no consideration of the analytic continuation of suitable dense sets in the state space $\mathscr{H}$ to the family $\mathscr{G}$.

C. Complete Set of States in Continued Spaces

If $\{|\lambda\rangle\}$ is the set of ideal eigenvectors for a self-adjoint, nonnegative (total Hamiltonian) operator so that

$$\int_0^\infty \Pi(\lambda)\, d\lambda = \int_0^\infty |\lambda\rangle\langle\lambda|\, d\lambda = 1; \qquad \langle\lambda|\mu\rangle = \delta(\lambda - \mu) \tag{4.19}$$

The vector

$$|\phi\rangle = \int_0^\infty \phi(\lambda)|\lambda\rangle\, d\lambda \tag{4.20}$$

is a vector in $\mathscr{H}$ if

$$\int_0^\infty |\phi(\lambda)|^2 \, d\lambda < \infty \tag{4.21}$$

If $\phi(\lambda)$ is analytic in λ in a suitable domain in the complex plane, we could deform the contour to write the vector as a vector in $\mathscr{G}$ (see Fig. 5):

$$|\phi\rangle = \int_C \phi(z) | z\rangle \, dz \tag{4.22}$$

The analytic continuation includes a simultaneous continuation of the bra vectors

$$\langle\psi| = \int_0^\infty \psi(\lambda)\langle\lambda| \, d\lambda \tag{4.23}$$

into a vector in $\tilde{\mathscr{G}}$:

$$\langle\widetilde{\psi}| = \int_C \widetilde{\psi(z)}\langle\widetilde{z}| \, dz \tag{4.24}$$

The additional closed contours C_1 and C_2 encountered in the continuation (see Fig. 6) are typical of poles and branch cuts. For resonance in scattering, we expect to find complex poles, but for multiparticle states involving unstable particles, we expect to have complex branch cuts. Although Fig. 6 shows only one pole and one pair of branch points in the finite complex plane, we may have more than one; and branch points may

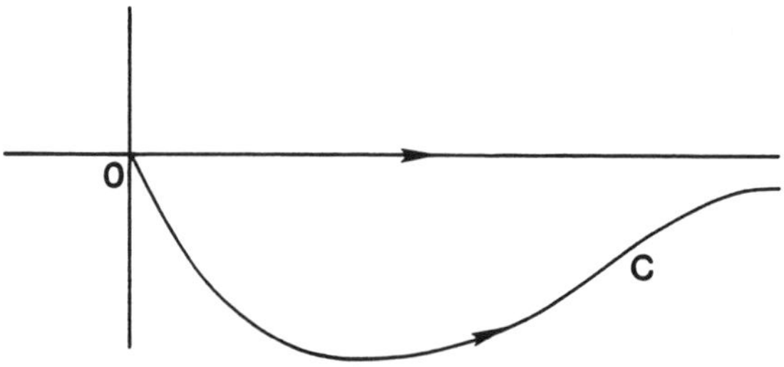

Figure 5. The z-plane contours defining vectors in $\mathscr{G}$.

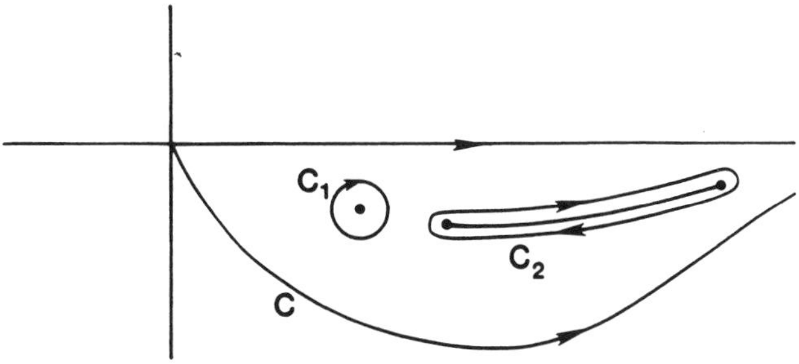

Figure 6. Possible singularities encountered and the modified contours.

move to infinity. The completeness identity (4.6) is modified to

$$\mathbb{1} = \int_C dz\,|z\rangle\langle\widetilde{z}| + \sum_{\text{poles}} |z_r\rangle\langle\widetilde{z}_r| + \int_{C_2} d\zeta\,|\zeta\rangle\langle\widetilde{\zeta}| \tag{4.25}$$

Furthermore, the scalar product remains unchanged in value:

$$\langle\widetilde{\psi}\,|\,\phi\rangle = \int_C \tilde{\psi}(z)\phi(z)\,dz + \sum_{\text{poles}} \tilde{\psi}(z_r)\phi(z_r) + \int_{C_2} \tilde{\psi}(\zeta)\phi(\zeta)\,d\zeta \tag{4.26}$$

Here and in Eq. (4.24), $\tilde{\psi}(z)$ is the analytic continuation of the function $\psi^*(z^*)$:

$$\tilde{\psi}(z) = \psi^*(z^*) \tag{4.27}$$

and the norm of $|\psi\rangle$ is given by $\langle\widetilde{\psi}\,|\,\psi\rangle$. If we have a definite state $\psi(\lambda)$ (which may be thought of as the created unstable particle state), the survival amplitude for the state is given by [36,41]

$$A(t) = \langle\widetilde{\psi}\,|\,e^{-iHt}\,|\,\psi\rangle = \mathrm{tr}(|\,\psi\rangle\langle\widetilde{\psi}\,|\,e^{-iHt}) \tag{4.28}$$

where H is the (total) Hamiltonian and can be expressed in the form of a Fourier integral:

$$A(t) = \int_0^\infty |\,\psi(\lambda)\,|^2 e^{-i\lambda t}\,d\lambda \tag{4.29}$$

This same survival amplitude can be computed in $\mathscr{G},\tilde{\mathscr{G}}$ if $|\,\psi\rangle$ is an analytic vector:

$$A(t) = \int_{C+C_1+C_2} dz\,\tilde{\psi}(z)\psi(z)e^{-izt} \tag{4.30}$$

If the analytically continued bilinear quantity is explicitly known, the pole and branch-cut contributions can be calculated. We do this when we consider solvable models like the Friedrichs–Lee [25,26] and the Cascade [44]. Suffice it to say that the survival amplitude can be defined for evolutions that are both forward *and backward* in time; and for all times the absolute value of the amplitude is bounded by unity.

For the generic case, the poles of the S-matrix coincide with the discrete states in the generalized completeness identity (4.25). However, the existence of a pole in the S-matrix is *neither sufficient nor necessary to have such additional discrete states in* $\mathscr{G}$. This is due to possible existence of redundant poles and of discrete states buried in the continuum. We discuss this further in another section.

D. Friedrichs–Lee Model States

A simple solvable model [2,25,26] is provided by a system with a discrete state and a one-dimensional continuum so that the vectors are of the form

$$[\eta, \phi(\omega)]^T = \Phi \tag{4.31}$$

with

$$\langle \Phi | \Phi \rangle = \eta^* \eta + \int d\omega \ \phi^*(\omega)\phi(\omega) \tag{4.32}$$

We choose a total Hamiltonian of the form

$$H(\eta, \phi(\omega))^T = \lambda[\eta, \phi(\omega)]^T$$

$$\lambda\eta = m_0 \eta + \int_0^\infty g^*(\omega')\phi(\omega')\, d\omega'$$

$$\lambda\phi(\omega) = \omega\phi(\omega) + g(\omega)\eta$$

Define the function

$$\alpha(\lambda) = \lambda - m_0 - \int_0^\infty \frac{g^*(\omega')g(\omega')}{\lambda - \omega'}\, d\omega' \tag{4.33}$$

If $\alpha(\lambda)$ has a real zero, it is for a negative value m [unless $g(\omega)$ vanishes some place in the interval $0 < \omega < \infty$]. If there is such a zero, there is a discrete eigenvalue m for the Hamiltonian H:

$$\phi_0(\omega) = \frac{g(\omega)}{m - \omega}\eta_0; \qquad \eta_0 = \left(\frac{d\alpha}{d\lambda}\bigg|_{\lambda = m} \right)^{-1/2}$$
$$H[\eta_0, \phi_0(\omega)]^T = m[\eta_0, \phi_0(\omega)]^T \tag{4.34}$$

There can be at most one zero. No such discrete state exists if

$$\alpha(0) = -m_0 + \int \frac{g^*(\omega')g(\omega')\, d\omega'}{\omega'} < 0 \tag{4.35}$$

However, if for some value $\lambda = M > 0$, we have the twin conditions

$$g(M) = 0; \qquad \alpha(M) = 0 \tag{4.36}$$

Then we can have a discrete state overlapped by the continuum.

There is a continuous spectrum $0 < \lambda < \infty$ and a corresponding continuum of scattering states which are ideal states with continuum normalization [82,85]:

$$\begin{gathered} |\Phi_\lambda\rangle = [\eta_\lambda, \phi_\lambda(\omega)]^T \equiv |\lambda\rangle \\ \eta_\lambda = \frac{g^*(\lambda)}{\alpha(\lambda + i\epsilon)}; \qquad \phi_\lambda(\omega) = \delta(\lambda - \omega) + \frac{g^*(\lambda)g(\omega)}{(\lambda - \omega + i\epsilon)\alpha(\lambda + i\epsilon)} \end{gathered} \tag{4.37}$$

These states satisfy the orthonormality and completeness relations

$$\begin{gathered} \langle m | m \rangle = 1, \qquad \langle m | \lambda \rangle = 0 \\ \langle \lambda | \lambda' \rangle = \delta(\lambda - \lambda') \end{gathered} \tag{4.38}$$

and

$$|m\rangle\langle m| + \int d\lambda\, |\lambda\rangle\langle\lambda| = \mathbb{1} \tag{4.39}$$

Here

$$|m\rangle = [\eta_0, \phi_0(\lambda)]^T \tag{4.40}$$

These calculations are already available in the literature and involve straightforward contour integration. If there is a discrete state buried in the continuum [86–88], Eqs. (4.36) and (4.37) show that there are two solutions at this value M: a discrete state of the form (4.34) with m replaced by M, and an ideal state with $\lambda = M$ which is a pure plane wave:

$$|M\rangle = \left(\frac{d\alpha}{d\lambda}\bigg|_{\lambda=M}\right)^{-1/2} \left[1, \frac{g(\omega)}{M - \omega}\right]^T \tag{4.41}$$

$$|M\rangle' = \left[0, \delta(\lambda - M) + \begin{matrix}\text{nonsingular} \\ \text{terms}\end{matrix}\right]^T \tag{4.42}$$

The state (4.41) would enter the completeness relation (4.39) and the orthonormality relations (4.38).

The S-matrix for the ideal scattering states reduces to a phase:

$$S(\lambda) = \alpha(\lambda - i\epsilon)/\alpha(\lambda + i\epsilon); \qquad 0 < \lambda < \infty \tag{4.43}$$

If $g(\omega)$ is analytic in ω, so is

$$g(\tilde{\omega}) = g'(\omega) = g^*(\omega^*) \tag{4.44}$$

Then the continuum ideal states $|\lambda\rangle$ can be replaced by complex eigenvalue ideal states denoted by the same symbol $|\lambda\rangle$, which have branch cuts along a different contour Γ beginning at 0 and ending at infinity. To see this, we consider the space of analytic functions in the region Δ bounded by Γ and the positive real axis for which the integral

$$\left|\int_\Gamma \phi^*(z^*)\phi(z)\, dz\right| < \infty \tag{4.45}$$

The spaces $\mathscr{G},\tilde{\mathscr{G}}$ consists of vectors $[\eta,\, \phi(z)]^T$ and $[\tilde{\eta},\, \widetilde{\phi(z)}]$ with such functions $\phi(z)$. We further require that these functions $\phi(z)$ vanish sufficiently fast at infinity so that

$$\int_0^\infty |\phi(\omega)|^2\, d\omega = \int_\Gamma \phi^*(z^*)\phi(z)\, dz \tag{4.46}$$

Note that the scalar product is between a vector in $\mathscr{G}$ and one in the dual space $\tilde{\mathscr{G}}$.

Along the contour Γ we can introduce a delta function $\delta(\lambda - z)$ defined by [24,42]

$$\int_\Gamma \phi(z)\delta(\lambda - z)\, dz = \phi(\lambda) \tag{4.47}$$

With this definition we can reinvestigate the eigenvalue problem

$$H[\eta,\, \phi(z)]^T = \lambda[\eta,\, \phi(z)]^T \tag{4.48}$$

with z along the contour Γ. Equation (4.48) implies

$$\begin{aligned}(\lambda - m_0)\eta &= \int_\Gamma g^*(z'^*)\phi(z')\, dz' \\ (\lambda - z)\phi(z) &= g(z)\eta\end{aligned} \tag{4.49}$$

The continuum ideal vectors have

$$\eta_\lambda = \frac{g^*(\lambda^*)}{\alpha(\lambda + i\epsilon)}$$

$$\phi_\lambda(z) = \delta(\lambda - z) + \frac{g^*(\lambda^*)g(z)}{(\lambda - z + i\epsilon)\alpha(\lambda + i\epsilon)} \tag{4.50}$$

$$\alpha(z) = z - m_0 - \int_\Gamma \frac{g^*(z'^*)g(z')\, dz'}{z - z'}$$

These are orthonormal; the computation follows the usual route. They are, together with the possible discrete state,

$$\eta_0 = [\alpha'(m)]^{-1/2}$$

$$\phi_0(z) = \frac{g(z)\eta_0}{m - z}$$

also complete, provided $\alpha(m) = 0$ for some $m < 0$.

In case $m_0 \gg 0$, there would be no discrete state $[\eta_0, \phi_0(z)]^T$. But if the contour Γ proceeds sufficiently far in the fourth quadrant, there would be a complex zero z_1 for $\alpha(z)$ and a discrete state with

$$\eta_1 = [\alpha'(z_1)]^{-1/2}$$

$$\phi_1(z) = \frac{g(z)\eta_1}{z_1 - z} \tag{4.51}$$

This state is orthogonal to the continuum states in $\mathcal{G}$ and enters as a discrete contribution to the completeness relation. Since $\alpha(z)$ is real analytic, if the contour Γ was in the upper half plane, there would be a zero z_1^* for $\alpha(z)$ and a corresponding state. In both cases, the discrete state remains fixed and contributes to the complete set of states or not according to whether Γ crosses z_1 (or z_1^*).

The demonstration of the completeness is the resolution of the identity in the form (see Fig. 7 for the contours defined.)

$$\mathbb{1} = \begin{cases} \displaystyle\int_\Gamma d\lambda\,|\lambda\rangle\langle\lambda| + |m\rangle\langle m| & m^* = m < 0,\ \alpha(m) = 0 \\ \displaystyle\int_{\Gamma'} d\lambda\,|\lambda\rangle\langle\widetilde{\lambda}| + |z_1\rangle\langle\widetilde{z}_1| & \alpha(z_1) = 0 \end{cases} \tag{4.52}$$

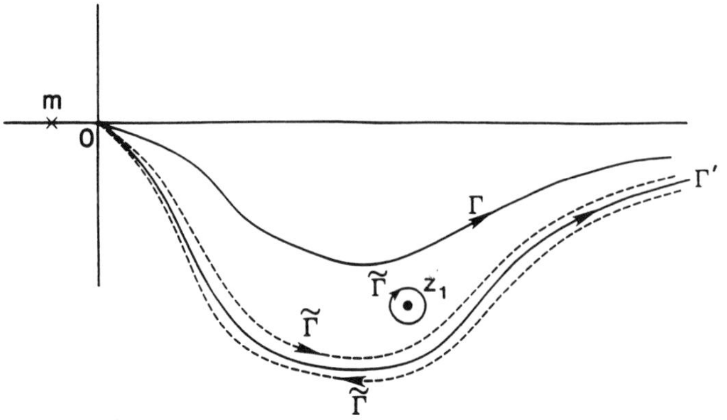

Figure 7. Contours Γ, Γ′, and Γ̃ for demonstrating completeness.

In doing the Γ or Γ′ integrals we have to compute, for example,

$$\int \phi^*_{\lambda^*}(z^*)\phi_\lambda(z')\, d\lambda = \delta(z - z') + \frac{g^*(z^*)g(z')}{(z' - z - i\epsilon)\alpha(z' - i\epsilon)}$$
$$+ \frac{g(z')g^*(z^*)}{(z - z' + i\epsilon)\alpha(z + i\epsilon)} + g^*(z^*)g(z')$$
$$\times \int_\Gamma \frac{g^*(\lambda^*)g(\lambda)\, d\lambda}{(\lambda - z - i\epsilon)(\lambda - z' + i\epsilon)\alpha(\lambda + i\epsilon)\alpha(\lambda - i\epsilon)} \tag{4.53}$$

The last term can be rewritten as a contour integral encasing the contour Γ because

$$g^*(\lambda^*)g(\lambda) = \frac{1}{2\pi i}\{\alpha(\lambda) - \alpha^*(\lambda^*)\} \tag{4.54}$$

so that the last term becomes

$$g^*(z^*)g(z)\frac{1}{2\pi i}\frac{1}{2\pi i}\int_\Gamma \frac{d\lambda}{(\lambda - z' + i\epsilon)(\lambda - z - i\epsilon)\alpha(\lambda)} \tag{4.55}$$

The poles at $\lambda = z' - i\epsilon$, $z + i\epsilon$ cancel the third and second terms respectively while the remaining contribution would be proportional to the residue at any pole of $1/\alpha(\lambda)$. Note that it is the zeros of $\alpha(z)$ that count, not the blow up of $g^*(z^*)g(z)$.

This conclusion is further demonstrated in the computation of the survival amplitude of the "unstable particle" state $[1, 0]^T$. Quite generally,

$$[(1, 0), e^{-iHt}(1, 0)^T] = \int \eta^*_{\lambda^*}\eta_\lambda e^{-i\lambda t}\, d\lambda \tag{4.56}$$

$$= \int_\Gamma e^{-i\lambda t} \frac{g^*(\lambda^*)g(\lambda)}{\alpha^*(\lambda^*)\alpha(\lambda)}\, d\lambda = \frac{1}{2\pi i}\int_{\tilde{\Gamma}} \frac{e^{-i\lambda t}\, d\lambda}{\alpha(\lambda)} \tag{4.57}$$

Again only the zeros of $\alpha(\lambda)$ contribute, not the singularities of $g^*(\lambda^*)g(\lambda)$. Any such pole of $g^*(\lambda^*)g(\lambda)$ is counterbalanced by a corresponding pole in $\alpha^*(\lambda^*)$.

Here we have acted as if poles are the only singularities encountered in the analytic continuation. But in many contexts there could be branch cuts. We discuss such a situation for the Cascade model.

E. Yamaguchi Potential Model States

A model related closely to the Friedrichs–Lee model is the separable potential model [54] which in its lowest relevant sector has a one-dimensional continuum. The states in $\mathscr{H}$ are, then, $L^2(0, \infty)$ functions:

$$\left\{\Phi: \int_0^\infty \phi^*(\omega)\phi(\omega)\, d\omega < \infty\right\}$$

We choose a total Hamiltonian of the form

$$(H\phi)(\omega) = \omega\phi(\omega) + \eta h(\omega) \int_0^\infty h^*(\omega')\phi(\omega')\, d\omega' \tag{4.58}$$

where $\eta^2 = 1$. Define the function

$$\beta(z) = 1 - \eta \int_0^\infty \frac{h^*(\omega')h(\omega')\, d\omega'}{z - \omega'} \tag{4.59}$$

If $\beta(z)$ has a real zero, it will arise for $\eta < 0$ at $z = z_0 < 0$. In that case there is a discrete solution:

$$\phi_0(\omega) = \frac{\eta h(\omega)}{z_0 - \omega}[\beta'(z_0)]^{-1/2}; \qquad \eta = -1; \qquad z_0 < 0 \tag{4.60}$$

There is a continuum of scattering states

$$\Phi_\lambda\colon \phi_\lambda(\omega) = \delta(\lambda - \omega) + \frac{\eta h^*(\lambda)h(\omega)}{(\lambda - \omega + i\epsilon)\beta(\lambda + i\epsilon)} \tag{4.61}$$

These ideal states satisfy orthonormality

$$\langle 0 | 0 \rangle = 1, \qquad \langle \lambda | 0 \rangle = 0$$
$$\langle \lambda | \lambda' \rangle = \delta(\lambda - \lambda')$$

and completeness

$$|0\rangle\langle 0| + \int d\lambda \, |\lambda\rangle\langle\lambda| = \mathbb{1} \tag{4.62}$$

Of course, if $\beta(z)$ has no zero, the discrete state $|0\rangle$ would be missing from this equation.

The S-matrix for the ideal scattering states reduces to a phase:

$$S(\lambda) = \beta(\lambda - i\epsilon)/\beta(\lambda + i\epsilon), \qquad 0 < \lambda < \infty \tag{4.63}$$

If $h(\omega)$ is analytic in ω, so is $h^*(\omega^*)$. Then we can continue the vector space $\mathscr{H}$ into $\mathscr{G}$ and get a spectrum along another contour Γ starting from the origin and going to infinity.

The dimensionless scattering amplitude (in $\mathscr{H}$) is given by

$$T(\omega) = \frac{\pi h(\omega)h^*(\omega)}{\beta(\omega + i\epsilon)} = \exp[i\theta(\omega)] \sin \theta(\omega) \tag{4.64}$$

where $\theta(\omega) = \arg \beta(\omega - i\epsilon)$ is the phase shift. If we choose nonrelativistic kinematics so that

$$\omega = k^2/2\mu \tag{4.65}$$

the more conventional scattering amplitude (with the dimension of a length) is given by

$$\begin{aligned} \mathscr{T}(k) &= \frac{\pi |h(\omega)|^2}{k\beta(\omega + i\epsilon)} = \frac{e^{i\theta(\omega)} \sin \theta(\omega)}{k} \\ &= [k \cot \theta(\omega) - ik]^{-1} \end{aligned} \tag{4.66}$$

which manifestly satisfies unitarity. The total (*s*-wave!) cross section is given by

$$\sigma(\omega) = \frac{4\pi}{k^2} \sin^2 \theta(\omega) \tag{4.67}$$

When analytic continuations are carried out, the scattering amplitude $T(\omega)$ is continued to yield

$$T(z) = \frac{\pi h(z) h^*(z^*)}{\beta(z + i\epsilon)}, \qquad z \text{ on } \Gamma \tag{4.68}$$

$T(z)$ so defined may have poles due to complex zeros of $\beta(z)$ or poles in $h(z)h^*(z^*)$. The latter do not correspond to extra physical states: they are "redundant poles" (see Section IV.G). If there are no complex zeros of $\beta(z)$, the completeness relation in the analytically continued space $\mathscr{G}$ is

$$\int_\Gamma dz \, |z\rangle \widetilde{\langle z|} = \mathbb{1} \tag{4.69}$$

The explicit expression for the ideal states $|z\rangle$ and the proof of the completeness and orthogonality are straightforward. In many contexts, there could be branch cuts. We discuss such a situation for the cascade model in Section VI.

F. Extended Spaces and Semigroup of the Time Evolution

We have so far formulated the passage from $\mathscr{H}$ to $\mathscr{G}$ as a correspondence between dense sets in $\mathscr{H}$ and $\mathscr{G}$. With this understanding, the basis in $\mathscr{G}$ is "the same" as in $\mathscr{H}$. Therefore, when we know that a pure exponential decay-time dependence is not possible for states in $\mathscr{H}$ (with a nonnegative spectrum for the total Hamiltonian), the same should also obtain for *corresponding states in* $\mathscr{G}$. Furthermore, because the time evolution (and regression) are implemented by a unitary family of linear operators realizing the time translation *group*, the same would also be true of the states in $\mathscr{G}$. A pure exponential decay or a Steiltjes integral over damped exponentials would then not be possible with states obtained by analytic continuation of physical states.

One can, however, ask what property has to be relaxed to realize an extended space $\tilde{\mathscr{H}}$ and its corresponding continuation $\tilde{\mathscr{G}}$ so that a semigroup of time evolutions can be realized. These semigroups would, generally, be realized by an isometry which is not, however, unitary. After all, an unrestricted Breit–Wigner resonance [4] with its Lorentz line shape does correspond to pure exponential decay (for positive time). We need to relax the positivity of energy and define states with all possible values of energy. In this case, we can realize semigroups of time evolution [45,46].

Let $\psi(\lambda)$ be a vector in a Hilbert space $\mathscr{H}$:

$$\int_0^\infty |\psi(\lambda)|^2 \, d\lambda = 1 \qquad \psi(\lambda) = 0 \qquad \lambda < 0 \tag{4.70}$$

We enlarge it into $\mathbb{H}_\pm$, where $\Psi(\lambda)$ is defined for negative values of λ also, in such a fashion that it is analytic in a half plane:

$$\Psi_\pm(z) = \frac{\mp}{2\pi i} \int_0^\infty d\lambda \, \frac{1}{\lambda - z \pm i\epsilon} \, \psi(\lambda) \tag{4.71}$$

These functions are *analytic* in the two *half planes* and their sum is equal to $\psi(\lambda)$:

$$\psi(\lambda) = \Psi_+(\lambda) + \Psi_-(\lambda) \tag{4.72}$$

On $\Psi_+(\lambda)$, the time evolution for positive times is realized by a contractive semigroup:

$$\Psi_+(z; t) = T_+(t)\Psi_+(z) = -\frac{1}{2\pi i} \int_0^\infty d\lambda \, e^{-i\lambda t} \frac{1}{\lambda - z + i\epsilon} \, \psi(\lambda) \tag{4.73}$$

$$\begin{aligned} T_+(t_1)T_+(t_2) &= T_+(t_1 + t_2) \qquad t_1, t_2 > 0 \\ T_+(t) = 0 \qquad t &< 0; \qquad T_+(0+) = 1 \end{aligned} \tag{4.74}$$

By the converse of a theorem of Titchmarsh [89]

$$\tilde{\Psi}_\pm(\tau) \equiv \int_{-\infty}^\infty \Psi_\pm(\lambda) e^{-i\lambda\tau} \, d\lambda = 0 \qquad \pm\tau < 0 \tag{4.75}$$

Then

$$T_+(t)\tilde{\Psi}_+(\tau) = \tilde{\Psi}_+(\tau + t) \qquad t > -\tau \tag{4.76}$$

$$T_+(t)\tilde{\Psi}_+(\tau) = 0 \qquad t < -\tau \tag{4.77}$$

Thus a semigroup evolution obtains on the half-plane analytic function $\Psi_+(\lambda)$. A similar conclusion obtains for the backward tracing of $\Psi_-(\lambda)$.

Given $\Psi_+(\lambda)$, we can continue it to a vector $\Psi_+(z)$ in $\tilde{\mathscr{J}}$ and the semigroup acts in $\tilde{\mathscr{J}}$ in the same fashion.

The functions $\Psi_+(z)$ are analytic in the half-plane by construction. They constitute the Hardy class of functions [90] which are square integrable along Re z for any negative imaginary part. *None* of this class *is a physical state* (expressible as linear combinations of states of nonnegative total energy). But many familiar unphysical states, like the Breit–Wigner function,

$$\Psi_+(\lambda) = \frac{\Gamma}{\sqrt{\pi}} \frac{1}{\lambda - \lambda_0 + i/2\Gamma} \tag{4.78}$$

are included in this Hardy class. In addition to such a single pole we could also have multiple poles and/or branch points. To obtain them, we can use a perfectly physical state obtained as a linear combination of states like Eq. (4.37), [for three-body case, see states like Eqs. (6.13) and (6.14) in Section VI] and carry out the linear maps (4.72) into the two Hardy class functions.

G. Redundant and Discrete States in the Continuum

For the model discussed in Section IV.F, when the contour Γ passes through $z = M_1$, the continuum wave function (4.47) exhibits singularity at $z = M_1$, a complex eigenvalue. There is, when the contour justifies it, a discrete eigenstate with eigenvalue M_1. The scattering amplitudes also have singularities (poles) at the same point. People often assume that the poles of the scattering amplitude correspond to unstable particles. It has, however, been known [91,92] that poles appear in the S-matrix (or the scattering amplitude) which do not correspond to discrete eigenstates of the Hamiltonian in $\mathscr{H}$. This is true of the (repulsive) exponential potential; and a number of phase-equivalent potentials [93,94] have been known for which some of the S-matrix poles correspond to bound (discrete) states and others do not. In the context of the Lee model and other such models, one could choose the poles to be redundant or genuine without changing the S-matrix. In the Lee model, this corresponds to the distinction between the zeros of the denominator function $\alpha(z)$ and the poles of the form factor $f^*(z^*)f(z)$. Nor are these redundant singularities restricted to being isolated poles; for example, the S-wave Yukawa potentials give a branch cut [95], but with no continuum of (ideal) states entering the description. In all such cases, the redundant singularities of the S-matrix do not correspond to states entering the complete set of states.

A similar situation obtains in the case of analytic continuation of the vector space $\mathscr{H}$ to $\mathscr{G}$. Consider the Lee model wave functions (4.50). They would develop singularities not connected with the spectrum of the Hamiltonian in $\mathscr{G}$ if the form factor $g(z)$ develops singularities. But these singularities do not give any contributions to the completeness identity because in these calculations we obtain the contour integrals involving $[1/\alpha(z)]$. The poles in $g^*(z^*)g(z)$ are matched by corresponding terms in $\alpha(z)$ and they disappear from the contour integral. As the contour Γ smoothly deforms itself, it is not snagged by singularities of $g^*(z^*)g(z)$. The same situation obtains for the Cascade model; only the zeros of $\alpha(z)$ contribute to the discrete state and only the branch cuts in $\gamma(\zeta)$ contribute to the scattering states involving an unstable particle.

A related phenomenon is that of states which contribute to the complete set of states located in the continuum but which do not contribute any singularity for the S-matrix [88]. This occurs when a zero of $\alpha(z)$ coincides with a zero of the form factor $g(z)$ as far as the Lee model is concerned. The spectrum is degenerate at this point M, $\alpha(M) = 0$ with a discrete state in $\mathscr{H}$ and an ideal state belonging to the continuum. In analytic continuation, we can have complex zeros of $\alpha(z)$ where the scattering amplitude vanishes; nevertheless, the complete set of states include these states. They also enter the computation of survival amplitudes (4.57).

For the Lee model, we choose a form factor $g^*(z^*)g(z)$ and an $\alpha(z)$ such that

$$\alpha(M_1) = 0; \qquad g^*(z^*)g(z) \sim (z - M_1)^2 G(z) \tag{4.79}$$

for some complex M_1. Then the scattering amplitude vanishes at this point

$$T(z) \sim (z - M_1)t(z) \tag{4.80}$$

The (ideal) state at this point is a "plane wave,"

$$\tilde{\eta}_1 = 0; \qquad \phi_1(z) = \delta(z - M_1) + \text{nonsingular terms} \tag{4.81}$$

(with no asymptotic diverging wave) which is degenerate in energy with the proper state in $\mathscr{G}$ with

$$\eta_1 = [\alpha'(M_1)]^{-1/2}; \qquad \phi_1(z) = \frac{g(z)\eta_1}{M_1 - z} \tag{4.82}$$

In a similar manner, if the form factors in the Cascade model have zeros along the cut beginning at the branch point μ_1, the scattering amplitude vanishes at these points on the branch cut, but the (ideal) states $|z\rangle$ in Eq. (6.14) beginning at μ_1 exist and contribute to the completeness (and to the survival amplitude for the unstable A particle).

Thus the S-matrix singularities and the spectrum of states are not necessarily in correspondence.

Along with redundant poles, we could also have redundant branch cuts from the "geometry of the potential." There will be no contribution from these to the completeness identity. Such branch cuts are familiar as the left-hand (and the short- and circle-) cuts in partial wave-dispersion relations.

H. Analytic Continuation of Survival Probability and Time-Reversal Invariance

1. *Analytic Continuation of Survival Probability*

The probability is the absolute value squared of the amplitude, which now involves the multiplication of two factors. One is $\langle\tilde{\psi}\,|\,\phi\rangle$, the inner product between the state in $\mathcal{G}$ and its dual in $\tilde{g}$. Both are defined along Γ. The other factor corresponds to complex conjugations, which is the inner product of the corresponding state in $\mathcal{G}^*$ and the dual state in $\tilde{\mathcal{G}}^*$ defined along Γ^*. For the analytic continuation of a probability function, there are two distinct pairs of vector spaces:

$$\mathcal{G},\tilde{\mathcal{G}} \quad \text{and} \quad \mathcal{G}^*,\tilde{\mathcal{G}}^*$$

For a discrete state $|M\rangle$ where $M = m - (i\Gamma/2)$, its time dependence is characterized by

$$|M, t\rangle = e^{-iHt}|M, 0\rangle = e^{-iMt}|M, 0\rangle \tag{4.83}$$

For the corresponding dual state in $\tilde{\mathcal{G}}$,

$$\langle\widetilde{M, t}\,| = (|\,M^*, t\rangle)^{\dagger} = e^{iMt}\langle\widetilde{M, 0}\,| \tag{4.84}$$

Their inner product is

$$\langle\widetilde{M, t}\,|\,Mt\rangle = e^{i(M-M)t}\langle\widetilde{M, 0}\,|\,M, 0\rangle = 1 \tag{4.85}$$

Consider the corresponding complex conjugate space. For the discrete state in $\mathcal{G}^*$,

$$|\,M^*t\rangle = e^{-iM^*t}|\,M^*, 0\rangle \tag{4.86}$$

and the $\tilde{\mathcal{G}}^*$ space,

$$\langle\widetilde{M^*, t}\,| \equiv \langle M, t\,| = e^{iM^*t}\langle M, 0\,| = e^{iM^*t}\langle\widetilde{M^*, 0}\,| \tag{4.87}$$

with the inner product

$$\langle \widetilde{M^*}, t \,|\, M^*, t\rangle = \langle M, t \,|\, M^*, t\rangle = 1 \tag{4.88}$$

2. *Time-Reversal Invariance*

Decay signifies irreversibility, but it is still relevant to investigate questions of time-reversal invariance. We recall some conventional wisdom on time reversal. It is a "kinematic" transformation, which is independent of the Hamiltonian or any other time evolution. Time reversal requires an antilinear correspondence in the primary space-state vectors. Under time reversal,

$$\psi(z, t) \xrightarrow{T} U_T \psi^*(z^*, -t) \tag{4.89}$$

where U_T is some suitable unitary operator. When we have "in" and "out" states, which are labeled by free particle momenta and helicities, under time reversal the states become respectively the "out" and "in" states, the momenta are reversed, and the helicities are unchanged. Although we do not use it in the following discussion, we also mention that for spinning objects, U_T is a rotation about the 2-axis by π:

$$U_T = \exp(i\pi J_2) \tag{4.90}$$

For internal symmetries like $SU(3)$ where 3 and $\bar{3}$ are distinct, the time reversal can be invoked only on the density operators $\psi\psi^\dagger$ rather than on the field operator ψ alone. The probabilities are sesquilinear in the amplitude (or absolute value square) and are always real. The time-reversal invariance predicts the equality between the probability and the corresponding time-reversed quantity. We recall that the survival amplitude is $\langle \widetilde{M}, 0 \,|\, M, t\rangle$. Applying the time-reversal operation, we have

$$\begin{gathered}
\langle M^*, 0 \,|\, M, t\rangle \xrightarrow{T} \langle M, 0 \,|\, M^*, -t\rangle^* = e^{-iMt}\langle M^*, 0 \,|\, M, 0\rangle = e^{-iMt} \\
\langle \widetilde{M}, 0 \,|\, M, t\rangle = e^{-iMt}\langle \widetilde{M}, 0 \,|\, M, 0\rangle \xrightarrow{T} (e^{-iM^*(-t)})^*\langle M, 0 \,|\, M^*, 0\rangle = e^{-iMt} \\
\langle \widetilde{M}, 0 \,|\, M, t\rangle^* = e^{+iMt}\langle \widetilde{M^*}, 0 \,|\, M^*, 0\rangle \xrightarrow{T} e^{+iM^*t}
\end{gathered} \tag{4.91}$$

Thus, the corresponding dependence of the time-reversed probability is given by

$$|\langle M^*, 0 \,|\, M, t\rangle|^2 \xrightarrow{T} |\langle M, 0 \,|\, M^*, -t\rangle|^2 = e^{-\Gamma t} \tag{4.92}$$

So that the survival amplitude involves the inner product of the state $|M, t\rangle$ in $\mathscr{G}$ with its dual state, $\langle \widetilde{M, 0}| = \langle M^*, 0|$ in $\tilde{\mathscr{G}}$, which leads to exponential decay. Also, for the complex conjugation of the inner product between $\mathscr{G}^*$ and $\tilde{\mathscr{G}}^*$ states, it again leads to an exponential decay.

I. Two Choices for Unstable Particle States

In our study of generalized quantum-state spaces, we have given an exposition of analytic continuation of state spaces, and the corespondence between dense sets of states in $\mathscr{H}$ and $\mathscr{G}$. For analytic Hamiltonians, the spectrum can be "analytically continued" in $\mathscr{G}$. The resolution of unity embodied in the completeness identity has alternate expressions. Incidentally, this is an example of reducible representations of the (time) translation group having different decompositions in which no component of one decomposition is equivalent to any component of the other. The notions of discrete states, continuous spectra, "in" and "out" states, and exact expressions for the (ideal) states all obtain for these generalized spaces.

One could take either of two views about what is an unstable particle. One is that it is a physical state of the system which is normalizable and which *ceases to exist as a discrete eigenstate of the total Hamiltonian.* If $|M\rangle$ denotes this normalized state, the survival amplitude is

$$A(t) = \langle M|e^{-iHt}|M\rangle = \int d\lambda\, e^{-i\lambda t}\langle M|\lambda\rangle\langle\lambda|M\rangle \tag{4.93}$$

Here λ is integrated along the positive real axis. This amplitude cannot ever be strictly exponential in t and is bounded in absolute value by unity for all t, positive or negative. It exhibits a Khalfin regime where it has an inverse power dependence and a Zeno regime where the departure of its absolute value from unity is quadratic in t. But for much of the intermediate region it is approximately exponential in $|t|$. One of the drawbacks of this picture of an unstable particle is that its survival amplitude does not furnish a representation of the time-translation group or semigroup. The unstable particle so defined is not "autonomous," it *ages.*

The other picture of the unstable particle is as a discrete state in the generalized space $\mathscr{G}$ and as such has a pure exponential dependence. The time evolutions form a semigroup (for $t > 0$) with the absolute value steadily decreasing exponentially. Such a state cannot have a counterpart physical state in $\mathscr{H}$. For negative values of t, the state tends to blow up. If we start from any state in $\mathscr{H}$ which can be continued into $\mathscr{G}$, the result so obtained would never be a pure discrete decaying state, but that plus remnants of a continuum. We could extend $\mathscr{H}$ to $\tilde{\mathscr{H}}$ by relaxing the spectral condition $H \geq 0$ and obtain a state in $\mathbb{H}_{\pm}$ as in Eq. (4.72); then we could obtain a semigroup evolution law (4.76, 4.77). We have also seen that both the time evolution of the decay process and that of the time-reversed process exhibit exponential decay. Although this choice appears to be elegant, it is deduced at the expense of giving up the lower boundedness of the energy spectrum. We consider it to be the less desirable choice.

Finally, we observe that the spaces $\mathscr{H}$ and $\mathscr{G}$ that we have used are distinct spaces though there is one-to-one correspondence between dense sets of analytic vectors in $\mathscr{H}$ and $\mathscr{G}$. This correspondence can be implemented by an intertwining operator $V : \mathscr{H} \to \mathscr{G}$ with its inverse $V^{-1} : \mathscr{G} \to \mathscr{H}$ given by the formal Steiltjes integral:

$$\begin{aligned} V(z, x) &= \int d\alpha\, \psi_\alpha(z)\psi_\alpha^*(x) \\ V^{-1}(z, x) &= \int d\alpha\, \psi_\alpha(x)\psi_\alpha^*(z^*) = \int d\alpha\, \psi_\alpha(x)\tilde{\psi}_\alpha(z) \end{aligned} \tag{4.94}$$

where $\{\psi_\alpha(x)\}$ is an analytic basis in $\mathscr{H}$ and $\{\psi_\alpha(z)\}$ its counterpart in $\mathscr{G}$. Any analytic operator, including the Hamiltonian in $\mathscr{H}$, has the counterpart in $\mathscr{G}$ defined by

$$A \to VAV^{-1} \tag{4.95}$$

These operators V, V^{-1} are intertwining between the spaces $\mathscr{H}$ and $\mathscr{G}$.

Two further remarks are in order. First, we can choose to concentrate on the eigenvalue equation being reduced to an equation for the unstable state alone by using one half of the equations to eliminate the daughter product amplitude. For the Fredrichs–Lee model,

$$\begin{aligned} (\lambda - M_0)\eta_0 &= \int_0^\infty f^*(\omega')\eta(\omega')\, d\omega' \\ (\lambda - \omega)\eta(\omega) &= f(\omega)\eta_0 \end{aligned} \tag{4.96}$$

For the discrete state, the second equation can be used to solve for $\eta(\omega)$ in terms of η_0:

$$\eta(\omega) = \frac{f(\omega)}{\lambda - \omega}\eta_0; \qquad \lambda < 0 \tag{4.97}$$

Then

$$(\lambda - M_0)\eta_0 = \int \frac{f^*(\omega')f(\omega')\, d\omega'}{\lambda - \omega'} \eta_0 \tag{4.98}$$

This is a nonstandard eigenvalue equation because the right-hand side is dependent on the eigenvalue. The solution is obtained by seeking the zeros of the function

$$\alpha(z) = z - m_0 - \int_0^\infty \frac{f^*(\omega')f(\omega')\, d\omega'}{\lambda - \omega'} \tag{4.99}$$

Note that the normalization of the state includes the continuum states also, so that instead of $|\eta_0| = 1$ we must choose

$$|\eta_0| = (\alpha'(M))^{-1/2} \tag{4.100}$$

If the subspace for which the solution is attempted is not one-dimensional, we would have a nonstandard matrix eigenvalue problem:

$$A\psi = \lambda\psi = F(\lambda)\psi \tag{4.101}$$

Such a situation obtains for the Kaon decay complex. The generic theory of such reduced nonstandard eigenvalue problem is due to Livsic [96,97].

The second remark contains improper models we have seen that the survival amplitude

$$A(t) = \langle \psi | e^{iHt} | \psi \rangle \tag{4.102}$$

can be expressed as a spectral integral

$$\int_0^\infty |\psi(\lambda)|^2 e^{-i\lambda t}\, d\lambda \tag{4.103}$$

with an absolute value no greater than unity. It is tempting to introduce an effective *nonself-adjoint* Hamiltonian K with the property

$$e^{-iKt} | \psi \rangle = A(t) | \psi \rangle \tag{4.104}$$

Since for a large class of dynamical models there is an extended region for which $A(t)$ is well approximated by a complex exponential

$$A(t) \sim e^{-iE_0 i - (1/2)\Gamma t} \tag{4.105}$$

one could consider

$$K = E_0 - \frac{i}{2}\Gamma \tag{4.106}$$

as the effective Hamiltonian. This would be very similar to the Livsic operator $F(\lambda)$ mentioned above. But if K is really thought of as describing the decaying system, we get into inconsistencies: to begin with we get complex eigenvalues *before* analytic continuations. Such complex poles ("in the physical sheet") violate general principles like causality. As pointed out by Peierls, the complex poles must be obtained only by analytic continuation. We see this in the Livsic decomposition, the function $\alpha(z)$ with a cut along the real axis has no complex pole, only its continuation has a pole. Lack of care in discussing this question leads to misleading statements even in current literature.

V. GENERALIZED MULTILEVEL QUANTUM SYSTEM

In Section VI we discussed the analytic continuation in the context of one level quantum system. In this section we apply the same approach to the multilevel and multichannel quantum system. In particle physics the neutral Kaon is the most familiar and a simple example of such a system. Therefore, in this section we also devote most of our attention to the neutral Kaon system. The wave functions we will be looking at, as in Section III, take on the general form labeled by running indices, so that it can readily be adapted to the multilevel, multichannel situation.

We saw in Section III that the Lee–Oehme–Yang (LOY) model makes use of the Breit–Wigner approximation as applied to the neutral Kaon system. In Section III, we also saw that within the LOY model, the K_L and K_S wave functions are superpositions of K^0 and $\bar{K}^0$, with

$$\Psi_L = N\begin{pmatrix} p \\ q \end{pmatrix}, \qquad \Psi_S = N\begin{pmatrix} p \\ -q \end{pmatrix} \tag{5.1}$$

Should one define the corresponding bra states to be the hermitian conjugate of the ket state, that is, $\langle K_\alpha | = | K_\alpha \rangle^\dagger$, one would arrive at

$$\langle K_S | K_S \rangle = \langle K_L | K_L \rangle = N^2(|p|^2 + |q|^2) \tag{5.2}$$

and

$$\langle K_S | K_L \rangle = N^2(|p|^2 - |q|^2) \tag{5.3}$$

One might ask why, if K_S and K_L are distinct eigenstates of the Hamiltonian, are the two states not "orthogonal," that is, $\langle K_S | K_L \rangle = 0$?

The answer is that the wave functions here are eigenfunctions of an effective Hamiltonian, namely, the operator "K" defined in Eq. (3.48). Although the total Hamiltonian of Eq. (3.19) is hermitian, the operator K is not; K is a 2×2 nonhermitian matrix. So the ψ_L and ψ_S given above are eigenfunctions with complex eigenvalues. The operation of complex conjugation takes the state with eigenvalue M into another state with the complex conjugated eigenvalue M^*. Thus, for a complex value of M, the product

$$\langle K_S | K_L \rangle \tag{5.4}$$

is an ill-defined product. This is not an inner product! This difficulty was recognized soon after the proposal of the LOY model. The resolution was found through working with the left eigenstates or the dual states. We denote the dual states by $\langle \tilde{K}_S |$ and $\langle \tilde{K}_L |$. Here the orthogonality relations should hold:

$$\langle \tilde{K}_S | K_L \rangle = 0, \qquad \langle \tilde{K}_L | K_S \rangle = 0$$

and we may choose

$$\langle \tilde{K}_S | K_S \rangle = \langle \tilde{K}_L | K_L \rangle = 1 \tag{5.5}$$

The notion of left eigenstates is well known. In the context of the neutral Kaon system, it was explained in detail by Sachs [70] over three decades ago. In his paper, Sachs worked in the same approximation as in the LOY model, where the K_L and K_S states are assumed to be the superposition of K^0 and $\bar{K}^0$. The continuum component of the wave functions is being neglected. In the theory presented in Section III, the continuum channel contribution is included explicitly. As we shall see, the inclusion of this piece leads to the exact orthogonality relation.

Our discussions in the remainder of this section are divided into four parts. In the first part, we recall the conventional solution of the theory as presented in Section III. In the second part, we demonstrate how the exact orthogonality relation alluded to above is obtained and show that when the continuum contribution is suppressed, it gives an approximate orthogonality relation. In the third part, we present the completeness and orthogonality properties of the analytically continued wave functions, which display the generalized spectrum of discrete states with complex eigenvalues together with the continuum states defined along a complex contour. In the last part, based on the analytically continued theory, we present a derivation of the refined version of the Bell–Steinberger relation [56].

A. Solution of Present Multilevel Model

In Section III, we saw that the continuum eigenfunctions take the form

$$\psi_\lambda = \begin{bmatrix} a_\lambda \\ b_\lambda(\omega) \end{bmatrix} \tag{5.6}$$

where

$$b_\lambda(\omega) = \delta(\lambda - \omega)I + \frac{g^\dagger(\omega)a_\lambda}{\lambda - \omega + i\epsilon} \tag{5.7}$$

and

$$Ka_\lambda = g \quad \text{with } K = \lambda I - m - G(\lambda) \tag{5.8}$$

where

$$\begin{aligned} G(\lambda + i\epsilon) &= \left\langle \frac{g(\omega)g^\dagger(\omega)}{\lambda - \omega + i\epsilon} \right\rangle \\ &= \int_0^\infty d\omega \, \frac{g(\omega)g^\dagger(\omega)}{\lambda - \omega + i\epsilon} \end{aligned} \tag{5.9}$$

If the discrete solution occurs at $\lambda = M$,

$$b_M = \frac{g^\dagger(\omega)a_M}{M - \omega + i\epsilon} \tag{5.10}$$

where a_M satisfies the equation at $\lambda = M$

$$\begin{aligned} Ka_\lambda = 0 \quad &\text{or} \quad [m + G(\lambda)]a_\lambda = \lambda a_\lambda \\ Ka_M = K(M)a_M &= (m + G(M))a_M = Ma_M \end{aligned} \tag{5.11}$$

If we identify K_L and K_S to be the second-sheet poles, we can define the unitarity cut in such a manner as to expose these poles (see Fig. 8). The

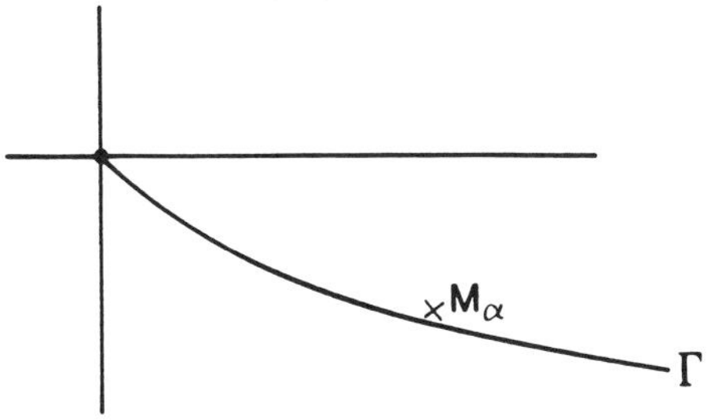

Figure 8. The contour Γ and the exposed pole at M_α.

corresponding analytically continued wave function is given by

$$\psi_\alpha = \begin{pmatrix} \eta_\alpha \\ \phi_{p\alpha} \end{pmatrix} = N_\alpha \begin{bmatrix} c_{k\alpha} \\ \dfrac{g_{pj}(\omega)c_{j\alpha}}{M_\alpha - \omega + i\epsilon} \end{bmatrix} \tag{5.12}$$

where $+i\epsilon$ serves as a reminder that the second sheet is now partially exposed and M_α is above the Γ-cut. The corresponding dual wave function of the discrete state at $\lambda = M_\beta$ is given by

$$\phi_\beta = (\chi_\beta, \zeta_\beta) = N_\beta \left(d_{\beta k}, \frac{d_{\beta k}\,\tilde{g}_{kp}}{M_\beta - \omega + i\epsilon} \right) \tag{5.13}$$

Here again, M_β is above the Γ-cut.

B. The Inner Product $\langle M_\beta^* \mid M_\alpha \rangle$

We denote the discrete eigenstate by K_L and K_S. Similar to the approach of Section III.A, we have

$$c_L = N_L \begin{pmatrix} p_L \\ q_L \end{pmatrix} \quad \text{and} \quad c_S = N_S \begin{pmatrix} p_S \\ -q_S \end{pmatrix} \tag{5.14}$$

except that p and q now depend on λ, which are evaluated at $\lambda = M_L$ and M_S. The N's are the normalization factors yet to be determined.

For the dual wave function, we proceed to solve for $d_\beta = (r, s)$ based on

$$(r, s)\begin{pmatrix} A_\lambda & B_\lambda \\ C_\lambda & A_\lambda \end{pmatrix} = \lambda(r, s) \tag{5.15}$$

Taking the transpose we have

$$\begin{pmatrix} A_\lambda & C_\lambda \\ B_\lambda & A_\lambda \end{pmatrix}\begin{pmatrix} r \\ s \end{pmatrix} = \lambda \begin{pmatrix} r \\ s \end{pmatrix} \tag{5.16}$$

Comparison with Eq. (3.48) reveals that, analogously to Eq. (3.50), it can be shown that

$$\left(\frac{r}{s}\right)^2 = \frac{C_\lambda}{B_\lambda}, \qquad \frac{r}{s} = \pm\sqrt{\frac{C_\lambda}{B_\lambda}} \tag{5.17}$$

In other words, for the K_L and K_S dual states,

$$d_L = N_L(q_L, p_L), \qquad d_S = N_S(-q_S, p_S) \tag{5.18}$$

1. *Orthonormality Relations*

The inner product of a discrete state labeled by α with another dual discrete state $\langle \beta |_D \equiv \langle \beta^* |$, is given by

$$\begin{aligned} \langle \beta^* | \alpha \rangle &= N_\alpha N_\beta \left[d_{\beta k}, \frac{d_{\beta j} \tilde{g}_{jp}(\omega)}{M_\beta - \omega + i\epsilon} \right] \left[\begin{matrix} c_{k\alpha} \\ \dfrac{g_{pj}(\omega) c_{j\alpha}}{M_\alpha - \omega + i\epsilon} \end{matrix} \right] \\ &= N_\alpha N_\beta \, d_{\beta k} \left[\delta_{kj} + \int_\Gamma d\omega \, \frac{\tilde{g}_{kp}(\omega) g_{pj}(\omega)}{(M_\beta - \omega)(M_\alpha - \omega)} \right] c_{j\alpha} \end{aligned} \tag{5.19}$$

The discontinuity of K across the Γ-cut can be read off from Eq. (5.11), and is found to be

$$K(\lambda + i\epsilon) - K(\lambda - i\epsilon) = 2\pi i \tilde{g}(\lambda) g(\lambda) \tag{5.20}$$

Thus, the integral in Eq. (5.19) can be deformed in the following manner: (see Fig. 9)

$$\begin{aligned} \int_\Gamma dw \, \tilde{g}_{kp}(\omega) g_{pj}(\omega) \cdots &= \frac{1}{2\pi k} \int_\Gamma dw \, [K_{kj}(\omega + i\epsilon) - K_{kj}(\omega - i\epsilon)] \cdots \\ &= \frac{-1}{2\pi k} \int_C dw \, K_{kj}(\omega) \cdots \end{aligned} \tag{5.21}$$

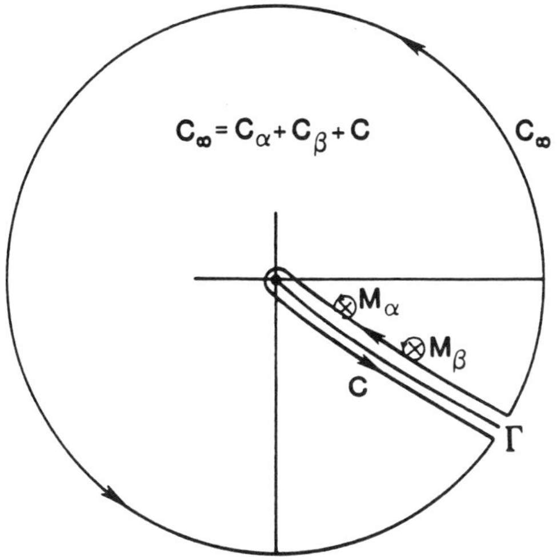

Figure 9. Relationship between C_∞ and those counterclockwise contours enclosed by C_∞.

where, as indicated in Fig. 9 the contour C wraps around the Γ-cut in a counterclockwise manner. The equations for the discrete solutions are

$$\begin{aligned} K_{kj}(M_\alpha)c_{j\alpha} &= 0 \quad \text{at } \lambda = M_\alpha \\ d_{\beta k} K_{kj}(M_\beta) &= 0 \quad \text{at } \lambda = M_\beta \end{aligned} \tag{5.22}$$

Inspection of the contours in Fig. 9 reveals

$$C_\infty = C_\alpha + C_\beta + C \tag{5.23}$$

The corresponding integrals are related by

$$I_C = -I_\alpha - I_\beta + I_\infty \quad \text{with } I_C = \frac{-d_\beta}{2\pi k} \int_C \frac{dw\ K(\omega)c_\alpha}{(M_\beta - \omega)(M_\alpha - \omega)} \tag{5.24}$$

Consider first the case $\beta \neq \alpha$. On account of Eq. (5.22),

$$I_\alpha = \frac{-1}{2\pi k} \int_{C_\alpha} \frac{dw}{(M_\beta - \omega)(M_\alpha - \omega)}\ d_\beta K(\omega)c_\alpha = 0 \tag{5.25}$$

Similarly,

$$I_\beta = 0 \tag{5.26}$$

From Eq. (3.31), the asymptotic behavior of $K(\omega)$ is

$$K_{kj}(\omega) \xrightarrow[\omega \to \infty]{} \omega \delta_{kj}(\omega) \tag{5.27}$$

$$\begin{aligned}(I_\infty)_{\beta\alpha} &= -\frac{1}{2\pi k} \int_{C_\infty} \frac{dw}{(M_\beta - \omega)(M_\alpha - \omega)} d_{\beta k}(\omega \delta_{kj}) c_{j\alpha} \\ &= -d_{\beta k}\, \delta_{kj}\, c_{k\alpha}\end{aligned} \tag{5.28}$$

Substituting Eqs. (5.25), (5.26), and (5.28) into Eq. (5.24), using the definition in Eq. (5.21), the inner product (5.19) becomes

$$\langle M_\beta^* \,|\, M_\alpha \rangle = N_\alpha N_\beta\, d_{\beta k} [\delta_{kj} + (-0 - 0 - \delta_{kj})] c_{j\alpha} = 0 \tag{5.29}$$

For the case $\beta = \alpha$,

$$\begin{aligned}\langle M_\alpha^* \,|\, M_\alpha \rangle = 1 &= N_\alpha^2\, d_{\alpha k} \left[\delta_{kj} + \int \frac{\tilde{g}_{kq}(\omega) g_{qj}(\omega)\, dw}{(M_\alpha - \omega)^2} \right] c_{\alpha j} \\ &= N_\alpha^2\, d_{\alpha k}\, K'_{kj}\, c_{j\alpha}\end{aligned} \tag{5.30}$$

with

$$K' = \left[\frac{dK}{d\lambda} \right]_{\lambda = M_\alpha}$$

Thus the normalization is given by

$$N_\alpha = [d_\alpha K' c_\alpha]^{-1/2} \tag{5.31}$$

2. *The "Overlap Function"*

The contribution to the overlap function from the discrete V components alone (i.e., V_1 and V_2 components) or the K^0 and $\bar{K}^0$ alone, is given by

$$\langle \beta^* \,|\, \alpha \rangle_V \equiv \sum_i \langle M_\beta^* \,|\, V_i \rangle \langle V_i \,|\, M_\alpha \rangle = N_\beta N_\alpha\, d_\beta\, c_\alpha \tag{5.32}$$

For $\alpha = K_L$ and $\beta = K_S$, using Eq. (5.18), the right-hand side of Eq. (5.32) becomes

$$\langle S^* \mid L \rangle_V = N_S N_L(-q_S p_S)\begin{pmatrix} p_L \\ q_L \end{pmatrix}$$

$$= N_S N_L(p_S q_L - q_S p_L) \tag{5.33}$$

Strictly speaking, since (p_L, q_L) and (p_S, q_S) are evaluated at M_L and M_S, $RHS \neq 0$. However, to the extent that the energy dependence of the coupling function $g(\omega)$ in the analytically continued Hamiltonian can be neglected, $p_L \approx p_S$, $q_L \approx q_S$, or $RHS \approx 0$.

This approximate result was discussed for instance in the work of Sachs [70]. Our contribution in this section is the demonstration of the exact orthogonality relation between the state α and its dual state β. More specifically, when the form-factor effect is taken into account, even though the discrete part alone (5.33) no longer vanishes, with the inclusion of the continuum contribution, the orthogonality relation holds exactly.

C. Continued Wave Functions and Continued Spectrum

Thus far we have looked at the discrete solutions in the analytically continued theory with the continuum states defined along the contour Γ. Hereafter, we refer to it as the "Γ-theory." The continuum states and the dual states are defined along the same contour Γ. We proceed to display the complete set of wave functions, including both the discrete states and the continuum states, and to investigate their orthonormality properties and completeness relations. Some of the calculations were given in Section V.B, and the remainder can be found in Appendices A–C of reference 55.

1. *Complete Set of Wave Functions*

1. *Discrete States.* From Eqs. (5.12) and (5.13), the wave functions of the discrete states and corresponding dual wave functions are given by

$$\Psi_\alpha \equiv \begin{pmatrix} \langle V_i \mid \alpha \rangle \\ \langle \widetilde{N\theta_p} \mid \alpha \rangle \end{pmatrix} \equiv \begin{pmatrix} \eta_\alpha \\ \phi_{p\alpha} \end{pmatrix} = N_\alpha \begin{bmatrix} c_{k\alpha} \\ \dfrac{g_{pj}(\omega) c_{j\alpha}}{M_\alpha - \omega} \end{bmatrix} \tag{5.34}$$

$$\widetilde{\Phi_\beta} \equiv (\langle \beta^* \mid V_i \rangle, \langle \beta^* \mid N\theta_p \rangle) \equiv (\chi_{\beta k}, \zeta_{\beta p})$$

$$= N_\beta \left[d_{\beta k}, \frac{d_{\beta k} \tilde{g}_{kp}(\omega)}{M_\beta - \omega} \right] \tag{5.35}$$

2. *Continuum States.* From Eqs. (5.6) and (5.7), the continuum wave functions and their dual wave functions are given by

$$\Psi_\lambda^r \equiv \begin{pmatrix} \langle V_i \mid \lambda r\rangle \\ \langle \widetilde{N\theta_p} \mid \lambda r\rangle \end{pmatrix} \equiv \begin{pmatrix} \eta_{k\lambda} \\ \phi_{p\lambda}^r \end{pmatrix} = \begin{bmatrix} a_{k\lambda} \\ \delta(\lambda-\omega)\delta_{rp} + \dfrac{g_{pj}(\omega)a_{j\lambda}}{\lambda-\omega+i\epsilon} \end{bmatrix} \tag{5.36}$$

$$\widetilde{\Phi_\lambda^r} \equiv (\langle \lambda^* r \mid V_i\rangle, \langle \lambda^* r \mid N\theta_p\rangle) \equiv (\chi_{\lambda k}, \zeta_{\lambda p}^r) = \left[\tilde{a}_{\lambda k}, \delta(\lambda-\omega)\delta_{pr} + \frac{\tilde{a}_{\lambda j}\tilde{g}_{jp}(\omega)}{\lambda-\omega-i\epsilon}\right] \tag{5.37}$$

From Eq. (5.8),

$$Ka = \tilde{g}, \qquad a = K^{-1}\tilde{g}, \qquad \tilde{a} = g\widetilde{K^{-1}} \tag{5.38}$$

2. *Orthonormality Relations*

The identity operator in the bare basis is

$$I = \mid V_i\rangle\langle V_i \mid + \int_\Gamma dw \mid N\theta_p(\omega)\rangle\langle N\theta_p(\omega^*) \mid \tag{5.39}$$

where summations over k and p are understood. The expected orthonormality relations are

$$\begin{aligned}\langle \beta^* \mid \alpha\rangle &= \langle \beta^* \mid V_i\rangle\langle V_i \mid \alpha\rangle + \int_\Gamma dw \langle \beta^* \mid N\theta_p\rangle\langle \widetilde{N\theta_p} \mid \alpha\rangle \\ &= \chi_{\beta k}\eta_{k\alpha} + \int_\Gamma dw\, \zeta_{\beta p}(\omega)\phi_{p\alpha}(\omega) = \delta_{\alpha\beta}\end{aligned} \tag{5.40}$$

$$\langle \lambda^*, r \mid \alpha\rangle = \chi_{\lambda k}\eta_{k\alpha} + \int_\Gamma dw\, \zeta_{\lambda p}(\omega)\phi_{p\alpha}(\omega) = 0 \tag{5.41}$$

$$\langle \beta^* \mid \lambda, r\rangle = \chi_{\beta k}\eta_{k\lambda} + \int dw\, \zeta_{\beta p}(\omega)\phi_{p\lambda}^r(\omega) = 0 \tag{5.42}$$

$$\langle \lambda^*, r \mid \mu, s\rangle = \chi_{\lambda k}^r\eta_{k\mu}^s + \int dw\, \zeta_{\lambda p}^r(\omega)\phi_{p\mu}^s(\omega) = \delta(\lambda-\mu)\delta_{rs} \tag{5.43}$$

The proof of Eq. (5.40) is given in the previous section [see Eqs. (5.29) and (5.30)]. The remaining relations are proved in Appendix A of reference 55.

3. *The Completeness Relations*

The spectrum in the analytic continued theory consists of the discrete states K_L and K_S at the complex energies M_L and M_S, respectively. This defines a space $\mathcal{G}$, where the identity operator is given by

$$I = |\alpha\rangle\langle\alpha^*| + \int_\Gamma d\lambda\, |\lambda r\rangle\langle\lambda^* r| \tag{5.44}$$

Again, summation over the discrete labels α and r is understood. The identity operator leads to following set of completeness relations:

$$\langle V_i|V_j\rangle = \eta_{k\alpha}\chi_{\alpha j} + \int_\Gamma d\lambda\ \eta_{k\lambda}\chi_{\lambda j} = \delta_{kj} \tag{5.45}$$

$$\langle \widetilde{N\theta}_p|V_i\rangle = \phi_{p\alpha}\chi_{\alpha k} + \int_\Gamma d\lambda\ \phi_{p\lambda}\chi_{\lambda k} = 0 \tag{5.46}$$

$$\langle V_i|N\theta_q\rangle = \eta_{k\alpha}\zeta_{\alpha q} + \int_\Gamma d\lambda\ \eta_{k\lambda}\zeta_{\lambda q} = 0 \tag{5.47}$$

$$\langle \widetilde{N\theta}_p(\omega)|N\theta_q(\omega')\rangle = \phi_{p\alpha}\zeta_{\alpha q} + \int_\Gamma d\lambda\ \phi_{p\lambda}\zeta_{\lambda q} = \delta(\omega-\omega')\delta_{pq} \tag{5.48}$$

The proofs of these relations are given in Appendices B and C of reference 55.

D. Derivation of the Bell–Steinberger Relation

The Bell–Steinberger relation [56] is usually associated with the unitarity relation. It is instructive to see how the corresponding relation arises within the present framework. We recall that the equation of the discrete solution is given by [see Eq. (5.11)]

$$K^{\alpha}_{kj}a_j = 0 \tag{5.49}$$

where

$$K_{kj}(\lambda) = \lambda\delta_{kj} - E_{kj}(\lambda) - m_{kj} \tag{5.50}$$

With analytic continuation one gets

$$E_{kj}(\lambda) = \int_\Gamma \frac{g^{+}_{kq}(\omega^*)g_{qj}(\omega)}{\lambda - \omega + i\epsilon} \tag{5.51}$$

We deform the unitarity cut running along the positive real axis to the contour Γ such that it "exposes" the discrete-state solution (see Fig. 9). In terms of the E-function, the discrete solution at $\lambda = M_\alpha$ is given by

$$[m_{kj} + E_{kj}(\lambda)]a_\lambda = M_\alpha a_\lambda \tag{5.52}$$

Taking the hermitian conjugate for the discrete solution at $\lambda = M_\beta$ gives

$$a_\lambda^+ [m^+ + E^+(\lambda)] = M_\beta^* a_\lambda^+ \tag{5.53}$$

But

$$[E(M_\beta)]^+ \int_{\Gamma^*} dw' \frac{g^+(\omega'^*)g(\omega')}{M_\beta^* - \omega' - i\epsilon} \equiv E(M_\beta^*) \tag{5.54}$$

where

$$[g^+(\omega^*)g(\omega)]^+ = g^+(\omega)g(\omega^*) = g^+(\omega'^*)g(\omega') \tag{5.55}$$

and $\omega' = \omega^*$ were used.

We assume each Yukawa coupling function in the Hamiltonian can be characterized by a coupling constant g_{pk} and a cutoff L_p. To evaluate $E(z + i\epsilon)$, when there is one discrete solution in the lower half plane, we choose the contour Γ such that it barely misses the point z. The principal value part

$$\begin{aligned} P[E(z)] &= \sum_p g_{jp}^+ g_{pk} \left[\int_0^{z-\delta} \frac{dw}{z - \omega + i\epsilon} + \int_{z+\delta}^{L_p} \frac{dw}{z - \omega + i\epsilon} \right] \\ &= \sum_p g_{jp}^+ g_{pk} \left[\ln \frac{z}{\delta} - \ln \frac{L_p}{\delta} \right] \\ &= \sum_p g_{jp}^+ g_{pk} \ln \frac{z}{L_p} \end{aligned} \tag{5.56}$$

Using the identity

$$\frac{1}{z - \omega \pm i\epsilon} = P \frac{1}{z - \omega} \mp i\pi\delta(z - \omega) \tag{5.57}$$

$$E(z + i\epsilon) = \sum_p g_{jp}^+ g_{pk} \ln\left(\frac{z}{L_p} e^{-i\pi} \right) \tag{5.58}$$

Assuming the bare mass matrix (m_{kj}) is hermitian, Eqs. (5.52) and (5.53) lead to

$$a^+_{\beta k}[E_{kj}(M^*_\beta) - E_{kj}(M_\alpha)]a_{j\alpha} = (M^*_\beta - M_\alpha)a^+_{\beta k}\, a_{k\alpha}$$
$$= \sum_p (a^+_{\beta j}\, g^+_{jp})(g_{pk}\, a_{k\alpha})\left[2\pi i + \ln \frac{M^*_\beta}{M_\alpha}\right] \quad (5.59)$$

The last equality is a refined version of the Bell–Steinberger relation which was deduced using the present theory.

For the Kaon system, both the mass and the width differences between K_L and K_S are small compared to the mean Kaon mass, that is,

$$\frac{M^*_\beta - M_\alpha}{M_\alpha} \ll 1 \quad (5.60)$$

or

$$\ln\left(\frac{M^*_\beta}{M_\alpha}\, e^{2\pi k}\right) \approx 2\pi i + \frac{M^*_\beta - M_\alpha}{M_\alpha} \approx 2\pi i \quad (5.61)$$

Denote

$$\text{“}\langle \beta \mid \alpha \rangle\text{”} \equiv a^+_{\beta k}\, a_{k\alpha}$$
$$\gamma^\alpha_p \equiv g_{pk}\, a_{k\alpha} \quad (5.62)$$
$$\gamma^{\beta+}_p \equiv a^+_{\beta k}\, g^+_{kp}$$

Equation (5.59) in the approximation of Eq. (5.61) is reduced to the original form of the Bell–Steinberger relation:

$$\text{“}\langle \beta \mid \alpha \rangle\text{”}(M^*_\beta - M_\alpha) = 2\pi i \sum_p \gamma^{\beta+}_p \gamma^\alpha_p \quad (5.63)$$

E. Summary

We have presented a theory for the neutral Kaon system based on the extended Lee model. The spectrum of the theory consists of the discrete states on the second sheet, which are the K_L and K_S states and the continuum states defined along a contour Γ. The spectrum spans the space $\mathcal{G}$. The bra states here are dual states of the ket states. For the discrete states, both the bra and ket states are at $\lambda = M$. For the continuum states, if the ket state is defined at $\lambda + i\epsilon$ along the upper lip of the contour Γ, the bra state is at $\lambda - i\epsilon$ along the lower lip of Γ.

Our analysis indicates that the nonvanishing of the "$\langle K_L | K_S \rangle$" in LOY theory is related to the fact that the quantity does not correspond to a properly defined amplitude. If the properly defined amplitude corresponds to the inner product between a state in the $\mathscr{G}$ space and a dual state in the $\tilde{\mathscr{G}}$ space, $\langle K_L^* | K_S \rangle$ is expected to vanish. As we see in Section V.B.1, it does.

Finally, based on our present theory, we derived a refined version of the Bell–Steinberger relation. The refinement differs from the original relation in the order of $O[(M_L^* - M_S)/M_S]$. Although this difference is insignificant for the neutral Kaon, $D^0\bar{D}^0$, $B^0\bar{B}^0$ systems, it still remains a challenge to look for quantum systems in nature where such correction does lead to a detectable effect.

VI. THE CASCADE MODEL

Up to this point, we confined our attention to two-body channels. In either the one-level system of Section IV or the multilevel system of Section V, the second-sheet singularities are simple poles. In this section we look at the quantum system which admits the decay into three-body channels. Here, in addition to second-sheet poles, there may also be second-sheet branch cuts. We consider a simple three-body model, namely, the cascade model, which is an exactly solvable model [44].

A. The Model

We consider a Hamiltonian system [44] where there are three classes of states for the unperturbed Hamiltonian; a particle A with bare energy M_0; a two-particle continuum with energy $\mu_0 + \omega$, $0 < \omega < \infty$; and a three-particle continuum with energy $\omega + \nu$, $0 \leq \omega, \nu < \infty$. We denote the amplitudes for these by η, $\phi(\omega)$, and $\psi(\omega, \nu)$ and the scalar product is given by

$$\eta_1^*\eta_2 + \int_0^\infty \phi_1^*(\omega)\phi_2(\omega)\, d\omega + \int_0^\infty \int \psi_1^*(\omega, \nu)\psi_2(\omega, \nu)\, d\omega\, d\nu < \infty \tag{6.1}$$

The vector space $\mathscr{H}$ of states is the completion of this vector space. The total Hamiltonian and eigenvalue equation are given by

$$\begin{bmatrix} M_0 & f^*(\omega') & 0 \\ f(\omega) & (\mu_0 + \omega)\delta(\omega - \omega') & g^*(\nu')\delta(\omega - \omega') \\ 0 & g(\nu)\delta(\omega - \omega') & (\omega + \nu)\delta(\omega - \omega')\delta(\nu - \nu') \end{bmatrix} \times \begin{bmatrix} \eta_\lambda \\ \phi_\lambda(\omega') \\ \psi_\lambda(\omega'\nu') \end{bmatrix} = \lambda \begin{bmatrix} \eta_\lambda \\ \phi_\lambda(\omega) \\ \psi_\lambda(\omega\nu) \end{bmatrix} \tag{6.2}$$

B. The Eigenstates

The energy eigenvalues are degenerate and infinitely degenerate once the three-particle channel becomes open. We can enumerate the (ideal) eigenstates of Eq. (6.2) in the following form:

$$
|\lambda n\rangle \equiv \begin{bmatrix} \eta_{\lambda n} \\ \phi_{\lambda n}(\omega) \\ \psi_{\lambda n}(\omega\nu) \end{bmatrix} = \begin{bmatrix} \dfrac{f^*(\lambda - n)}{\alpha(\lambda + i\epsilon} \, \dfrac{g^*(n)}{\gamma(n + i\epsilon)} \\[2ex] \dfrac{g^*(n)\delta(\lambda - \omega - n)}{\gamma(\lambda - \omega + i\epsilon)} + \dfrac{f(\omega)\eta_{\lambda n}}{\gamma(\lambda - \omega + i\epsilon)} \\[2ex] \delta(\nu - n)\delta(\lambda - \omega - n) + \dfrac{g(\nu)}{\lambda - \omega - \nu + i\epsilon}\, \phi_{\lambda n}(\omega) \end{bmatrix} \tag{6.3}
$$

where $0 \leq n \leq \lambda < \infty$, and

$$
\begin{aligned}
\alpha(z) &= z - M_0 - \int_0^\infty \frac{f^*(\omega')f(\omega')}{\gamma(z - \omega' + i\epsilon)} \\
\gamma(z) &= z - \mu_0 - \int_0^\infty \frac{g^*(\nu)g(\nu)}{z - \nu + i\epsilon}\, d\nu
\end{aligned} \tag{6.4}
$$

If there is a real value μ such that

$$
\gamma(\mu) = 0; \qquad \gamma' = \left.\frac{\partial \gamma(z)}{\partial z}\right|_{z=\mu} \tag{6.5}
$$

there exists a two-particle, one-parameter family:

$$
|\tau\rangle = \begin{bmatrix} \eta_\tau \\ \phi_\tau(\omega) \\ \psi_\tau(\omega\nu) \end{bmatrix} = \begin{bmatrix} \dfrac{f^*(\tau - \mu)}{\sqrt{\gamma'}\,\alpha(\tau + i\epsilon)} \\[2ex] \dfrac{1}{\sqrt{\gamma'}}\,\delta(\tau - \mu - \omega) + \dfrac{f(\omega)}{\gamma(\tau - \omega + i\epsilon)}\,\eta_\tau \\[2ex] \dfrac{g(\nu)}{\tau - \omega - \nu + i\epsilon}\,\phi_\tau(\omega) \end{bmatrix} \tag{6.6}
$$

Note that λ and τ vary over ranges differing by μ so that

$$
0 < \lambda, \qquad (\tau - \mu) < \infty
$$

If there is a real value M such that

$$
\alpha(M) = 0; \qquad \alpha' = \left.\frac{\partial \alpha(z)}{\partial z}\right|_{z=M} \tag{6.7}
$$

then there exists a discrete state

$$|\lambda n\rangle \equiv \begin{bmatrix} \eta_{\lambda n} \\ \phi_{\lambda n}(\omega) \\ \psi_{\lambda n}(\omega v) \end{bmatrix} = \frac{1}{\sqrt{\alpha'}} \begin{bmatrix} 1 \\ \dfrac{f(\omega)}{\gamma(M-\omega)} \\ \dfrac{g(v)f(\omega)}{\gamma(M-\omega)(M-\omega-v)} \end{bmatrix} \tag{6.8}$$

C. Orthonormality Relations

These states are (ideal) normalized. By a straightforward calculation, they can be shown to be mutually orthogonal. We can also show them to be complete. The best way is to compute $\iint d\omega'\, dv'\, \psi^*(\omega'v')\psi(\omega'v')$ and convert it into a contour integral. If there are zeros of $\gamma(z)$ they will compensate the one-parameter continuum and so on, and we may obtain

$$\begin{gathered} \langle M \,|\, M\rangle = 1, \qquad \langle M \,|\, \tau\rangle = 0, \qquad \langle M \,|\, \lambda n\rangle = 0 \\ \langle \tau' \,|\, \tau\rangle = \delta(\tau' - \tau), \qquad \langle \tau' \,|\, \lambda n\rangle = 0 \\ \langle \lambda' n' \,|\, \lambda n\rangle = \delta(\lambda - \lambda')\delta(n - n') \end{gathered} \tag{6.9}$$

and

$$\begin{gathered} \iint \psi_{\lambda n}(\omega'v')\psi^*_{\lambda n}(\omega v)\, d\lambda\, dn + \int \psi_\tau(\omega'v')\psi^*_\tau(\omega v)\, d\tau \\ + \psi_0(\omega'v')\psi^*_0(\omega v) = \delta(\omega - \omega')\delta(v - v') \\ \iint \psi_{\lambda n}(\omega'v')\phi^*_{\lambda n}(\omega)\, d\lambda\, dn + \int \psi_\tau(\omega'v')\phi^*_\tau(\omega)\, d\tau + \phi_0(\omega'v')\phi^*_0(\omega) = 0 \\ \iint \psi_{\lambda n}(\omega'v')\eta^*_{\lambda n}\, d\lambda\, dn + \int \psi_\tau(\omega'v')\eta^*_\tau\, d\tau + \psi_0(\omega'v')\eta^*_0 = 0 \qquad (6.10) \\ \iint \phi_{\lambda n}(\omega')\phi^*_{\lambda n}(\omega)\, d\lambda\, dn + \int \phi_\tau(\omega')\phi^*_\tau(\omega)\, d\tau + \phi_0(\omega')\phi^*_0(\omega) = \delta(\omega' - \omega) \\ \iint \phi_{\lambda n}(\omega')\eta^*_{\lambda n}\, d\lambda\, dn + \int \phi_\tau(\omega')\eta^*_\tau\, d\tau + \phi_0(\omega')\eta^*_0 = 0 \\ \iint \eta_{\lambda n}\eta^*_{\lambda n}\, d\lambda\, dn + \int \eta_\tau\eta^*_\tau\, d\tau + \eta_0\eta^*_0 = 1 \end{gathered}$$

D. Continuation of Scattering Amplitudes and Unitarity Relations

To study analytic continuation [55] with complex branch cuts, we choose M_0 and μ_0 sufficiently positive so that there is no real zero for $\gamma(z)$ or $\alpha(z)$. Then the only states in $\mathscr{H}$ which are (ideal) eigenstates are $|\lambda n\rangle$ and these are complete in the sense of Eq. (6.10). The S-matrix elements are

$$\langle \lambda n, \text{out} | \lambda' n', \text{in} \rangle = \delta(\lambda - \lambda') \cdot \{\delta(n - n') + 2iT(n, n'; \lambda)\} \tag{6.11}$$

$$T(n, n'; \lambda) = -\pi \left\{ \alpha(\lambda + i\epsilon)\eta_{\lambda n}\eta_{\lambda n'} + \frac{g^*(n)g(n)}{\gamma(n + i\epsilon)}\,\delta(n - n') \right\} \tag{6.12}$$

Both the S- and T-matrix elements considered as a function of λ can be viewed as analytic functions of (complex) energy z with a branch cut $0 < z < \infty$. Because by hypotheses $\gamma(\zeta)$ has no real zero, we would find a complex zero at μ_1 in the lower half plane as we deform the branch cut from that along the positive real axis to the appropriate contour in the fourth quadrant. This pole induces a branch cut in $T(n, n'; \lambda)$ from μ_1 to infinity along a contour of our choice. So we can have, as illustrated in Fig. 10, the choice of the contours Γ_1, $\Gamma_2 + \Gamma_2'$, or $\Gamma_3 + \Gamma_3' + \Gamma_3''$. For $\Gamma_2 + \Gamma_2'$, we have the complex branch cut beginning at μ_1. For $\Gamma_3 + \Gamma_3' + \Gamma_3''$, we have the complex branch cut beginning at μ_1 and the pole at M_1.

These analytic properties signal the possibility of analytic continuation of the space $\mathscr{H}$ into $\mathscr{G}$. For the contour Γ_1, we get the complete set of states $|z, \zeta\rangle$:

$$|z, \zeta\rangle = \left\{ \begin{array}{l} \dfrac{f^*(z^* - \zeta^*)g^*(\zeta^*)}{\alpha(z + i\epsilon)\gamma(\zeta + i\epsilon)} \\[2ex] \dfrac{g^*(\zeta^*)\delta(z - \zeta - \xi)}{\gamma(z - \xi + i\epsilon)} + \dfrac{f(\xi) \cdot f^*(z^* - \zeta^*)g^*(\zeta^*)}{\alpha(z + i\epsilon)\gamma(\zeta + i\epsilon)\gamma(z - \xi + i\epsilon)} \\[2ex] \delta(\zeta - \nu)\delta(z - \xi - \nu) + \dfrac{g(\nu)}{z - \xi - \nu + i\epsilon} \\[2ex] \quad \times \left[\dfrac{g^*(\zeta^*)\delta(z - \zeta - \xi)}{\gamma(z - \xi + i\epsilon)} + \dfrac{f(\xi)f^*(z^* - \zeta^*)g^*(\zeta^*)}{\alpha(z + i\epsilon)\gamma(\zeta + i\epsilon)\gamma(z - \xi + i\epsilon)} \right] \end{array} \right\} \tag{6.13}$$

where z lies on the contour Γ_1 and we may choose $\xi + \nu$, ζ, and ν also to lie on this contour. By a lengthy but straightforward calculation using the conversion of open contour integrals into closed contour integrals, we can show that Eq. (6.13) constitutes a complete (ideal) orthonormal system. Neither the zeros of α nor of γ are in the complex plane cut along Γ_1 and,

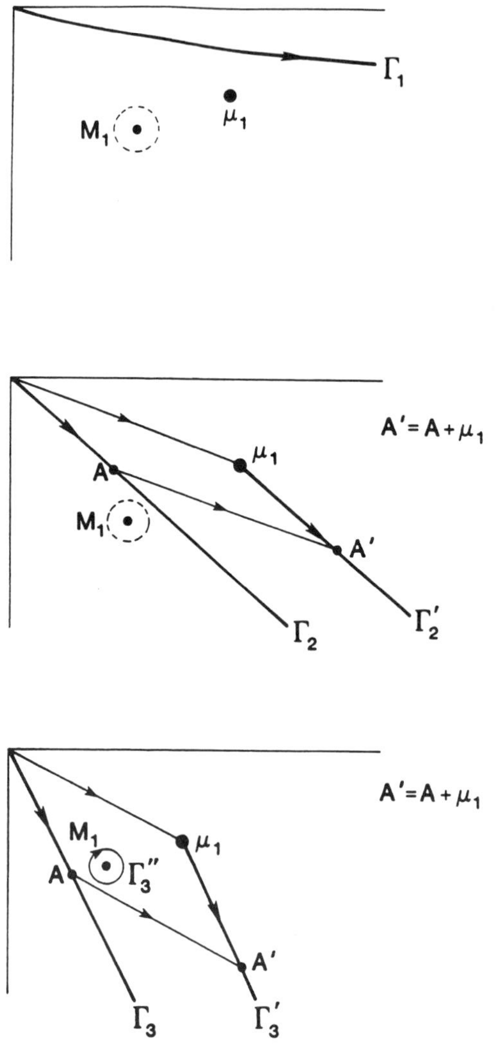

Figure 10. Spectra and contours for the cascade model with $M_0 \gg \mu_0 \gg 0$.

consequently, the closed-contour integrals do not enclose any of the related singularities.

If we choose the contour Γ_2, we have crossed the branch point at μ_1. This branch point "snags" the closed contour over which we integrate and completeness is restored only by including the generalized (ideal) states

$$|y\rangle = \begin{cases} \dfrac{f^*(y^* - \mu_1)}{\sqrt{\gamma_1'}\alpha(y + i\epsilon)} \\ \dfrac{1}{\sqrt{\gamma_1'}}\delta(y - \mu_1 - \xi) + \dfrac{f(\xi)}{\gamma(y - \xi + i\epsilon)} \cdot \dfrac{f^*(y^* - \mu_1)}{\sqrt{\gamma_1'}\alpha(y + i\epsilon)} \\ \dfrac{g(v)}{y - \xi - v + i\epsilon}\left[\dfrac{1}{\sqrt{\gamma_1'}}\delta(y - \mu_1 - \xi) + \dfrac{f(\xi)}{\gamma(y - \xi + i\epsilon)}\dfrac{f^*(y^* - \mu_1)}{\sqrt{\gamma_1^*}\alpha(y + i\epsilon)}\right] \end{cases} \tag{6.14}$$

with

$$\gamma_1' = \left.\frac{\partial\gamma(\zeta)}{\partial\zeta}\right|_{\zeta = \mu_1} \tag{6.15}$$

Here y and $\xi + \mu_1$ are along Γ_2 and ξ lies on Γ_2', which is obtained from Γ_2 by displacing it by the fixed complex number μ'. The states $|y\rangle$ and $|z, \zeta\rangle$ in Eqs. (6.14) and (6.13) now form a complete set. The contour Γ_2' is the spectrum of the "unstable" particle B (which has now become a "stable particle"!), scattering a θ particle with energy ξ. In addition to the generalized unitarity relation along Γ_2', this scattering also obeys

$$T(\zeta, \zeta'; z) - T^*(\zeta^*, \zeta'^*; z^*) = \int_{\Gamma_2} d\gamma''\, T^*(\zeta''^*, \zeta^*; z^*)T(\zeta'', \zeta'; z) \tag{6.16}$$

the unitarity relation

$$T(\xi) - T^*(\xi^*) = T^*(\xi^*)T(\xi) \tag{6.17}$$

along Γ_2'. There is a technical point here. For the definition of the continued wave functions, the contour Γ_2' is chosen through the "parallel-transport" prescription stated above. However, for the continued unitarity relation, it can be shown that it is no longer necessary to be confined to the parallel transported contour Γ_2'.

In the context of the continuation of wave functions, further deformation of the contour does alter the states $|\tau\rangle$. When z and $\xi + \zeta$ are along the contour Γ_3, τ is along $\Gamma_3' = \Gamma_3 + \mu_1$ (see Fig. 10). It could also uncover the discrete state $|M_1\rangle$ with

$$|M_1\rangle = \frac{1}{\sqrt{\alpha_1'}}\begin{bmatrix} 1 \\ \dfrac{f(\xi)}{\gamma(M_1 - \xi)} \\ \dfrac{g(\nu)f(\xi)}{\gamma(M_1 - \xi)(M_1 - \xi - \nu)} \end{bmatrix} \tag{6.18}$$

which then needs to be included in the completeness relation.

Unitarity relations for the T-matrix are energy-local relations [98] and as such *do not mix the unstable- and stable-particle scattering.*

VII. SUMMARY AND CONCLUSIONS

Let us recapitulate some of the points considered in this chapter. The Breit–Wigner approximation has been the phenomenological framework for the description of unstable states and it predicts a pure exponential decay. There are several shortcomings to this approach. The resonance is associated with a pair of complex conjugate poles on the physical sheet; this violates "causality." *Viewing* the Breit–Wigner model as a continuous spectrum violates the semiboundedness condition, which, in turn, leads to the violation of the second law of thermodynamics. Thus it is necessary to describe unstable quantum systems by going beyond the Breit–Wigner approximation, not only for minor technical corrections but for a conceptually satisfactory formulation.

Our discussions have been divided into two parts. In the first part, we see that insisting on the semiboundness of the spectrum cause the time evolution of an unstable quantum system to deviate from strict exponential decay both in the very small and the very large time region. In the neutral Kaon-type system in the very small and large t regions, there is a regeneration effect between K_L and K_S states.

From the study of solvable models, we saw that departure from exponential law, with the present experimental limits of time resolution, is numericaly insignificant. Nevertheless, we find it useful for the sake of conceptual clarity to pursue a consistent generalized quantum mechanical framework for the description of unstable states. The predictions of this framework coincide with the Breit–Wigner approximation in the bulk of the exponential decay region and at the same time allow extension to the very small and very large time regions. This is analogous to the formulation of the relativistic theory in the nonrelativistic domain, which allows for natural extrapolation to the relativistic domain.

With this in mind, by means of analytic continuation, we identify an unstable particle state as a discrete state in the generalized space $\mathcal{G}$ with complex energy eigenvalue. Here the continuum states are defined along some complex contour and the inner product and transition amplitudes are defined between states in $\mathscr{G}$ and its dual state in the corresponding dual space $\tilde{G}$.

The Breit–Wigner approximation [4] was introduced in the 1930s. A systematic and rigorous approach began with the paper by Sudarshan, Chiu, and Gorini [42], which proposed the notion of generalized quantum states leading to a consistent treatment of an unstable quantum particle as a complex eigenvalue solution of the operator in $\mathscr{G}$ associated with a hermitian Hamiltonian in $\mathscr{H}$. The analytic continuation of this program was carried out for various models, demonstrating that this approach can indeed be implemented consistently in various models. Within this framework, a resonance pole is a bona fide eigenstate of the continuation of a hermitian Hamiltonian with complex energy eigenvalues. We have applied the same generalized framework to scattering problems. The analytically continued scattering amplitudes and the extended unitarity relations were presented. The generalized framework provides the essential ingredient needed for a consistent description of the scattering process involving resonances.

The present formalism of dual space differs from the rigged Hilbert space theory, which also deals with dual spaces. But the dual spaces are the primary entities here. Some earlier papers in the literature claiming time asymmetry obtained their results by introducing unphysical states with energies unbounded from below.

The present approach is in one sense the completion of Heisenberg's program to make dynamics out of directly measured quantities like spectral frequencies and intensities augmented by resonance positions and widths; and in another sense a further generalization of the Dirac formalism of quantum theory in terms of ket and bra vectors. It is instructive that these old ideas contain the germs of many modern developments [45].

REFERENCES

1. Gamow, G., *Z. Phys.* **51**, 204 (1928).
2. Dirac, P. A. M., *Proc. Roy. Soc. London* **114**, 243 (1927).
3. Weisskopf, V. F. and E. P. Wigner, *Z. Phys.* **63**, 54 (1930).
4. Breit, G. and E. P. Wigner, *Phys. Rev.* **49**, 519 (1936).
5. Bohm, A., *Boulder Lect. Theor. Phys.* **9A**, 255 (1966).
6. Bohm, A., *J. Math. Phys.* **21**, 1040 (1980); **22**, 2313 (1981); *Physica* **124A**, 103 (1984).
7. Bohm, A., *Quantum Mechanics: Foundations and Applications* (Springer-Verlag, Berlin, 1986).
8. Bohm, A. and M. Gadella, in *Dirac Kets, Gamow Vectors and Gelfand Triplets*, A. Bohm and J. D. Dollard (Eds.) Springer Lecture Notes in Physics, Vol. 348 (Springer-Verlag, Berlin, 1989).

9. Fonda, L., G. C. Ghirardi, A. Rimini, and T. Weber, *Nuovo Cimento* **15A**, 689 (1973); *Rep. Prog. Phys.* **41**, 587 (1978).
10. Ghirardi, G. C., C. Omero, T. Weber, and A. Rimini, *Nuovo Cimento* **A52**, 421 (1979).
11. Cho, Gi-chol, H. Kasai, and Y. Yamaguchi, *The Time Evolution of Unstable Particles*, Tokai University preprint, TKU-HEP93/04.
12. Fermi, E., *Nuclear Physics* (University of Chicago Press, Chicago, 1950), p. 142.
13. Bohr, N., *Nature* **137**, 344 (1936).
14. Kapur, P. L. and R. Peierls, *Proc. R. Soc. London*, **A166**, 277 (1938).
15. Peierls, R. E., *Proc. R. Soc. London* **253A**, 16 (1960).
16. Matthews, P. T. and A. Salam, *Phys. Rev.* **112**, 283 (1958).
17. Matthews, P. T. and A. Salam, *Phys. Rev.* **115**, 1079 (1959).
18. Siegert, A. J. F., *Phys. Rev.* **56**, 750 (1939).
19. Wheeler, J., *Phys. Rev.* **52**, 1107 (1937).
20. Peierls, R. E., *Proceedings of the Glasgow Conference*, Glasgow, Scotland, 1954, pp. 296–299.
21. van Kampen, N., *Phys. Rev.* **91**, 1267 (1953).
22. Glaser, V. and G. Källen, *Nucl. Phys.* **2**, 706 (1956).
23. Höhler, G., *Z. Phys.* **152**, 546 (1958).
24. Nakanishi, M., *Progr. Theor. Phys.* **19**, 607 (1958).
25. Lee, T. D., *Phys. Rev.* **95**, 1329 (1954).
26. Friedrichs, K. O., *Commun. Pure Appl. Math.* **1**, 361 (1948).
27. Moshinsky, M., *Phys. Rev.* **84**, 525 (1951); see also H. M. Nussenzweig, "Moshinsky Functions, Resonances and Tunneling," in *Symmetries in Physics*, A. Frank and B. Wolf (Eds.) (Springer-Verlag, Berlin, 1992), p. 294.
28. Winter, R. G., *Phys. Rev.* **123**, 1503 (1961).
29. Frey, B. B. and E. Thiele, *J. Chem. Phys.* **48**, 3240 (1968).
30. Levy, M., *Nuovo Cimento* **30**, 115 (1959).
31. Williams, D. N., *Commun. Math. Phys.* **21**, 314 (1971).
32. Fleming, G., *Nuovo Cimento* **A16**, 232 (1973).
33. Khalfin, L., *Sov. Phys. JETP* **6**, 1053 (1958).
34. Khalfin, L., *JETP Lett.* **8**, 65 (1968).
35. Paley, R. and N. Wiener, *Fourier Transform in Complex Domain* (Providence, RI, 1934), Theorem XII.
36. Misra, B. and E. C. G. Sudarshan, *J. Math. Phys.*, **18**, 756 (1977).
37. Tasaki, S., T. Petrosky, and I. Prigogine, *Physica* **A173**, 175 (1991); see also [38].
38. Petrosky, T. and I. Prigogine, *Proc. Nat. Acad. Sci. U.S.A.* **93**, 9393 (1993).
39. Antoniou, I. and I. Prigogine, *Physica* **A192**, 443 (1993).
40. Prigogine, I., *From Being to Becoming: Time & Complexity in Physical Sciences* (W. H. Freeman, San Francisco, 1980).
41. Chiu, C. B., E. C. G. Sudarshan, and B. Misra, *Phys. Rev.* **D16**, 520 (1977).
42. Sudarshan, E. C. G., C. B. Chiu, and V. Gorini, *Phys. Rev.* **D18**, 2914 (1978).
43. Sudarshan, E. C. G. and C. B. Chiu, *Phys. Rev.* **D47**, 2602 (1993).
44. Chiu, C. B., E. C. G. Sudarshan, and G. Bhamathi, *Phys. Rev.* **D46**, 3508 (1992).

45. Sudarshan, E. C. G., *Phys. Rev.* **A50**, 2006 (1994).
46. Sudarshan, E. C. G., *J. Math. Phys. Sci.* **25**(5/6), 1 (1996).
47. Jenkin, F. A. and H. E. White, *Fundamentals of Optics* (McGraw-Hill, New York, 1957).
48. Mead, C., *Phys. Rev.* **116**, 359 (1958).
49. Mehra, J. and E. C. G. Sudarshan, *Nuovo Cimento* **11B**, 215 (1972).
50. Chiu, C. B. and E. C. G. Sudarshan, *Phys. Rev.* **D42**, 3712 (1990).
51. Lee, T. D., R. Oehme, and C. N. Yang, *Phys. Rev.* **106**, 340 (1957).
52. Khalfin, L., USSR Academy of Sciences, Steklov Mathematical Institute, Leningrad, LOMI Reports E-6-87 and E-7-87, 1987 (unpublished).
53. Khalfin, L., University of Texas at Austin, CPT Report No. 211, 1990 (unpublished).
54. Yamaguchi, Y., *Phys. Rev.* **95**, 1628 (1954); see also *Progr. Theor. Phys. Suppl.* **7**, 1 (1958).
55. Chiu, C. B. and E. C. G. Sudarshan, "Theory of the neutral Kaon system," in *A Gift of Prophecy*, E. C. G. Sudarshan (Ed.) (World Scientific, Singapore, 1994), p. 81.
56. Bell, J. S. and J. Steinberger, *Proceedings of the International Conference on Elementary Particles*, Oxford, 1965, p. 195. See also R. E. Marshak, Riazuddin and C. P. Ryan, *Theory of Weak Interactions in Particle Physics* (Wiley, New York, 1969), Section 6.5.A.
57. Desgasperis, A., L. Fonda, and G. C. Ghirardi, *Nuovo Cimento* **21A**, 471 (1974).
58. Rau, J., *Phys. Rev.* **129**, 1880 (1963).
59. Peres, A., *Am. J. Phys.* **48**, 931 (1980).
60. Fleming, G., *Phys. Lett.* **125B**, 287 (1982).
61. Valanju, P. C., Ph.D. Dissertation, University of Texas, Austin, Texas, 1980.
62. Valanju, P. C., B. Chiu, and E. C. G. Sudarshan, *Phys. Rev.* **D21**, 1304 (1980).
63. Itano, W. M., D. J. Heinzen, J. J. Bollinger, and D. J. Wineland, *Phys. Rev.* **A41**, 2295 (1990).
64. Ekstein, H. and A. J. F. Siegert, *Ann. Phys. (NY)* **68**, 509 (1971).
65. Norman, E. B., S. B. Gazes, S. G. Crane, and D. A. Bennett, *Phys. Rev. Lett.* **60**, 2246 (1988).
66. Capra, F., *The Tao of Physics* (Shambhala, Berkeley, CA, 1975), p. 25.
67. *Proceedings of the Oxford Colloquium on Multiparticle Dynamics* (Oxford University, Oxford, UK, 1975), p. 577.
68. Bialas, A., *Proceedings of Topical Conference on Electronuclear Physics with Internal Targets*, G. Arnold (Ed.) (World Scientific, Singapore, 1990), p. 65 and refrences quoted there.
69. Gell-Mann, M. and A. Pais, *Phys. Rev.* **97**, 1387 (1955).
70. Sachs, R. G., *Ann. Phys.* **22**, 239 (1963).
71. Kenny, B. G. and R. G. Sachs, *Phys. Rev.* **D8**, 1605 (1973).
72. Dirac, P. A. M., *Principles of Quantum Mechanics*, 4th ed. (Clarendon, Oxford, 1954), p. 206, Eq. (52).
73. Mandl, F., *Quantum Mechanics* (Pergamon Press, New York, 1954).
74. von Neumann, J., *Mathematical Foundation of Quantum Mechanics* (Springer-Verlag, Berlin, 1932; Princeton University Press, Princeton, NJ, 1955).
75. Roberts, J. F., *J. Math. Phys.* **7**, 1097 (1966).
76. Segal, I. E., *Ann. Math.* **48**, 930 (1947).
77. Gelfand, I. M. and G. F. Shilov, *Generalized Functions*, Vols. II and IV (Academic, New

York, 1967).

78. Antoine, J. P., *J. Math. Phys.* **10**, 53 (1969); **10**, 2276 (1969).
79. Kuriyan, J. G., N. Mukunda, and E. C. G. Sudarshan, *Comm. Math. Phys.* **8**, 204 (1968); *J. Math. Phys.* **9**, 12 (1968).
80. Parravicini, G., V. Gorini, and E. C. G. Sudarshan, *J. Math. Phys.* **21**, 2208 (1980).
81. Yamaguchi, Y., *J. Phys. Soc. Jpn.* **57**, 1525 (1988); **57**, 3339 (1988); **58**, 4375 (1989); **60**, 1545 (1991).
82. Sudarshan, E. C. G., in "Relativistic Particle Interactions, Notes by V. Teplitz," *Proceedings of the 1961 Brandeis University Summer Institute* (Benjamin, New York, 1962).
83. Hu, N., *Phys. Rev.* **74**, 131 (1948).
84. Sakurai, J. J., *Modern Quantum Mechanics* (Addison-Wesley, New York, 1985).
85. Heisenberg, W., *Nucl. Phys.* **4**, 532 (1957).
86. Wigner, E. P. and J. von Neumann, *Z. Phyusik* **30**, 465 (1929).
87. Simon, B., *Commun. Pure Appl. Math.* **22**, 531 (1967).
88. Sudarshan, E. C. G., *Field Theory, Quantizations and Statistical Physics*, E. Terapegui (Ed.) (Reidel, Dordrecht, Holland, 1981).
89. Titchmarsh, E. C., *Theory of Fourier Integrals* (Oxford University Press, Oxford, UK, 1937).
90. Rosenblum, M. and J. Rovnyak, *Hardy Classes and Operator Theory* (Oxford University Press, New York, 1985).
91. Ma, S. T., *Phys. Rev.* **69**, 668 (1946); **71**, 195 (1947).
92. Biswas, S. N., T. Pradhan, and E. C. G. Sudarshan, *Nucl. Phys.* **B50**, 269 (1972).
93. Bargman, V., *Rev. Mod. Phys.* **21**, 488 (1949).
94. Newton, R. G., *J. Math. Phys.* **1**, 319 (1960).
95. Wu, T. Y. and T. Ohmura, *Quantum Theory of Scattering* (Prentice-Hall, Englewood Cliffs, NJ, 1962).
96. Livsic, M. S., *Sov. Phys. Dokl.* **1**, 620 (1956); *Sov. Phys. JETP* **4**, 91 (1957).
97. Howland, J. S., *J. Math. Anal. Appl.* **50**, 415 (1975).
98. Gleeson, A. M., R. J. Moore, H. Rechenberg, and E. C. G. Sudarshan, *Phys. Rev.* **D4**, 2242 (1971).

RESONANCES AND DILATATION ANALYTICITY IN LIOUVILLE SPACE

ERKKI J. BRÄNDAS

Department of Quantum Chemistry, Uppsala University, Uppsala, Sweden

CONTENTS

ABSTRACT

We consider a general formulation of a complex system within the subdynamics framework. The derivations are suggested from recent work using the complex scaling method (CSM), of quantum mechanical systems with dilatation analytic perturbations in Liouville space. Our approach is mathematically similar to the Prigogine subdynamics decomposition of the Liouville operator for systems having absolutely continuous spectra. However, the question of irreversibility and the approach to equilibrium are somewhat different. A specific point is the occurrence of Jordan blocks in the dilated equations, and they are suggested to be connected with microscopic self-organization. Applications with respect to both classical and quantum situations are considered in some detail.

Advances in Chemical Physics, Volume XCIX, Edited by I. Prigogine and Stuart A. Rice.
ISBN 0-471-16526-3

I. INTRODUCTION

Fundamental physics has usually been developed according to the very large, as in astro- or space physics, or to the very small, as in particle, nuclear, high-energy, and quantum physics. However, a new overlapping component has emerged in modern natural sciences, namely, the very complex [1,2], involving, for example, chaos, self-organization, nonlinear dynamics, and nonequilibrium phenomena. The field of Complexity is rapidly becoming a target for contemporary research and its relevance to many fundamental questions is currently on the rise.

The fundamental nature of complex phenomena is something we see every day, yet the simplifications, introduced to define a physical system, lead us most of the time into a limited "linear thinking." The novelty and importance of a more basic viewpoint has been particularly emphasized and developed by the Brussels School [1] under the leadership of Prof. I. Prigogine. Further attention has been given to the field through the European network of *nonlinear phenomena and complex systems* organized by Prof. G. Nicolis.

Obviously, the present direction penetrates fundamental areas like material science, big science, biophysics, information technology, and meteorology and the bridge between generic and strategic research is becoming stronger every day. Recently [3], a microscopic formulation of subdynamics, based on the complex scaling method (CSM) [4] and a fundamental hierarchy of reduced density matrices [5], in the extreme configuration [6], prompted the notion of a coherent–dissipative system [7] with many specific and surprising applications. The idea to look for dissipative structures in physicochemical systems was originally supported by EG-SCIENCE and concerns today, for example,

Water and ionic liquids, proton transfer processes, anomalous H_2O/D_2O conductances as well as ionic conductances of molten alkali chlorides

Condensed matter, high-temperature superconductivity (HTSC) and the fractional quantum Hall effect (FQHE), the spin dynamics in *Gd* far above the Curie point, and conductance background effects in high-quality tunnel junctions

Biology—proton dynamics in DNA and related evolutionary consequences

Accelerator physics—for example, in connection with spontaneous and stimulated emission related to the electron cyclotron maser [8].

See also [9] for a review of earlier work. In this chapter, we briefly describe the theoretical development leading to a simple Liouville equation for a dissipative (open) system. Since the formulation emphasizes complexity rather than a quantal or classical angle, Planck's constant is not used in the general equations, except when a particular choice is made in favor of the latter. We thus show that a particular (super)operator

$$\hat{\mathscr{P}} = (\omega_0 \tau - i)\hat{\mathscr{I}} + \hat{\mathscr{J}} \tag{1.1}$$

and its propagator

$$\exp(-i\hat{\mathscr{P}}t/\tau) \tag{1.2}$$

can be obtained from general (first) principles. In Eqs. (1.1) and (1.2), ω_0 is the resonance frequency and t is the time, scaled in relation to τ, the corresponding phenomenological relaxation lifetime. Further notation involves $\hat{\mathscr{I}}$ as the unit operator, and $\hat{\mathscr{J}}$, representing the interaction with the environment. The latter has a very curious property (see below) and it has no counterpart in standard linear equilibrium dynamics. One may also think of the product $\omega_0 \tau$ as a Q value of a "resonating cavity," $\tau^{-1} = \omega_S$ as a mismatching frequency, and ω_0^{-1} as the fast-time variable. Obviously this Q value could be very large compared to unity and hence one might question the importance of the perturbation $\hat{\mathscr{J}}$. Nevertheless, the operator $\hat{\mathscr{J}}$ will induce new dynamical features and connects microscopic subdynamics with self-organization.

We first indicate how Eqs. (1.1) and (1.2) can be obtained from first principles. We refer to CSM (Appendix A), the occurrence of Jordan blocks (Appendix B), the relevant algebraic structure (Appendix C), and the analytic structure suggested by dilatation analyticity (Appendix D). In so doing we also prove some additional theorems for nonabelian representations not published before. The electron cyclotron maser is used as an example of a system where subsequent formulations of classical and quantum formulations coexist; but applications in condensed matter, as complex systems in this context, are also emphasized.

II. CORRELATED TRANSITIONS OF FERMIONS

To obtain a more concrete understanding of the implications of Eqs. (1.1) and (1.2), we first discuss the problem of correlated fermions. The result of this exercise suggests a possible form of the general Liouvillian in Eq. (1.1). This will also lead to a compact formulation of spontaneous and stimulated transitions and a general microscopic formulation of self-organization.

Since fermions obey the Pauli principle, a quantum formulation must incorporate Fermi statistics. In condensed-matter situations, quantum correlations between electrons, for example, play a fundamental role. In other situations, as for instance in accelerator physics, a beam of electrons, may contain “classical” particles at relativistic energies. In both cases, the particles obey in principle a Liouville equation for the phase-space density matrix, ρ, of the form

$$i\frac{\partial\rho}{\partial t} = \hat{L}\rho \tag{2.1}$$

As usual, $\hat{L}$ is the commutator with the Hamiltonian ($\hbar$ should be inserted unless we deal directly with energies in terms of frequencies) in the quantum case and $i\hat{L}$ is the Poisson bracket in the classical domain.

To develop the microdynamical picture we will, as mentioned above, first treat the case of correlated fermions. In passing, however, we want to point out the importance of finding a consistent transition from the quantum to the classical regime. The classical situation of gyrating electrons in a constant magnetic field perturbed by a suitable time-dependent electromagnetic (resonator) field is of particular interest here, because the theoretical formulation can be made both quantum mechanically and classically.

First, we quote some general results from the interactions and the correlations in a many-body fermionic system. It is a well-known fact that, owing to the Pauli principle, quantum correlations manifest themselves in the appearance of occupation numbers [5] fulfilling $0 \leq \lambda^{(1)} \leq 1$. For weakly correlated particles, these occupation numbers are close to 1, but in strongly correlated situations, for example, in high-T_c superconductivity (HTSC), or in the FQHE, these occupancies may deviate significantly from unity.

Under optimal conditions, the fermionic statistics may lead to the so-called extreme case of maximum coherence, see Coleman [6], which could further develop into Yang's off-diagonal long-range order (ODLRO) [10]. We want to emphasize that these correlations can extend over macroscopic dimensions in condensed-matter applications (cf. HTSC and FQHE). In the domain of accelerator physics, however, such extensions are not usually seen, but it should be further investigated whether such effects could appear, for instance, during bunching in cyclotron maser experiments. In general, quantum effects are entering accelerator physics, for example, through laser cooling and the fascinating new area of crystalline beams. Keep in mind, therefore, that general correlations (not to be confused with the concept of interactions) may play a more active role than before, for example, in storage-ring physics, owing to the statistical (fermionic or bosonic) character of the participating particles.

In the correlated (extreme) case [6], the occupation number is given by

$$\lambda^{(1)} = \frac{N}{2r} \tag{2.2}$$

where N is the number of fermions and $2r$ is the number of available spin orbitals. Equivalently, r is the rank of the fermion (geminal) pair subspace. It is now a simple task to set up a "correlation matrix" which describes quantum mechanically the excitation of an electron from one level to another as correlated with or stimulated by other electronic excitations [3]. The off-diagonal element is simply given by the product of the probability of the correlated electron being in the appropriate spin orbital, that is, $N/2r$, with the probability of the other spin orbital being empty, that is, $(N/2r) \cdot [(2r - N)/2(r - 1)]$. The detailed statistical analysis have been carried out in [3], so we do not repeat it here.

It seems perhaps a contradiction to discuss fermions, while at the same time having Landau oscillators and levels in mind when we discuss, for example, the cyclotron maser concept (CMC) [11–15], which refers to bosonic entities. There is of course no problem here if we realize that we are considering fermion pair subspaces and associated second-order, reduced-density matrices, and that fermion pairs behave as (quasi) bosons. Analogously, the two-dimensional projection of gyrating electrons, which leads to the (highly degenerate) Landau oscillator formulation, again subscribes to boson statistics. In both cases, there is a "hidden" fermionic origin of the problem, which cannot be forgotten, yet the transition to the classical regime follows accordingly, as discussed below.

The analogy, which is carried out in more detail in the next section, implies (cf. HTSC and FQHE) that the structure, exhibited by correlated electrons, can be transferred directly to a (relativistic) Landau oscillator picture.

Returning to the "correlation matrix" discussed above, it is a simple matter to diagonalize it. This leads to a large eigenvalue λ_L and a small one λ_S given by

$$\lambda_S = \frac{N(N-2)}{4r(r-1)}, \qquad \lambda_L = \frac{N}{2} - (r-1)\lambda_S \tag{2.3}$$

For the identification with Coleman's extreme case or with Yang's concept of ODLRO, see [3,7]. Note that in the oscillator picture, referred to above, the large eigenvalue λ_L is directly related to the stimulated part of the transition leading to amplification through ODLRO.

For our present purpose, we start with a localized basis of electron pair functions of rank r given by $\mathbf{h} = |h_1, h_2, \ldots, h_r\rangle$. It may here be chosen as a real localized set of spin-paired reference determinants, but in general it is important to realize that a dynamic formulation necessitates transformations to complex Gamow-like representations (see [3] for more details). To arrive at the nonabelian structure, (see Appendix B), we introduce a delocalized coherence basis **g** and a complex correlation basis **f** via

$$|\mathbf{h}\rangle\mathbf{B} = |\mathbf{g}\rangle = |g_1, g_2, \ldots, g_r\rangle \tag{2.4a}$$

and

$$|\mathbf{h}\rangle\mathbf{B}^{-1} = |\mathbf{f}\rangle = |f_1, f_2, \ldots, f_r\rangle \tag{2.4b}$$

with **B** given by

$$\sqrt{(r)}\cdot\mathbf{B} = \begin{pmatrix} 1 & \omega & \omega^2 & \cdots & \omega^{r-1} \\ 1 & \omega^3 & \omega^6 & \cdots & \omega^{3(r-1)} \\ \cdots & \cdots & \cdots & \cdots & \cdots \\ 1 & \omega^{2r-1} & \omega^{2(2r-1)} & \cdots & \omega^{(r-1)(2r-1)} \end{pmatrix} \tag{2.4c}$$

and $\omega = \exp[i(\pi/r)]$. The transformations, Eq. (2.4a–c), were derived by Reid and Brändas [16] and their properties were directly connected with the appearance of Jordan blocks.

The (relevant part of) the reduced second-order density matrix for the correlated electron problem reads [3,6,7]

$$\Gamma^{(2)} = \Gamma_{\mathrm{L}}^{(2)} + \Gamma_{\mathrm{S}}^{(2)} = \lambda_{\mathrm{L}}|g_1\rangle\langle g_1| + \lambda_{\mathrm{S}}\sum_{k=1}^{r}\sum_{l=1}^{r}|h_k\rangle\left(\delta_{kl} - \frac{1}{r}\right)\langle h_l| \tag{2.5}$$

with $Tr\{\Gamma^{(2)}\} = N/2$ i.e. the correlation matrix is normalized to the number of fermionic pairs. Utilizing the transformation above we can also write

$$\Gamma^{(2)} = \Gamma_{\mathrm{L}}^{(2)} + \Gamma_{\mathrm{S}}^{(2)} = \lambda_{\mathrm{L}}|g_1\rangle\langle g_1| + \lambda_{\mathrm{S}}\sum_{k=2}^{r}|g_k\rangle\langle g_k| \tag{2.6}$$

In passing, we mention that we have neglected an uninteresting "unpaired" contribution to the full $\Gamma^{(2)}$ in the trace relation above, but here this technicality is not explicitly pursued [3], except to point out its importance in connection with the general problem of information and entropy content of the evolving system.

It is interesting to note that, for a given value of r (i.e., $2r$ much greater than N), it follows from Eqs. (2.2)–(2.5) that λ_{L} first "grows" linearly with N until the quadratic term [see Eq. (2.3)] begins contributing. The parabola "ends" for $2r = N$, where coherence and delocalization (in the sense given above) no longer persists [3]. We have put the words "grows" and "ends" within quotation marks to issue a warning that changes in N and/or r are not to be interpreted as in linear theory, that is, to see what happens with the system when certain parameters vary. Here the formulation is *manifestly nonlinear* and therefore N and/or r are not at "our disposal" because they simply reflect that all parts of the system are intrinsically coupled as a dissipative structure.

We have shown how Eq. (2.5), via complex dilations [4,17,18] and thermalization (e.g., see [3,7,16] and Appendix C), assumes a Jordan block structure which for Γ_S takes the simple form (Appendix B)

$$\Gamma_S = \lambda_S \sum_{k=1}^{r-1} |f_k\rangle\langle f_{k+1}| = \lambda_S J \tag{2.7}$$

with the important property

$$J^r = 0, \qquad J^{r-1} \neq 0 \tag{2.8}$$

As mentioned above, we may take our reference space as real and let all dynamic complexities be invoked through effective reduced Hamiltonians (and/or Liouvillians) and associated partitioning techniques. The dilation transformation that lies behind the new nonunitary picture was given a noncontradictory Liouville formulation within subdynamics by Obcemea and Brändas [19] and it was further developed in [3,7,16]. We must remember, however, that in this formulation the Liouvillian is no longer self-adjoint and hence there is the possibility of a spontaneous breaking of time reversibility. We will also see what consequences the present extension has for the Liouville phase space density or rather the associated space of reduced-density matrices.

III. SPONTANEOUS AND STIMULATED TRANSITIONS

Since our aim is to derive the Liouville equation corresponding to Eq. (1.1), which is valid not only in the quantum domain, we will interchangably proceed to classical limits when appropriate. We will also couple the formulation to a suitably defined impedance or gain function. To obtain the relevant Liouville equation, we consider the physical picture related to the CMC, that is, in terms of correlated relativistic Landau scillators. By analogy, the transfer of the structure of the correlation matrix displayed in Section II suggests a superoperator framework. The crucial assumption is thus that spontaneous and stimulated emission display coherent properties similar to ODLRO.

We denote the density matrix corresponding to a transition (system absorbs) between the Landau levels $i + 1$ and i by $\|h_{k(i)}\rangle\rangle = |\{i + 1\}(k)\rangle\langle i(k)|$, where k denotes a particular transition involving Landau levels i and $i + 1$, characterized by the angular frequency $\omega_{i+1,i} \approx \omega_L$, and given by the eigenvalue (of the "zero-order" Liouvillian) $\hbar\omega_{i+1,i} = E_{i+1} - E_i$.

As before, we obtain a superoperator transformation between the transition matrices $\|\mathbf{h}\rangle\rangle$, $\|\mathbf{f}\rangle\rangle$, and $\|\mathbf{g}\rangle\rangle$ via the $r \times r$ transformation matrices $\mathbf{B}^{-1}$ and $\mathbf{B}$, respectively. In the basis $\|\mathbf{f}\rangle\rangle$, the transitions are correlated, and in $\|\mathbf{g}\rangle\rangle$ they are coherent, according to the properties of respective transformations. For example, $\|g_1\rangle\rangle$ describes simultaneous coherent transitions by the frequency $\omega_{i+1,i}$, while $\|f_1\rangle\rangle$ represents transitions that are phase correlated through the factors $\exp[(i\pi/r)(k-1)]$, $k = 1, 2, \ldots, r$. Note also that $\langle\langle\mathbf{g}\|$ and $\langle\langle\mathbf{f}\|$ describe coherent and correlated emissions with the eigenvalues $\hbar\omega_{i,i+1} = E_i - E_{i+1}$ of the corresponding (zero-order) Liouvillian.

Absorption and emission are therefore intrinsically connected through the bra-ket correlated transformations based on $\|\mathbf{h}\rangle\rangle\mathbf{B}$ and $\|\mathbf{h}\rangle\rangle\mathbf{B}^{-1}$. We also remark that the usual transformation property for time reversal (i.e., $t \to -t$; $\hat{L} \to -\hat{L}$), does not work here, because we have an intrinsic coupling with the environment through the new terms (see below) obtained from the dissipative dynamics.

In the present formulation, the scalar product becomes

$$\langle\langle h_{k(i)}\|h_{l(j)}\rangle\rangle = Tr\{h_{k(i)}^{\dagger} h_{l(j)}\} \tag{3.1}$$

In general, one must use the generalized scalar product for the $\|\mathbf{h}\rangle\rangle$ complex

$$\langle\langle h_{k(i)}^{*}\|h_{l(j)}\rangle\rangle = Tr\{h_{k(i)}^{*\,\dagger} h_{l(j)}\} \tag{3.1$'$}$$

Since our chosen reference basis is real, we do not dwell on this complication here. Suffice it to say that meaningful scalar products should in general be taken between components of the two conjugate Hilbert spaces and all vectors in the bra position should have a complex conjugate sign to show that we take a component from the dual space; see [16] for details. Note that here we are treating non-self-adjoint Liouvillians with real reference spaces, and that unitary transformations in this context could lead to nonsymmetric Jordan blocks [16]. Because the density matrix in the extreme case develops a large component and a small one, it is clear that the corresponding superselection sectors become entangled during thermalization. The occurrence of Jordan blocks (see Appendix B) thus implies important consequences for the dissipative dynamics.

We now construct the relevant Liouvillian $\hat{L}$ commensurate with the Jordan block structure. Note that we will obtain a time-independent effective Liouvillian even if the original Liouvillian contains time-dependent parts; we return to this problem below (see also Appendix D). In analogy with the previous section we write [note that Γ here means the resonance width related to the characteristic (collision or phase debunching) lifetime of the state via $\hbar/\Gamma = \tau$, not to be confused with the density matrix that has indices S or L]

$$\hat{L} = (z_L - i\Gamma)\hat{I} + \hbar\omega_S \hat{J} \tag{3.2a}$$

$$\hat{I} = \sum_{k=1}^{r} \|g_k\rangle\!\rangle\langle\!\langle g_k\| = \sum_{k=1}^{r} \|f_k\rangle\!\rangle\langle\!\langle f_k\|, \qquad \hat{J} = \sum_{k=1}^{r-1} \|f_k\rangle\!\rangle\langle\!\langle f_{k+1}\| \tag{3.2b}$$

In Eq. (3.2a) $z_L = \hbar\omega_L$, with ω_L being the characteristic angular frequency associated with a specific transition in question and ω_S to be determined. One possibility is to define a relevant time scale through $\hbar\omega_S = \Gamma$. This choice follows from the uncertainty relation and it has some experimental justification [20].

Another choice would be to deduce a relation [from inspection of Eq. (2.6)], that is, that the large component (in the identity operator) scales with the Jordan block component as

$$\omega_S = \omega_L \cdot \frac{\lambda_S}{\lambda_L} \tag{3.3}$$

which means that the quotient between ω_S and ω_L can be expressed in terms of N, the number of particles in the system, and r, the dimension of the pair space, see Eq. (2.3). This might be relevant in connection with spontaneous and stimulated emission. Both choices may be mutually compatible, and we assume that

$$\omega_S = 1/\tau \tag{3.4}$$

where $\tau = \hbar/\Gamma$ is the microscopic time scale commensurate with the relevant relaxation time (see Appendix C).

Our density matrix is now properly specified. Although this construction appears quite ad hoc, our intention is clearly not to derive the details of the full Liouvillian, but rather to see how the second term in Eq. (3.2a) arises from first principles and to demonstrate the subsequent consequences with respect to its Liouvillian properties.

In the discussion of the exact underlying principles for this construction, we must realize that the Liouvillian in, for example, the CMC case, will be time dependent through a perturbing electromagnetic rf field. Of course one might attempt to derive appropriate equations from general partitioning techniques while continuing to analyze the nonlinear equations via well-defined regularizations or dilations; this complication can be dealt with (see Appendix D). Here we briefly express the abstract resolvent equations for a general Liouvillian referring to a particular time scale, where explicit time dependencies on a shorter scale have been averaged out.

The full nonlinearity of the problem can now be explicitly specified, giving the precise condition for the characteristic angular frequency as well as the one determing $\Gamma = \hbar/\tau$. This follows from the unique analytic continuation of the Liouville problem:

$$\{\hat{P}\hat{\mathscr{L}}\,\hat{P} + \hat{\Psi}(z) - z\hat{\mathscr{I}}\}\|f_l\rangle\rangle = 0; \qquad l = 1, 2, \ldots, r \tag{3.5a}$$

in which $\hat{P} = \sum_{k=1}^{r} \|g_k\rangle\rangle\langle\langle g_k\| = \sum_{k=1}^{r} \|f_k\rangle\rangle\langle\langle f_k\|$ with $\hat{P} + \hat{Q} = \hat{\mathscr{I}}$ and

$$\hat{\Psi}(z) = -\hat{P}\hat{\mathscr{L}}\,\hat{Q}(\hat{Q}\hat{\mathscr{L}}\,\hat{Q} - z\hat{\mathscr{I}})^{-1}\hat{Q}\hat{\mathscr{L}}\,\hat{Q} \tag{3.5b}$$

In the equations above, the Liouvillian $\hat{\mathscr{L}}$ describes (in principle) all physical mechanisms, including correlations, relativistic effects, and so on. Further, $\hat{\mathscr{I}}$ is the complete unit operator and Eqs. (3.5a and b) give the condition for ω_L and Γ, including in principle all the nonlinearities of the problem. The construction in Eq. (3.2a) obtains from Eq. (3.5) by the choice $\hat{P} = \hat{I}$ and $\hat{L} = \hat{P}\hat{\mathscr{L}}\,\hat{P} + \hat{\Psi}(z_L - i\Gamma)$. See also [19,21] for more details.

Before we continue to demonstrate how the associated dynamics of the present model evolve, we return to the particular situation experienced in CMC [12–15]. At first, we note the enormous degeneracies of the Landau oscillator problem, which, however, are coupled with decreasing level spacings as energy increases. This implies that stimulated absorption and emission must be correlated, so that excited oscillators in some specified energy levels should be phase correlated with the corresponding oscillators, already situated at higher energies with appropriate deexcitation properties, for stimulated emission to occur. Obviously, this collective effect could lead to a decrease of energy spread and thus organization. The intriguing possibility that appears here is the following. The "classical" Maxwellian probability for an appropriately defined excitation reads

$$p = \exp - (h\nu/k_B T)$$

where $h\nu$ is the frequency multiplied by Planck's constant, k_B is Boltzmann's constant, and T is the relevant temperature. We have put "classical" within quotation marks because the famous Einstein relation between spontaneous and stimulated emissions involves the intensity $I(\nu)$ and the famous A and B

coefficients, that is,

$$\frac{AI(\nu)}{B} = \frac{1}{\exp(h\nu/k_{\mathrm{B}}T) - 1} = \delta \tag{3.6a}$$

is equivalent to the (quantum) Planck radiation law. An interesting analogy with the previous derivations occurs if we consider the correlation matrix [cf. discussions in connection with Eq. (2.2)] with diagonal elements p and off-diagonal elements $p(1-p)$, where the latter corresponds to the phase-correlated coupling between a particular oscillator being excited and another one being deexcited.

It is a remarkable coincidence that this simple correlation matrix has the same algebraic structure as the one appearing in condensed matter (see Section II), albeit for a weakly correlated situation. Thus we obtain

$$\lambda_{\mathrm{L}} = sp - (s-1)p^2, \qquad \lambda_{\mathrm{S}} = p^2 \tag{3.6b}$$

where in the weakly correlated regime $s \approx r \approx N/2$. Since Eq. (3.3) seems compatible with Eq. (3.4), one obtains

$$\omega_{\mathrm{S}}/\omega_{\mathrm{L}} = \delta/s \approx 2 \cdot \delta/N \tag{3.6c}$$

Equations (3.2)–(3.6) lead to a general impedance or gain function (cf. the classical formulation of CMC) [11,13–15], at the same time coupling the formulation, through the present dissipative subdynamics model, to an organization or cooling property.

IV. SELF-ORGANIZATION AND DYNAMIC CORRELATION PATTERNS

We now show how the Liouvillian obtained in the previous section directly connects a suitable gain function with organization through the mechanism of spontaneous and stimulated transitions. We begin with the general relation between the propagator and resolvent ($t > 0$ and C^+ contour in the upper half plane and $\hbar = 1$, see [19] for details):

$$e^{-i\hat{L}t} = \frac{i}{2\pi}\int_{C^+} (z - \hat{L})^{-1} e^{-izt}\, dz \tag{4.1}$$

Equation (4.1) is convenient when working directly with angular frequencies and inverse lifetimes. For the remainder of this section, however, we insert $\hbar$ again to connect with the Liouvillian of the previous section [e.g., Eq.

(3.2a)]. Regarding the problem of time scales, it is easy to see that one can replace the left-hand side of Eq. (4.1) with a general expression corresponding to a more complicated correlation function (see Appendix D). Through appropriate dispersion relations (see Appendix D), a relevant time-independent operator can be constructed so that Eq. (4.1) and, in principle, Eq. (3.5), can be expressed without contradictions, albeit within a generalized analytical context.

Thus, leaving this technicality aside as solved in principle, the resolvent straightforwardly becomes

$$(z = \hat{L})^{-1} = \frac{1}{z - (\hbar\omega_{\mathrm{L}} - i\Gamma)}\,\hat{I} + \frac{\hbar\omega_{\mathrm{S}}}{(z - (\hbar\omega_{\mathrm{L}} - i\Gamma))^2}\,\hat{J} + \cdots \frac{(\hbar\omega_{\mathrm{S}})^{r-1}}{(z - (\hbar\omega_{\mathrm{L}} - i\Gamma))^r}\,\hat{J}^{r-1} \tag{4.2}$$

and the propagator is

$$\exp\left(-\frac{i\hat{L}t}{\hbar}\right) = \exp(-i\omega_{\mathrm{L}}\,t)\exp\left(-\frac{\Gamma t}{\hbar}\right)\cdot\exp(-i\omega_{\mathrm{S}}\,\hat{J}t)\hat{I} \tag{4.3}$$

with

$$\exp(-i\omega_{\mathrm{S}}\,\hat{J}t) = \left[1 - i\omega_{\mathrm{S}}\,t\hat{J}\cdots + \frac{(-i\omega_{\mathrm{S}}\,t)^{r-1}}{(r-1)!}\,\hat{J}^{r-1}\right]$$

The formula can be obtained either from simple matrix algebra or from the general resolvent propagator expression above [22].

Let us briefly look at the transitions described by $\| f_1 \rangle\!\rangle$ correlated with the transitions $\| f_k \rangle\!\rangle$ characterized by the operator $\hat{J}^{k-1}$ with $k = 2, 3, \ldots, r$. Note that the characteristic (collision) lifetime τ is still given by $\hbar/\Gamma$. The time rule for, let us say, transitions involving $\| f_1 \rangle\!\rangle$ and $\| f_k \rangle\!\rangle$ is given by (note that the present resonance model is only realistic or relevant "in between" small and large times, i.e., does not describe the evolution for very short or very large times compared to multiples $\hbar/\Gamma$)

$$N(t) = \left| \langle\!\langle f_1 \| \exp\left(-\frac{i\hat{L}t}{\hbar}\right) \| f_k \rangle\!\rangle \right| \propto t^{k-1}\exp\left(-\frac{\Gamma t}{\hbar}\right) \tag{4.4a}$$

which yields

$$dN \propto t^{k-2}\left(k - 1 - \frac{\Gamma t}{\hbar}\right)\exp\left(-\frac{\Gamma t}{\hbar}\right)\cdot dt \tag{4.4b}$$

Equations (4.4) show that $dN > 0$ for $t < (k - 1) \cdot (\hbar/\Gamma)$, $k = 2, 3, \ldots, r$. Thus the occurrence of Jordan blocks in the Liouville picture leads to increase of $N(t)$ for the times specified above, that is, an increase of number of particles in the correlated state given by $\| f_1 \rangle\!\rangle$ by correlated transitions from all other states $\| f_k \rangle\!\rangle$, $k = 2, \ldots, r$. In general, there may be a certain balance between increase and decrease in coherence, correlation, and dissipation. Because there is an overall decay out of $\hat{I}$ given by

$$N(t) = \left| \langle\!\langle f_1 \| \exp\left(-\frac{i\hat{L}t}{\hbar} \right) \| f_1 \rangle\!\rangle \right| \propto \exp\left(-\frac{\Gamma t}{\hbar} \right) \tag{4.5a}$$

yielding the standard exponential decay rule

$$dN \propto \left(-\frac{\Gamma}{\hbar} \right) \exp\left(-\frac{\Gamma t}{\hbar} \right) \cdot dt \tag{4.5b}$$

the summing up of all contributions or transitions to (or from) the correlated state $\| f_1 \rangle\!\rangle$ may lead to dissipative structures which, during certain time spans of order $r\tau$ have increasing order, and in this sense we may speak of self-organization on a microscopic level "created" by the Jordan block term in $\hat{L}$. In the following section, we explain the explicit relation between these Jordan blocks and general impedance or gain functions.

V. CONNECTION WITH THE IMPEDANCE–GAIN FUNCTION

A typical gain function in the cyclotron maser problem was derived quantum mechanically by Schneider [12]. Based on Schneider's work, Ikegami [15] expanded the expression for the radiated net transfer of power in the quantum mechanical linear theory as well as attempted a derivation based on classical expressions involving specifically obtained friction terms [23].

He obtained, for particles gyrating in a magnetic field undergoing radiative transitions leading to a change of the transverse energy, the rate formula

$$\frac{d\gamma_\perp^*}{dt^*} \propto \tau \frac{2ax}{(1 + x^2)^2}$$

where $a = \omega_L^* \tau^* (\gamma_\perp^* - 1)$, $x = \tau^*(\omega_L^* - \omega)$, $\omega_L^* = \omega_c^*/\gamma_\perp^*$ is the relativistic cyclotron frequency, $\tau^* = 1/\nu^*$ with ν^* being the phase bunching frequency, and finally $\gamma_\perp^*$ is the relativistic energy factor. The "constant" above involves the energy-flow density (Poynting vector), the classical radius of the particles divided by $m_0 c$, with m_0 being the rest mass and c the speed of light.

We may also compare the Schneider expression where $a = Q(W/m_0 c^2)$, $Q = \omega_0 \tau$, and $x = (\omega_{i+1,i} - \omega)\tau$, with τ being the characteristic time, W the kinetic energy of the electron, and ω_0 the cyclotron frequency.

From Eq. (3.5) we obtain the nonlinear equations determining $\hbar\omega_L - i\Gamma$. The analytic continuation—necessary to obtain the complex solutions corresponding to the full subdynamics—rests on the simple dispersion relation (see [19] and Appendix D for details)

$$\Psi(\hbar\omega + i0) = \mathscr{P}[\hat{P}\mathscr{L}\hat{Q}(\hbar\omega - \hat{Q}\mathscr{L}\hat{Q})^{-1}\hat{Q}\mathscr{L}\hat{P}] - i\pi\delta[\hat{P}\mathscr{L}\hat{Q}(\hbar\omega - \hat{Q}\mathscr{L}\hat{Q})\hat{Q}\mathscr{L}\hat{P}]$$

with $\mathscr{P}(\Psi)$ and $\delta(\Psi)$ denoting the principal part and the delta function contribution of Ψ respectively.

Using the Obcemea–Brändas construction [19], the $\delta(\Psi)$ contribution can be retrieved from the corresponding Hamiltonian dynamics via a dispersion relation for $\bar{H}(E + i0) = H + HG(E + i0)H$, where $G(E + i0)$ is a suitably defined reduced resolvent and $H = H_0 + V$ with V a suitable optical potential [24]. For a sufficiently monochromatic beam of wave packets φ with kinetic energies given by H_0, one can identify $\bar{H}(E + i0)$ with the t-matrix, that is, $\bar{H}(E + i0) = H_0 + V + VG(E + i0)V = H_0 + t(E + i0)$ with G being the full resolvent, $G = (E + i0 - H)^{-1}$ and $t = 1/2\pi(\hat{S} - \mathscr{I})$, with $\hat{S}$ being the scattering matrix.

Eigenvalues of $\bar{H}(E)$ are then given by $E_i(E) - i\epsilon(E)$ with

$$E_i(E) = \langle \varphi_i | H_0 + V + \mathscr{P}[V(E - H)^{-1}V] | \varphi_i \rangle$$

and

$$\epsilon_i(E) = \pi\langle \varphi_i | V\delta(E - H)V) | \varphi_i \rangle$$

Further, the construction [19] amounts to finding an analytic continuation based on $\hat{L}$ with eigenvalues

$$E_i(E) - E_j(E) - i[\epsilon_i(E) + \epsilon_j(E)]$$

We note that $\epsilon_i(E) = \pi\{-\Im t_{\varphi_i\varphi_i}\} \propto \sum_\gamma \sigma_{i\to\gamma}$, where $\sigma_{i\to\gamma}$ is the cross section for the transition $i \to \gamma$ and the sum is the total cross section.

We conclude, therefore, that the absorption curve for the electron beam in the experimental setup [23] should follow a Lorentzian curve as described above in this section. However, the fundamental nonlinearities of this problem require additional work in the degeneracy space $\hat{I}$. Under "normal" conditions, that is, with no Jordan blocks present, the dispersion relation for $(z - \hat{L})^{-1}$ should give the projector associated with $\hat{P}$ [see Eq.

(3.5)], and hence only trivial heating of the particle beam is possible. In the present case, however, we need to study the resolvent $(z - \hat{L})^{-1}$, including the Jordan block term, in order to find the appropriate evolution commensurate with the degenerate root manifold spanned by $\|\mathbf{f}\rangle\rangle$. Thus, we obtain from Eq. (4.2)

$$(z - \hat{L})^{-1} = \frac{\tau}{\hbar}\left\{\frac{1}{(y+i)}\hat{I} + \frac{\omega_S \tau}{(y+i)^2}\hat{J} + \cdots \right. \tag{5.1}$$

with $y = \tau(z/\hbar - \omega_L)$. Intuitively, we might conclude that we only have to put $y = -x$, with x defined in connection with the previous identification with Schneider's gain curve [12]. Even if the contributions to the rate (see below) look similar in nature, there are some important differences. First, it is important to realize that ω_L, obtained from eq. (3.5), *is not equivalent to* Ikegami's ω_L^* or the gyrofrequency. Our ω_L is the energy parameter divided by $\hbar$ and thus contains relativistic corrections, with the leading terms obtained from the relativistic Schrödinger (or Klein–Gordon) equation, small collective effects, and possibly contains a weak interaction with the longitudinal degrees of freedom. Furthermore, there are higher-order terms in the resolvent expansion, Eqs. (4.2) and (4.3) which might completely blur out the simple and clear-cut information produced by a "well-behaved" gain function. It is nevertheless an interesting coincidence that such a gain-like function indeed can be obtained here. To see that consistent gains can be obtained, which have the desired domain of negative absorption, we briefly return to our scattering theoretic review.

Since the cross section (or associated spectral density) refers to taking the imaginary parts of the factors in front of $\hat{I}, \hat{J}, \ldots, \hat{J}^r$ (the correct spectral density would of course be derived from a correlation function of the type discussed in Appendix D, but from the mathematical point of view the present development is essentially correct), we can directly find the contributions, that is, from the first term,

$$\frac{d\gamma_\perp^*}{dt^*} \propto \langle\langle f_1 \| \Im\{-(z - \hat{L}\} \| f_1 \rangle\rangle = \frac{\tau}{\hbar}\frac{1}{(y^2+1)} \tag{5.2}$$

and from the second term (and similarly for higher orders),

$$\frac{d\gamma_\perp^*}{dt^*} \propto \langle\langle f_1 \| \Im\{-(z - \hat{L})^{-1}\} \| f_2 \rangle\rangle = \frac{\tau}{\hbar}\left\{\left(\frac{2\omega_S \tau \cdot y}{(y^2+1)^2}\right)\right\} \tag{5.3}$$

We note that for $\omega_L > z/\hbar$, or $y < 0$, we obtain a negative absorption, which is characteristic of a gain function. We can also compare with Schneider's and Ikegami's expressions and it might be tempting to estimate the value of $a = \omega_S \tau$, determined from Eq. (3.3), that is,

$$a = \omega_S \tau = \omega_L \tau \cdot 2\delta/N$$

which is related to Schneider's relativistic factor $\omega_0 \tau \cdot W/m_0 c^2$ (W is the kinetic energy of the electrons) or Ikegami's parametrization.

The gain function has been a key quantity in accelerator physics, because it suggests the possibility to improving the quality of the accelerated beam. It is important, however, to realize that this property may not coincide with cooling, because

1. It does not guarantee organization or increase of phase space density.
2. It may be blurred out by higher-order terms in Eq. (5.1).

Nevertheless, the absolute size of $a = \omega_S \tau$ may, in the present context [note that (3.4) can be overcome by $\tau \rightarrow r\tau$ from (4.4)], still be of fundamental importance for any cooling effects to appear. This conclusion is only indirectly drawn from Eq. (5.3), because $|a|^2$ appears in the proportionality factor of Eq. (4.4), see also Eq. (4.3). We do not discuss this issue further here, except to refer to recent discussions on the topic [8,15,28,29].

VI. CONCLUSION

The present approach has also been applied to condensed-matter situations, where quantum effects are prominent. These topics are not explicitly discussed here; see, for example, [3,9] for some earlier applications. The point we want to make in this final section concerns the transition from classical to quantum and within this framework we have emphasized the cyclotron maser concept as a convenient example where both aspects seem compatible. We have derived a Liouvillian operator in Section IV, whose time evolution was explicitly carried out and interpreted in terms of an organizational property (e.g., see, Eqs. (4.3)–(4.5)). We also demonstrated how the underlying microscopic characteristics of our formulation, discussed in Appendices A–D, tied the formulation to a suitable impedance–gain function. Obviously there is a strong connection between this formulation and quantum mechanical principles. Nevertheless, the formulation incorporates the classical regime also (see Section V).

To combine these aspects we present the equations in the "classical" form promised in the introduction. This is simply done by introducing dimensionless quantities in the standard way, that is,

$$\hat{\mathscr{P}} = \frac{\tau}{\hbar} \hat{L} \tag{6.1}$$

and

$$y = \tau(\omega - \omega_0) \tag{6.2}$$

It follows directly from Eqs. (6.1) and (6.2) and the properties of $\hat{\mathscr{J}}$ that the propagator can be written as, $n = r$

$$\exp(-i\hat{\mathscr{P}}t/\tau) = \exp(-i\omega t)\exp(-t/\tau)\sum_{k=0}^{n-1}(-it/\tau)^k \frac{1}{k!}\hat{\mathscr{J}}^k \tag{6.3}$$

and

$$(\omega\tau\hat{\mathscr{I}} - \hat{\mathscr{P}})^{-1} = \sum_{k=1}^{n}(y+i)^{-k}\hat{\mathscr{J}}^{k-1} \tag{6.4}$$

As in the introduction $\omega_L = \omega_0$ is the resonance frequency, ω_0^{-1} is the short time scale, τ is the phenomenological time scale, and $\omega_S = (\tau)^{-1}$, the "mismatch frequency" that may lead to chaotic behavior or organization, depending on the situation. Planck's constant can be included without problems when quantum conditions have to be explicitly considered.

Thus the operator $\hat{\mathscr{J}}$ "condenses" r singular points into one higher-order singularity. In principle, this set of points could have an infinite number of elements condensed to an essential singularity; it may be speculated whether one could interpret this as the sought-after connection between chaos in the classical and quantum domains.

APPENDIX A. THE COMPLEX SCALING METHOD

In this appendix, we give a brief review on the CSM. Most of the present material can be found in reference [7].

A key quantity in the CSM formulation is the scaling operator

$$U(\theta) = \exp(iA\theta) \tag{A.1}$$

where

$$A = \frac{1}{2}\sum_{k=1}^{N}[\vec{p}_k\vec{x}_k + \vec{x}_k\vec{p}_k] \tag{A.2}$$

is the generator of the scaling transformation, θ is a parameter, which may be real or complex, and $\vec{x}_k$ and $\vec{p}_k$ are the coordinate and momentum of particle k. If $\theta \in R$, then U is a unitary operator, defining the dilation group, with $\mathscr{D}(U) = \mathscr{H}$, which effectuates the scaling

$$U(\theta)\Phi(x_1, \ldots, x_N) = \exp\left(\frac{3N}{2}\theta\right)\Phi(e^{\theta}x_1, \ldots, e^{\theta}x_N) \tag{A.3}$$

By considering complex scalings, that is, $U(\eta)$ with $\eta = |\eta| e^{i\theta}$ (note that our parameter in U now refers to the whole analytic parameter η rather than the dilation group parameter θ), Balslev and Combes [4] proved that for certain classes of Hamiltonians, that is, so-called dilation analytic Hamiltonians, the continuous spectrum changed under the complex deformation and became rotated down -2θ in the complex energy plane, thereby opening up sectors on the "unphysical" Riemann sheet, where possible finite dimensional "resonance eigenvalues" would appear. They also demonstrated the nonexistence of singularly continuous spectra in these cases, as well as invariance of the exposed spectrum (including bound states) to variations of θ.

Without going into too many mathematical details, we will consider $\eta = |\eta| e^{i\theta}$, with $0 \leq \theta < \theta_0$. Generally θ_0 depends on the potential, for instance, in the Coulomb case no limit is invoked, while in some other cases, as shown below, it is natural to have $\theta_0 \leq \pi/4 - \delta$. We also introduce Ω as

$$\Omega = \{\eta, |\arg(\eta)| < \theta_0\} \tag{A.4}$$

Furthermore, we make the decomposition $\Omega = \Omega^+ \cup \Omega^- \cup R$, where the real axis $R = R^+ \cup R^- \cup \{0\}$ partitions Ω into its upper and lower parts in the complex plane. In what follows we assume that the interaction V is a sum of two-body interactions V_{ij} such that V_{ij} is Δ_{ij}-compact in $L^2(R^3)$, and furthermore that $V_{ij}(\eta) = U_{ij}(\eta)V_{ij}U_{ij}^{-1}(\eta)$, $\eta \in R^+$, has a compact analytic extension to Ω^+. The definition of the dilatation analytic family of operators $H(\eta)$ is then given by

$$H(\eta) = U(\eta)HU^{-1}(\eta), \qquad \eta \in R^+ \tag{A.5}$$

and

$$H(\eta) = \eta^{-2}T + V(\eta), \qquad \eta \in \Omega^+ \tag{A.6}$$

In the expression above it is notable that $H(\eta)$ is obtained in two steps—first via a unitary transformation to a scaled representation and thereafter by an analytic continuation to Ω^+. Although this is a mathematically rigorous procedure it is preferable, as we will see below, to consider the unbounded similarity transformation $U(\eta)$, $\eta \in \Omega^+$, directly. We then introduce the domain $\mathcal{N}(\Omega)$ as the subset

$$\mathcal{N}(\Omega) = \{\Phi, \Phi \in \mathcal{H};\ U(\eta)\Phi \in \mathcal{H};\ \eta \in \Omega\} \tag{A.7}$$

We furthermore complete the space with respect to the norm

$$\int_{-\theta_0}^{+\theta_0} \|U(\eta)\Phi\|_{L^2}^2\, d\theta = \|\Phi\|_{N\theta_0}^2 \tag{A.8}$$

Since we also want the first and second partial derivatives to satisfy Eq. (A.8), it is natural to define the spaces $\mathcal{N}_{\theta_0}^{(i)}$, $i = 0, 1, 2$, analogously. We are now in the position to make an alternative definition of the analytic family $H(\eta)$:

$$H(\eta) = U(\eta)HU^{-1}(\eta); \qquad \mathcal{D}(UHU^{-1}) = \mathcal{N}_{\theta_0}^{(2)} \tag{A.9}$$

and

$$H(\eta) = \eta^{-2}T + V(\eta); \qquad \mathcal{D}(UHU^{-1}) \rightarrow \mathcal{D}(T) \tag{A.10}$$

Note that $\mathcal{N}_{\theta_0}^{(i)}$, $i = 0, 1, 2$ are Hilbert spaces, which allow $H(\eta)$ to be interpreted as a similarity transformation of the self-adjoint unscaled operator H. We also note that $U(\eta)$ exhibits the "star-unitary" property

$$U^{\dagger}(\eta^{\star}) = U^{-1}(\eta) \tag{A.11}$$

which can easily be proven from

$$\langle\Phi|\Psi\rangle = \langle\Phi(\eta^{\star})|\Psi(\eta)\rangle = \langle\Phi|U^{\dagger}(\eta^{\star})U(\eta)|\Psi\rangle \tag{A.12}$$

which holds for all $\Phi, \Psi \in \mathcal{N}_{\theta_0}$.

In the definition presented above, leading of course to the same analytic family $H(\eta)$, we can see that the two steps involved in Eqs. (A.9) and (A.10) are different in comparison to Eqs. (A.5) and (A.6). Here the first step consists of restricting $\mathcal{H}$ to a smaller domain $\mathcal{N}_{\theta_0}^{(2)}$ for which the scaling $U(\eta)$ is defined for all η such that $\arg(\eta) < \theta_0$. After making the complex rotation, that is, changing the parameter θ in e^{θ} from real to complex, one completes the subset (A.7) (dense in $\mathcal{H}$) to $\mathcal{D}(T)$, which means that completion is made with respect to the "standard" L^2-norm for the functions and its first and second partial derivatives.

One can understand the abstract transformations discussed above in a very simple way. Consider, for example, the integral

$$I(b) = \int_0^{+\infty} e^{-br}\, dr = \frac{1}{b} \tag{A.13}$$

whose simple analytic form trivially allows analytic continuation to negative values of b. In the latter case, the integral is not convergent even if $I(b) = 1/b$ for all $b \neq 0$. If $b \neq 0$ is complex (or negative), so that the integrand in Eq. (A.1) does not vanish for $r \to \infty$, one may consider defining a complex integration path with θ sufficiently large that

$$I = \lim_{R \to \infty} \int_0^{Re^{i\theta}} e^{-br}\, dr = \frac{1}{b} \tag{A.14}$$

Hence the complex path can be used to explicitly compute a numerical value of the integral I even when its analytical form is unknown.

Thus we see that complex scaling can be used to extend quantum mechanics beyond its conventional domain in that it assigns a well-defined meaning to vectors and matrix elements with respect to operators (resolvents), requiring the existence of a scaled representation for some nonreal η.

Before we continue, we want to emphasize that we have implicitly assumed that $V\Psi \in \mathcal{N}_{\theta_0}$ for $\Psi \in \mathcal{N}_{\theta_0}^{(2)}$. Even if this is a restriction on V, one can usually circumvent this problem by considering more general deformations such as exterior scaling. These are, however, technical points that will not alter our general considerations, conclusions, or physical interpretation.

With the preceding development in mind, it is easy to avoid the following paradoxical situation. If we consider the resolvent operator (see below), one may ask whether the following relation with $\eta \in \Omega^+$, $\lambda^\star \neq \lambda$,

$$\langle \Phi | (H - \lambda I)^{-1} | \Psi \rangle = \langle \Phi(\eta^\star) | (H(\eta) - \lambda I)^{-1} | \Psi(\eta) \rangle \tag{A.15}$$

holds for all $\Phi, \Psi \in \mathcal{N}_{\theta_0}$. The question arises because the left-hand side of Eq. (A.15) is always finite for λ complex, while the right-hand side may become infinite if λ is a complex eigenvalue of the complex rotated operator $H(\eta)$, which is a possibility according to the Balslev–Combes theorem. The paradox has been discussed at some length in several circumstances, (e.g., see [7]) so we will not repeat it here, except to recommend that the reader attempt to find his/her own explanation.

It is also clear that we have considered the Hilbert spaces $\mathcal{N}^{(i)}_{\theta_0}$, $i = 0, 1, 2$ in order to find convenient domains for the unbounded (in $\mathcal{H}$) similitude $U(\eta)$, $\eta \in \Omega$. After appropriate deformations, completion with respect to the standard L^2 norm is made. In this manner, we arrive at the formal eigenvalue relation ($\eta = |\eta| e^{i\theta}$):

$$H(\eta)\Psi(\eta) = \epsilon(\eta)\Psi(\eta), \qquad \eta \in \Omega^+ \tag{A.16}$$

with arg η sufficiently large to uncover the resonance ϵ. The conjugate of Eq. (A.16) becomes

$$\overline{H(\eta)}\,\overline{\Psi(\eta)} = \overline{\epsilon(\eta)}\,\overline{\Psi(\eta)}, \qquad \eta \in \Omega^- \tag{A.17}$$

with the involution $\overline{A(\eta)} = A^\star(\eta^\star)$ being introduced to include also (complex) optical potentials. Obviously Eqs. (A.16) and (A.17) motivate the construction

$$\langle \overline{\Psi(\eta^\star)} \,|\, H(\eta) \,|\, \Psi(\eta) \rangle = \epsilon(\eta) \langle \overline{\Psi(\eta^\star)} \,|\, \Psi(\eta) \rangle \tag{A.18}$$

from which stationary variational principles can be derived in almost the same fashion as in ordinary quantum mechanics, with the important distinction that the extremum property of the principle has been lost. Note also that if H satisfies $H^\star = H$, we can assume, *from the variational point of view*, that $\Psi^\star = \Psi$, and then Eq. (A.18), defining a trial $\epsilon(\eta)$ based on a trial $\Psi(\eta)$ and corresponding to $\overline{\Psi(\eta^\star)} = \Psi(\eta)^\star$, becomes complex symmetric (cf. Gantmacher's theorem [16,24]). However, in many cases it may be useful to analyze the wave function in terms of nonreal spherical harmonics, $Y_{l,m}(\vartheta, \varphi)$, or other convenient nonreal representations. The complex symmetry is thus a convenience which may be imposed without restricting the formulation. We impose it below for simplicity.

Since η is sometimes used primarily as a numerical convergence factor, we may replace the explicit η-dependence for the quantity A by replacing $A(\eta)$ by A^c. Hence we write

$$H^c\Psi^c = \epsilon\Psi^c \tag{A.19}$$

and for the complex conjugate equation,

$$H^{c\star}\Psi^{c\star} = \epsilon^\star\Psi^{c\star} \tag{A.20}$$

and

$$\langle \Psi^{c\star} | H^c | \Psi^c \rangle = \epsilon \langle \Psi^{c\star} | \Psi^c \rangle \tag{A.21}$$

Even if this simple notation is very appealing in that almost any standard quantum mechanical technique can be taken over if appropriately modified, one should note that it results in a formulation that goes beyond conventional mechanics "on the real axis". The most direct consequence is, as mentioned, the appearance of complex resonance eigenvalues and associated Gamow vectors, see Fig. 1.

A closer study of all generalized spectral properties of the complex deformed problem shows that these eigenstates essentially deflate the (generalized) spectral density giving, in an asymptotic sense, a decomposition of the continuum into resonances and background. This has the important consequence that each Gamow vector represents a well-defined section of the continuous spectrum associated with the unscaled self-adjoint problem. In other words, one can say that *the Gamow representation condensates an infinite dimensional Hilbert space associated with a particular spectral part of the continuum into a finite dimensional linear space of suitable*

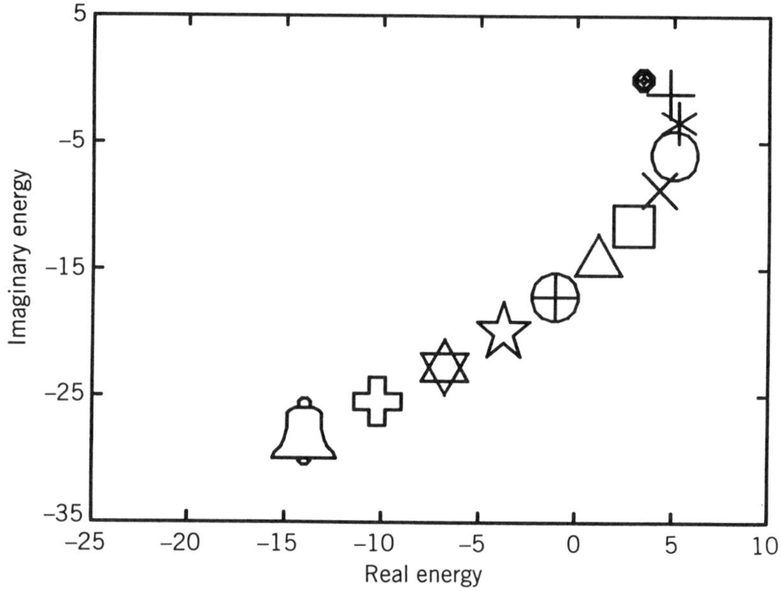

Figure 1. "Individualistic" complex resonances for a dilatation analytic potential containing an exponential barrier, see [26] for more details.

Gamow vectors. This condensation has played a fundamental role in our theoretical work and applications, especially regarding two fundamental problems:

1. Since the general case may lead to matrices that cannot be diagonalized, we must be able to incorporate so-called Jordan blocks into our theory (this turns out to be a blessing in disguise).
2. We need to introduce these Gamow-like resonances into a noncontradictory formulation of the appropriate Liouville equation (this turns out to be nontrivial).

The consequences of the CSM extension to dissipative subdynamics are quite surprising and have led to specific applications in concrete physical situations [3,7–9,16,20,26].

APPENDIX B. THE JORDAN BLOCK

In this appendix, we briefly discuss *how it may happen that a matrix cannot be diagonalized* and what a Jordan block means in this connection.

Consider the simple (complex) symmetric matrix **Q** given by

$$\mathbf{Q} = 1/2\begin{pmatrix} 1 & -i \\ -i & -1 \end{pmatrix} \tag{B.1}$$

By inspection we see that the square of Eq. (B.1) is the zero matrix, but because the rank (the number of linearly independent vectors that constitute the matrix) is 1, $\mathbf{Q} \neq 0$. Hence **Q** is similar to

$$\mathbf{C} = \begin{pmatrix} 0 & 1 \\ 0 & 0 \end{pmatrix}$$

that is, there exists an invertible matrix **B**, such that the triangular matrix **C** is related to the symmetric **Q** via

$$\mathbf{Q} = \mathbf{B}^{-1}\mathbf{C}\mathbf{B} \tag{B.2}$$

Rather than giving an explicit form for the 2×2 case shown above, we present a simple general solution. We also show that the vectors that constitute **B** or $\mathbf{B}^{-1}$ have some specific physical properties that may be directly related to the extreme case or Yang's ODLRO.

To generalize the discussion above, we focus on the classical canonical form of any finite dimensional matrix. We assume that the reader is familiar with the simple fact that a matrix with distinct eigenvalues (corresponding to one-dimensional linearly independent eigenvectors) can always be transformed, via a similarity transformation, like Eq. (B.2), to the diagonal form. The problem arises when degeneracies appear. If the matrix represents a normal operator (i.e., an operator that commutes with its own adjoint), one can also prove diagonalizability of the matrix. However, for general matrices, this is no longer true, and blocks of various dimensions may appear in the matrix, corresponding to a particular degenerate eigenvalue. The largest dimension of such a Jordan block defines the Segrè characteristic for this particular degenerate eigenvalue, and this will be a key quantity in what follows.

Because one can easily prove that any matrix can be transformed into a triangular form, it is obvious that any general matrix can also be transformed into a complex symmetric form provided a similarity transformation of type (B.2) exists between the triangular and the symmetric block. It suffices to focus on a particular degenerate eigenvalue, and without restriction we can put this eigenvalue equal to zero in the analysis. Assuming that the corresponding Segrè characteristic is s, the dimension of the largest Jordan block of the (zero) degenerate eigenvalue, it is clear that the rank of this nondiagonalizable block must be $s - 1$. The fact that the matrix, taken to the power $s - 1$, is different from zero (but equals zero for higher powers) means that the nondiagonalizable block can always be transformed to a classical canonical (Jordan) form with 1s in the entries above or below the diagonal.

In a previous study, Reid and Brändas [16] found a simple complex symmetric form of this Jordan block:

$$q_{kl} = \left(\delta_{kl} - \frac{1}{s}\right) \exp\left(i\pi \frac{k + l - 2}{s}\right) \tag{B.3}$$

where s is the dimensionality of the Jordan block and

$$1 \leq k, \qquad l \leq s \tag{B.4}$$

with the property that the matrix in Eq. (B.3) is similar to

$$\mathbf{C} = \begin{pmatrix} 0 & 1 & 0 & \cdots & 0 \\ \vdots & \ddots & \ddots & \cdots & \vdots \\ & & \ddots & \ddots & 0 \\ \vdots & & & \ddots & 1 \\ 0 & \cdots & & \cdots & 0 \end{pmatrix} \tag{B.5}$$

One can further prove that the matrix defined in Eq. (B.3) ($s = n$), is similar to Eq. (B.5) by considering the matrix **B** given by $[\omega = \exp(i\pi/n)]$

$$\mathbf{B} = \frac{1}{\sqrt{n}} \begin{pmatrix} 1 & \omega & \omega^2 & \cdots & \omega^{n-1} \\ 1 & \omega^3 & \omega^6 & \cdots & \omega^{3(n-1)} \\ \cdots & \cdots & \cdots & \cdots & \cdots \\ 1 & \omega^{2n-1} & \omega^{2(2n-1)} & \cdots & \omega^{(n-1)(2n-1)} \end{pmatrix} \tag{B.6}$$

or equivalently by $b_{kl} = \omega^{(2k-1)(l-1)}$. One realizes that **B** is unitary and then it is simple to compute and demonstrate that $\mathbf{Q} = \mathbf{B}^{-1}\mathbf{C}\mathbf{B}$.

Since Jordan blocks of various orders may occur in the classical canonical form, we also give the formulas for $\mathbf{Q}^r$ for $r = 1, 2, 3, \ldots, n-1$. Note that **Q** defined by

$$\mathbf{Q}_{kl} = \omega^{(k+l-2)}\left(\delta_{kl} - \frac{1}{n}\right)$$

should be zero if taken to the nth power. Direct evaluation of $\mathbf{Q}^r$ gives, $(r < n)$

$$(\mathbf{Q}^r)_{kl} = \omega^{r(k+l-2)}[\delta_{kl} - (\mathbf{R}^r)_{kl}] \tag{B.7}$$

with

$$(\mathbf{R}^r)_{kl} = \begin{cases} \dfrac{1}{n}\sin\left[\dfrac{\pi r(l-k)}{n}\right] \Big/ \sin\left[\dfrac{\pi(l-k)}{n}\right] & k \neq 1 \\ \dfrac{r}{n} & k = 1 \end{cases} \tag{B.8}$$

which, because of the underlying shift property, bears a resemblance to angular momentum algebra.

The transformation above is quite interesting in its own right, and it shows that a real triangular matrix and a complex symmetric matrix are connected through a unitary transformation. This particular form is of fundamental importance in the following appendix.

APPENDIX C. THE COHERENT–DISSIPATIVE ENSEMBLE

Here we describe how CSM is incorporated into the algebraic structure commensurate with dissipative dynamics. We assume associative and convex structures.

The starting point is, as before, the equation of motion for the density matrix or phase, space density ρ given by the Liouville–von Neumann equation

$$i\hbar \frac{\partial \rho}{\partial t} = \hat{L}\rho \tag{C.1}$$

which, together with some given initial conditions, defines a unitary time-reversible evolution. Note that we have inserted $\hbar$ for completeness and further indicated the superoperator $\hat{L}$ with a caret to distinguish it from the Schrödinger differential operator, as in the Hamiltonian formulation.

Our specific interest is the density matrix ρ, or its reduced system operator projections, which are defined later. The transition to the classical domain is not considered explicitly here. We first show how the CSM can be incorporated, which necessitates using a bit of algebra.

To allow for different choices of representation we define the superoperator $\hat{P}$ by

$$\hat{P} = A\rangle\langle + \rangle\langle B \tag{C.2}$$

where the choice of A and B will be made below and $\rangle\langle$ is short for the "bra" and "ket" components of the density matrix ρ. Thus Eq. (C.2) is equivalent to

$$\hat{P}\rho = A\rho + \rho B$$

and from this follows

$$e^{\hat{P}} = e^{A}\rangle\langle e^{B} \tag{C.3}$$

or

$$e^{\hat{P}} = e^{A}\rho e^{B}$$

and from Eq. (C.1) that

$$\partial\rho = dA\rho + \rho dB \tag{C.4}$$

There are basically two choices of A, B, and $\hat{P}$ that correspond to physically meaningful realizations. Note that we are now focusing on dilation analytic extension or a CSM framework for the Gamow-type solutions of Eq. (A.19) (see previous paragraph).

1. The first choice of operators concerns the time evolution of the Liouville equation [19]. One obtains $\hat{P} = -i\hat{L}t$ and $A = -iHt = B^{\dagger}$, which leads to

$$e^{-i\hat{L}t}\rangle\langle = |e^{-iHt}\rangle\langle e^{-iHt}| \tag{C.5}$$

and

$$i\frac{\partial \varrho}{\partial t} = \hat{L}\varrho \tag{C.6}$$

For a particular component, using the CSM construction [19], one obtains

$$\varrho_{kl}^{c} = |\Psi_k^c\rangle\langle\Psi_l^c| \tag{C.7}$$

and

$$H^c\Psi_k^c = \epsilon_k \Psi_k^c \tag{C.8}$$

with

$$\epsilon_k = E_k - i\epsilon_k \tag{C.9}$$

and the following eigenvalue relation:

$$e^{-i\hat{L}^c t}\varrho_{kl}^c = e^{\{-i(E_k - E_l) - (\epsilon_k + \epsilon_l)\}t}\varrho_{kl}^c \tag{C.10}$$

This choice is characterized by the fact that Eq. (C.10) contains energy differences, although the widths are added. Note also that ϱ in Eq. (C.7) for $k = l$ is *not* a projector, that is, $\varrho^2 \neq \varrho$, which makes the interpretation fundamentally different in connection with direct probability interpretations.

2. The second choice concerns the Boltzmann factor containing $\beta = 1/k_B T$, where k_B is the Boltzmann constant and T is the absolute temperature. Here the representation obtains from $\hat{P} = -\beta\hat{L}$ and $A = -\beta H = B$, satisfying

$$-\frac{\partial \rho}{\partial \beta} = \hat{L}_B \rho \tag{C.11}$$

with the eigenvalue relation (note the complex conjugate sign in the bra position)

$$e^{-\beta\hat{L}_{B^c}}|\Psi_k^c\rangle\langle\Psi_l^{c\star}| = e^{-\beta\{(E_k+E_l)-i(\epsilon_k+\epsilon_l)\}}|\Psi_k^c\rangle\langle\Psi_l^{c\star}| \tag{C.12}$$

Note that $\rho_{kk}^c = |\Psi_k^c\rangle\langle\Psi_k^{c\star}|$ here *is a (non-self-adjoint) projector* (i.e., $\rho^{c2} = \rho^c$), while $\rho^{c\dagger} \neq \rho^c$. In addition, we also see that *both widths and energies are added* in Eq. (C.12). In what follows we let $\beta \to \beta/2$ so that our formulation agrees with the standard definition of the Boltzmann factor, that is,

$$e^{-(\beta/2)\hat{L}_{B^c}}\rho_{kk}^c = e^{-\beta\epsilon_k}\rho_{kk}^c \tag{C.13}$$

We observe that $\hat{L}$ in Eq. (C.6) is defined by

$$\hat{L} = H\rangle\langle - \rangle\langle H \tag{C.14}$$

while $\hat{L}_B$ in Eq. (C.11) is given by

$$\hat{L}_B = H\rangle\langle + \rangle\langle H \tag{C.15}$$

and the fundamental difference between ϱ and ρ in the two formulations is shown above.

Within this algebraic structure, it is straightforward to apply CSM to $\Gamma_S^{(2)}$, Eq. (7), utilizing the second choice (C.11–13). The relevant matrix element (see [7] for more details) contains the factor

$$\exp\left[\frac{\beta}{2}(\epsilon_k + \epsilon_l)\right]\left(\delta_{kl} - \frac{1}{n}\right) \tag{C.16}$$

If the quantization condition below is satisfied, for example,

$$\pi\frac{(k-1)}{n} = \frac{\beta}{2}\epsilon_k \tag{C.17}$$

we obviously find that $\Gamma_S^{(2)}$ in Eq. (2.5) in the thermalization process assumes the form of a Jordan block of order r. Also $\Gamma_L^{(2)}$ becomes a Jordan block of order 2. Furthermore, the two superselection sectors (a superselection rule operates between two subspaces, superselection sectors, if there are neither spontaneous transitions between their state vectors nor any measurable quantities with finite matrix elements between their state vectors) corresponding to the large and small components of $\Gamma^{(2)}$ become connected through the thermalization. It also follows that the largest relaxation time, $\tau_{\rm rel}$, obtains from Eq. (C.17) with $r_{\min}$ the minimum size, that is,

$$\tau_{\rm rel} = r_{\min} \cdot \frac{\hbar}{4\pi k_B T} \tag{C.18}$$

In deriving Eq. (C.18) from Eq. (C.17), we have made use of the standard relation

$$\tau_k = \frac{2\hbar}{\epsilon_k}$$

Finally, we want to emphasize (see also Appendix A) that complex scaling leads to a certain asymptotic decoupling of the localized resonance state from the background. This is also very easy to visualize from the physical point of view. By scaling the coordinates so that the appropriate outgoing Gamow waves become square integrable, we find that the environment will gradually be shielded from the system in a manner precisely given by the CSM deformation. The incorporation of complex scaling into the present superoperator picture therefore gives, both in a physical as well as a mathematical sense, a valid subdynamical picture where the correlations from the environment can be preserved as much as needed for the dissipative state to "survive."

Although complex scaling exhibits the abovementioned localization properties, it is important to remember (see Appendix B) that irreducible nondiagonal structures, so-called Jordan blocks, may appear in contrast to the conventional formulation "on the real axis." This is a desired "complication" because it will, in fact, be associated with the self-organization of new coherence patterns of a nonlocal, nonlinear quantum statistical origin.

APPENDIX D. THE CORRELATION FUNCTION

Here we present some results on the general transforms between a correlation function and the corresponding spectrum.

We begin by giving a simple account of the relations between the resolvent and the evolution operator. Despite our emphasis on density matrices and the Liouville equation, we give some of our formulas in the Hamiltonian form, but the replacement of the Hamiltonian H by a general superoperator should in most cases be quite straightforward. In the most general case, however, and in the transition to the classical domain, the analytic problems must be controlled specifically.

We temporarily put $\hbar = 1$, which means that the Hamiltonian (or equivalently the Liouvillian) expresses energies (or energy differences) in frequency units. We keep the time directions (as well as associated analyticity requirements for the resolvent) and therefore define the *retarded–advanced* ($\pm$) evolution operator and the associated resolvents as

$$\mathscr{G}^{\pm}(t) = \mp i\vartheta(\pm t)\exp(-iHt) \tag{D.1}$$

$$\mathscr{G}(z) = (zI - H)^{-1} \tag{D.2}$$

where $\vartheta(x) = 1$ for $x \geq 0$ and zero for $x < 0$. The advantage of using the retarded–advanced propagators is obvious from the viewpoint of the associated resolvents, because they are related through the Fourier–Laplace transforms

$$\mathscr{G}^{\pm}(t) = \frac{1}{2\pi}\int_{C^{\pm}} \mathscr{G}(z)\exp(-izt)\,dz \tag{D.3}$$

$$\mathscr{G}(z) = \int_{-\infty}^{+\infty} \mathscr{G}^{\pm}(t)\exp(izt)\,dt. \tag{D.4}$$

The contour $C^{\pm}$ runs in the upper (+) and lower (−) complex half plane, respectively, from $-\infty$ to $+\infty$. With the initial condition (replace Ψ with the appropriate density matrix in the Liouville case)

$$\Psi(t) = \Psi_0 \qquad \text{for } t = 0$$

and rule of evolution given by

$$\Psi^{\pm}(t) = \pm i\mathscr{G}^{\pm}(t)\Psi_0$$

one obtains the *inhomogeneous* differential equations

$$\left(i\frac{\partial}{\partial t} - H\right)\mathscr{G}^{\pm}(t) = \delta(t)$$

and

$$(z - H)\Psi(z) = \begin{cases} +i\Psi_0 & z \text{ in upper half plane, } t > 0 \\ -i\Psi_0 & z \text{ in lower half plane, } t < 0 \end{cases}$$

The interpretation of a complex resonance (cf. discussions around so-called Gamow waves) usually evolves as follows. For $t > 0$, one obtains, for example, the probability amplitude

$$a(t) = \langle \Psi_0 | \Psi^+(t) \rangle = +i \langle \Psi_0 | \mathcal{G}^+(t) | \Psi_0 \rangle$$

$$a(t) = \langle \Psi_0 | \frac{i}{2\pi} \int_{C+} (zI - H)^{-1} \exp(-izt)\, dz \,| \Psi_0 \rangle \tag{D.5}$$

from which one deduces via a quick detour into the "unphysical Riemann sheet" that the contribution from simple poles is

$$a(t) = \sum_k |\langle \Psi_0 | \Psi_k \rangle|^2 \exp(-iE_k t) + \text{background contributions} \tag{D.6}$$

Since a self-adjoint operator only exhibits a real spectrum, the true interpretation of Eq. (D.6) should be that of a complex amplitude resting inside the circle in the complex plane, defined by $|a(t)| \leq a_{\max}$, for some $a_{\max}$. However, if one introduces the concept of a *complex resonance eigenvalue* and assumes that there is, in some sense, an expansion formula of the conventional "spectral resolution type," one obtains (superficially), with $\epsilon_k = E_k - i(\Gamma_k/2)$ (and $R_k = \langle \Psi_0 | \Psi_k \rangle$) using the residue theorem on Eq. (D.5),

$$|a(t)|^2 = |R_k|^2 \exp - \Gamma_k t + \cdots \tag{D.7}$$

The precise meaning of Eq. (D.6) is usually made rather vaguely. First, the so-called interference terms are neglected, which may or may not be important, there is also the problem of small times and long time tails, which do not subscribe to the formula above. Nevertheless, it is possible to identify the evolution according to Eq. (D.6) for some relevant time intervals of some given prepared system (as in radioactive decay). Despite the hand-wavy nature of the present description and the nonrigorous mathematical setting, we can already introduce the concept of a *level width*,

$$-2\Im \epsilon_k = \Gamma_k \tag{D.8}$$

through our knowledge of the complex eigenvalue given by CSM. The contour integration can basically be carried out rigorously.

From these simple equations it is possible to derive the general relations between a correlation function $C(t)$ and the corresponding spectrum $Q(\omega)$. In analogy with the retarded/advanced formulation above,

$$g^{\pm}(t) = \mp i\vartheta(\pm t) C(t) \tag{D.9}$$

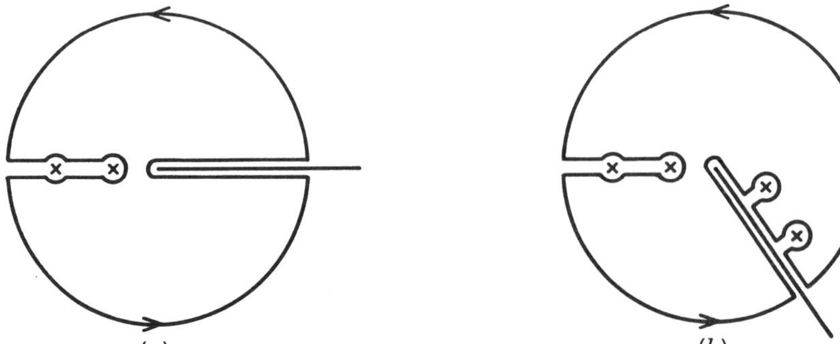

Figure 2. Integration contour for the Cauchy representation of the Nevanlinna function displaying the spectrum of a differential operator (left), and the integration contour for the "deformed" case, obtained by CSM. In the latter case, one finds a "rotated" continuum and also complex resonance poles (right).

Replacing the averaged Green's function in Eq. (D.4), or resolvent in Eq. (5.1), by a function $-f(z)$ with simple analyticity requirements, for example, Kramers–Kronig type dispersion relations or Nevanlinna character {a mapping of respective (complex) half-planes into each other plus the requirement of Cauchy analyticity; see also discussion in [27]}, one finds

$$z \to E \pm i0 \Rightarrow f(z) \to \Im f^{\pm}(E) = \pm \Im f(E) \tag{D.10}$$

with

$$g^{\pm}(t) = \frac{1}{2\pi} \int_{C\pm} [-f(z)] \exp(-izt)\, dz \tag{D.11}$$

and

$$-f(z) = \int_{-\infty}^{+\infty} g^{\pm}(t) \exp(izt)\, dt \tag{D.12}$$

and finally

$$\frac{1}{\pi} \int_{-\infty}^{+\infty} \Im f(\omega) \exp(-i\omega t)\, d\omega = C(t) \tag{D.13}$$

Using the notation $Q(\omega) = (1/\pi)\Im f(\omega)$, one can write

$$\int_{-\infty}^{+\infty} Q(\omega)\exp(-i\omega t)\,d\omega = C(t) \tag{D.14}$$

and

$$\frac{1}{2\pi}\int_{-\infty}^{+\infty} C(t)\exp i\omega t\,dt = Q(\omega) \tag{D.15}$$

However, the occurrence of Jordan blocks in the underlying operator $\hat{L}$, containing $\hat{\mathscr{J}}$, introduces more complicated dispersion relations and non-Nevanlinna behavior, yet the algebra corresponding to Eqs. (6.3) and (6.4) can be worked out simply and analyzed. We do not go into the various analyticity conditions needed for possible continuation into the complex plane, except to point out that careful investigation of the underlying operators may throw additional light on the associated dynamical behavior in complex systems. For general integration contours see Fig. 2a and 2b.

ACKNOWLEDGMENTS

I thank the Swedish Natural Science Research Council and the Swedish Research Council for Engineering Sciences for financial support. I also thank M. Höghede for discussions related to formula (B.7–8). In particular I am thankful to Prof. H. Ikegami for his never failing enthusiasm to communicate and provoke new ideas and Prof. S. Kullander for stimulating cross-disciplinary interaction at its best.

REFERENCES

1. I. Prigogine, *From Being to Becoming. Time and Complexity in the Physical Sciences* (Freeman, New York, 1980).
2. G. Nicolis and I. Prigogine, *Exploring Complexity* (Freeman, New York, 1989).
3. E. Brändas and C. A. Chatzidimitriou-Dreismann, *Int. J. Quant. Chem.* **40**, 649 (1991).
4. E. Balslev and J. M. Combes, *Commun. Math. Phys.* **22**, 280 (1971).
5. P. O. Löwdin, *Phys. Rev.* **97**, 1474 (1955).
6. A. J. Coleman, *Rev. Mod. Phys.* **35**, 668 (1963); *J. Math. Phys.* **6**, 1425 (1965); see also *Density Matrices and Density Functionals, Proceedings of the A. John Coleman Symposium*, R. Erdahl and V. H. Smith, Jr. (eds.) (Riedel, Dordrecht, 1987).
7. E. J. Brändas and C. A. Chatzidimitriou-Dreismann, Lecture Notes in Physics **325**, 486 (1989).
8. E. J. Brändas, "Organization of particle beams—A possible new cooling concept," in *Proceedings of the Workshop on Beam Cooling and Related Topics*, Montreaux, October 4–8, 1993.
9. C. A. Chatzidimitriou-Dreismann, *Adv. Chem. Phys.* **80**, 201 (1991).

10. C. N. Yang, *Rev. Mod. Phys.* **34**, 694 (1962).
11. R. Q. Twiss, *Aust. J. Phys.* **11**, 564 (1958).
12. J. Schneider, *Phys. Rev. Lett.* **2**, 504 (1959).
13. P. Sprangle and A. T. Drobot, *IEEE Trans. Microwave Theory Tech.* **25**, 528 (1977).
14. J. L. Hirshfield and J. M. Wachtel, *Phys. Rev. Lett.* **12**, 533 (1964).
15. H. Ikegami, *Phys. Rev. Lett.* **64**, 1737 (1990); *ibid.* 2593 (1990).
16. C. E. Reid and E. J. Brändas, *Lecture Notes in Physics*, **325**, 475 (1989).
17. C. van Winter, *J. Math. Anal.* **47**, 633 (1974).
18. B. Simon, *Ann. Math.* **97**, 247 (1973).
19. CH. Obcemea and E. Brändas, *Ann. Phys.* **151**, 383 (1983).
20. C. A. Chatzidimitriou-Dreismann and E. J. Brändas, *Physica* **201C**, 340 (1992).
21. I. Prigogine, C. George, F. Henin, and L. Rosenfeld, *Chem. Scripta* **4** (1973).
22. E. Brändas in *Quantum Science Methods and Structure. A Tribute to Per-Olov Löwdin*, J-L. Calais, O. Goscinski, J. Linderberg, and T. Öhrn (eds.) (Plenum New York, 1976), p. 381; see also E. Brändas and P. Froelich, *Int. J. Quant. Chem.* **S17**, 113 (1983).
23. H. Ikegami, *Phys. Scripta* **48**, 32 (1993); "Coherent Microwave Cooling of Electron and Ion Beams," in *Proceedings of the Workshop on Beam Cooling and Related Topics*, Montreaux, October 4–8, 1993.
24. D. A. Micha and E. Brändas, *J. Chem. Phys.* **55**, 4792 (1971).
25. F. R. Gantmacher, *The Theory of Matrices*, Vol. II (Chelsea Publishing, New York, 1959).
26. E. J. Brändas (ed.), *Proceedings from the Nobel Satellite Symposium on Resonances and Microscopic Irreversibility*, *Int. J. Quant. Chem.* **46**, (3) (1993).
27. E. Engdahl, E. Brändas, M. Rittby, and N. Elander, *Phys. Rev. A* **37**, 3777 (1988).
28. J. L. Hirschfield and G. S. Park, *Phys. Rev. Lett.* **66**, 2312 (1991).
29. D. J. Larson, *Phys. Rev. Lett.* **68**, 133 (1992).

TIME, IRREVERSIBILITY, AND UNSTABLE SYSTEMS IN QUANTUM PHYSICS

E. EISENBERG

Department of Physics, Bar-Ilan University, Ramat-Gan, Israel

L. P. HORWITZ

Department of Physics, Bar-Ilan University, Ramat-Gan, and School of Physics, Raymond and Beverly Sackler Faculty of Exact Sciences, Tel-Aviv University, Ramat-Aviv, Israel

CONTENTS

ABSTRACT

The recently developed quantum theory utilizing the ideas and results of Lax and Phillips for the description of scattering and resonances, or unstable systems, is reviewed. The framework for the construction of the Lax–Phillips theory is given by a functional space which is the direct integral over time of the usual quantum mechanical Hilbert spaces, defined at each t. It has been shown that quantum scattering theory can be formulated in this way. The theory of Lax and Phillips, however, also obtains a simple relation between the poles of the S-matrix and the spectrum of the generator of the semigroup corresponding to the reduced motion of the resonant

Advances in Chemical Physics, Volume XCIX, Edited by I. Prigogine and Stuart A. Rice.
ISBN 0-471-16526-3

state. In this Chapter, we show that to obtain such a relation in the quantum mechanical case, the evolution operator must act as an integral operator on the time variable. The structure required appears naturally in the Liouville space formulation of the evolution of the state of the system. The resulting S-matrix is a function, as an integral operator, of $t - t'$ (i.e., homogeneous), and the semigroup is contractive. A physical interpretation of this structure may be introduced, from which we obtain a quantitative description of the expected age of a created system, and the expected time of decay of an unstable system. The superselection rule which distinguishes between the unstable system and its decay products is realized in this way. It is also shown that, from this point of view, one has a natural mechanism for the dynamical mixing of the quantum mechanical states as observed by means of time-translation invariant operators. In particular, this provides a model for certain types of irreversible processes, and for the measurement process for closed and open systems.

I. INTRODUCTION

The unstable quantum system is an important example of irreversible phenomena in nature. Such systems, ranging from excited atomic states to short-lived elementary particles, are characterized by what is generally observed to be an irreversible evolution. These phenomena cause us to seek an explanation for such processes from first principles. Moreover, since most of the decay processes are observed experimentally to obey an exponential decay law, one expects this behavior to follow from very general assumptions.

In this section, we give some historical background of efforts to describe the unstable system. In the next section, we review the main ideas of Lax–Phillips scattering theory and show how it can emerge from the quantum theory. In Section III, we show that a general law of evolution leads to an S-matrix of Lax–Phillips form or which the singularities are associated with the spectrum of the generator of a semigroup which describes the evolution of the unstable states. In Section IV, we discuss the connection of Lax–Phillips theory to some aspects of measurement theory. In Section V, using related ideas, we show that mixing of states can occur (even in closed systems) by a similar mechanism in the Liouville space formulation of both classical and quantum mechanics.

The description of irreversible evolution in the quantum theory has been described by the addition of non-Hermitian terms to the Hamiltonian, such that it has complex eigenvalues, and the induced evolution is nonunitary. Structures of this type were originally introduced by Gamow [1] who studied the effect of assigning complex eigenvalues to the energy spectrum, and hence introduced a kind of generalized eigenvector. Wu and Yang [2] parameterized the K-meson decay in this way. In this method, the non-Hermitian terms in the Hamiltonian are introduced phenomenologically, and may only indirectly be associated with some known interaction terms. In addition, the interpretation of the generalized eigenvectors, and their relation to some wave function which describes a definite state of the system, are not at all clear.

In 1930, Weisskopf and Wigner [3], introduced an alternative approach to the decay problem. According to their approach, the evolution takes place in a Hilbert space which is a direct sum of two subspaces: that of the decaying states and that of decay products. These two subspaces are stable under the "free" evolution induced by H_0, but are combined linearly under the full evolution induced by $H = H_0 + V$. In this Hilbert space, the evolution is unitary, and hence its generator (i.e., the Hamiltonian), is self-adjoint. The decay is described as the probability flow from the subspace of the decaying states to its complement, the subspace of the decay products. Weisskopf and Wigner considered the most simple model, in which there is only one decaying state ψ, and a continuum of states corresponding to the decay channel. They studied perturbatively what has become known as the *survival amplitude*,

$$A(t) = (\psi, e^{-Ht}\psi) \tag{1.1}$$

which is the probability amplitude for the system to remain in the discrete state until time t. Horwitz and Marchand extended this model for systems in which the unstable-state subspace is two-dimensional, applied it to the decay of the K^0-meson [4], and then extended the treatment to the most general case [5]. They formulated the decay problem as the evolution of the full unitary evolution projected into the subspace of the unstable states, and unified the mathematical treatment of this problem with that of scattering resonances. In the following discussion, we describe this approach and pose critical problems, motivating the development of a more general theory.

The Hilbert space that corresponds to an unstable system $\mathscr{H}$ may be represented as the direct sum of two Hilbert spaces that correspond to the state space of the decaying system and the decay products. Let us denote the projection operators on these two subspaces P and $\bar{P}$, such that $P + \bar{P} = 1$. For the decay problem, the basic quantity is the *reduced motion*

$$U'(t) = PU(t)P \tag{1.2}$$

where $U(t) = e^{-iHt}$, which governs the time evolution of the subspace $P\mathscr{H}$ of the unstable states. From this one can derive the decay law of the unstable states. If $\{\phi_i\}$ is an orthonormal basis of $P\mathscr{H}$, the probability that an unstable state ϕ, which exists at time $t = 0$, is in the subspace $P\mathscr{H}$ of unstable states at time t is given by

$$p(t) = \sum_i |(\phi_i, U(t)\phi)|^2 \tag{1.3}$$

Another way of writing this quantity is

$$p(t) = Tr_P(U'^{\dagger}(t)U'(t)P_\phi) \tag{1.4}$$

where P_ϕ is the projection on the subspace spanned by the initial state $\phi \in P\mathscr{H}$.

The total evolution operator $U(t) = e^{-iHt}$ and the resolvent $R(z) = (z - H)^{-1}$ are related to each other by the (inverse) Laplace transform

$$U(t) = \frac{1}{2\pi i}\oint R(z)e^{-izt}\,dz \tag{1.5}$$

where the integration contour is around the spectrum of H. If we project this operator into the subspace $P\mathscr{H}$, we can obtain a similar relation which expresses the reduced motion $U'(t)$ in terms of the *reduced resolvent* $R'(z) = PR(z)P$:

$$U'(t) = \frac{1}{2\pi i}\oint R'(z)e^{-izt}\,dz \tag{1.6}$$

Using this relation, one can derive an evolution equation for the reduced motion, which permits us to examine the behavior of this function in different time regimes. For this purpose, let us use the identity $zR(z) = 1 + HR(z)$ and project it on the two (orthogonal) subspaces $P\mathscr{H}$ and $\bar{P}\mathscr{H}$. Doing so, we obtain the coupled system

$$\begin{aligned} zR'(z) - P &= PHR'(z) + PH\bar{P}R(z)P \\ z\bar{P}R(z)P &= \bar{P}HR'(z) + \bar{P}H\bar{P}R(z)P \end{aligned} \tag{1.7}$$

and eliminating $\bar{P}R(z)P$,

$$zR'(z) - P = PHR'(z) - PH\bar{P}(z - \bar{P}H\bar{P})^{-1}\bar{P}HR'(z) \tag{1.8}$$

Taking the inverse Laplace transform of Eq. (1.8), and using the convolution theorem and Eq. (1.6), one obtains the following integro-differential equation for the reduced motion:

$$i \frac{d}{dt} U'(t) = PHPU'(t) - i \int_0^t d\tau\, PH\bar{P}e^{-i\bar{P}H\bar{P}\tau}\bar{P}HPU'(t-\tau) \tag{1.9}$$

The latter equation is called the *Master Equation for the reduced motion*, and applies not only to decay systems, but in general to evolutions obtained from a unitary group $\{U(t)\}$ by projection into a subspace [6]. This equation is particularly useful for small times where a solution by iteration is highly convergent [7]. In the limit of very small times, the second term of Eq. (1.9) is negligible and $U'(t)$ evolves unitarily.

That the decay rate vanishes for $t \to 0$ can also be seen directly from Eq. (1.4) (provided that the Hamiltonian is defined on $P\mathscr{H}$):

$$\begin{aligned} \frac{d}{dt} p(t) &= \frac{d}{dt} Tr_P(PU^{\dagger}(t)PU(t)P_{\phi})\Big|_{t=0} \\ &= Tr_P(iPHU^{\dagger}(t)PU(t)P_{\phi} - iPU^{\dagger}(t)PHU(t)P_{\phi}\Big|_{t=0} \\ &= iTr_P(PHP_{\phi} - PHP_{\phi}) = 0 \end{aligned} \tag{1.10}$$

This reflects the so-called $O(t^2)$ short-time behavior, which leads to the prediction of special quantum effects such as the quantum Zeno effect [8].

If the term $\bar{P}H\bar{P}$ vanishes, one obtains an algebraically soluble model [9,10] which has been very helpful in investigating the analytic properties of the theory of unstable systems. The generalized states (elements of a Gel'fand triple [11]), providing exact exponential decay, have been studied in detail in this model [12,13].

It is not difficult to see that an irreversible evolution must be described by a semigroup [14] (for the reversible case this is a group induced by a unitary transformation). However, it can be shown generally that the reduced motion, as described above, cannot generate a semigroup [15].

An operator family $U'(t)$ on the subspace $\mathscr{K}$ is called *of positive type and contractive* if the following conditions hold:

$$U'^{\dagger}(-t) = U'(t), \qquad U'(0) = I_{\mathscr{K}} \tag{1.11}$$

According to the theory of extensions of Hilbert spaces [16], one may construct a Hilbert space $\mathscr{H}$ (which is an extension of $\mathscr{K}$), and a continuous group $U(t)$ of unitary operators on it, such that $U'(t)$ is the contraction of $U(t)$ to the subspace $\mathscr{K}$. Such an extension is unique if one requires $\mathscr{H}$ to be spanned by $U(t)\mathscr{K}$.

The physical interpretation of this theorem is that, knowing the subspace of unstable states and the contracted evolution in it, one can construct uniquely the Hilbert space of the unstable states and the decay products, and the generator of the evolution in this space, which is the Hamiltonian.

If we require $U'(t)$ to be strongly contractive, that is,

$$s - \lim_{t \to \infty} U'(t) = 0 \tag{1.12}$$

it can be shown [15,16] that $U'(t)$ may be a semigroup only if the spectrum of the Hamiltonian H is all the real line $\mathscr{R}$. When $\mathscr{K}$ is finite-dimensional, one can prove further that if $U'(t)$ is a semigroup, there are states in $\mathscr{K}$ with infinite energy, and for states with finite energy the decay rate vanishes at $t = 0$. These conditions cannot all be achieved for Schrödinger systems in which the Hamiltonian is semibounded and the subspace contains only finite energy states. Hence this theorem is, in fact, a no-go theorem, which proves that an exact semigroup law cannot be obtained for ordinary quantum-mechanical systems.

We have briefly discussed Weisskopf–Wigner theory, and seen the difficulties that arise. According to this theory, the decay law is not exponential, and the exponential, and the experimentally observed exponential law is approximated only for times not too short and not too long. Moreover, we have seen that the description of the decay process in the framework of contraction of unitary evolution, in the usual quantum-mechanical Hilbert space, to a subspace, does not lead to semigroup behavior, and therefore this is not an irreversible process in the full sense (i.e., there are regeneration effects). The $O(t^2)$ short-time behavior for the contracted evolution also indicates deviations from the semigroup law, leading to effects such as the Quantum Zeno effect, which may occur in some physical systems; one would like to have a more general theory in which this effect may or may not occur, according to the dynamical laws.

Techniques of analytic continuation of the reduced resolvent to the second Riemann sheet, resulting in an exact exponential decay, lead to the concept of generalized vectors. The physical interpretation of these generalized states, and their relation to vectors of the Hilbert space which describe well-defined states of the system, are not at all clear. In particular, there are many ways of representing the space in which the pole is a (generalized) eigenvalue, for example, by defining the space by analytic continuation through the cut in the complex energy plane, or by analytic continuation of dilations [12]. One would also like to have a theory in which the states associated with the unstable system have a consistent physical interpretation.

There is, furthermore, another, perhaps more fundamental, problem associated with the general method of Wigner and Weisskopf—the expression (1.1) for the survival amplitude implicitly assumes the existence of a linear superposition

$$e^{-iHt}\psi = A(t)\psi + \chi(t) \tag{1.13}$$

where $\chi(t)$ represents the decayed system and $(\psi, \chi(t)) = 0$. In general, this linear superposition does not correspond to any physical situation in our experience; a short-lived particle, for example, is seen as either the particle before the decay or the decay products at a certain time, which cannot be predicted. This linear superposition does not correspond to the object that we see experimentally in such a process.

These features are essentially related to the attempt to describe an unstable system in a framework more suitable to the description of reversible phenomena. In what follows we show another approach to irreversible phenomena which attempts to solve these difficulties.

II. THE QUANTUM UNSTABLE SYSTEM: A DIRECT INTEGRAL SPACE DESCRIPTION

In Section I, we briefly discussed the traditional methods of dealing with unstable states in the framework of quantum theory and pointed out some of the difficulties in such methods. However, within the scope of quantum theory, there is a framework more suitable to the description of deterministic, irreversible phenomena. It is based on the use of a direct integral Hilbert space, which we describe in the following.

The characterization of a system undergoing an irreversible process cannot, in principle, be specified at a given instant in time. In fact, the physical quantities describing such processes involve time measurements (i.e., measurements of the time at which certain defined phenomena occur). Therefore, information about the decay to be deduced from the state is associated with its distribution in time, which is an essential property of the system, as is the location or momentum of a quantal particle. The time variable is, from this point of view, an internal degree of freedom of the system, which provides a framework for the description of interactions which can influence the structure of the state. A more detailed discussion of interpretation is given in a later section.

The dynamical evolution of the system involves a change in its internal structure, including its distribution in t, along with other observables characterizing the state. This evolution, parameterized by the laboratory time τ (which is not a dynamical variable), is defined on a Hilbert space with t in its measure space (e.g., with Lebesgue measure), with the norm given by

$$\int \|\psi_t^\tau\|^2 \, dt = \|\psi^\tau\|^2$$

where the norm in the integral is taken as the norm in $\mathscr{H}_t$, a member of a family of auxiliary Hilbert spaces (all isomorphic) defined for each t.

The theory of Lax and Phillips [17], designed for systems of hyperbolic differential equations describing the scattering of electromagnetic or acoustic waves, for example, and the Floquet theory [18] for periodic, time-dependent, quantum-mechanical problems are examples of such a structure. Piron [14] has shown that methods of this type are applicable to the general, time-dependent, quantum-mechanical problem. Recently, Flesia and Piron [19] have shown that scattering problems in quantum theory can be put in the form of Lax–Phillips theory (Horwitz and Piron [20] have discussed its applicability to the problem of the unstable system) by forming a direct integral of the quantum mechanical Hilbert spaces $\mathscr{H}_t$ over t in order to construct a larger space $\mathscr{H}$ which includes t in its measure space. Their approach to this kind of structure is as follows.

Horwitz and Piron [20] point out that it has been known since 1952 [21] that physical systems cannot always be described by a single Hilbert space. In general, they are described by a direct sum of Hilbert spaces for which the matrix elements of any observable between states belonging to different components vanish, that is, there is a superselection rule between these components. There may be many superselection rules in physics. In particular, in the description of the state of a system, the time plays the role of a continuous superselection rule. The system, therefore, may be described by a family of Hilbert spaces indexed by t [22]. It is useful to consider this family of Hilbert spaces as a functional space [23]. The Lebesgue measure is chosen for the weights of the orthogonal direct integral. Therefore, this functional space is a Hilbert space, which we call the direct integral space, of the form

$$\bar{\mathscr{H}} = \int_{\oplus} \mathscr{H}_t \, dt = L^2(-\infty, \infty; \mathscr{H}) \tag{2.1}$$

Each component in the direct integral (2.1) is a copy of a single Hilbert space $\mathscr{H}$ (which is isomorphic to the Schrödinger quantum-mechanical Hilbert space). Scalar products in $\bar{\mathscr{H}}$ have the form

$$(f, g) = \int (f_t, g_t)_{\mathscr{H}} \, dt \tag{2.2}$$

and the norm squared is

$$\|f\|_{\mathscr{H}}^2 = \int \|f_t\|_{\mathscr{H}}^2 \, dt \tag{2.3}$$

It is instructive to see how this method arises from a procedure in classical mechanics. Consider Hamilton's equation of motion,

$$\frac{dq}{dt} = \frac{\partial H}{\partial p}, \qquad \frac{dp}{dt} = -\frac{\partial H}{\partial q} \tag{2.4}$$

where the Hamiltonian $H(p, q, t)$ depends on time. Defining the time as a new variable and the energy E of outside sources as its conjugate momentum, the new Hamiltonian [24]

$$K(q, p, t, E) = E + H(q, p, t) \tag{2.5}$$

leads to the equivalent equations

$$\begin{aligned} \frac{dq}{d\tau} &= \frac{\partial H}{\partial p}, \qquad \frac{dp}{d\tau} = -\frac{\partial H}{\partial q} \\ \frac{dt}{d\tau} &= 1, \qquad \frac{dE}{d\tau} = -\frac{\partial H}{\partial t} \end{aligned} \tag{2.6}$$

in which K is independent of the time τ, and the time t and the new time τ have the same rate. Thus, the procedure of lifting the time to be a dynamical variable is used to transform a time-dependent problem to a time-independent one.

We have claimed before that the observation of irreversible processes involves time measurements. Consequently, one would like to have a *time operator*, that would correspond to the measured time. It is well known [25] that the canonical quantization rule, $[A, B] = i$, implies that the spectrum of both A and B is the whole real line. Therefore, the desired time operator T, whose canonical conjugate is the generator of translations in time, i.e., the generator of the evolution, cannot be defined in the framework of single-Hilbert space where the generator of the evolution is the Hamiltonian which is semibounded. In the direct integral space, however, such a time operator can be defined naturally as

$$(T\psi)_t = t\psi_t \tag{2.7}$$

To complete the construction of the theory in the framework of the direct integral space, one should specify the evolution law. In this section, we define the evolution of a system described by $\psi \in \bar{\mathscr{H}}$ as the ordinary Hilbert space unitary evolution combined with translation along the t-axis [19],

$$\psi_{t+r}^{\tau} = W_t(\tau)\psi_t^0 \tag{2.8}$$

Since $W_t(\tau)$ represents an evolution, it follows that

$$W_{t+\tau_1}(\tau_2)W_t(\tau_1) = W_t(\tau_1 + \tau_2) \tag{2.9}$$

Equation (2.8) is a representation (in fact, the t-representation) of a unitary evolution in $\bar{\mathscr{H}}$, which we denote by $U(\tau)$, that is,

$$\psi_{t+\tau}^{\tau} = (U(\tau)\psi)_{t+\tau} = W_t(\tau)\psi_t^0 \tag{2.10}$$

The $U(\tau)$ are then unitary operators:

$$\begin{aligned}(U(\tau)\psi, U(\tau)\phi)_{\bar{\mathscr{H}}} &= \int (W_t(\tau)\psi_t, W_t(\tau)\phi_t)_{\mathscr{H}}\, dt \\ &= \int (\psi_t, \phi_t)_{\mathscr{H}}\, dt = (\psi, \phi)_{\bar{\mathscr{H}}}\end{aligned} \tag{2.11}$$

Furthermore, they form a one-parameter group, that is,

$$\begin{aligned}(U(\tau_1)U(\tau_2)\phi)_{t+\tau_1+\tau_2} &= W_{t+\tau_1}(\tau_2)W_t(\tau_1)\phi_t = W_t(\tau_1 + \tau_2)\phi_t \\ &= (U(\tau_1 + \tau_2)\phi)_{t+\tau_1+\tau_2}\end{aligned} \tag{2.12}$$

Since the $U(\tau)$ form a one-parameter group of unitary operators, it follows that if its action is continuous, it has a self-adjoint generator, K, where

$$K = s - \lim_{\tau \to 0} \frac{1}{\tau}(U(\tau) - I) \tag{2.13}$$

It has been shown by Piron [14] that if K, $-i\partial_t$, and $K + i\partial_t$ have a common dense domain on which they are essentially self-adjoint, then H, defined as the self-adjoint extension of $K + i\partial_t$, is a decomposable operator on $\bar{\mathscr{H}}$, that is, $(H\psi)_t = H_t\psi_t$. Moreover, if $\| W_t(\tau)\phi_t \|_{\bar{\mathscr{H}}}$ is measurable in t and τ, K is unitarily equivalent to $-i\partial_t$, and hence its spectrum is absolutely continuous over all the real axis. The unitary transformation providing this equivalence is of the form

$$(R(t_0)\phi)_t = W_t(t_0 - t)\phi_t = W_{t0}^{-1}(t - t_0)\phi_t \tag{2.14}$$

This result follows from the application of the definition of the operators:

$$\begin{aligned}(R(t_0)e^{-iK\tau}R^{-1}(t_0)\phi)_t &= W_{t0}^{-1}(t - t_0)(e^{-iK\tau}R^{-1}(t_0)\phi)_t \\ &= W_{t0}^{-1}(t - t_0)W_{t-\tau}(\tau)(R^{-1}(t_0)\phi)_{t-\tau} \\ &= W_{t0}^{-1}(t - t_0)W_{t-\tau}(\tau)W_{t0}(t - t_0)\phi_{t-\tau}\end{aligned}$$

But according to the composition law (2.9), this is

$$(R(t_0)e^{-iK\tau}R^{-1}(t_0)\phi)_t = \phi_{t-\tau} \tag{2.15}$$

so that

$$R(t_0)KR^{-1}(t_0) = -i\partial_t \tag{2.16}$$

and therefore the spectrum of K is all the real line.

Misra, Courbage, and Prigogine [26] have shown that the existence of an evolution generator with unbounded spectrum is necessary for the existence of an entropy operator M, with simple properties, that is, that the rate of change of the entropy, D, is compatible with the entropy itself, thus $[D, M] = 0$. This result follows when we construct the expectation value of the evolution generator in states defined as $e^{iMs}\psi$, where s is an arbitrary parameter. Since

$$\frac{d}{ds}(e^{iMs}\psi, He^{iMs}\psi) = i(e^{iMs}\psi, [H, M]e^{iMs}\psi) \tag{2.17}$$

and $D = i[H, M]$, the expression on the right-hand side is just $(e^{iMs}\psi, De^{iMs}\psi)$; from the commutation relation $[D, M] = 0$, it follows that it is independent of s. Integrating Eq. (2.17) over s, one finds

$$(e^{iMs}\psi, He^{iMs}\psi) = (\psi, D\psi)s + \text{constant} \tag{2.18}$$

Hence, taking s to any arbitrary value, we see that H must be unbounded (in particular, from below), unless $(\psi, D\psi) = 0$, in which case the entropy is constant.

However, if the spectrum of H is unbounded from below (and absolutely continuous), then there exists a time operator. In this case, the theory can be put in correspondence with a theory of evolution in a larger Hilbert space. As discussed in the next chapter, the formulation of quantum dynamical problems in the Liouville space [27] forms a natural framework for this type of structure.

In the following, we describe how Flesia and Piron [19] applied Lax–Phillips formalism to quantum systems with the help of the direct integral space which we have just discussed.

Lax–Phillips theory [17] assumes the existence of a one-parameter unitary group of evolution on a Hilbert space $\bar{\mathscr{H}}$, and incoming and outgoing subspaces $\mathscr{D}_-$ and $\mathscr{D}_+$ such that

$$\begin{aligned} U(\tau)\mathscr{D}_+ &\subset \mathscr{D}_+ \qquad \text{for all } \tau > 0 \\ U(\tau)\mathscr{D}_- &\subset \mathscr{D}_- \qquad \text{for all } \tau < 0 \\ \bigcap_\tau U(\tau)\mathscr{D}_\pm &= \{0\} \\ \overline{\bigcup_\tau U(\tau)\mathscr{D}_\pm} &= \bar{\mathscr{H}} \end{aligned} \tag{2.19}$$

where τ is the evolution parameter identified with the laboratory time. It follows from a theorem of Sinai [28] that $\bar{\mathscr{H}}$ can be foliated in such a way that it can be represented as a family of (auxiliary) Hilbert spaces in the form $L^2(-\infty, +\infty; \mathscr{H}_t)$ over Lebesgue measure in t, and all the $\mathscr{H}_t$ are isomorphic (we therefore sometimes refer to these spaces simply as $\mathscr{H}$). The scalar product in $\bar{\mathscr{H}}$ is given by

$$(f, g) = \int_{-\infty}^{\infty} (f_t, g_t)_{\mathscr{H}_t}\, dt \tag{2.20}$$

Lax and Phillips show that there are unitary operators W_+^{-1}, W_-^{-1} which map the elements of $\bar{\mathscr{H}}$ into representations—called the outgoing and incoming translation representations—for which the evolution is translation in t. The subspaces $\mathscr{D}_+$, $\mathscr{D}_-$ correspond to the sets of functions with, in these representations, support in semiinfinite segments of the positive and negative t-axis respectively. They define the S-matrix abstractly as the map from the incoming translation representation to the outgoing one (i.e., $S = W_+^{-1}W_-$). This map is defined up to unitary transformations on the auxiliary spaces $\{\mathscr{H}_t\}$ and refers to the equivalence classes for which the incoming and outgoing representations have the property that the evolution is represented by translation.

In the quantum theory, one constructs the space $\bar{\mathscr{H}}$ by taking the direct integral of the quantum mechanical Hilbert spaces over all values of the time t with Lebesgue measure. The form of the theory adopted by Flesia and Piron [19] distinguishes the elements of these equivalence classes, and constructs an S-matrix which maps the auxiliary space in the incoming translation representation to the auxiliary space of the outgoing one. In the model that they use to illustrate the structure, this map corresponds to a preasymptotic form of the S-matrix of the usual scattering theory. Their model assumes that the subspaces $\mathscr{D}_+$, $\mathscr{D}_-$ are represented in the "free" representation, for which the free evolution is translation, by $L^2(-\infty, \rho_-; \mathscr{H})$, $L^2(\rho_+, \infty; \mathscr{H})$, respectively. In the limit in which the interval between the two semiinfinite regions of support tends to infinity, their S-matrix becomes the usual S-matrix.

Under these conditions they identify the wave operators $W_\pm^{-1}$ with the operators $R(\rho_\pm)$, defined, according to Eq. (2.14), as

$$(R(\rho_\pm)\phi)_t = (W_\pm^{-1}\phi)_t = W_t(\rho_\pm - t)\phi_t = W_{\rho\pm}^{-1}(t - \rho_\pm)\phi_t \tag{2.21}$$

Therefore, the S-matrix becomes (we denote by FP the Flesia–Piron form)

$$(S^{\mathrm{FP}}(\rho_-, \rho_+)\phi)_t = (R(\rho_+)R(\rho_-)^{-1}\phi)_t = W_{\rho_-}(\rho_+)\phi_t \tag{2.22}$$

When the model for the evolution is taken to be the interaction picture evolution, that is,

$$W_t(\tau) = e^{iH_0(t+\tau)}e^{-iH\tau}e^{-iH_0t} \tag{2.23}$$

the S-matrix takes the form

$$S^{\mathrm{FP}}(\rho_-, \rho_+) = e^{iH_0\rho_+}e^{-iH\rho_+}\,e^{iH\rho_-}\,e^{-iH_0\rho_-} \tag{2.24}$$

In the limit $\rho_+ \to \infty$, $\rho_- \to -\infty$, this becomes $\Omega_+^\dagger \Omega_-$ (in the sense of a bilinear form on a dense set) which defines the S-matrix of the usual scattering theory.

Lax and Phillips define the operator

$$\mathscr{Z}(\tau) = P_+ U(\tau) P_- \tag{2.25}$$

on $\bar{\mathscr{H}}$, where $P_\pm$ is the projection on the orthogonal complement of $\mathscr{D}_\pm$. This operator vanishes on $\mathscr{D}_\pm$ and maps the subspace

$$\mathscr{K} = \bar{\mathscr{H}} \ominus (\mathscr{D}_+ \oplus \mathscr{D}_-) \tag{2.26}$$

into itself. These mappings form a semigroup [17], that is, for $\tau_1, \tau_2 \geq 0$,

$$\mathscr{Z}(\tau_1)\mathscr{Z}(\tau_2) = \mathscr{Z}(\tau_1 + \tau_2) \tag{2.27}$$

and this semigroup is strongly contractive—for each $\phi \in \mathscr{K}$ and any ϵ, there exists a τ_ϕ such that

$$\|\mathscr{Z}(\tau)\phi\|_{\bar{\mathscr{H}}} < \epsilon \tag{2.28}$$

for $\tau > \tau_\phi$. It can be shown that under the conditions of Eq. (2.19), $\mathscr{Z}(\tau)$ is just the unitary evolution $U(\tau)$ projected into the subspace $\mathscr{K}$. Because the states which lie in the subspaces $\mathscr{D}_\pm$, in the case of scattering, describe the incoming and outgoing waves that are not influenced by the interaction, the states that lie in $\mathscr{K}$ describe the unstable states (i.e., resonances of the scattering). From this point of view, the Lax–Phillips semigroup is analogous to the reduced motion discussed in the previous section. The direct integral space method, solves, therefore, the problem of deriving an exact semigroup law for the reduced motion and provides a realization of the unitary dilation of Nagy and Foias [16] [note that, because the generator of the evolution in $\bar{\mathscr{H}}$ has an absolutely continuous spectrum $(-\infty, \infty)$, this result does not contradict the no-go theorem discussed above].

Lax and Phillips prove that the S-matrix is a multiplicative operator in the spectral representation of K (which is the Fourier transform of the translation representation), that is,

$$(S\psi)_\sigma = S(\sigma)\psi_\sigma$$

and the eigenvalues of the generator of the semigroup $\mathscr{Z}(\tau)$ correspond to the singularities of the analytic continuation of $S(\sigma)$. The eigenstates corresponding to these eigenvalues are analogous to the generalized eigenstates found in the framework of Wigner and Weisskopf, as discussed in Section I. Thus, the S-matrix contains all the information about the unstable states. It can be seen, however, from Eq. (2.22) [and, explicitly, in Eq. (2.24)] that the S-matrix obtained from a model in which the evolution is given in the form (2.8) has no t-dependence, and hence its spectral representation is trivial. We discuss this problem and its resolution in the next section.

III. CONSTRUCTION OF THE NONTRIVIAL S-MATRIX

As we have seen, the Lax–Phillips theory is a natural framework for describing irreversible processes, and it may be applied to the quantum theory, using the direct integral space method, as suggested by Flesia and Piron.

There is, however a fundamental problem in their construction. As mentioned at the end of the previous section, the relation between the Lax–Phillips semigroup (which corresponds to the reduced motion) and the

S-matrix is established in terms of the spectral representation of the generator of the evolution, that is, the Fourier transform of the translation representation. However, the operator S^{FP} of Eq. (2.22) does not depend on t and, therefore, the Fourier transform, in terms of which the relation to the Lax–Phillips semigroup can be established, has only a trivial structure as a function of σ:

$$\tilde{S}^{\mathrm{FP}}(\sigma) = (e^{iH_0\rho_+}e^{-iH\rho_+}e^{iH\rho_-}e^{-iH_0\rho_-})\delta(\sigma) \tag{3.1}$$

Hence one can not obtain, under these conditions, a description of a decaying system.

In fact, from the point of view taken by Lax and Phillips, such an S-matrix, relating elements of an equivalence class (corresponding to unitary maps of the auxiliary spaces), corresponds to no scattering. To see this, we prove in the following that, in this case, in the *outgoing* translation representation, $\mathscr{D}_-$ is represented by $L^2(-\infty, \rho_-; \mathscr{H})$; that is, in this representation $\mathscr{D}_-$ has a definite support property; therefore, by definition, this representation is also *incoming*. Up to an isomorphic mapping of the auxiliary spaces (used by Flesia and Piron [19,20] to represent the scattering process), the Lax–Phillips S-matrix which relates these incoming and outgoing translations is therefore trivial.

Let $\psi \in \mathscr{D}_-$; then, in the free representation, $\psi \in L^2(-\infty, \rho_-; \mathscr{H})$. The outgoing representor of ψ is then given by

$$\psi_t^{\mathrm{out}} = (W_+^{-1}\psi)_t = (R(\rho_+)\psi)_t = W_t(\rho_+ - t)\psi_t \tag{3.2}$$

Since ψ_t vanishes for $t > \rho_-$, so does $(W_+^{-1}\psi)_t$, and hence the set $\mathscr{D}_-^{\mathrm{out}}$ of incoming states in the outgoing representation satisfies

$$\mathscr{D}_-^{\mathrm{out}} \subset L^2(-\infty, \rho_-; \mathscr{H}) \tag{3.3}$$

The opposite direction of the demonstration is similar. Let us assume now that $\psi^{\mathrm{out}} \in L^2(-\infty, \rho_-; \mathscr{H})$, and consider ϕ defined by $\phi = W_+ \psi^{\mathrm{out}}$, that is,

$$\phi_t = W_{\rho+}(t - \rho_+)\psi_t^{\mathrm{out}} \tag{3.4}$$

Since ψ_t^{out} vanishes when $t > \rho_-$, ϕ_t has the same property, and hence $\phi \in \mathscr{D}_-$ and $\psi^{\mathrm{out}} \in W_+^{-1}\mathscr{D}_-$ or $L^2(-\infty, \rho_-; \mathscr{H}) \subset \mathscr{D}_-^{\mathrm{out}}$. We conclude from this result and Eq. (3.3) that

$$\mathscr{D}_-^{\mathrm{out}} = L^2(-\infty, \rho_-; \mathscr{H}) \tag{3.5}$$

and hence the outgoing representation coincides with the incoming one.

In fact, for any system in which the evolution may be written as $\psi^{\tau}_{t+\tau} = W_t(\tau)\psi_t$, the incoming and outgoing representations coincide. To see this, we construct the incoming (outgoing) translation representations using the free representation. Let us assume that there is a representation in which both $\mathscr{D}_{\pm}$ have definite support properties, but the evolution is not necessarily just translation. The free representation, which in the absence of interaction is both incoming and outgoing, for example, should have this property. Let us define a new representation in the following way. Take f_0 to be in the free representation, and define its outgoing image:

$$f_+ = W_+^{-1} f_0 = s - \lim_{\tau \to \infty} U_0(-\tau)U(\tau) f_0 \tag{3.6}$$

First, we show that it belongs to an outgoing representation. Since $U(\tau)$ is the full evolution in the free representation, f_+ evolves according to

$$f_+(\tau) = W_+^{-1} U(\tau) W_+ f$$

However,

$$\begin{aligned} W_+^{-1} U(\tau) &= s - \lim_{\tilde{\tau} \to \infty} U_0(-\tilde{\tau})U(\tilde{\tau})U(\tau) \\ &= s - \lim_{\tilde{\tau}+\tau \to \infty} U_0(\tau)U_0(-\tilde{\tau} - \tau)U(\tilde{\tau} + \tau) = U_0(\tau)W_+^{-1} \end{aligned} \tag{3.7}$$

and, therefore, $f_+(\tau) = U_0(\tau) f_+(0)$, that is, the evolution is translation.

We now prove the second condition for the outgoing representation, $\mathscr{D}_+^{\text{out}} = \mathscr{D}_+^{\text{free}} = L^2(\rho_+, \infty; \mathscr{H})$. This follows when the evolution is $\psi^{\tau}_{t+\tau} = W_t(\tau)\psi_t$, because in this case, for any positive τ,

$$(U_0(-\tau)U(\tau)\psi)_t = (U(\tau)\psi)_{t+\tau} = W_t(\tau)\psi_t \tag{3.8}$$

that is, there is only a unitary change in the little space $\mathscr{H}$ with no translation, which implies that $U_0(-\tau)U(\tau)L^2(\rho_+, \infty; \mathscr{H}) = L^2(\rho_+, \infty; \mathscr{H})$, and therefore it is true in the limit $\tau \to \infty$. In a similar way we define

$$W_- = s - \lim_{\tau \to -\infty} U_0(-\tau)U(\tau) \tag{3.9}$$

to build the incoming representation.

Let us now further assume that the evolution is of the form $\psi^{\tau}_{t+\tau} = W_t(\tau)\psi_t$, and that there is given a free representation such that

$$\mathscr{D}_+ = L^2(\rho_+, \infty; \mathscr{H}) \qquad \mathscr{D}_- = L^2(-\infty, \rho_-; \mathscr{H}) \tag{3.10}$$

and check the properties of the subspace $\mathscr{D}^{\text{out}}_- = W^{-1}_+\mathscr{D}_-$. Let $\{\psi_t\} \in \mathscr{D}_-$, and τ be some positive number. Then, one has

$$(U(\tau)\psi)_{t+\tau} = W_t(\tau)\psi_t \tag{3.11}$$

The right-hand side vanishes for $t > \rho_-$ and, thus $(U(\tau)\psi)_t = 0$ for $t > \rho_- + \tau$. We therefore have

$$(U_0(-\tau)U(\tau)\psi)^{\text{free}}_t = 0 \qquad \text{for } t > \rho_- \tag{3.12}$$

Since this is true for every positive τ, it is true in the limit $\tau \to \infty$, and therefore

$$\{\psi_t\}^{\text{out}} \in L^2(-\infty, \rho_-; H) \tag{3.13}$$

and $\{\psi_t\} \in \mathscr{D}_-$.

The inverse direction uses the same type of argument, and it follows that

$$\mathscr{D}^{\text{out}}_- = L^2(-\infty, \rho_-; \mathscr{H}) \tag{3.14}$$

Therefore, as before, the outgoing translation representation is also incoming and the S-matrix is the identity.

The assumption of an evolution law which is pointwise on the time axis of the measure space of $\bar{\mathscr{H}}$ therefore does not lead to the full structure of Lax–Phillips theory, that is, an S-matrix which has a nontrivial Fourier transform. This result is highly significant for the theory of unstable systems. Although an axiomatization of the structure of such systems has not yet been carried out, it is clear, as noted above, that they cannot be characterized at a particular moment in time. The second law of thermodynamics, in particular, is a statement about time evolution which is intrinsic to irreversible processes; it is stronger than the condition for the unitary evolution of reversible systems, for which the only requirement is conservation of probability (and once differentiability for the element of the equivalence class of L^2 functions which must satisfy the Schrödinger equation). In the latter case, models are based on correspondence with a classical analog with Hamiltonian dynamics. For the irreversible system, one might expect a classical analog for the dynamical evolution, which contains some representations of correlation in time [29], such as a Boltzmann-type equation, encompassing, in particular, the thermodynamics of the second law.

Although the generalization of Lax–Phillips theory by Flesia and Piron [19] provides a new point of view for scattering theory, we see that to extend the theory further to include a description of the evolution of an unstable system, it is necessary to generalize the law of evolution to that of a nontrivial integral operator over the time.

The most general linear evolution law has the form

$$(U(\tau)\psi)_{t+\tau} = \int_{-\infty}^{+\infty} W_{t,t'}(\tau)\psi_{t'}\, dt' \tag{3.15}$$

We show that this type of evolution, which goes beyond the formulation of Flesia and Piron [19] and Floquet theory [24], can correspond to unitary evolution in $\bar{\mathscr{H}}$ with a nontrivial S-matrix for which the singularities of its Fourier transform are associated with the spectrum of the generator of the Lax–Phillips semigroup. As we show in Section V, the form of the evolution law (3.15) has a natural realization in Liouville space. For mathematical and conceptual purposes, and possibly for actual physical applications, we wish, however, to study first the condition for which the full structure of the Lax–Phillips theory is applicable in the framework of ordinary Hilbert space theory as well. In this framework, one can understand the structure of Eq. (3.15) by examining the class of physical systems actually studied by Lax–Phillips (i.e., hyperbolic systems), as for electromagnetic scattering theory. Their foliation of the Hilbert space of solutions along the time axis, leads to the existence of a nontrivial semigroup behavior. According to our remark following Eq. (3.5), it follows that the evolution law must be of the type of Eq. (3.15). The action of the Green's function for a hyperbolic system requires an integral over t, introducing the correlations necessary for the construction of a nontrivial Lax–Phillips theory.

In the framework of the ordinary quantum theory, one generally constructs Hamiltonian dynamical models with Hamiltonian functions which are either time-independent or depend on time in a pointwise manner. One can understand the time parameter entering this construction as given in terms of a particular representation for the time operator which can be defined in the larger Hilbert space $\bar{\mathscr{H}}$. In this representation, the subspaces $\mathscr{D}_-$ and $\mathscr{D}_+$, in general, may not have definite support properties. The transformation to a representation in which they have definite support properties (as in the "free" representation in the Flesia–Piron theory) may result in a representation for the Hamiltonian which is not decomposable, inducing a nonpointwise evolution, as in Eq. (3.15). Such transformations were discussed, for example, by Friedrichs [9]. We use such a representation in the following.

Assuming that the evolution is represented by Eq. (2.10) (i.e., pointwise on the t-axis), Flesia and Piron proved that the generator K of the evolution on $\bar{\mathscr{H}}$ is of the form $-i\partial_t$ plus a decomposable operator. This follows from the application of Trotter's formula [30],

$$e^{-iH\tau} = s - \lim_{n\to\infty} (e^{-\partial_t \tau/n} e^{-iK\tau/n})^n \tag{3.16}$$

and the assumption that the evolution is generated pointwise as in Eq. (1.4). One examines

$$\begin{aligned}(e^{-\partial_t \tau/n} e^{-iK\tau/n} f)_t &= (e^{-iK\tau/n} f)_{t+\tau/n} \\ &= (U(\tau/n) f)_{t+\tau/n} = W_t(\tau/n) f_t\end{aligned} \tag{3.17}$$

where the first half follows from the action of translation and the last from the assumption that the evolution is represented by an operator acting pointwise on the t-axis. Taking the limit of the sequence, one obtains

$$(e^{-iH\tau} f)_t = s - \lim_{n\to\infty} (W_t(\tau/n))^n f_t \tag{3.18}$$

which is clearly decomposable.

However, if the evolution acts as a nontrivial kernel on the time variable, as in Eq. (3.15), one obtains

$$(e^{-\partial_t \tau/n} e^{-iK\tau/n} f)_t = (e^{-iK\tau/n} f)_{t+\tau/n} = \int W_{t,t'}(\tau/n) f_{t'}\, dt' \tag{3.19}$$

Applying this operator again n times, one obtains

$$((e^{-\partial_t \tau/n} e^{-iK\tau/n})^n f)_t = \int W_{t,t_1}(\tau/n) W_{t_1,t_2}(\tau/n) \cdots W_{t_{n-1},t_n}(\tau/n) f_{t_n}\, dt_1\, dt_2 \cdots dt_n \tag{3.20}$$

This n-fold convolution converges, because the product (3.16) converges if $-i\partial_t$, K, and $i\partial_t + K$ have a common dense domain. It is clear that the right-hand side is, in general, not decomposable.

The operators $W_{tt'}(\tau)$ must satisfy some conditions if the $U(\tau)$ are to form a one-parameter unitary group. Since

$$\begin{aligned}(U(\tau_1+\tau_2)\psi)_{t+\tau_1+\tau_2} &= \int W_{t,t'}(\tau_1+\tau_2)\psi_{t'}\,dt' \\ &= (U(\tau_1)U(\tau_2)\psi)_{t+\tau_1+\tau_2} \\ &= \int W_{t+\tau_2,t'}(\tau_1)(U(\tau_2)\psi)_{t'}\,dt' \\ &= \iint W_{t+\tau_2,t'}(\tau_1)W_{t'-\tau_2,t''}(\tau_2)\psi_{t''}\,dt'\,dt''\end{aligned}$$

must be true for arbitrary ψ, we require the relation

$$W_{t,t'}(\tau_1+\tau_2) = \int W_{t+\tau_2,t''+\tau_2}(\tau_1)W_{t'',t'}(\tau_2)\,dt'' \tag{3.21}$$

From the property $U^{-1}(\tau) = U(-\tau) = U^\dagger(\tau)$, that is,

$$(\psi, U(\tau)\phi)_{\bar{\mathscr{H}}} = (U(-\tau)\psi, \phi)_{\bar{\mathscr{H}}}$$

one finds

$$\int\left(\psi_t, \int W_{t-\tau,t'}\,\phi_{t'}\,dt'\right)_{\mathscr{H}} dt = \int\left(\int W_{t+\tau,t'}(-\tau)\psi_{t'}\,dt', \phi_t\right)_{\mathscr{H}} dt$$

Because this is true for arbitrary ψ and ϕ, we obtain

$$W_{t_1-\tau,t_2}(\tau) = W_{t_2+\tau,t_1}(-\tau)^\dagger$$

or

$$W_{t,t'}(-\tau) = W_{t'-\tau,t-\tau}(\tau)^\dagger \tag{3.22}$$

The two conditions, Eqs. (3.21) and (3.22), ensure that the evolution is unitary. If we use the chain property, putting $\tau_1 = -\tau_2 = \tau$ in Eq. (3.21), we have

$$\int dt''\,W_{t-\tau,t''-\tau}(\tau)W_{t'',t'}(-\tau) = W_{t,t'}(0) - \delta(t-t') \tag{3.23}$$

With Eq. (3.22), Eq. (3.23) becomes

$$\int W_{t,t''}(\tau)W_{t',t''}(\tau)^{\dagger}\,dt'' = \delta(t-t')$$

Let us now study, for this general evolution, some properties of the S-matrix,

$$(S\psi)_t = \int S_{t,t'}\psi_{t'}\,dt'$$

and show that in this general case the S-matrix must have the form $S_{t,t'} = S(t-t')$. Using the definitions $S = W_{+}^{-1}W_{-}$, where

$$W_{\pm} = s - \lim_{\tau\to\pm\infty} U(-\tau)U_0(\tau)$$

we find

$$(S\psi)_t = s - \lim_{\tau_1,\tau_2\to\infty} (U_0(-\tau_1)U(\tau_1)U(\tau_2)U_0(-\tau_2)\psi)_t$$

But

$$\begin{aligned}(U_0(-\tau_1)U(\tau_1+\tau_2)U_0(-\tau_2)\psi)_t &= (U(\tau_1+\tau_2)U_0(-\tau_2)\psi)_{t+\tau_1}\\ &= \int W_{t+\tau_1,t'}(\tau_1+\tau_2)(U_0(-\tau_2)\psi)_{t'}\,dt'\\ &= \int W_{t+\tau_1,t'}(\tau_1+\tau_2)\psi_{t'+\tau_2}\,dt'\\ &= \int W_{t+\tau_1,t'-\tau_2}(\tau_1+\tau_2)\psi_{t'}\,dt'\end{aligned}$$

and, therefore, the matrix elements of S are

$$\begin{aligned}S_{t,t'} &= s - \lim_{\tau_1,\tau_2\to\infty} W_{t+\tau_1,t'-\tau_2}(\tau_1+\tau_2)\\ &= s - \lim_{\tau_1',\tau_2'\to\infty} W_{t-t'+\tau_1',-\tau_2'}(\tau_1'+\tau_2') = S(t-t')\end{aligned} \tag{3.24}$$

(where $\tau_1' = \tau_1 + t'\tau_2' = \tau_2 - t'$). This is a very important property of the S-matrix, according to which, when one goes to the spectral representation

$\hat{\psi}_\sigma = \int e^{-i\sigma t}\psi_t \, dt$, the S-matrix takes the simple form

$$\hat{S}_{\sigma,\sigma'} = \frac{1}{2\pi}\int e^{-i\sigma t} S_{t,t'} e^{i\sigma' t'} \, dt \, dt' = \delta(\sigma - \sigma')\hat{S}(\sigma)$$

where

$$\hat{S}(\sigma) = \int e^{-i\sigma t} S(t) \, dt \tag{3.25}$$

that is, in this basis the S-matrix is diagonal, and the S-operator is multiplication on the subspaces, labeled by σ, of $\{\mathscr{H}_\sigma\}$, the set of (isomorphic) Hilbert spaces which are the Fourier dual to the set $\{\mathscr{H}_t\}$. This result can also be obtained by looking at the definition of the S-matrix,

$$S = s - \lim_{\tau_1, \tau_2 \to \infty} U_0(-\tau_1)U(\tau_1 + \tau_2)U_0(-\tau_2)$$

from which it follows that

$$SU_0(\tau) = U_0(\tau)S$$

Since $U_0(\tau)$ is the translation operator, one obtains the result $[S, i\partial_t] = 0$ (which corresponds to the usual result of scattering theory $[S, H_0] = 0$). It follows from this commutation relation that $S_{t,t'} = S(t - t')$.

Since $U(\tau)$ is a continuous group of unitary operators, one can write it in the form $U(\tau) = e^{-iK\tau}$. We now find the form of the S-matrix in terms of the generator K. The generator may be calculated from the equation $i\partial_\tau U(\tau) = KU(\tau)$. Looking at the components of this equation one has

$$i\partial_\tau(U(\tau)\psi)_t = (KU(\tau)\psi)_t \tag{3.26}$$

On the other hand, by the definition of $U(\tau)$,

$$\begin{aligned} i\partial_\tau[U(\tau)]_t &= i\partial_\tau \int W_{t-\tau,t'}(\tau)\psi_{t'} \, dt' \\ &= -i\partial_t \int W_{t-\tau,t'}(\tau)\psi_{t'} \, dt' + i\partial_\tau \int W_{\tilde{t},t'}(\tau)\psi_{t'} \, dt' \end{aligned} \tag{3.27}$$

where we define $\tilde{t} = t - \tau$ with the implication that ∂_τ operates only on the argument τ of $W_{\tilde{t},t'}(\tau)$ which is explicitly displayed. Let us write the generator in the (general) form, writing the t-derivative explicitly to take into account the form of Eq. (3.27),

$$(K\psi)_t = -i\partial_t \psi_t + \int \kappa_{t,t'} \psi_{t'}\, dt'$$

so that

$$(KU(\tau)\psi)_t = -i\partial_t \int W_{t-\tau,t'}(\tau)\psi_{t'}\, dt' + \iint \kappa_{t,t'} W_{t'-\tau,t''}(\tau)\psi_{t''}\, dt'\, dt'' \quad (3.28)$$

Comparing this result to Eq. (3.27) one obtains

$$i\partial_t \int W_{\tilde{t},t'}(\tau)\psi_{t'}\, dt' = \iint \kappa_{t,t'} W_{t'-\tau,t''}(\tau)\psi_{t''}\, dt'\, dt''$$

and since this equation holds for arbitrary ψ,

$$i\partial_\tau W_{\tilde{t},t''}(\tau) = \int \kappa_{t,t'} W_{t'-\tau,t''}(\tau)\, dt'$$

that is,

$$i\partial_\tau W_{t,t'}(\tau) = \int \kappa_{t+\tau,t''+\tau} W_{t'',t'}(\tau)\, dt'' \quad (3.29)$$

This differential equation determines the evolution operators W in terms of the generator K, and may formally be expanded in a series (convergent for sufficiently small κ). The first terms in the series are

$$W_{t,t'}(\tau) = \delta(t-t') - i\int_0^\tau \kappa_{t+\tau',t'+\tau'}\, d\tau'$$
$$-\frac{1}{2}\int_0^\tau d\tau'' \int_0^\tau d\tau' \int dt''\, T[\kappa_{t+\tau'',t''+\tau''}\kappa_{t+\tau',t''+\tau'}] + \cdots$$

where T implies the τ-ordered product. Using the formula (3.6), we obtain a perturbative formula for S in the form

$$\begin{aligned}
S_{t,t'} &= s - \lim_{\tau_1,\tau_2\to\infty}\Bigg(\delta(t-t') - i\int_0^{\tau_1+\tau_2}\kappa_{t-t'+\tau-\tau_2,\tau-\tau_2}\,d\tau \\
&\quad - \frac{1}{2}\int_0^{\tau_1+\tau_2} d\tau \int_0^{\tau_1+\tau_2} d\tau' \\
&\quad \times \int dt''\, T[\kappa_{t-t'-\tau_2+\tau,t''+\tau}\kappa_{t-t'-\tau_2+\tau',t''+\tau'}]\Bigg) + \cdots \\
&= s - \lim_{\tau_1,\tau_2\to\infty}\Bigg(\delta(t-t') - i\int_{-\tau_2}^{\tau_1}\kappa_{t-t'+\tau,\tau}\,d\tau \\
&\quad - \frac{1}{2}\int_{-\tau_2}^{\tau_1} d\tau \int_{-\tau_2}^{\tau_1} d\tau' \int dt''\, T[\kappa_{t-t'+\tau,t''+\tau}\kappa_{t''+\tau',\tau'}]\Bigg) + \cdots \\
&= \delta(t-t') - i\int_{-\infty}^{\infty}\kappa_{t-t'+\tau,\tau}\,d\tau \\
&\quad - \frac{1}{2}\int_{-\infty}^{\infty} d\tau \int_{-\infty}^{\infty} d\tau' \int dt''\, T[\kappa_{t-t'+\tau,t''+\tau}\kappa_{t''+\tau',\tau'}] + \cdots
\end{aligned} \tag{3.30}$$

It is interesting to examine one particular case for which Eq. (3.29) can be solved exactly. Consider the case in which the perturbation κ is of the form $\kappa_{t,t'} = \kappa(t-t')$. Let us take the Fourier transform of Eq. (3.29) with respect to t and t'.

$$i\partial_\tau W_{\sigma,\sigma'}(\tau) = \int e^{i\sigma\tau}\kappa_{\sigma,\sigma''}e^{-i\sigma''\tau}W_{\sigma'',\sigma'}(\tau)\,d\sigma'' \tag{3.31}$$

Since $\kappa_{t,t'} = \kappa_{t-t'}$, one obtains

$$\kappa_{\sigma,\sigma'} = \tilde{\kappa}(\sigma)\delta(\sigma-\sigma') \tag{3.32}$$

Using Eq. (3.32) in Eq. (3.31), one obtains

$$i\partial_\tau W_{\sigma,\sigma'}(\tau) = \tilde{\kappa}(\sigma)W_{\sigma,\sigma'}(\tau)$$

from which follows

$$W_{\sigma,\sigma'}(\tau) = e^{-i\tilde{\kappa}(\sigma)\tau}\delta(\sigma-\sigma') \tag{3.33}$$

Taking the inverse transform of Eq. (3.33) we get

$$W_{t,t'}(\tau) = \frac{1}{2\pi}\int d\sigma\, d\sigma'\, e^{i\sigma t} e^{-i\tilde{\kappa}(\tau)\tau}\delta(\sigma - \sigma')e^{-i\sigma' t'} = \frac{1}{2\pi}\int d\sigma\; e^{i\sigma(t-t')}e^{-i\tilde{\kappa}(\sigma)\tau} \tag{3.34}$$

Using Eqs. (3.24) and (3.34), one obtains

$$S_{t,t'} = s - \lim_{\tau\to\infty}\int_{-\infty}^{\infty} d\sigma\; e^{i\sigma(t-t')}e^{i\sigma\tau}e^{-i\tilde{\kappa}(\sigma)\tau} \tag{3.35}$$

where $\tau = \tau_1 + \tau_2$, and it follows that

$$\tilde{S}(\sigma) = s - \lim_{\tau\to\infty} e^{i(\sigma - \tilde{\kappa}(\sigma))\tau} \tag{3.36}$$

If there is a subset in $\bar{\mathscr{H}}$ for which this limit exists, one may replace it by the limit in the sense of distributions (i.e., pointwise in σ). We now show that the limit in the sense of distributions is either trivial, $\tilde{S}(\sigma) = 0$, or it is not meromorphic as a function of σ. Therefore, either the strong limit does not exist, or it is not meromorphic (since the strong limit has unity norm when it exists, it cannot vanish identically). Assuming that the limit exists as a distribution,

$$\tilde{S}(\sigma) = s - \lim_{\tau\to\infty} e^{i(\sigma-\tilde{\kappa}(\sigma))\tau} = \lim_{\epsilon\to\infty} \epsilon\int_0^{\epsilon} e^{i(\sigma-\tilde{\kappa}(\sigma))\tau}e^{-\epsilon\tau}\, d\tau = \lim_{\epsilon\to 0}\frac{i\epsilon}{\sigma - \tilde{\kappa}(\sigma) + i\epsilon} \tag{3.37}$$

This expression obviously vanishes for $\sigma \neq \tilde{\kappa}(\sigma)$, and is 1 where they are equal; therefore, it is clear that either it vanishes everywhere, or it is not a meromorphic function.

We now consider the structure of the Lax–Phillips semigroup

$$\mathscr{Z}(\tau) = P_+ U(\tau) P_- \tag{3.38}$$

It is shown by Lax–Phillips that because $\{U(\tau)\mathscr{D}_+\}$ is dense in $\bar{\mathscr{H}}$, $\mathscr{Z}(\tau)$ is strongly contractive [17]. We show now that under the general evolution (3.15), the semigroup is still contractive, as for the Flesia–Piron case [19,20]. Let us calculate the generator of the semigroup B. We use the free-translation representation in which both $\mathscr{D}_\pm$ have definite support properties. In this representation,

$$\mathscr{Z}(\tau) = P_+ U(\tau) P_- = E(\rho) U(\tau)(I - E(0)) \tag{3.39}$$

where $E(t)$ is the spectral resolution corresponding to T_0, the free-time operator (the conjugate of K_0 which is, in the free translation representation, $-i\partial_t$). Then, the generator (in the subspace $\mathscr{K}$) of $\mathscr{Z}(\tau)$ is

$$\begin{aligned} B &= i \lim_{\tau \to \infty} \frac{\mathscr{Z}(\tau) - I_{\mathscr{K}}}{\tau} = i \lim_{\tau \to 0} \frac{E(\rho)(I - iK\tau)(I - E(0)) - I_{\mathscr{K}}}{\tau} \\ &= E(\rho) K (I - E(0)) = P_+ K P_- \end{aligned} \tag{3.40}$$

Note that one cannot apply the imprimitivity relations directly here because K is not conjugate to T_0. According to the requirements on $\mathscr{D}_\pm$, the matrix elements of κ between states from $\mathscr{D}_-$ to $\mathscr{D}_+$, or $\mathscr{D}_\pm$ to $\mathscr{K}$, vanish and, therefore,

$$B = P_+ K_0 P_- + \kappa_{\mathscr{K}} \tag{3.41}$$

An operator B is called dissipative [31,32] if

$$-i((\phi, B\phi) - (B\phi, \phi)) \leq 0 \tag{3.42}$$

for all ϕ in the domain of B. Since $\kappa_{\mathscr{K}}$ is self-adjoint, only the first term determines whether the operator is dissipative, that is, this property does not depend on the perturbation. As shown by Horwitz and Piron [20], the operator $P_+ K_0 P_-$ is, in fact, dissipative. It is known [32] that $\mathscr{Z}(\tau)$ is a contractive semigroup if and only if its generator is dissipative. It therefore follows, independently of (self-adjoint) interaction, that the semigroup $\mathscr{Z}(\tau)$ is contractive. We see from this [20] the *essential mechanism of Lax–Phillips theory*. The nonself-adjointness of $P_+ K_0 P_-$ corresponds to the restriction of $-i\partial_t$ to a finite interval, so that, in fact, the operator has imaginary eigenvalues. In the presence of interaction (nontrivial κ), these eigenvalues emerge as the actual eigenvalues of B, corresponding to the singularities of $S(\sigma)$.

The direct integral space provides a framework as a functional space for quantum mechanics in which the Nagy–Foias construction can be realized, that is, for which unitary evolution can be restricted to a contractive semigroup. We now introduce an extension of the conceptual framework which considers the set $\{\psi_t\}$, corresponding to the Lax–Phillips vector ψ, as an ensemble of the same type, for example, as $\{\psi(x)\} \in \mathscr{H}$, where x is a point of the spectrum of the position observable, in the usual form of the quantum theory. In concluding this section, we investigate some consequences of this interpretation.

In particular, we discuss some properties of the time operator and the realization of the superselection rule in time. In the next section, we discuss the possibility of decoherence in $\mathscr{H}$ induced by the unitary evolution in $\mathscr{H}$.

In fact, there are three distinct types of time operator. One, which we call the incoming time operator T^{in}, provides a spectral family in terms of which the incoming representation can be constructed, and in which functions in $\mathscr{D}_-$ have definite support and functions in $\bar{\mathscr{H}}$ evolve by translation. In this representation, the norm of the evolving states in $L^2(-\infty, 0; \mathscr{H})$ must decrease. After sufficient laboratory time τ passes, the states evolve to $\mathscr{D}_+$, and in the outgoing representation, provided by the spectral family of the outgoing time operator T^{out}, they have definite support in $L^2(\rho, \infty; \mathscr{H})$. The mapping of functions in the incoming representation to the outgoing representation is provided by the Lax–Phillips S-matrix, and the time operators are related by

$$T^{\text{out}} = S T^{\text{in}} S^{\dagger} \tag{3.43}$$

The third type of time operator corresponds to the "free" representation and is related to T^{in}, T^{out} by the Lax–Phillips wave operators. The spectral family for this operator provides the "standard" representation (analogous to Dirac's choice of "standard" spectral families), which we have used above.

There is an interval, in general, when the system is in interaction, and its state is neither in $\mathscr{D}_-$ nor $\mathscr{D}_+$. The expectation value of the operator T^{in} in the state ψ^{τ} projected into $\mathscr{K} \oplus \mathscr{D}_+$ (corresponding to the projection P_-) can be interpreted as the interaction interval. If the system in interaction is considered as an unstable particle (a resonance), this interval is its *age* after creation at $t = 0$. This interpretation follows from that of T^{in}, that is,

$$\frac{dT^{\text{in}}}{d\tau} = i[K, T^{\text{in}}] = 1 \tag{3.44}$$

Hence, the expectation value of T^{in}, corresponding to the support properties of the states in $\mathscr{D}_-$ in the incoming representation, moves with the laboratory time τ. The expectation value of T^{in} then moves out of $(-\infty, 0)$. The expectation value of T^{in} in the state $P_- \psi^{\tau}$ is

$$\langle T^{\text{in}} \rangle_{\tau} = \int t \,|{}_{\text{in}}\langle t \,|\, P_- \psi^{\tau} \rangle|^2 \, dt \tag{3.45}$$

Here $|_{\text{in}}\langle t | P_- \psi^\tau \rangle|^2$ is the probability density for the age t at time τ, an intrinsic dynamical property of the system. The positive value that the expectation value develops corresponds to the average age.

Similarly, when an unstable system decays, it moves to the subspace $\mathscr{D}_+$, the subspace of outgoing states. In the outgoing representation, these states have support in $L^2(\rho, \infty; \mathscr{H})$. The operator T^{out} satisfies

$$\frac{dT^{\text{out}}}{d\tau} = i[K, T^{\text{out}}] = 1 \tag{3.46}$$

and its expectation value goes with the laboratory time. Hence, for an unstable system which decays, the expectation value of $T^{\text{out}} - \rho$ in states in $\mathscr{D}_+$ is the time after decay, that is,

$$\text{time after decay} = \int (r - \rho) |_{\text{out}}\langle t | (1 - P_+) \psi^\tau \rangle|^2 \, dt \tag{3.47}$$

where $|_{\text{out}}\langle t | (1 - P_+) \psi^\tau \rangle|^2$ is the probability density at each τ that the system has decayed at laboratory time $t - \rho$ (for τ sufficiently early, this quantity vanishes). Similarly, the expectation value of $\rho - T^{\text{out}}$ in states in $\mathscr{K} \oplus \mathscr{D}_-$ is the average time interval for the system to decay, that is,

$$\text{time interval to decay} = \int (\rho - t) |_{\text{out}}\langle t | P_+ \psi^\tau \rangle|^2 \, dt \tag{3.48}$$

where, $|_{\text{out}}\langle t | P_+ \psi^\tau \rangle|^2$ is the probability density at each τ, that the system will decay at laboratory time $\rho - t$. We then understand the subspace $\mathscr{K}$ as corresponding to the unstable system.

An unstable system must be characterized by (at least) *two* time operators; if there were only one, in terms of which both the age and the time of decay are described, incoming and outgoing representations would coincide (up to isomorphisms of $\mathscr{H}$), as discussed above, and the Lax–Phillips S-matrix would be unity. In this representation, both $\mathscr{D}_-$ and $\mathscr{D}_+$ have definite support properties and the evolution is represented by translation. Hence, the sum of the age and time of decay would necessarily be constant. This is not consistent with the observation of known unstable systems, for which the time of decay, given the time of creation of the system, is not definite. Therefore, our treatment, which introduces two time operators, is necessary for the construction of a physically consistent theory.

The Lax–Phillips description of an unstable system developed here has the following important characteristics:

1. A state in $\mathscr{K}$ is indistinguishable from any other in $\mathscr{K}$ by its support property in t, if detected according to operators that are independent of t, that is, the time of decay associated with these states cannot be determined by measurements of time-independent observables. This characteristic is consistent with our experience, in which one cannot predict the time of decay, or distinguish different stages of development of the undecayed system.

2. The structure of the theory is somewhat similar to the Wigner–Weisskopf idea, in that a subspace is associated with the decaying system. The decay of the system is also associated with the probability flow out of the subspace. However, in Wigner–Weisskopf theory, the process of decay is represented as a continuous evolution from the original unstable state to the final state through a changing linear superposition. In the Lax–Phillips, theory the expectation value of an observable which is decomposable in the free or outgoing representations, where $\mathscr{D}_+$ has definite support properties, necessarily reduces to the sum of the expectation values in the subspaces $\mathscr{K} \oplus \mathscr{D}_-$ and in the subspace $\mathscr{D}_+$ (the decay products).

There is, therefore, an exact superselection rule for measurements of the system by means of such decomposable operators.

IV. MIXING OF STATES UNDER THE LAX–PHILLIPS EVOLUTION

Recently, Machida and Namiki [33] proposed a measurement theory based on a direct integral space of many continuous Hilbert spaces and a continuous superselection rule. As noted by Tasaki et al. [34], although Machida and Namiki had some success, their theory has a conceptual difficulty—the apparatus is described by many Hilbert spaces, but the system corresponds to a single Hilbert space, as in conventional theory. Thus, one needs to specify the boundary between the system and the apparatus. As discussed by von Neumann, this is impossible.

In this section, we investigate the possibility of using the quantum version of Lax–Phillips theory as discussed above to solve this problem by describing both the system and the apparatus by a direct integral space of the form of Eq. (2.1).

In the direct integral space $\bar{\mathscr{H}}$, the most general operator $\hat{A}$ takes the form

$$(\hat{A}\psi)_t = \int dt' \, A_{t,t'} \, \psi_{t'} \tag{4.1}$$

where $A_{t,t'}$ is an operator from $\mathscr{H}_{t'}$ to $\mathscr{H}_t$. However, if the operator is self-adjoint in $\bar{\mathscr{H}}$, the foliation may be changed such that the operator is decomposable, that is,

$$(\hat{A}\psi)_t = A_t \psi_t \tag{4.2}$$

Moreover, any (time-dependent) observable $A(t)$ defined in the usual quantum Hilbert space $\mathscr{H}$ can be lifted naturally to the direct integral space $\bar{\mathscr{H}}$ as follows:

$$(\hat{A}\psi)_t = A(t)\psi_t \tag{4.3}$$

For any such decomposable self-adjoint operator in the direct integral space, we define an "expectation value" (consistent with our discussion in Section III) as

$$\langle \hat{A} \rangle_\psi = \frac{(\psi, \hat{A}\psi)_{\bar{\mathscr{H}}}}{(\psi, \psi)_{\bar{\mathscr{H}}}} = \frac{\int dt(\psi_t, A_t \psi_t)_{\mathscr{H}}}{\int dt(\psi_t, \psi_t)_{\mathscr{H}}} \tag{4.4}$$

This definition is a natural generalization of the expectation value in the conventional quantum mechanics. Indeed, for a state,

$$(\psi^\epsilon)_t = \sqrt{\frac{1}{\pi} \frac{\epsilon}{(t - t_0)^2 + \epsilon^2}}\, \psi_0 \tag{4.5}$$

(where ψ_0 is in $\mathscr{H}$), the average value of the operator $\hat{A}$ of Eq. (4.3) is given by

$$\begin{aligned} \langle A \rangle_{\psi^\epsilon} &= \int dt(\psi_0, A(t)\psi_0)_{\mathscr{H}} \left[\frac{1}{\pi} \frac{\epsilon}{(t - t_0)^2 + \epsilon^2} \right] \\ &\times \left\{ \int dt(\psi_0, \psi_0)_{\mathscr{H}} \left[\frac{1}{\pi} \frac{\epsilon}{(t - t_0)^2 + \epsilon^2} \right] \right\}^{-1} \\ &\to \frac{\int dt(\psi_0, A(t)\psi_0)_{\mathscr{H}}\, \delta(t - t_0)}{\int dt(\psi_0, \psi_0)_{\mathscr{H}}\, \delta(t - t_0)} = \frac{(\psi_0, A(t_0)\psi_0)_{\mathscr{H}}}{(\psi_0, \psi_0)_{\mathscr{H}}} \end{aligned} \tag{4.6}$$

for $\epsilon \to 0$, clearly the usual quantum mechanical expectation value.

We wish to show now that a vector in the direct integral space (which we refer to as a Lax–Phillips state) can represent both pure and mixed states in the usual sense.

Most measurement processes are concerned with measurements of observables that are time-independent in the Schrödinger picture. Therefore, if two different Lax–Phillips states give the same expectation value for all time-independent observables, these two states are essentially indistinguishable. In this sense, we define the following:

1. A Lax–Phillips vector $\psi \in \mathscr{H}$ is called "pure-like" if there exists a pure state

$$\rho_0 = \phi_0 \phi_0^*, \qquad \phi_0 \in \mathscr{H}$$

such that

$$\langle \hat{A} \rangle_\psi = \mathrm{Tr}\, \rho_0 A = (\phi_0, A\phi_0) \tag{4.7}$$

for every element of the algebra of bounded linear operators associated with the spectral families of the time-independent observables* on the original space $\mathscr{H}$.

2. A Lax–Phillips vector is called "mixed-like" if no such (pure) ρ_0 exists.

We now show that $\psi = \{\psi_t\} \in \mathscr{H}$ is pure-like if and only if it has the form

$$\psi_t = f(t)\phi_0 \tag{4.8}$$

The proof is as follows (we take $\int dt \|\psi_t\|_{\mathscr{H}}^2 = 1$ henceforth):

$$\langle \hat{A} \rangle_\psi = \int dt(\psi_t, A\psi_t) = w(A) \tag{4.9}$$

is a convex linear functional of A. Consider a sequence of projection operators P_n which converge to some projection P in operator norm. Then

$$w(P_n) = \int dt(\psi_t, P_n \psi_t) = \int dt \|P_n \psi_t\|^2 \leq 1 \tag{4.10}$$

is a positive sequence converging to $w(P) = \int dt \|P\psi_t\|^2$. Hence $w(P)$ is a continuous linear functional on the projection operators on $\mathscr{H}$, and Gleason's theorem (e.g., see [20]) assures, for a Hilbert space of ≥ 3 real

* We wish to emphasize that what is meant is *explicit* time dependence in the Schrödinger picture; we do not refer here to the dynamical time dependence that may arise in the Heisenberg picture if A is not a constant of the motion.

dimensions, that there exists a density operator ρ_ψ (for which Tr $\rho_\psi = 1$, Tr $\rho_\psi^2 \leq 1$, $\rho_\psi \geq 0$) such that

$$w(A) = \text{Tr}(\rho_\psi A) \tag{4.11}$$

If $\rho_\psi = \rho_0 = \phi_0 \phi_0^*$, a pure state in $\mathscr{H}$, the condition that must be satisfied is (ρ_0 is a time-independent bounded self-adjoint operator, in fact, a projection), by Eq. (4.10),

$$w(\rho_0) = \text{Tr } \rho_0^2 = 1 \tag{4.12}$$

and, therefore, from Eq. (4.9), we must have

$$\int dt \, |(\psi_t, \phi_0)|^2 = 1 \tag{4.13}$$

By the Schwartz inequality in $\mathscr{H}$ ($\|\phi_0\|_{\mathscr{H}} = 1$),

$$|(\psi_t, \phi_0)|^2 \leq \|\psi_t\|_{\mathscr{H}}^2 \tag{4.14}$$

and $\int dt \|\psi_t\|_{\mathscr{H}}^2 = 1$, the right-hand side of Eq. (4.13) is an upper bound on the integral. To achieve this upper bound, ψ_t must be proportional to ϕ_0, that is, Eq. (4.8) must hold, and $\int dt \, |f(t)|^2 = 1$.

We remark that the lift $\hat{\rho}_\psi$ of the time-independent operator ρ_ψ defined in Eq. (4.11) (or of ρ_0), is, according to Eq. (4.3), defined by

$$(\hat{\rho}_\psi \psi)_t = \rho_\psi \psi_t$$

for $\psi_t \in \mathscr{H}$. The operator-valued (on $\mathscr{H}$) kernel of $\hat{\rho}_\psi$ in $\bar{\mathscr{H}}$ in the t representation is, therefore, formally of the form

$$\langle t \,|\, \hat{\rho}_\psi \,|\, t' \rangle = \delta(t - t')\rho_\psi$$

so that clearly the trace of $\hat{\rho}_\psi$ in $\bar{\mathscr{H}}$ does not exist. Our discussion has been primarily with the definition of *states on $\mathscr{H}$ induced by the vectors of $\bar{\mathscr{H}}$*.

We now discuss the possibility of decoherence, or the evolution from pure- to mixed-like states. First, we consider the Schrödinger evolution for a time-dependent Hamiltonian. We then study a more general evolution, for which we obtain stronger results. The solution of the time-dependent

Schrödinger equation can always be written formally as $\psi_t = U(t, t')\psi_{t'}$, where $U(t, t')$ satisfies the chain property $U(t, t')U(t', t'') = U(t, t'')$, and can be expressed in terms of the integral of a time-ordered product. We define $W_t(\tau) = U(t + \tau, t)$, and lift the evolution to $\bar{\mathscr{H}}$ as follows:

$$\psi^{\tau}_{t+\tau} = W_t(\tau)\psi_t \tag{4.15}$$

where $W_t(\tau)$ is given by (T implies the time-ordered product)

$$W_t(\tau) = T(e^{-i\int_t^{t+\tau} H(t')\,dt'}) \tag{4.16}$$

For this kind of time-evolution, we obtain

$$\langle \hat{A} \rangle_{\psi} = \int dt (W_t(\tau)\psi_t, AW_t(\tau)\psi_t)_{\mathscr{H}} \tag{4.17}$$

where we have taken the normalization as unity. For the pure-like state introduced in Eq. (4.8), we then have

$$\langle \hat{A} \rangle_{\psi} = \int dt\, |f(t)|^2 (W_t(\tau)\phi_0, AW_t(\tau)\phi_0)_{\mathscr{H}} \tag{4.18}$$

It follows from our previous argument that the effective state corresponding to Eq. (4.18) [in the sense of Eqs. (4.9) and (4.11)] is mixed-like if $W_t(\tau)\phi_0 \neq W_{t'}(\tau)\phi_0$ [i.e., the state ρ_{ψ} induced from $\psi^{\tau}_{t+\tau} = W_t(\tau)\psi_t = f(t)W_t(\tau)\phi_0$ is not pure in $\mathscr{H}$].

The evolution operator $W_t(\tau)$, in a full-evolution model, does not depend on t if the Hamiltonian is time independent. In this case, $W_t(\tau) = W(\tau) = e^{-iH\tau}$, and

$$\begin{aligned}\langle \hat{A} \rangle_{\psi} &= \left(\int dt\, |f(t)|^2\right)(W(\tau)\phi_0, AW(\tau)\phi_0)_{\mathscr{H}} \\ &= (W(\tau)\phi_0, AW(\tau)\phi_0)_{\mathscr{H}}\end{aligned} \tag{4.19}$$

so that the corresponding state is pure-like. If the Hamiltonian does not depend on time explicitly, a pure-like state remains pure-like, and no apparent decoherence (in the state induced in $\mathscr{H}$) arises. If the Hamiltonian depends on time explicitly, the states induced in $\mathscr{H}$ do not, in general, maintain their purity and decoherence may take place. For the interaction picture model of Flesia and Piron [19], decoherence may occur. If, however, consistently with the interaction picture model, one takes for the time-independent observables their corresponding interaction picture forms, no decoherence takes place. In this case, the result is, of course, independent of the choice of picture.

As we shall see in a concrete example, the degree of decoherence depends not only on the time dependence of the Hamiltonian, but also on the initial states.

A. Example

Tasaki et al. [34] considered a simple example described by the following Hamiltonian:

$$H(t) = -\frac{\Omega_0}{2}\sigma_z + \frac{\Omega}{2}[\sigma_+ e^{i\Omega_0 t} + \sigma_- e^{-i\Omega_0 t}] \tag{4.20}$$

where σ_i are the Pauli matrices:

$$\sigma_z = \begin{pmatrix} 1 & 0 \\ 0 & -1 \end{pmatrix} \qquad \sigma_+ = \begin{pmatrix} 0 & 1 \\ 0 & 0 \end{pmatrix} \qquad \sigma_- = \begin{pmatrix} 0 & 0 \\ 1 & 0 \end{pmatrix}$$

The Hilbert space for this model is the two-dimensional complex space $\mathscr{H} = \mathbf{C}^2$. It is easy to derive the evolution operator $W_t(\tau)$ corresponding to the Hamiltonian $H(t)$. One obtains

$$W_t(\tau) = u(\tau)\left\{\cos\frac{\Omega}{2}\tau - i\sin\frac{\Omega}{2}\tau(\sigma_+ e^{i\Omega_0 t} + \sigma_- e^{-i\Omega_0 t})\right\} \tag{4.21}$$

where the operator $u(\tau)$ is given by

$$u(\tau) = \begin{bmatrix} \exp\left(i\frac{\Omega_0}{2}\tau\right) & 0 \\ 0 & \exp\left(-i\frac{\Omega_0}{2}\tau\right) \end{bmatrix} \tag{4.22}$$

The direct integral space for this model is given by $L^2(-\infty, \infty; \mathbf{C}^2)$. We wish to study now the time evolution of the pure-like state given by Eq.

(4.8). From Eqs. (4.15) and (4.21) we have

$$
\begin{aligned}
\rho_{\psi_P} &= u(\tau) \int dt\,|f(t)|^2 \left\{\cos\left(\frac{\Omega}{2}\tau\right) - i\sin\left(\frac{\Omega}{2}\tau\right)(\sigma_+ e^{i\Omega_0 t} + \sigma_- e^{-i\Omega_0 t})\right\} |\psi_0\rangle \\
&\quad \times \langle\psi_0| \left\{\cos\left(\frac{\Omega}{2}\tau\right) + i\sin\left(\frac{\Omega}{2}\tau\right)(\sigma_+ e^{i\Omega_0 t} + \sigma_- e^{-i\Omega_0 t})\right\} u^\dagger(\tau) \\
&= u(\tau)\Bigg[\cos^2\left(\frac{\Omega}{2}\tau\right)|\psi_0\rangle\langle\psi_0| + \sin^2\left(\frac{\Omega}{2}\tau\right)\sigma_+|\psi_0\rangle\langle\psi_0|\sigma_- \\
&\quad + \sin^2\left(\frac{\Omega}{2}\tau\right)\sigma_-|\psi_0\rangle\langle\psi_0|\sigma_+ \\
&\quad + \Bigg\{iF(\Omega_0)\cos\left(\frac{\Omega}{2}\tau\right)\sin\left(\frac{\Omega}{2}\tau\right)(|\psi_0\rangle\langle\psi_0|\sigma_+ - \sigma_+|\psi_0\rangle\langle\psi_0|) \\
&\quad + \sin^2\left(\frac{\Omega}{2}\tau\right)F(2\Omega_0)\sigma_+|\psi_0\rangle\langle\psi_0|\sigma_+ + h.c.\Bigg\}\Bigg]u^\dagger(\tau) \\
&= u(\tau)W(\tau,\alpha)|\psi_0\rangle\langle\psi_0|W^\dagger(\tau,\alpha)u^\dagger(\tau) + u(\tau)\Bigg\{i[F(\Omega_0) - e^{i\alpha}] \\
&\quad \times \cos\left(\frac{\Omega}{2}\tau\right)\sin\left(\frac{\Omega}{2}\tau\right)(|\psi_0\rangle\langle\psi_0|\sigma_+ - \sigma_+|\psi_0\rangle\langle\psi_0|) \\
&\quad + \sin^2\left(\frac{\Omega}{2}\tau\right)[F(2\Omega_0) - e^{2i\alpha}]\sigma_+|\psi_0\rangle\langle\psi_0|\sigma_+ + h.c.\Bigg\}u^\dagger(\tau)
\end{aligned}
\tag{4.23}
$$

where $F(\omega)$ is the Fourier transform of $|f(t)|^2$,

$$
F(\omega) \equiv \int dt\,|f(t)|^2 e^{i\omega t} \tag{4.24}
$$

α is an arbitrary real number (to be chosen for convenience), and the operator $W(\tau, \alpha)$ is given by

$$
W(\tau,\alpha) = \cos\left(\frac{\Omega}{2}\tau\right) - i\sin\left(\frac{\Omega}{2}\tau\right)(e^{i\alpha}\sigma_+ + e^{-i\alpha}\sigma_-) \tag{4.25}
$$

and *h.c.* stands for the Hermitian conjugate of the first part in brackets. The first term clearly keeps the purity. Therefore, if we can choose α such that the remaining terms vanish, the time evolution does not destroy the purity of the state. The necessary and sufficient condition is the existence of a real

α satisfying

$$F(\Omega_0) = e^{i\alpha}, \qquad F(2\Omega_0) = e^{2i\alpha} \tag{4.26}$$

As an example, if $|f(t)|^2$ is the Gaussian form,

$$|f(t)|^2 = \frac{1}{2\sqrt{\pi}\,\Delta} \exp\left[\frac{(t - t_0)^2}{4\Delta^2}\right] \tag{4.27}$$

we obtain

$$F(\omega) = e^{i\omega t_0} e^{-(\omega\Delta)^2} \tag{4.28}$$

This function clearly cannot satisfy the conditions (4.26), but in order to minimize the terms in Eq. (4.23) that destroy the purity, we may take α to be $\alpha = \Omega_0 t_0$. Then we have

$$\begin{aligned}\rho_{\psi_P} &= u(\tau)W(\tau, \Omega_0 t_0)|\psi_0\rangle\langle\psi_0| W^\dagger(\tau, \Omega_0 t_0)u^\dagger(\tau) \\ &+ u(\tau)\left\{ ig(\Omega_0)e^{i\Omega_0 t_0} \cos\left(\frac{\Omega}{2}\tau\right) \sin\left(\frac{\Omega}{2}\tau\right)(|\psi_0\rangle\langle\psi_0|\sigma_+ - \sigma_+|\psi_0\rangle\langle\psi_0|) \right. \\ &\left. + \sin^2\left(\frac{\Omega}{2}\tau\right) g(2\Omega_0)e^{2i\Omega_0 t_0}\sigma_+|\psi_0\rangle\langle\psi_0|\sigma_+ + h.c. \right\} u^\dagger(\tau)\end{aligned} \tag{4.29}$$

where the function $g(\omega) = |F(\omega)| - 1 = \exp[-(\omega\Delta)^2] - 1$ describes the initial state dependence of the degree of decoherence. Strictly speaking, because $g(\Omega_0)$ and $g(2\Omega_0)$ are different from zero, decoherence takes place irrespective of the value of $\Delta(\neq 0)$. However, if the g terms are very small, the first term dominates, and the state ρ_{ψ_P} corresponds to an almost pure state. In short, we find for the Gaussian example, that when the initial state is well localized on the t-axis compared with the time scale of the change of the Hamiltonian, (i.e., $\Omega_0 \Delta \ll 1$), the state ρ_{ψ_P} remains practically pure. Otherwise, decoherence takes place.

We now generalize the discussion above to the generalized evolution introduced in the previous section. In particular, we find that even for *closed* systems, mixing of pure states is possible, when the interaction is not local on the time axis.

As we have seen, the generator K of the evolution $U(\tau)$ described by Eq. (3.15), may, in general, be written in the form

$$(K\psi)_t = -i\partial_t \psi_t + \int \kappa_{t,t'} \psi_{t'}\, dt' \tag{4.30}$$

When the kernel κ is in the form

$$\kappa_{t,t'} = \kappa_{t-t'} \tag{4.31}$$

it follows that (here $-\partial_t$ stands for the operator on $\bar{\mathscr{H}}$ which is represented as a derivative in the t-representation)

$$[\kappa, -i\partial_t] = 0 \tag{4.32}$$

Therefore, the system described by a generator of this form is closed in the sense that it is invariant to translations on the time axis, that is,

$$[K, -i\partial_t] = 0 \tag{4.33}$$

As in the previous section [Eq. (3.34)], this kind of interaction leads to an evolution operator of the form

$$W_{t,t'}(\tau) = \frac{1}{2\pi} \int e^{i(t-t')\sigma} e^{-i\kappa(\sigma)\tau}\, d\sigma = W_{t-t'}(\tau) \tag{4.34}$$

where $\kappa(\sigma)$ is the Fourier transform of $\kappa_{t-t'}$ with respect to $t - t'$.

Now, consider the most general form of pure state, $\psi_t = f(t)\psi_0$. The time evolution of such a state is

$$(\psi^\tau)_{t+\tau} = \int W_{t,t'}(\tau)\psi_0 f(t')\, dt' \tag{4.35}$$

For an evolution operator of the form (4.34), it follows that

$$\begin{aligned} (\psi^\tau)_{t+\tau} &= \int W_{t-t'}(\tau)\psi_0 f(t')\, dt' \\ &= \int W_{t'}(\tau)\psi_0 f(t-t')\, dt' \end{aligned} \tag{4.36}$$

This corresponds, for every t, to a superposition of the states $W_{t'}(\tau)\psi_0$, but, in general, for each t, the weights are different, and we conclude that the state may be mixed by the evolution. The purity of the state will be conserved if and only if all the states $W_{t'}(\tau)\psi_0$ are the same up to a factor which is a function of t' (and τ; the discussion that follows is, however, for each τ). We now prove that this occurs for *any* ψ_0 if and only if $\kappa_{t-t'} = \kappa\delta(t - t')$, where κ is some constant operator.

Let us assume that the state remains pure under evolution, that is,

$$W_t(\tau)\psi_0 = \alpha_t \psi_1 \tag{4.37}$$

for any arbitrary ψ_0 and corresponding ψ_1. Let $\{\phi_n\}$ be a complete orthonormal set in $\mathscr{H}$; then for each τ,

$$W_t(\tau)\phi_n = \alpha_t \psi_n = \alpha_t \sum_n \beta_{mn} \phi_m \tag{4.38}$$

and, therefore,

$$(\phi_m, W_t(\tau)\phi_n) = \beta_{mn} \alpha_t \tag{4.39}$$

Hence,

$$W_t(\tau) = \alpha_t W(\tau) \tag{4.40}$$

where

$$(\phi_m, W(\tau)\phi_n) = \beta_{mn} \tag{4.41}$$

Taking the Fourier transform of Eq. (4.40) one obtains

$$\tilde{W}_\sigma(\tau) = \tilde{\alpha}(\sigma) W(\tau) \tag{4.42}$$

However, from Eq. (4.34) it follows that

$$\tilde{W}_\sigma(\tau) = e^{-i\kappa(\sigma)\tau} \tag{4.43}$$

We now show that $W(\tau)$ has an inverse. As we have seen, the evolution operators satisfy the relation

$$\int W_{t,t''}(\tau) W_{t',t''}(\tau)^\dagger \, dt'' = \delta(t - t') \tag{4.44}$$

It follows from Eq. (4.44) and (4.40) that

$$\int \alpha_{t-t''} \alpha^*_{t'-t''} \, dt'' \, W(\tau) W(\tau)^\dagger = \delta(t - t') \tag{4.45}$$

and, therefore,

$$\int \alpha_{t-t''}\alpha^*_{t'-t''}\,dt'' = \lambda\delta(t-t'), \qquad \lambda \neq 0 \tag{4.46}$$

so that

$$\lambda W(\tau)W(\tau)^\dagger = 1 \tag{4.47}$$

It follows from Eqs. (3.23) and (3.22), by shifting t, t' and taking $\tau \to -\tau$, that the conjugate can appear on the first instead of the second factor in Eq. (4.44); thus $\lambda W(\tau)^\dagger W(\tau) = 1$ (λ must be real) as well, that is,

$$W^{-1}(\tau) = \lambda W(\tau)^\dagger \tag{4.48}$$

Hence, W^{-1} exists. Then, from Eqs. (4.42) and (4.43),

$$\tilde{W}_\sigma(\tau_1)\tilde{W}_\sigma(\tau_2)^{-1} = e^{-i\kappa(\sigma)(\tau_1-\tau_2)} = W(\tau_1)W(\tau_2)^{-1} \tag{4.49}$$

independently of σ; hence,

$$\kappa(\sigma) = \text{const.} \Rightarrow \kappa_{t-t'} = \kappa\delta(t-t') \tag{4.50}$$

Thus, we realize that pure states remain pure if and only if condition (4.50) is satisfied, which is exactly the case of a time-independent, pointwise Hamiltonian.

We therefore see that a generalized evolution of the form (3.15) may lead to mixing of pure states without assuming that the system is open or nonconservative, in the sense that $W_{t,t'}(\tau)$ may be of the form $W_{t-t'}(\tau)$, as discussed after Eq. (4.30). As found in Section III, in the framework of Lax–Phillips theory, the relation between the singularities of the S-matrix and the spectrum of the generator of the semigroup can be obtained only from the more general evolution, which indicates that the origin of irreversibility may be found in such structures.

In conclusion, we have seen that the Lax–Phillips theory provides a description of the quantum states which admits the possibility of decoherence not only for time-dependent Hamiltonian systems, but also for closed (but not Hamiltonian in the original Hilbert space) systems. Therefore, Machida–Namiki theory can be formulated naturally in this framework, and it is not necessary to specify the limit between the system and the measuring apparatus.

V. INTRINSIC DECOHERENCE IN CLASSICAL AND QUANTUM EVOLUTION

It has long been emphasized by Prigogine and his co-workers [27] that the natural description for the evolution of a system with many degrees of freedom is that of the evolution of the density matrix ρ, through the Liouville equation,

$$i\frac{d\rho}{dt} = [H, \rho] \tag{5.1}$$

The density matrix ρ ($\rho \geq 0$, $\mathrm{Tr}\,\rho = 1$) has the property that $\mathrm{Tr}\,\rho^2 \leq 1$, where the equality is attained only for a pure state. In general, one considers the space of Hilbert–Schmidt operators A for which

$$\mathrm{Tr}\,A^*A < \infty \tag{5.2}$$

the positive (normalized) elements of such a space correspond to the physical states, the density matrices. On this space, the commutator with the Hamiltonian H defines a linear operator $\mathscr{L}$, called the Liouvillian, for which

$$i\frac{d\rho}{d\tau} = \mathscr{L}\rho \tag{5.3}$$

where one assumes that $\mathscr{L}$ is self-adjoint in the Liouville space. The spectrum of the Liouvillian is, in general, continuous in $(-\infty, \infty)$, and hence there may exist an operator T conjugate to $\mathscr{L}$, such that

$$[T, \mathscr{L}] = i \tag{5.4}$$

Suppose T is self-adjoint and has the spectral representation

$$T = \int t'\, dE(t') \tag{5.5}$$

It follows from the commutation relation (5.4) that

$$e^{i\mathscr{L}\tau}Te^{-i\mathscr{L}\tau} = T + \tau$$

or

$$e^{i\mathscr{L}\tau}\, dE(t')e^{-i\mathscr{L}\tau} = dE(t' - \tau) \tag{5.6}$$

that is, $\mathscr{L}$, T, and $dE(t')$ form an imprimitivity system [35]. With this, we see that the spectral family of the operator T shifts with $\mathscr{L}$ in the same way as the time evolution of the state ρ [in Eq. (5.3)], and we may therefore identify T as the "time operator."

In particular, for a Hamiltonian of the form of the sum of an unperturbed operator H_0 and a perturbation V, that is $H = H_0 + V$, the corresponding Liouvillian is

$$\mathscr{L} = \mathscr{L}_0 + \mathscr{L}_I \tag{5.7}$$

Now suppose we consider the "time operator" T_0, conjugate to $\mathscr{L}_0$; it satisfies

$$[T_0, \mathscr{L}_0] = i$$

Then, in the spectral representation of T_0,

$$_0\langle t\,|[T_0, \mathscr{L}_0]|\,t'\rangle_0 = i\delta(t - t')$$

or

$$(t - t')_0\langle t\,|\,\mathscr{L}_0\,|\,t'\rangle_0 = i\delta(t - t') \tag{5.8}$$

It follows that

$$_0\langle t\,|\,\mathscr{L}_0\,|t'\rangle_0 = -i\partial_t\,\delta(t - t') \tag{5.9}$$

Hence,

$$_0\langle t\,|\,\mathscr{L}\,|\,t'\rangle_0 = -i\partial_t\,\delta(t - t') + {}_0\langle t\,|\,\mathscr{L}_I\,|\,t'\rangle_0 \tag{5.10}$$

where the last term is, in general, not diagonal.

We see, therefore, that the Liouville space formulation of dynamics provides a physical example of a structure in which the evolution law (for which the evolution parameter τ corresponds to the laboratory time) is a nontrivial kernel (nondecomposable) on the time axis and, hence, a Lax–

Phillips system which may have an S-matrix with nontrivial analytic properties.

In this section, we show that the existence of a time operator in the Liouville space provides a natural and consistent mechanism for the decoherence of physical states, that is, that pure states become mixed during the evolution, both for quantum and classical systems. The Hamiltonian evolution of states in classical mechanics is known by the Liouville theorem to be nonmixing, that is, to preserve the entropy of the system [36]. The same property holds for the quantum evolution as well, and follows from the unitarity of the evolution operator. This has been an obstacle to the consistent description of irreversible processes from first principles [37]. The use of techniques such as coarse-graining or truncation to achieve a realization of the second law does not follow from basic dynamical laws, and is fundamentally not consistent with the underlying Hamiltonian dynamical structure [38].

To show that the Liouville space provides a natural framework for such mixing to develop, we first make some definitions. The notion of a pure state is defined by means of expectation values of observables, that is, a state is called "pure" if the expectation value of each observable in this state is equal to the corresponding expectation value computed with respect to some well-defined wave function. We wish to weaken this condition, and require such an equality only for a t-independent subset of observables (to be defined precisely below). One obtains all the physical information concerning this subset of observables from an *effective* state resulting from the reduction of the full state by integration over the degree of freedom which is not relevant for this subset (i.e., the spectrum of the time operator). We call this reduced state the *physical state*. We show that there exist mixed states for which the effective physical state is pure and denote them as "effectively pure." These states may become effectively mixed during the evolution of the system. We formulate these ideas in the framework of the quantum Liouville space, and consider their application to classical mechanics. We also utilize a simple example to illustrate this mechanism.

The kernel representing a Hilbert–Schmidt operator A on the original Hilbert space of n degrees of freedom, $\langle \mathbf{k} | A | \mathbf{k}' \rangle$, where $\mathbf{k}$ consists of n parameters, corresponds to the function $A(\mathbf{k}, \mathbf{k}') \equiv \langle \mathbf{k}, \mathbf{k}' | A \rangle$ representing the vector A of the Liouville space. We then change variables from $\mathbf{k}, \mathbf{k}'$ to t, the spectrum of T, and $(2n - 1)$ other independent parameters β. This transformation is defind by a kernel $K(t, \beta | \mathbf{k}, \mathbf{k}')$ such that

$$A(t, \beta) \equiv \langle t, \beta | A \rangle = \int K(t, \beta | \mathbf{k}, \mathbf{k}') \langle \mathbf{k}, \mathbf{k}' | A \rangle \, d\mathbf{k} \, d\mathbf{k}' \tag{5.11}$$

and, in particular, for the density operator ρ (positive A),

$$\rho_t(\beta) \equiv \langle t, \beta \,|\, \rho \rangle = \int K(t, \beta \,|\, \mathbf{k}, \mathbf{k}')\langle \mathbf{k}, \mathbf{k}' \,|\, \rho \rangle \, d\mathbf{k} \, d\mathbf{k}' \tag{5.12}$$

In what follows, we use the time operator $T \equiv T_0$ conjugate to the unperturbed Liouville operator, which is defined according to the decomposition [39]

$$\mathscr{L} = \mathscr{L}_0 + \mathscr{L}_I \tag{5.13}$$

that is, on a suitable domain,

$$[T, \mathscr{L}_0] = i \tag{5.14}$$

It is clear from Eq. (5.14) that

$$(e^{-i\mathscr{L}_0\tau}A)(t, \beta) = A(t + \tau, \beta) \tag{5.15}$$

Under the free evolution, the representation of A on the Liouville space undergoes translation $t \to t + \tau$, so that t acquires the meaning of a label for the free translation in time.

Using this new basis, the expectation value of an observable is written as

$$\langle A \rangle_\rho = \mathrm{Tr}(A\rho) = \int \rho_t(\beta) A(t, \beta) \, dt \, d\beta \tag{5.16}$$

where

$$A(t, \beta) = \int K(t, \beta \,|\, \mathbf{k}, \mathbf{k}')\langle \mathbf{k}, \mathbf{k}' \,|\, A \rangle \, d\mathbf{k} \, d\mathbf{k}' \tag{5.17}$$

If A belongs to the subset of t-independent operators, that is, $A(t, \beta) \equiv A(\beta)$, then from Eq. (5.16) it follows that

$$\langle A \rangle = \mathrm{Tr}(A\rho) = \int \hat{\rho}(\beta) A(\beta) \, d\beta \tag{5.18}$$

where $\hat{\rho}$ is defined as

$$\hat{\rho}(\beta) \equiv \int dt \; \rho_t(\beta) \tag{5.19}$$

We therefore see that with respect to the set of t-independent observables, all of the information available in the state is contained in $\hat{\rho}$. It follows from Eq. (5.15) that t-independent observables commute with the free Hamiltonian H_0. In this case, clearly the asymptotic form of the observable A (in Heisenberg picture) exists if the wave operator for the scattering theory exists, that is,

$$\begin{aligned}\lim_{\tau\to\pm\infty} e^{-i\mathscr{L}\tau}A &= \lim_{\tau\to\pm\infty} U(\tau)^{-1}AU(\tau)\\ &= \lim_{\tau\to\pm\infty} U(\tau)^{-1}U_0(\tau)AU_0(\tau)^{-1}U(\tau)\\ &= \Omega_{\pm} A\Omega_{\pm}^{-1} = A_{\pm} \end{aligned} \tag{5.20}$$

where $U(\tau)$ is the full evaluation operator, and $U_0(\tau)$ is that of the unperturbed evolution. The t-independent observables therefore correspond to the asymptotic variables in a scattering theory [40].

Ludwig [41] has emphasized that measurements on a quantum system are made by means of the detection of signals corresponding to observables which are operationally on a semiclassical or classical level. These measurable signals, which characterize the state, are the properties propagating to the detectors, and are therefore asymptotic variables (i.e., ξ-independent). We do not argue that observables which are time dependent in the Heisenberg picture (such as the electromagnetic field) play no role. These operators may even be useful for calculations of measurable quantities, and their expectation values can be evaluated using, for example, the Schwinger–Keldysh technique [42]. However, from a physical point of view, based on the abovementioned theoretical arguments on the nature of measurement, only functions of these observables that have asymptotic limits (e.g., the free-number density and the momentum of an electromagnetic field) provide for experimental measurement. Measurements carried out upon an evolving system involve, in fact, interactions with apparatus which are essentially asymptotic (e.g., magnetic fields far from an electron beam, or the e-ν or photon signal from the pions in the final state of K-meson decay). These asymptotic observables determine the structure of the state, and hence (with a sufficient number of such measurements) can be used to define the nature of the evolution, that is whether a pure state tends to a mixed state. We thus conclude that the subset of ξ-independent observables corresponds to all the experimentally accessible measurements, and is, therefore, the subset of observables which can be used to characterize experimentally the structure of a physical state.

Note that $\hat{\rho}$ is not simply related to the density matrix of the system, but is given by the integral of Eq. (2.12) over the variable t. In fact, the unit operator on the original Hilbert space is represented by

$$1(t, \beta) = \int K(t, \beta \,|\, \mathbf{k}, \mathbf{k}')\delta(\mathbf{k} - \mathbf{k}')\, d\mathbf{k}\, d\mathbf{k}'$$

$$= \int K(t, \beta \,|\, \mathbf{k}, \mathbf{k})\, d\mathbf{k} \tag{5.21}$$

Now, $\mathscr{L}_0$ annihilates the unit operator, that is,

$$(e^{-i\mathscr{L}_0 t}1)(\mathbf{k}, \mathbf{k}') = 1(\mathbf{k}, \mathbf{k}') \tag{5.22}$$

so that, according to Eq. (5.15), $1(t, \beta)$ is independent of t [i.e., $1(t, \beta)$]. The function $1(\beta)$ is, moreover, invariant under all automorphisms of the algebra of observables that leave $t(\mathbf{k}, \mathbf{k}')$ invariant. We discuss the properties of the representations provided by t, β in more detail elsewhere [43]. For our present purpose, we note that

$$\mathrm{Tr}\, \rho = \int \rho(t, \beta) 1(t, \beta)\, dt\, d\beta$$

$$= \int \hat{\rho}(\beta) 1(\beta)\, d\beta = 1 \tag{5.23}$$

so that $1(\beta)$ provides the appropriate measure for what we have called the physical state.

A state $\hat{\rho}$ is called *effectively pure* if there exists a wave function ψ such that for every t-independent observable A,

$$\langle \psi \,|\, A \,|\, \psi \rangle = \langle A \rangle_{\hat{\rho}} = \int \hat{\rho}(\beta) A(\beta)\, d\beta \tag{5.24}$$

The form of $\hat{\rho}(\beta)$ can be determined by the measurement of all the t-independent observables. As shown below, in the basis of (generalized) eigenstates of the free Hamiltonian, $\hat{\rho}(\beta)$ must be represented as a sum of bilinear functions over equal energy subspaces. The state $\hat{\rho}(\beta)$ is effectively pure if and only if the coefficients of these bilinear functions are factorizable.

If ρ is pure in the usual sense (i.e., $\mathrm{Tr}\, \rho^2 = 1$), the condition (5.24) holds for any observable, and therefore the resulting $\hat{\rho}$ is effectively pure. However, it is clear that the reduction, Eq. (5.19), is not one to one, and therefore each $\hat{\rho}$ corresponds to an *equivalence class* of states in Liouville space. Even if only one of these states is pure, $\hat{\rho}$ would be effectively pure, because it does not distinguish between elements of the equivalence class. We thus see that strict purity implies effective purity but not the opposite, that is, *even mixed states may appear to be physically pure.*

We wish to show now that while unitarity excludes the possibility of mixing of pure states, mixing of effectively pure states is still possible. Generally, in the presence of interaction, the full Liouvillian takes the form [from Eq. (5.10)]

$$\langle t|\mathscr{L}|t'\rangle = -i\delta_t\delta(t-t') + \langle t|\mathscr{L}_I|t'\rangle \tag{5.25}$$

where the second term is, in general, not diagonal, but rather acts as an integral operator on t. The resulting evolution is also of an integral operator structure and takes the form (we call the evolution parameter τ to distinguish it from the spectrum of the T-operator)

$$\rho_t^\tau = \int W_{t,t'}(\tau)\rho_{t'}^0\,dt' \tag{5.26}$$

where the operator $W_{t,t'}(\tau)$ acts only on the β dependence.

For simplicity, we use the Fourier transform representation

$$\begin{aligned} \rho(\alpha,\beta) &= \int e^{-it\alpha}\rho_t(\beta)\,dt \\ \bar{W}_{\alpha,\alpha'}(\tau) &= \int e^{-it\alpha}e^{it'\alpha'}W_{t,t'}(\tau)\,dt\,dt' \end{aligned} \tag{5.27}$$

Note that $\hat{\rho}(\beta) = \rho(\alpha,\beta)|_{\alpha=0}$, and, therefore,

$$\hat{\rho}^\tau(\beta) = \rho^\tau(\alpha,\beta)\Big|_{\alpha=0} = \int \bar{W}_{0,\alpha'}(\tau;\beta,\beta')\rho(\alpha',\beta')\,d\alpha'\,d\beta' \tag{5.28}$$

The initial effective purity of ρ provides information only on its $\alpha = 0$ component while the other components may be effectively mixed, but, as we see from Eq. (5.28), during the evolution the $\alpha = 0$ component develops contributions from the other components, and therefore it may become mixed. The states keep their effective purity, in general, only if $\bar{W}_{0,\alpha'} \sim \delta(\alpha')$.

We are now in a position to characterize an effectively pure state more explicitly. Since α is the Fourier dual of the variable t, it follows from Eq. (5.3) that α is the spectrum of the unperturbed Liouvillian $\mathscr{L}_0$. Hence, the $\alpha = 0$ component of a state $\sum c(\mathbf{k},\mathbf{k}')\psi_\mathbf{k}\psi_{\mathbf{k}'}^*$, for a basis $\{\psi_\mathbf{k}\}$, which are (generalized) eigenfunctions of H_0, is the partial sum over the terms for which the (unperturbed) energy eigenvalues associated with $\psi_\mathbf{k}$ and $\psi_{\mathbf{k}'}$ are equal. For a pure state corresponding to $\psi = \sum a(\mathbf{k})\psi_{\mathbf{k}}$, $c(\mathbf{k}, \mathbf{k}') = a(\mathbf{k})a(\mathbf{k}')^*$ is factorizable. An effectively pure state coincides with the $\alpha = 0$ component of a pure state, and therefore satisfies this factorizability condition in the equal-energy subspaces. However, this condition in the equal-energy subspaces does not imply its general validity (for $\alpha \neq 0$). Hence, an effectively pure state is associated with an equivalence class which includes mixed states as well.

Note that if H_0 is nondegenerate, the effective purity condition holds trivially for every state (diagonal elements of the density matrix in H_0 representation are positive definite), and the evolution cannot induce mixing.

A. Example

We wish to consider now a simple concrete example to illustrate the ideas presented above. Consider the evolution of a particle in three dimensions in the presence of a screened Coulomb (Yukawa) potential. The matrix elements of the free Liouvillian are given by (we take $2m = 1$)

$$\langle \mathbf{k}_1, \mathbf{k}_2 | \mathscr{L}_0 | \mathbf{k}_3, \mathbf{k}_4 \rangle = \delta^3(\mathbf{k}_1 - \mathbf{k}_3)\delta^3(\mathbf{k}_2 - \mathbf{k}_4)(\mathbf{k}_2^2 - \mathbf{k}_1^2) \tag{5.29}$$

We change the variables in Liouville space from $(\mathbf{k}_1, \mathbf{k}_2)$ to $(\alpha, \beta, \Omega_1, \Omega_2)$ by the transformation

$$\alpha = \mathbf{k}_2^2 - \mathbf{k}_1^2, \qquad \beta = \mathbf{k}_2^2 + \mathbf{k}_1^2 \tag{5.30}$$

and Ω_1, Ω_2 are the angle variables of the momenta $\mathbf{k}_1$, $\mathbf{k}_2$, respectively. We denote the set of variables β, Ω_1, Ω_2 by $\bar{\beta}$. In this new basis the matrix elements of the free Liouvillian are given by

$$\langle \alpha, \bar{\beta} | \mathscr{L}_0 | \alpha', \bar{\beta}' \rangle = \alpha\delta(\alpha - \alpha')\delta(\bar{\beta} - \bar{\beta}') \tag{5.31}$$

The variable α defined by this change of basis coincides with the α of our general discussion above.

As mentioned before, effectively pure states are mixed during the evolution unless $\bar{W}_{0,\alpha'} \sim \delta(\alpha')$. We therefore look at the evolution operators induced by the perturbation to see whether this is the case. The matrix elements of the interaction Liouvillian are given by

$$\langle \mathbf{k}_1, \mathbf{k}_2 | \mathscr{L}_I | \mathbf{k}_3, \mathbf{k}_4 \rangle = \delta^3(\mathbf{k}_1 - \mathbf{k}_3)\tilde{V}_{\mathbf{k}_2 - \mathbf{k}_4} - \delta^3(\mathbf{k}_2 - \mathbf{k}_4)\tilde{V}_{\mathbf{k}_1 - \mathbf{k}_3} \tag{5.32}$$

where $\tilde{V}_{\mathbf{k}}$ is the Fourier transform of the potential V, taken at the point $\mathbf{k}$.

For the screened Coulomb potential,

$$V(r) = \frac{Ae^{-\mu r}}{\mu r} \tag{5.33}$$

where $\tilde{V}_{\mathbf{k}}$ is given by

$$\tilde{V}_{\mathbf{k}} = \frac{4\pi A}{\mu(\mathbf{k}^2 + \mu^2)} \tag{5.34}$$

and the matrix element takes the form

$$\langle \mathbf{k}_1, \mathbf{k}_2 | \mathscr{L}_I | \mathbf{k}_3, \mathbf{k}_4 \rangle = \frac{4\pi A}{\mu}\left[\frac{\delta^3(\mathbf{k}_1 - \mathbf{k}_3)}{(\mathbf{k}_2 - \mathbf{k}_4)^2 + \mu^2} - \frac{\delta^3(\mathbf{k}_2 - \mathbf{k}_4)}{(\mathbf{k}_1 - \mathbf{k}_3)^2 + \mu^2}\right] \tag{5.35}$$

Changing the variables to (α, β), one obtains

$$\begin{aligned}
&\langle \alpha, \bar{\beta} | \mathscr{L}_I | \alpha', \bar{\beta}' \rangle \\
&\quad \equiv \mathscr{L}_I(\alpha, \alpha', \bar{\beta}, \bar{\beta}') \\
&\quad = \frac{64\pi A}{\sqrt{2\mu}\,\mu}\left\{\frac{\left[\delta(\beta - \alpha - \beta' + \alpha')\dfrac{1}{\sqrt{\beta - \alpha}}\right]\delta(\Omega_1, \Omega_3)}{\beta + \beta' + \alpha + \alpha' - 2\sqrt{(\beta + \alpha)(\beta' + \alpha')}\,B(\Omega_2, \Omega_4) + \mu^2}\right. \\
&\qquad \left. - \frac{\left[\delta(\beta + \alpha - \beta' - \alpha')\dfrac{1}{\sqrt{\beta + \alpha}}\right]\delta(\Omega_2, \Omega_4)}{(\beta + \beta' - (\alpha + \alpha') - 2\sqrt{(\beta - \alpha)(\beta' - \alpha')}\,B(\Omega_1, \Omega_3) + \mu^2}\right\}
\end{aligned} \tag{5.36}$$

where $B(\Omega_1, \Omega_2)$ is defined by

$$B(\Omega_1, \Omega_2) = \sin\theta_1 \sin\theta_2 \cos(\phi_1 - \phi_2) + \cos\theta_1 \cos\theta_2 \tag{5.37}$$

Therefore, it is clear that the kernel $\mathscr{L}_I(\alpha, \alpha', \bar{\beta}, \bar{\beta}')$ is *not* of the form $\delta(\alpha - \alpha')\hat{A}(\bar{\beta}, \bar{\beta}')$ and that the evolution operators also do not have this form. In particular, for weak interactions, first-order perturbation theory gives

$$\bar{W}_{0,\alpha}(\tau) = \delta(\alpha) - i\tau\mathscr{L}_I(0, \alpha, \bar{\beta}, \bar{\beta}') + O(\tau^2 A^2) \tag{5.38}$$

where the second term induces mixing.

B. Application to Classical Mechanics

The method that we have described above applies as well to the formulation of classical mechanics on a Hilbert space defined on the manifold of phase space which was introduced by Koopman [44] and used extensively in statistical mechanics [38]. Misra [26] has shown that dynamical systems which admit a Lyapunov operator necessarily have an absolutely continuous spectrum; therefore, one can construct a time operator on the classical

Liouville space for such systems. We identify the variables **k**, **k**′ with the variables of the classical phase space, and consider the trace as an integral over this space. The expectation value of a t-independent operator defines a reduced-density function in the form (5.19). Because a pure state is defined by a density function concentrated at a point of the phase space, a state that is effectively pure must have the form $\hat{\rho}(\beta) = \delta(\beta - \beta_0)$. The equivalence class associated with this reduced density contains mixed states as well, such as $\rho(t, \beta) = \delta(\beta - \beta_0) f(t)$, corresponding to a nonlocalized function on the phase space (**k**, **k**′). The structure of the theory, and the conclusions we have reached, are therefore identical to those of the quantum case.

VI. CONCLUSIONS

The Lax–Phillips theory assumes the existence of a unitary evolution $U(\tau)$ on a Hilbert space $\bar{\mathscr{H}}$, and incoming and outgoing subspaces satisfying the conditions of Eq. (2.19). Under these conditions, incoming (outgoing) representations of $\bar{\mathscr{H}}$ of the form $\bar{\mathscr{H}} = L^2(\mathscr{R}; \mathscr{H})$ are defined, where $\mathscr{H}$ is an auxiliary Hilbert space, for which the evolution is represented by translation and the incoming (outgoing) subspace has definite support properties. The unitary transformation between these representations is called the S-matrix, and is of a multiplicative form in the spectral representation of the generator of the evolution (which is the Fourier transform of the outgoing representation). Lax and Phillips (17) define a (strongly contractive) semigroup in the form $\mathscr{Z}(\tau) = P_+ U(\tau) P_-$, where $P_\pm$ is the projection on the orthogonal complement of $\mathscr{D}_\pm$. The main result is that the eigenvalues of the generator of the semigroup, which determine the lifetime for the contraction, correspond to the singularities (of the analytic continuation) of the S-matrix in the spectral representation of the generator of the evolution.

The physical system Lax–Phillips had primarily in mind is scattering of waves (electromagnetic or acoustic) from an obstacle. For this system, they identified the incoming and outgoing subspaces with the subspaces of incoming and outgoing waves, respectively. Thus, the abstract Lax–Phillips semigroup corresponds in this case to the contracted evolution of waves in the region of the obstacle. It originally appeared that the ideas of the Lax–Phillips theory could not be extended to the quantum theory [14] because the generator of translations, which are the representation of the evolution in the incoming and outgoing representation, has a spectrum on the whole real line, and the generator of the evolution for the quantum case is semibounded for most cases.

Flesia and Piron [19] have shown that the use of the larger Hilbert space formulation of quantum mechanics solves this problem because the spectrum of the generator of the evolution in the larger space is indeed the whole real line. In this formulation the evolution takes place in a direct integral Hilbert space, over the usual quantum mechanical Hilbert spaces, indexed by t, and the evolution is according to a new evolution parameter τ. In this framework, they have calculated the Lax–Phillips S-matrix, and shown that it has the form $(S\psi)_t = S_t \psi_t$, where S_t is a map from $\mathscr{H}$ to $\mathscr{H}$, independent of t, and corresponds to the usual scattering-theory S-matrix in the limit of the gap between the subspaces $\mathscr{D}_\pm$ going to ∞.

However, this theory has a fundamental difficulty. As mentioned above, the relation between the S-matrix and the semigroup is established in terms of the Fourier transform of the S-matrix. However, because S_t is independent of t, $S_t \equiv S_0$, the Fourier transform of S is trivial, that is $S(\sigma) = S_0\,\delta(\sigma)$, and is not a meromorphic function of σ. Therefore, this theory cannot use the full structure of the Lax–Phillips theory, and in particular, is not applicable to the description of unstable systems.

The Flesia–Piron construction can, however, be extended to *nondecomposable* evolution. As discussed in Section III, this kind of evolution occurs naturally in the framework of the Liouville space; it may also correspond to action at a distance as a consequence of Green's function, or to a pointwise Hamiltonian in another representation, which becomes nondecomposable in the representation in which $\mathscr{D}_\pm$ have definite support properties.

We have shown for this general evolution law that the S-matrix is indeed multiplicative in the Fourier transform basis, but, in general, is not trivial. Furthermore, the semigroup is contractive, independently of the form of the nondecomposable (self-adjoint) perturbation. One finds, in this construction, an exact superselection rule separating the states of the unstable system and the decay products.

We have reviewed the application of Lax–Phillips theory to the idea of Machida and Namiki [33] for the transition to a mixed state during the measurement [34]. Such a transition can consistently be realized in the framework of the Lax–Phillips theory, with no necessity for distinguishing the system from the measurement apparatus. The measurement is characterized by expectation values in the Lax–Phillips pure state of time-independent observables; if the Lax–Phillips state contains nontrivial time dependence (i.e., not simply factorizable), the resulting measurement appears as that of a mixed state, resulting in one or another state with some probability, but not a linear superposition.

Extending this idea to the generalized evolution, we have found that decoherence takes place even for closed systems, that is, systems which are invariant for translations along the t-axis, (excluding generators of the form $-i\partial_t + H$, where H is t-independent). Therefore, Namiki's measurement theory can be applied for closed systems also. This may be considered a first step toward a theory of the measurement process.

As discussed in Section V, it has also recently been shown, following similar ideas, that the Liouville space, which provides a natural framework for the realization of Lax–Phillips theory in its most general form, also admits the construction of a mechanism for the decoherence of physical states. This mechanism is contained in the formation of equivalence classes of states with respect to the measurement of operators which are independent of the Liouville time, for example, corresponding to asymptotic observables. It is shown that an effective pure state may actually be mixed, and its evolution will lead, in general, to an effective mixed state. The evolution of such equivalence classes, rather than the evolution of a particular state, therefore becomes a proper subject of study.

Lockhart and Misra [45] have discussed the notion of dynamical evolution (following the unitary or measure-preserving standard mathematical models of evolution) and *physical evolution*, for which the evolution is realized as a dissipative semigroup consistent with the second law of thermodynamics. In fact, they state that the central problem of quantum measurement theory is the reconciliation of the Schrödinger evolution with the statistical evolution caused by the measurement process generally referred to as the "collapse of the wave packet." Criteria are given by Lockhart and Misra [45], in which a detailed model of irreversible processes is developed, for systems that admit irreversible behavior. It may be possible to imbed these ideas into the framework of the Lax–Phillips theory.

In this review, we have discussed states in a space that includes time as a variable, with an additional parameter that corresponds to the laboratory clock. The interpretation of such a time variable (subject, as well, to transformations from one representation to another) is not at all trivial; the a priori distribution of its values varies systematically under the Lax–Phillips or Liouville evolution, with the evolution of the state. The realization of a nontrivial Lax–Phillips theory, as well as the structure of the Liouville evolution, impose a dynamical role for this variable, as would be expected to be necessary for the description of a physical *process*, reversible or irreversible. Some progress, as we have pointed out, has been made in understanding this essential aspect of the theory, and it is hoped that future work will bring further insight.

ACKNOWLEDGMENTS

One of us (LPH) wishes to thank C. Piron for many discussions, and is grateful for his hospitality in Geneva during several visits, and to the Swiss National Science Foundation for its support. He also thanks the International Solvay Institutes for Physics and Chemistry in Brussels, where this work was completed, for the opportunity to visit Brussels, and I. Prigogine for his hospitality, his emphasis on the essential importance of the Liouville space, and his continuing interest in the development of the ideas we have discussed here. We thank our colleagues in Brussels—I. Antoniou, B. Misra, and S. Tasaki—for helpful discussions and comments which contributed significantly to many of the results that we have presented.

REFERENCES

1. G. Gamow, *Zeits. f. Phys.* **51**, 204 (1928).
2. T. D. Lee, R. Oehme, and C. N. Yang, *Phys. Rev.* **106**, 340 (1957); T. T. Wu and C. N. Yang, *Phys. Rev. Lett.* **13**, 380 (1964).
3. V. F. Weisskopf and E. P. Wigner, *Zeits. f. Phys.* **63**, 54 (1930); **65**, 18 (1930).
4. L. P. Horwitz and J. P. Marchand, *Helv. Phys. Acta* **42**, 1039 (1969).
5. L. P. Horwitz and J. P. Marchand, *Rocky Mt. J. Math.* **1**, 225 (1973).
6. R. Zwanzig, *Physica* **30**, 1109 (1964).
7. N. Bleistein, H. Neumann, R. Handelsman, and L. P. Horwitz, *Nuovo Cim.* **41A**, 389 (1977).
8. B. Misra and E. C. G. Sudarshan, *J. Math. Phys.* **18**, 756 (1977); C. Chiu, E. C. G. Sudarshan, and B. Misra, *Phys. Rev.* **D16**, 520 (1979).
9. K. O. Friedrichs, *Comm. Pure Appl. Math.* **1**, 361 (1948).
10. T. D. Lee, *Phys. Rev.* **95**, 1329 (1956).
11. I. M. Gel'fand and G. E. Shilov, *Generalized Functions*, Vol. 4 (Academic Press, New York, 1968), p. 77ff., English translation.
12. L. P. Horwitz and I. M. Sigal, *Helv. Phys. Acta* **51**, 685 (1980); W. Baumgartel, *Math. Nachr.* **75**, 133 (1978). See also G. Parravicini, V. Gorini, and E. C. G. Sudarshan, *J. Math. Phys.* **21**, 2208 (1980); A. Bohm, *The Rigged Hilbert Space and Quantum Mechanics*, Springer Lecture Notes on Physics, Vol. 78 (Springer, Berlin, 1978); A. Bohm, *Quantum Mechanics: Foundations and Applications* (Springer, Berlin, 1986); A. Bohm, M. Gadella, and G. B. Mainland, *Am. J. Phys.* **57**, 1103 (1989); T. Bailey and W. C. Schieve, *Nuovo Cim.* **47A**, 231 (1978).
13. M. Reed and B. Simon, *Methods of Modern Mathematical Physics*, Vol. 1 (Academic Press, New York, 1972), p. 148.
14. C. Piron, *Foundations of Quantum Physics* (Benjamin/Cummings, Reading, MA, 1976).
15. L. P. Horwitz, J. P. Marchand, and J. LaVita, *J. Math. Phys.* **12**, 2537 (1971); D. Williams, *Comm. Math. Phys.* **21**, 314 (1971).
16. C. Foias and B. Sz. Nagy, *Acta Sci. Math.* **23**, 106 (1962); F. Riesz and B. Sz. Nagy, *Functional Analysis*, 2nd ed., Appendix by B. Sz. Nagy (Dover, New York, 1990), English translation.
17. P. D. Lax and R. S. Phillips, *Scattering Theory* (Academic Press, New York, 1967).
18. See, for example, L. E. Reichl, *The Transition to Chaos in Conservative Classical Systems: Quantum Manifestation* (Springer, New York, 1992), and references therein; I. M. Sigal, personal communication.
19. C. Flesia and C. Piron, *Helv. Phys. Acta* **57**, 697 (1984).
20. L. P. Horwitz and C. Piron, *Helv. Phys. Acta* **66**, 694 (1993).
21. G. C. Wick, A. S. Wightman, and E. P. Wigner, *Phys. Rev.* **88**, 101 (1952).
22. C. Piron, *Mecanique Quantique, Bases et Applications* (Presses polytechnique et universitaires romandes, Lausanne, 1991), p. 61.
23. I. M. Gel'fand and N. Ya. Vilenkin, *Generalized Functions*, Vol. 4, p. 110 ff. (Academic Press, New York, 1964).

24. J. S. Howland, *Springer Lecture Notes in Physics*, Vol. 130 (Springer, Berlin, 1979), p. 163.
25. For example, K. Gottfried, *Quantum Mechanics*, Vol. 1 (Benjamin, New York, 1966), p. 248.
26. B. Misra, *Proc. Natl. Acad. Sci.* **75**, 1627 (1978); B. Misra, I. Prigogine, and M. Courbage, *Proc. Natl. Acad. Sci. U.S.A.* **76**, 4768 (1979).
27. C. George, *Physica* **65**, 277 (1973); I. Prigogine, C. George, F. Henin, and L. Rosenfeld, *Chem. Scripta* **4**, 5 (1973); T. Petrosky, I. Prigogine, and S. Tasaki, *Physica* **A173**, 175 (1991); T. Petrosky and I. Prigogine, *Physica* **A175**, 146 (1991), and references therein.
28. I. P. Cornfeld, S. V. Formin, and Ya. G. Sinai, *Ergodic Theory* (Springer, Berlin, 1982).
29. For example, E. Kogan and M. Kaveh, *Phys. Rev.* **B46**, 10636 (1992).
30. H. F. Trotter, *Pac. J. Math.* **8**, 887 (1958).
31. M. Reed and B. Simon, *Methods of Modern Mathematical Physics*, Vol. 3 (Academic Press, New York, 1979), p. 236.
32. E. B. Davies, *Quantum Theory of Open Systems* (Academic Press, London, 1976), p. 103.
33. S. Machida and M. Namiki, *Prog. Theor. Phys.* **63**, 1457 (1980); **63**, 1833 (1980); M. Namiki and S. Pascazio, *Phys. Rev.* **A44**, 39 (1991).
34. S. Tasaki, E. Eisenberg, and L. P. Horwitz, *Found. Phys.* **24**, 1179 (1994).
35. G. W. Mackey, *Induced Representations of Groups and Quantum Mechanics* (Benjamin, New York, 1968).
36. See, for example, J. Yvon, *Correlations and Entropy in Classical Statistical Mechanics* (Pergamon Press, Oxford, 1969).
37. See, for example, I. E. Antoniou and I. Prigogine, *Physica* **A192**, 443 (1993).
38. I. Prigogine, *Non-equilibrium Statistical Mechanics* (Wiley, New York, 1961).
39. The Liouville operator is defined by

$$e^{-i\mathscr{L}\tau}\rho = e^{-iH\tau}\rho e^{iH\tau}$$

and $\mathscr{L}_0$ by

$$e^{-i\mathscr{L}_0\tau}\rho = e^{-iH_0\tau}\rho e^{iH_0\tau}$$

Then, $\mathscr{L}_I \equiv \mathscr{L} - \mathscr{L}_0$. We thank I. Antoniou for a communication on this point, as well as for a discussion of the interpretation of Eq. (5.15).
40. W. O. Amrein, P. Martin, and B. Misra, *Helv. Phys. Acta* **43**, 313 (1970).
41. G. Ludwig, *Foundations of Quantum Mechanics I* (Springer Verlag, New York, 1982); *Foundations of Quantum Mechanics II* (Springer Verlag, New York, 1983).
42. See, for example, E. Lifshitz and L. P. Pitaevskii, *Physical Kinetics* (Pergamon Press, Oxford, 1981), Chapter 10.
43. E. Eisenberg and L. P. Horwitz, *Phys. Rev.* **A52**, 70 (1995).
44. B. Koopman, *Proc. Natl. Acad. Sci. U.S.A.* **17**, 315 (1931).
45. C. M. Lockhart and B. Misra, *Physica* **136A**, 47 (1986).

QUANTUM SYSTEMS WITH DIAGONAL SINGULARITY

I. ANTONIOU

International Solvay Institute for Physics and Chemistry, Brussels, and Theoretische Natuurkunde, Free University of Brussels, Brussels, Belgium

Z. SUCHANECKI

International Solvay Institute for Physics and Chemistry, Brussels, Belgium, and Hugo Steinhaus Center and Institute of Mathematics, Wroclaw Technical University, Wroclaw, Poland

CONTENTS

ABSTRACT

The work of the Brussels–Austin group on irreversibility over the last years has shown that quantum large Poincaré systems lead to states and observables with diagonal singularity. States with diagonal singularity include microcanonical equilibrium and they provide the natural framework for the discussion of quantum irreversible processes. This formulation actually

Advances in Chemical Physics, Volume XCIX, Edited by I. Prigogine and Stuart A. Rice.
ISBN 0-471-16526-3

defines an extension of quantum theory beyond the conventional Hilbert space framework and logic. We characterize the algebra of observables and the states with diagonal singularity and derive the Pauli master equation without resorting to the thermodynamic limit.

I. INTRODUCTION

States with diagonal singularity were introduced in classical and quantum statistical physics by I. Prigogine and co-workers [1–4] in their establishment of the master equation. These states are actually the entities that approach equilibrium for large nonintegrable Poincaré systems. States with diagonal singularity are linear functionals over the algebra of operators with diagonal singularity introduced by van Hove [5,6] in his perturbation analysis of large quantum systems.

The very large size of the system manifests itself in the continuous nature of the unperturbed energies. Van Hove supplemented the diagonal singularity property with the initial random-phase condition in order to derive the Pauli Master equation [7,8]. The irreversible equations, such as the Boltzmann, Fokker–Planck, and Pauli equations, arise as first-order approximations in the subdynamics decomposition of Liouville operator [9–12]. The subdynamics decomposition leads moreover to a spectral representation of the evolution of large Poincaré (nonintegrable) systems which includes the resonances and diffusion parameters in the spectrum [13–16]. These spectral decompositions are meaningless in the conventional Hilbert space topology but they acquire meaning in terms of weaker topologies like the Rigged Hilbert Space [15–16]. The key point in the subdynamics construction is the projection onto the diagonal part of states or observables. This projection is usually performed through the thermodynamic limit because the direct calculation on the continuous spectrum in the conventional formulation suffers from divergences. The reason for the divergences is that the continuum labeled basis $|\alpha\rangle$ of the space of wave functions

$$\langle\alpha\,|\,\alpha'\rangle = \delta(\alpha - \alpha')$$

cannot be lifted to a product basis for the operators. Indeed, the dyadic basis $|\alpha\rangle\langle\alpha'|$ gives rise to meaningless expressions in the continuous case because

$$\mathrm{tr}\{(|\alpha\rangle\langle\alpha|)^{\dagger}(|\alpha'\rangle\langle\alpha'|)\} \equiv (\alpha\alpha\,|\,\alpha'\alpha')_{\mathrm{HS}} = \delta(\alpha - \alpha')\delta(\alpha - \alpha')$$

where $(\ |\ )_{\mathrm{HS}}$ is the scalar product in the Hilbert–Schmidt space $\mathscr{H} \otimes \mathscr{H}^{\times}$.

The problem of the thermodynamic limit is one of the reasons why "the derivations of van Hove is by no means rigorous" as Hugenholtz states in

Festschrift Leon van Hove [17]. Moreover, the conventional Hilbert–Schmidt formulation does not allow us to define the microcanonical equilibrium and therefore the approach to equilibrium. The reason is that the identity operator, although bounded, is not in the Hilbert–Schmidt space $\mathcal{H} \otimes \mathcal{H}^{\times}$ because

$$\mathrm{tr}\, I = \sum_{\nu} \langle u_{\nu} | u_{\nu} \rangle = +\infty$$

for any orthonormal basis (u_{ν}) of the Hilbert space $\mathcal{H}$.

However, the extension of the Hilbert–Schmidt algebra to an algebra with identity [18,19]

$$\mathbf{C} \cdot I \oplus (\mathcal{H} \otimes \mathcal{H}^{\times})$$

together with the diagonal singularity property holds the clue for a proper formulation of quantum theory which allows for

1. A clear definition of states and observables with the diagonal singularity
2. A definition of the projections onto the diagonal and off-diagonal parts without resorting to the thermodynamic limit
3. The construction of continuous biorthonormal basis for states and observables

We show in this work that following the algebraic approach to dynamical systems we can construct a natural extension of quantum theory which fulfills requirements 1–3 and gives meaning to the resulting spectral decompositions of the Liouville operator which are intertwined with the master equation. Therefore, the master equation is not in contradiction with quantum mechanics but requires an extension of the conventional formulation.

The main idea of the algebraic approach is that the states ρ correspond to the normalized positive linear functionals $(\rho|$ over a suitably chosen algebra of operators representing the observables of the dynamical system. The value of the functional $(\rho|$ for an operator A is the expectation value of the observable A in the state ρ:

$$(\rho | A) \equiv \langle A \rangle_{\rho}$$

The algebraic generalization of quantum theory was pioneered by Segal [20], who assumed that the observables form a C*-algebra of bounded operators. C*-algebras were applied in the 1960s to local quantum field theory and statistical physics. For a general introduction to the methodology and results, we refer to Emch [21], Bratteli and Robinson [22], and

Haag [23]. In fact, Emch concludes his paper [21] with the remark that the algebraic approach can be of definite use in the problem of Poincaré invariants.

However, the algebra of operators with the diagonal singularity turns out to be an involutive Banach algebra which is not a C*-algebra. This more general structure allows us to incorporate the states with diagonal singularity which cannot be described in terms of the conventional [24–26] or generalized [16,24,27] formulation of quantum theory.

The additional possibilities for extension at the level of densities (the so-called Liouville space formulation) have been emphasized repeatedly by Prof. I. Prigogine and co-workers [9,14,28–31]. The states with diagonal singularity are nonlocal, they include the microcanonical equilibrium, and they describe the approach to equilibrium of large quantum systems. These properties of states with diagonal singularity are discussed in detail by Petrosky and Prigogine in the same volume [31].

First (Section II) we discuss the continuum-labeled bases of large quantum systems which define the diagonal parts of operators. Based on this discussion we define the operators with diagonal singularity (Section III) and the states with diagonal singularity (Section IV). After a discussion on operators acting on observables and on states (Section V) we formulate the time evolution (Section VI) and the spectral decomposition of the Liouville operator (Section VII). The intrinsic irreversibility of systems with a nonvanishing collision operator is illustrated by the relation with the Pauli master equation (Section VIII).

II. COMPLETE SYSTEMS OF COMMUTING OBSERVABLES AND COMMON EIGENVECTORS

We consider quantum systems associated with a separable Hilbert space $\mathscr{H}$ with a left conjugate linear and right linear scalar product $\langle \mid \rangle$. The Hamiltonian operator H can be decomposed into a diagonal part H_0 and an off-diagonal part or perturbation V:

$$H = H_0 + \lambda V$$

The operator H_0 is diagonal with respect to the noninteracting entities $|\alpha\rangle$. The eigenvectors $|\alpha\rangle$ are the common eigenvectors of a complete system of commuting observables [24–27] and provide a representation of the system. The actual choice of the complete set of commuting observables is based upon physical considerations. Let us remark in passing that for every chosen, countable, complete set of commuting self-adjoint operators A_1, $A_2, \ldots$ on the separable Hilbert space $\mathscr{H}$, there always exists a generalized orthonormal basis of common eigenvectors $|\alpha\rangle \equiv |\alpha_1\alpha_2 \cdots)$ labeled by the spectral values $\alpha \equiv (\alpha_1, \alpha_2, \ldots)$ in the spectrum $\sigma = \sigma(A_1) \times \sigma(A_2) \times \cdots \subseteq \mathbf{R} \times \mathbf{R} \times \cdots$ of the complete set $A_1, A_2, \ldots$.

The eigenvectors $\langle\alpha|$ are linear functionals living in the dual $\Phi^\times$ of a dense nuclear subspace Φ of the Hilbert space $\mathscr{H}$. The existence of the nuclear subspace Φ was proved by Maurin [32,33] under the additional assumption that the intersection of the domains of the powers of the operators A_i^n, $n = 1, 2, \ldots$ is dense. This assumption was shown to be unnecessary by de Dormale and Gautrin [34], who extended the work of Jauch and Misra [35] and Prugovecki [26].

The labeling quantum numbers α are discrete in the case of atoms and oscillators or continuous in the case of scattering, fields, and large systems.

In the case of potential scattering, $|\alpha\rangle \equiv |k_1, k_2, k_3\rangle \equiv |\mathbf{k}\rangle$ are the plane waves describing the asymptotic states. The complete system consists of the three components of the momentum operator.

The perturbation λV (λ is a dimensionless parameter) couples the noninteracting degrees of freedom $|\alpha\rangle$. In the case of integrable systems, the perturbation just gives rise to new independent conserved entities associated with the total Hamiltonian H:

$$H = \sum_a z_a |a\rangle\langle a|$$

However, if resonances are present, the interaction cannot be reincorporated into any conventional noninteracting conserved entities*. The main idea of the Brussels school is that in such cases the natural noninteracting entities are not conserved. In fact, the subdynamics decomposition [9–13] amounts to a generalized spectral decomposition of the evolution operator with complex eigenvalues associated with resonances [14–17]. The construction is particularly appropriate for large Poincaré (nonintegrable) systems that are classical or quantum mechanical.

III. OPERATORS WITH DIAGONAL SINGULARITY

In the representation provided by continuous bases like $|\alpha\rangle$, van Hove encountered [1,2] operators represented by the kernels

$$A_{\alpha\alpha'} \equiv \langle\alpha|A|\alpha'\rangle = A_\alpha^d \,\delta(\alpha - \alpha') + A_{\alpha\alpha'}^c$$

The operator A is said to have diagonal singularity with respect to the basis $|\alpha\rangle$ if the diagonal part A_α^d does not vanish. The diagonal function A_α^d has

* We thank Prof. C. George for this expression.

no singularities while the off-diagonal part $A^c_{\alpha\alpha'}$ may have certain weaker singularities as in the case of continuous degeneracies appearing in scattering or in coupled fields. In such cases, one may decompose further the off-diagonal part $A^c_{\alpha\alpha'}$. Here we assume the simplest case where $A^c_{\alpha\alpha'}$ are kernels of compact operators. The reason is that the states on the algebra of compact operators correspond to trace class operators, because the dual of the algebra $\mathscr{B}^\infty_{\mathscr{H}}$ of compact operators is isomorphic to the trace class operators $\mathscr{B}^1_{\mathscr{H}}$. Tracial states are the traditional states of quantum statistical physics [25–27].

A reasonable choice for the algebra of operators with diagonal singularity is the sum

$$\mathscr{A}^d + \mathscr{B}^\infty_{\mathscr{H}}$$

The diagonal part is the maximal Abelian von Neumann algebra generated by the chosen complete system of commuting observables [36,34]. The algebra $\mathscr{A}^d$ is a Banach algebra with respect to the norm

$$\|A^d\| \equiv \sup_{\psi \neq 0} \frac{\|A\psi\|_H}{\|\psi\|_H} \qquad \text{any } A^d \in \mathscr{A}^d \tag{1}$$

If some observables correspond to unbounded self-adjoint operators, as in the case of potential scattering, the algebra $\mathscr{A}^d$ is generated by their spectral projections, which are bounded operators.

The operator H_0 as well as any other unbounded operator are affiliated with the algebra $\mathscr{A}^d$, if their spectral projections are included in $\mathscr{A}^d$. In the case of unbounded operators, Antoine et al. [37] considered more general structures like V^*-algebra instead of maximal Abelian von Neumann algebras.

The nondiagonal part $\mathscr{B}^\infty_{\mathscr{H}}$ is the C^*-algebra [19] of compact operators on the Hilbert Space $\mathscr{H}$. The algebras $\mathscr{A}^d$ and $\mathscr{B}^\infty_{\mathscr{H}}$ have no common elements, because the diagonal operators are noncompact. Indeed, suppose that

$$Af(\alpha) = a(\alpha)f(\alpha)$$

where $a(\alpha)$ is a bounded function, $a(\alpha) \neq 0$ on a set of positive measure. Therefore, there is a number $\gamma > 0$ such that the set

$$\sigma_0 \equiv \{\alpha : |a(\alpha)| > \gamma\}$$

has a positive measure. Since our measure is nonatomic, the space $\mathscr{L}^2_{\sigma 0}$ is infinite dimensional. Thus there is an orthonormal basis $\{e_n\}$ in $\mathscr{L}^2_{\sigma 0}$. Let us extend the functions $\{e_n\}$ on the whole σ, putting $e_n(\alpha) = 0$ for $\alpha \notin \sigma_0$. Then

$\|e_n\|_{\mathscr{L}_\sigma{}^2} = \|e_n\|\mathscr{L}^2_{\sigma 0}$ and $\{e_n\}$ is an orthonormal sequence in $\mathscr{L}^2_\sigma$. Moreover, we have

$$\|Ae_n - Ae_m\|^2_{\mathscr{L}_\sigma{}^2} = \int_{\sigma 0} d\alpha\,|a(\alpha)|^2\,|e_n(\alpha) - e_m(\alpha)|^2 \geq \gamma^2\|e_n - e_m\|^2\mathscr{L}^2_{\sigma 0} = 2\gamma^2$$

This implies that the sequence $\{Ae_n\}$ does not contain any convergent subsequence, that is, that the operator A cannot be compact.

Consequently, any operator A in the algebra is uniquely represented as the sum of the diagonal and off-diagonal parts:

$$A = A^d + A^c$$

The algebra $\mathscr{A}$ of operators with diagonal singularity is therefore the direct sum

$$\mathscr{A} = \mathscr{A}^d \oplus \mathscr{B}^\infty_{\mathscr{H}}$$

A natural norm on $\mathscr{A}$ which captures the physical properties of these operators can be defined as follows:

$$\|A\|_{\mathscr{A}} = \|A^d\| + \|A^c\|$$

where $\|\cdot\|$ is the operator norm defined previously (1).

The properties of the algebra $\mathscr{A}$ are summarized by the following

Theorem

1. *$\mathscr{A}$ is an involutive Banach algebra of bounded operators which includes the identity operator.*
2. *$\mathscr{A}$ is not a C^*-algebra.*

Proof

a. $\mathscr{A}$ is clearly a Banach space.

b. The product of two operators A, B in $\mathscr{A}$ is also an element of $\mathscr{A}$

$$\begin{aligned} AB &= (A^d + A^c)(B^d + B^c) \\ &= A^dB^d + A^cB^c + A^cB^d + A^dB^c = (AB)^d + (AB)^c \end{aligned}$$

where $(AB)^d \equiv A^dB^d$, the diagonal part of AB, and $(AB)^c \equiv A^cB^c + A^cB^d + A^dB^c$, the nondiagonal part of AB. The products A^cB^c, A^cB^d, and A^dB^c are compact operators because the product of any bounded operator with compact operators is a compact operator [38].

c. The identity operator is clearly included in $\mathscr{A}^d$.
d. The continuity of the product is guaranteed by the inequality

$$\|AB\|_{\mathscr{A}} \leq \|A\|_{\mathscr{A}} \|B\|_{\mathscr{A}}$$

The proof of this inequality is straightforward:

$$\begin{aligned}\|AB\|_{\mathscr{A}} &= \|A^dB^d\| + \|A^cB^c + A^cB^d + A^dB^c\| \\ &\leq \|A^d\| \|B^d\| + \|A^cB^c\| + \|A^cB^d\| + \|A^dB^c\| \\ &\leq \|A^d\| \|B^d\| + \|A^c\| \|B^c\| + \|A^c\| \|B^d\| + \|A^d\| \|B^c\| \\ &= (\|A^d\| + \|A^c\|)(\|B^d\| + \|B^c\|) \\ &= \|A\|_{\mathscr{A}} \|B\|_{\mathscr{A}}\end{aligned}$$

In the proof we have used the facts that both $\mathscr{A}^d$ and $\mathscr{B}^{\infty}_{\mathscr{H}}$ are Banach algebras and that the triangle inequality is valid for the usual operator norm $\|\cdot\|$.
e. $\mathscr{A}$ is not a C^*-algebra. To see this one should check the nonvalidity of the complete regularity condition [18,19]

$$\|A^{\dagger}A\|_{\mathscr{A}} = \|A\|^2_{\mathscr{A}}$$

Putting in this equality $A = A^d + A^c$ we get

$$\|(A^d + A^c)^{\dagger}(A^d + A^c)\|_{\mathscr{A}} = \|A^d + A^c\|^2_{\mathscr{A}}$$

which is equivalent to

$$\|A^{d\dagger}A^d + (A^{d\dagger}A^c + A^{c\dagger}A^d + A^{c\dagger}A^c)\|_{\mathscr{A}} = \|A^d + A^c\|^2_{\mathscr{A}}$$

or

$$\|A^{d\dagger}A^d\| + \|A^{d\dagger}A^c + A^{c\dagger}A^d + A^{c\dagger}A^c\| = (\|A^d\| + \|A^c\|)^2$$

Then, using the *-property of the operator norm $\|\cdot\|$, we get

$$\|A^{d\dagger}A^c + A^{c\dagger}A^d + A^{c\dagger}A^c\| = 2\|A^d\| \|A^c\| + \|A^c\|^2$$

In particular, putting $A^d = I$ and $A^c = K$, where I is the identity and K is a compact operator, we have the equality

$$\|K + K^{\dagger} + K^{\dagger}K\| = 2\|K\| + \|K\|^2 \tag{2}$$

Therefore, it is enough to show that there is a compact operator K for which Eq. (2) does not hold. Let us consider an orthonormal basis $\{e_n\}$ in $\mathscr{L}_\sigma^2$ and the operator K defined as follows:

$$Ke_1 = \alpha e_2$$

$$Ke_n = 0 \qquad \text{for } n = 2, 3, \ldots$$

Without loss of generality, we can reduce our considerations to the operator

$$K = \begin{pmatrix} 0 & \alpha \\ 0 & 0 \end{pmatrix}$$

acting on two-dimensional Euclidean space, where $\alpha > 0$. Then

$$K^\dagger = \begin{pmatrix} 0 & 0 \\ \alpha & 0 \end{pmatrix}$$

and

$$K^\dagger K = \begin{pmatrix} 0 & \alpha^2 \\ 0 & 0 \end{pmatrix}$$

Using an estimation for the operator norm in finite dimensional Euclidean spaces (see Eq. (4.16), [39]) we get

$$\|K + K^\dagger + K^\dagger K\| = \left\| \begin{pmatrix} \alpha^2 & \alpha \\ \alpha & 0 \end{pmatrix} \right\| \leq \alpha + \alpha^2$$

On the other hand,

$$\|K\| = \left\| \begin{pmatrix} 0 & \alpha \\ 0 & 0 \end{pmatrix} \right\| = \alpha$$

which gives

$$2\|B\| + \|B\|^2 = 2\alpha + \alpha^2$$

Therefore, we have, instead of Eq. (2),

$$\alpha + \alpha^2 < 2\alpha + \alpha^2$$

which means that for such a K as above equality (2) does not hold.

The generalized basis $|\alpha\rangle$ for $\mathscr{H}$ leads to a natural basis for the algebra $\mathscr{A}$ constructed as follows. The diagonal part $\mathscr{A}^d$ can be expanded in terms of operators $|\alpha) \equiv |\alpha)^d \equiv |\alpha\rangle\langle\alpha|$:

$$A^d = \int_\sigma d\alpha \, A^d_\alpha |\alpha)$$

where A^d_α is an essentially bounded measurable function [26,34,35] on the spectrum σ. For the definition and properties of the W^*-algebra $\mathscr{L}^\infty_\sigma$ of essentially bounded functions on σ, see Dunford and Schwartz [40].

Physicists usually expect that the off-diagonal compact part can be "expanded" in terms of the operators $|\alpha\alpha') \equiv |\alpha : \alpha') \equiv |\alpha\rangle\langle\alpha'|$, that is, that A^c can be represented by a kernel $A^c_{\alpha\alpha'}$ on the product space $\sigma \times \sigma$:

$$A^c = \int_{\sigma \times \sigma} d\alpha \, d\alpha' \, A^c_{\alpha\alpha'} |\alpha\alpha')$$

where $A^c_{\alpha\alpha'}$ is the kernel function on the product space $\sigma \times \sigma$. In this case, we have

$$A = \int_\sigma d\alpha \, A_\alpha |\alpha) + \int_{\sigma \times \sigma} d\alpha \, d\alpha' \, A^c_{\alpha\alpha'} |\alpha\alpha') \tag{3}$$

Let us remark before concluding this section that

1. The basis $\{|\alpha), |\alpha\alpha')\}$ may be used for the expansion of unbounded operators which are not in the algebra $\mathscr{A}$. In such cases, we have more general structures like the partial involutive algebras [41,42]. For example, we can consider the algebra of operators with the property that the off-diagonal part has a measurable kernel, that is, the off-diagonal part is an integral operator [43]. In fact, $\mathscr{A}^d$ and the algebra of integral operators have no common elements. See Appendix A for the proof. However, the algebra of integral operators is too big and unless we restrict the admissible measurable kernels, we may have operators with very narrow domains. For example, if the kernel is a square integrable function, the off-diagonal part is a Hilbert–Schmidt operator. There are conditions which guarantee that an integral operator is compact [44]. However, there are compact operators that cannot be represented by a measurable kernel [43]. We shall not discuss further the possible choices for kernels because it is, in our opinion, a strictly model-dependent question.

2. A different choice for the complete system of commuting observables gives rise to a different generalized basis $|\alpha\rangle$ for the Hilbert space $\mathscr{H}$ and

therefore to a different algebra $\mathscr{A}$. However, both bases are related by a unitary operator which induces a Banach algebra isomorphism.

IV. STATES WITH DIAGONAL SINGULARITY

The states ρ of the system correspond to the normalized positive linear functionals $(\rho|$ over the algebra $\mathscr{A}$, that is, they satisfy the properties

1. Linearity

$$(\rho | z_1 A_1 + z_2 A_2) = z_1(\rho | A_1) + z_2(\rho | A_2)$$

2. Positivity

$$(\rho | A^\dagger A) \geq 0$$

3. Normalization

$$(\rho | I) = 1$$

Because $\mathscr{A}$ is an involutive Banach algebra, the states ρ are bounded with norm [18,19]

$$\|\rho\| = (\rho | I) = 1$$

In this paper, we adopt the following bra/ket notation in connection with the states/observables duality:

$$\rho(A) \equiv (\rho | A) \equiv \langle A \rangle_\rho$$

The bracket $(\rho | A)$ represents the expectation value of the observable A in the state ρ. The bras $(\rho|$ are to be understood as linear functionals over the algebra $\mathscr{A}$ of observables. The reader should not confuse the bras and kets $(\cdot|$ and $|\cdot)$ expressing the state/observables duality with the Dirac [24] bracket notation $\langle\cdot|$ and $|\cdot\rangle$ for the linear and antilinear functionals representing the wave functions [27]. Our bras $(\cdot|$ and kets $|\cdot)$ notation is just Dirac's notation in the Liouville space.

The states ρ form a convex subset of the Banach dual $\mathscr{A}^\times$ of the algebra $\mathscr{A}$:

$$\mathscr{A}^\times = \mathscr{A}^{d\times} \oplus (\mathscr{B}^\infty_{\mathscr{H}})^\times = \mathscr{A}^{d\times} \oplus \mathscr{B}^1_{\mathscr{H}}$$

Any state ρ has therefore a diagonal and an off-diagonal part, ρ^d and ρ^c, correspondingly

$$\rho = \rho^d + \rho^c$$

with

$$(\rho^d \mid A) = (\rho^d \mid A^d)$$
$$(\rho^c \mid A) = (\rho^c \mid A^c)$$

for any operator $A \in \mathscr{A}$.

As the dual space of $\mathscr{B}^{\infty}_{\mathscr{H}}$ is the space of trace class $\mathscr{B}^{1}_{\mathscr{H}}$, the off-diagonal states ρ correspond to the tracial functionals

$$(\rho^c \mid A) = \operatorname{tr} \hat{\rho}\hat{A}$$

where $\hat{\rho}$ is the density operator representing the state ρ. The norm of the dual space $\mathscr{A}^{\times}$ is therefore

$$\|\rho\| = \max\{(\rho^d \mid I), \operatorname{tr} \hat{\rho}\} \tag{4}$$

The proof is based on the theorem that gives the norm of the dual of the direct sum of Banach spaces. The theorem is stated and proved in Appendix B.

The dual space $(\mathscr{L}^{\infty}_{\sigma})^{\times}$ is isomorphic to the space of complex additive measures of bounded total variation and absolutely continuous with respect to a suitable [40] extension $(\tilde{\mathscr{B}}, \tilde{\mu})$ of the Lebesgue measure structure $(\mathscr{B}, \mu)$ associated with the spectrum σ. The diagonal states ρ^d are represented therefore in terms of these normalized positive measures or their Radon–Nikodym derivatives ρ^d_α.

The diagonal elements ρ^d_α represent the probability densities associated with the pure states $|\alpha\rangle$, whereas the off-diagonal elements $\rho^c_{\alpha\alpha'}$ describe correlations or coherent superpositions of the pure states $|\alpha\rangle$ and $|\alpha'\rangle$.

For observables A which admit kernel representation (3), the expectation value in the state ρ is written

$$(\rho \mid A) = \langle A \rangle_\rho = \int_\sigma d\alpha\, \rho^d_\alpha A^d_\alpha + \int_{\sigma \times \sigma} d\alpha\, d\alpha'\, \rho^{c*}_{\alpha\alpha'} A^c_{\alpha\alpha'} \tag{5}$$

In view of the duality relation (4), we define the action of the states ρ on the basis $|\alpha)$, $|\alpha\alpha')$ as follows

$$(\rho\,|\,\alpha) \equiv \rho_\alpha^{d*} = \rho_\alpha^d$$
$$(\rho\,|\,\alpha\alpha') \equiv \rho_{\alpha\alpha'}^{c*}$$

These formulas suggest the definition of a dual basis $(\alpha| \equiv (\alpha\!:\!\alpha|$, $(\alpha\alpha'| \equiv (\alpha\!:\!\alpha'|$ for the space of states

$$(\alpha\,|\,A) \equiv (\alpha\,|\,A^d) \equiv A_\alpha^d$$
$$(\alpha\alpha'\,|\,A) \equiv (\alpha\alpha'\,|\,A^c) \equiv A_{\alpha\alpha'}^c$$

The duality relation (4) suggests the expansion of any state ρ

$$(\rho| = \int_\sigma d\alpha\, \rho_\alpha^d(\alpha| + \int_{\sigma\times\sigma} d\alpha\, d\alpha'\, \rho_{\alpha\alpha'}^{c*}(\alpha\alpha'|$$

Lemma. *The bases* $\{(\alpha|, (\alpha\alpha'|\}$, $\{|\beta), |\beta\beta')\}$ *form a biorthonormal dual pair, that is, the following properties hold:*

$$(\alpha\,|\,\beta) = \delta(\alpha - \beta) \tag{a}$$
$$(\alpha\,|\,\beta\beta') = 0 \tag{b}$$
$$(\alpha\alpha'\,|\,\beta) = 0 \tag{c}$$
$$(\alpha\alpha'\,|\,\beta\beta') = \delta(\alpha - \beta)\delta(\alpha' - \beta') \tag{d}$$

Proof. The linear functionals $(\alpha|$ are positive and therefore bounded [18]. As formula (4) shows, $|\beta)$ picks the function representing the diagonal part of any state ρ. The diagonal part of the state $(\rho| = (\alpha|$ is represented by the measure concentrated on α with the generalized Radon–Nikodym derivative $\delta_\alpha(\beta) = \delta(\alpha - \beta)$ [38]. Such measures in fact live in the dual space $(\mathscr{L}_\sigma^\infty)^\times$ [39, p. 86, Problem 8]. Therefore, $(\alpha\,|\,\beta) = \delta(\alpha - \beta)$.

As $(\alpha\,|\,B^c) = 0$ for all off-diagonal operators B^c, (b) expresses just the natural extension of the functional $(\alpha|$ to the generalized basis $|\beta\beta') = |\beta\rangle\langle\beta'|$. Indeed, the generalized operator $|\beta\beta') \equiv |\beta\rangle\langle\beta'|$ is the weak limit of the sequence $|\beta_\nu\rangle\langle\beta'_\nu|$ in $\Phi\otimes\Phi$ such that

$$|\beta\rangle = \mathrm{w}-\lim\,|\beta_\nu\rangle$$
$$|\beta'\rangle = \mathrm{w}-\lim\,|\beta'_\nu\rangle$$

We have

$$(\alpha \mid \beta\beta') = \mathrm{w} - \lim(\alpha \mid \beta_\nu \beta'_\nu) = 0$$

Formulas (c) and (d) are similarly the natural extensions of the functionals $(\alpha\alpha' \mid$ to the basis $\mid \beta)$, $\mid \beta\beta')$

$$\begin{aligned}(\alpha\alpha' \mid \beta) &= \mathrm{w} - \lim(\alpha\alpha' \mid \beta_\nu) = 0\\ (\alpha\alpha' \mid \beta\beta') &= \mathrm{w} - \lim(\alpha\alpha' \mid \beta_\nu \beta'_\nu)\\ &= \mathrm{w} - \lim\langle \alpha \mid \beta_\nu\rangle\langle \beta'_\nu \mid \alpha'\rangle\\ &= \delta(\alpha - \beta)\delta(\alpha' - \beta')\end{aligned}$$

The normalized positive linear functionals over the algebra $\mathscr{A}$ include not only the conventional quantum states but also the states with diagonal singularity which cannot be discussed in terms of the Hilbert space formulation. In fact, the reason for the algebraic extension [20–23] of quantum theory is the possibility to discuss states of physical interest which cannot be formulated in terms of the conventional Hilbert space language. For example, the dual $\mathscr{B}_{\mathscr{H}}^{\times}$ of the von Neumann algebra $\mathscr{B}_{\mathscr{H}}$ of bounded operators on the Hilbert space $\mathscr{H}$ includes not only the density operators that live in the trace class space $\mathscr{B}_{\mathscr{H}}$ but also states that vanish on the compact operators $\mathscr{B}_{\mathscr{H}}^{\infty}$. In fact, we have the isomorphism

$$\mathscr{B}_{\mathscr{H}}^{\times} \cong (\mathscr{B}_{\mathscr{H}}^{\infty})^{\perp} \oplus \mathscr{B}_{\mathscr{H}}^{1}$$

This result was proved by Dixmier [45, p. 51]. The dual of our algebra has a similar structure with $\mathscr{B}_{\mathscr{H}}^{\times}$, although it contains a smaller number of states:

$$\mathscr{A}^{\times} = (\mathscr{A}^{d})^{\times} \oplus \mathscr{B}_{\mathscr{H}}^{1}$$

Nevertheless, $\mathscr{A}^{\times}$ is rich enough to contain interesting and physically relevant states which we characterize below. First let us add, however, a few remarks about kernels of tracial operators.

Observe that a tracial operator on $\mathscr{L}_{\sigma}^{2}$ is also a Hilbert–Schmidt operator; thus it is an integral operator with a kernel from $\mathscr{L}_{\sigma\times\sigma}^{2}$. Generally, a kernel $k_{\alpha\alpha'}$ of an integral operator K is a function defined almost everywhere with respect to the planar measure. This means, in particular, that on the diagonal, the values $k_{\alpha\alpha}$ may have no meaning or can be defined arbitrarily without changing the operator. However, as shown in Appendix C, if K is additionally tracial, any orthonormal basis in $\mathscr{L}^{2}$ determines a function $\tilde{k}_{\alpha\alpha'} = k_{\alpha\alpha'}$ almost everywhere such that $\tilde{k}_{\alpha\alpha'}$ is defined a.e. and

$$\mathrm{tr}(K) = \int_\sigma d\alpha \, \tilde{k}_{\alpha\alpha}$$

Moreover, the values of $\tilde{k}_{\alpha\alpha'}$ on the diagonal do not depend almost everywhere on the choice of the basis in $\mathscr{L}^2$. We call such $\tilde{k}_{\alpha\alpha'}$ a *natural kernel* of the tracial operator K.

Let us assume that any tracial operator ρ^c which appears in the following has a natural integral kernel $\rho_{\alpha\alpha'}$. The states of the algebra are now characterized as follows:

1. The positive linear functionals $\rho \in \mathscr{A}^\times$ should fulfill the condition

$$\|\rho^c\| \leq \|\rho^d\|$$

so that

$$\|\rho\| = (\rho \,|\, I) = (\rho^d \,|\, I) = \max\{\|\rho^d\|, \|\rho^c\|\}$$

in view of Eq. (4) and the theorem in Appendix B. Note also that the algebra $\mathscr{A}^d$ is isomorphic to $\mathscr{L}^\infty_\sigma$. Thus the diagonal part ρ^c of ρ is represented by an integrable function ρ^d_α and its norm is the $\mathscr{L}^1$-norm. Since the norm of ρ^d is its trace, the inequality above, which guarantees the positivity of ρ can be written as

$$\int_\sigma d\alpha \, \rho^c_{\alpha\alpha} \leq \int_\sigma d\alpha \, \rho^d_\alpha$$

2. The pure states are represented by the wavefunctions ψ in $\mathscr{H}$ such that for any pure state ρ_ψ,

$$(\rho_\psi \,|\, A) = \langle \psi \,|\, A\psi \rangle$$

This condition implies that

$$\rho^d_\alpha = |\psi_\alpha|^2 \tag{6}$$

$$\rho^c_{\alpha\alpha'} = \psi_\alpha \psi^*_{\alpha'} \tag{7}$$

Although the corresponding projector $|\psi\rangle\langle\psi|$ is a tracial state, that is, $\rho^d = 0$, we interpret Eqs. (6) and (7) as an embedding of the state $|\psi\rangle\langle\psi|$ in the space $\mathscr{A}^\times$ and we do not distinguish between the states ρ satisfying (6) and (7) and the one-dimensional projections $\mathscr{A}^\times$.

3. The states ρ in $\mathscr{A}^\times$ with

$$\rho^d_\alpha = \rho^c_{\alpha\alpha} \tag{8}$$

represent the density operators $\hat{\rho}$ in $\mathscr{B}^1_{\mathscr{H}}$ with

$$\rho_{\alpha\alpha'} = \langle \alpha | \hat{\rho} | \alpha' \rangle \tag{9}$$

which is the natural kernel of the tracial operator $\hat{\rho}$ discussed above and in Appendix C.

Proof

$$\begin{aligned}
(\rho | A) &= \int_\sigma d\alpha\, \rho^d_\alpha A^d_\alpha + \int_{\sigma^2} d\alpha\, d\alpha'\, \rho^c_{\alpha\alpha'} A^c_{\alpha\alpha'} \\
&= \int_\sigma d\alpha\, \rho_{\alpha\alpha'} (A^d_\alpha \delta(\alpha - \alpha') + A^c_{\alpha\alpha'}) \\
&= \int_{\sigma^2} d\alpha\, d\alpha' \langle \alpha | \hat{\rho} | \alpha' \rangle \langle \alpha | A | \alpha' \rangle \\
&= \operatorname{tr} \hat{\rho} A
\end{aligned}$$

The dual pair $|\alpha)$, $|\alpha\alpha')$, $(\beta|$, $(\beta\beta'|$ of Lemma 1 allows us to give meaning to the continuous biorthonormal basis in Liouville space. For example, the one-particle free Hamiltonian

$$H = \frac{1}{2m} \hat{P}^2$$

where $\hat{P}$ is the momentum operator has the spectral decomposition

$$H = \int_{-\infty}^{\infty} dp\, \frac{p^2}{2m} |p\rangle\langle p|$$

where $P|p\rangle = p|p\rangle$.

The basis for the algebra of observables is

$$|p) \equiv |p\rangle\langle p|$$

$$|pp') \equiv |p\rangle\langle p'|$$

The dual basis $(k|$, $(kk'|$ is defined by

$$(k|p) = \delta(k-p)$$
$$(k|pp') = 0$$
$$(kk'|pp') = \delta(k-k')\delta(p-p')$$
$$(kk'|p) = 0$$

The diagonal state $(k|$ has well-defined energy and momentum and it is normalized:

$$(k|\hat{H}) = (k|\int_{-\infty}^{\infty} dp\, \frac{p^2}{2m}\, |p) = \int_{-\infty}^{\infty} dp\, \frac{p^2}{2m}\, \delta(k-p) = \frac{k^2}{2m}$$

$$(k|\hat{P}) = (k|\int dp\, p\, |p) = \int dp\, p\delta(k-p) = k$$

$$\|(k|\| = (k|I) = (k|\int dp\, |p) = 1$$

The microcanonical and canonical equilibrium states for the evolution generated by H_0 have a natural description within our formalism:

$$(\rho_{\text{micro}}| = \frac{1}{\mathcal{N}_{\text{micro}}} \int d\alpha\, \delta(\omega_\alpha - E)(\alpha|$$

$$(\rho_{\text{can}}| = \frac{1}{\mathcal{N}_{\text{can}}} \int d\alpha\, e^{-\beta\omega_\alpha}(\alpha|$$

where $\mathcal{N}_{\text{micro}}$, $\mathcal{N}_{\text{can}}$ are the normalization constants determined by the condition

$$(\rho_{\text{micro}}|I) = (\rho_{\text{can}}|I) = 1$$

We see that ρ_{micro} and ρ_{can} cannot be described in terms of the conventional formulation of quantum physics because they are purely diagonal states.

V. OPERATIONS ON STATES AND OBSERVABLES—SUPEROPERATORS

Respecting the tradition of statistical physics, we work on states, although some calculations may be performed on observables through the (state | observable) duality, when convenient. The action of a linear operator U on the state ρ satisfies the duality relation

$$(U\rho \mid A) = (\rho \mid VA)$$

for all observables A. The operator V acting on the observables A is the predual of U, that is, $V = U^{+}$ or equivalently U is the dual of V, that is, $U = V^{\times}$. This is just the duality between the Schrödinger and the Heisenberg pictures. We discuss below several examples of linear operators on states, known as linear operations or superoperators.

1. The projector P_d on the diagonal part of the states is defined as follows:

$$(P_d \rho \mid = \int d\alpha(\rho \mid \alpha)(\alpha \mid \equiv (\rho^d \mid$$

or

$$(P_d \rho \mid A) = \int d\alpha(\rho \mid \alpha)(\alpha \mid A) = (\rho \mid P_d A)$$

for any observable A. To avoid heavy notation, we used the same symbol for the projection superoperators on states and on observables because the action is formally the same, although we understand that they act on different spaces. This is just the generalization of the self-adjointness of the projections acting on the Hilbert–Schmidt space.

2. The projector P_c on the off-diagonal part of the states is

$$(P_c \rho \mid = \int_{\sigma \times \sigma} d\alpha \, d\alpha'(\rho \mid \alpha\alpha')(\alpha\alpha' \mid \equiv (\rho^c \mid$$

or

$$(P_c \rho \mid A) = \int_{\sigma \times \sigma} d\alpha \, d\alpha'(\rho \mid \alpha\alpha')(\alpha\alpha' \mid A) = (\rho^c \mid A) = (\rho \mid P_c A)$$

for any observable A.

3. The Liouville operator L on the states ρ is defined by the relation

$$(L\rho \mid A) \equiv (\rho \mid [H, A]) \equiv (\rho \mid LA)$$

for all observables A.

4. Scalar operators zI, z complex, on the states ρ satisfy the relation

$$(z\rho \mid A) = (\rho \mid z^* A) = z^*(\rho \mid A)$$

5. The eigenvalue problem for superoperators is formulated as follows. The eigenvectors of the operator U satisfy the equation

$$Uf_\nu = z_\nu f_\nu$$

understood as

$$(Uf_\nu \mid A) = (z_\nu f_\nu \mid A) = z_\nu^*(f_\nu \mid A)$$

for all observables A. The left eigenvectors of U are the eigenoperators of the predual operator V, that is,

$$VF_\nu = z_\nu F_\nu$$

to be understood as

$$(\rho \mid VF_\nu) = (\rho \mid z_\nu F_\nu)$$

or

$$(U\rho \mid F_\nu) = z_\nu(\rho \mid F_\nu)$$

for all states ρ. In case the eigenvectors of U form a biorthonormal pair (|), the operator U has the spectral decomposition

$$\sum_\nu z_\nu \mid F_\nu)(f_\nu \mid$$

to be understood as

$$(U\rho \mid A) = (\rho \mid VA) = \sum_\nu z_\nu(\rho \mid F_\nu)(f_\nu \mid A)$$

for all states ρ and observables A.

6. Any operator U can be represented in terms of the biorthonormal dual pair $(\alpha|$, $(\alpha\alpha'|$, $|\beta)$ and $|\beta\beta')$ as follows:

$$(U\rho\,|\,A) = (\rho\,|\,VA) = \int_{\sigma 2} d\alpha\, d\beta (\rho\,|\,\alpha)(\alpha\,|\,V\,|\,\beta)(\beta\,|\,A)$$
$$+ \int_{\sigma 3} d\alpha\, d\beta\, d\beta' (\rho\,|\,\alpha)(\alpha\,|\,V\,|\,\beta\beta')(\beta\beta'\,|\,A)$$
$$+ \int_{\sigma 3} d\alpha\, d\alpha'\, d\beta (\rho\,|\,\alpha\alpha')(\alpha\alpha'\,|\,V\,|\,\beta)(\beta\,|\,A)$$
$$+ \int_{\sigma 4} d\alpha\, d\alpha'\, d\beta\, d\beta' (\rho\,|\,\alpha\alpha')(\alpha\alpha'\,|\,V\,|\,\beta\beta')(\beta\beta'\,|\,A)$$

VI. TIME EVOLUTION

In the Heisenberg picture, the observables A evolve according to the group of automorphisms of the algebra $\mathscr{A}$ induced by the Schrödinger equation:

$$V_t A \equiv e^{iHt} A e^{-iHt} = e^{iLt} A$$

The Heisenberg evolution equation is

$$\partial_t A = i[H, A] = iLA$$

where $LA = [H, A] = HA - AH$ is the Liouville operator.

In the Schrödinger picture, the states ρ evolve according to the dual evolution

$$U_t = V_t^{\times}$$

or

$$(U_t \rho\,|\,A) = (\rho\,|\,V_t A)$$

The Liouville–von Neumann evolution equation for states is

$$(\partial_t \rho\,|\,A) = (\rho\,|\,i[H, A]) = (-iL\rho\,|\,A)$$

If the state ρ is tracial, the evolution is implemented by the evolution on the Hilbert space and L is the Liouville-von Neumann operator of the conventional formulation of quantum statistical physics [22,23,25].

The Liouville operator may lead out of the algebra $\mathscr{A}$. Indeed, for Hamiltonians of the form

$$H = H_0 + V$$

with

$$H_0 = \int_\sigma d\alpha\, \epsilon_\alpha |\alpha\rangle\langle\alpha|$$

$$V = \int_{\sigma\times\sigma} d\alpha\, d\alpha'\, V_{\alpha\alpha'} |\alpha\rangle\langle\alpha'|$$

the Liouville operator acts on the algebra as

$$\begin{aligned} LA = [H, A] &= [H_0, A^c] + [V, A^d] + [V, A^c] \\ &= \int_{\sigma^2} d\alpha\, d\alpha' (\epsilon_\alpha - \epsilon_{\alpha'}) A^c_{\alpha\alpha'} |\alpha\alpha') \\ &\quad + \int_{\sigma^3} d\alpha\, d\alpha'\, d\alpha''\, A^c_{\alpha\alpha'} [V_{\alpha''\alpha} |\alpha''\alpha') - V_{\alpha'\alpha''} |\alpha\alpha'')] \\ &\quad + \int_{\sigma^2} d\alpha\, d\alpha'\, A^d_\alpha [V_{\alpha'\alpha} |\alpha'\alpha) - V_{\alpha\alpha'} |\alpha\alpha')] \end{aligned}$$

or

$$\begin{aligned} L = &\int_{\sigma^2} d\alpha\, d\alpha' (\epsilon_\alpha - \epsilon_{\alpha'}) |\alpha\alpha')(\alpha\alpha'| \\ &+ \int_{\sigma^3} d\alpha\, d\alpha'\, d\alpha'' [V_{\alpha''\alpha} |\alpha''\alpha')(\alpha\alpha'| - V_{\alpha'\alpha''} |\alpha\alpha'')(\alpha\alpha'|] \\ &+ \int_{\sigma^2} d\alpha\, d\alpha' [V_{\alpha'\alpha} |\alpha'\alpha)(\alpha| - V_{\alpha\alpha'} |\alpha\alpha')(\alpha|]. \end{aligned}$$

It is now clear that if $V_{\alpha'\alpha}$ contain singularities, as in the case of the van Hove condition discussed in Section VIII, L leads out of the algebra and we have to extend the algebra accordingly.

The solution of the eigenvalue problem may lead to a spectral decomposition of the Liouville operator,

$$L = \sum_\nu z_\nu |F_\nu)(f_\nu|$$

in the sense

$$(L\rho| = \sum_{\nu} z_{\nu}(\rho | F_{\nu})(f_{\nu}|$$

with f_{ν}, f_{ν} a biorthonormal dual pair:

$$(f_{\nu} | F_{\nu'}) = \delta_{\nu\nu'}$$

The evolution of states can be decomposed in terms of the eigen-components

$$(e^{-iLt}\rho| = \sum_{\nu} e^{iz_{\nu}t}(\rho | F_{\nu})(f_{\nu}|$$

Our extended formulation allows z_{ν} to be complex. We expect that the imaginary part represents decay rates or diffusion coefficients. In the conventional Hilbert–Schmidt formulation of course z_{ν} is real and $f_{\nu} \equiv F_{\nu}$ because L is a self-adjoint operator on the Hilbert–Schmidt space.

VII. NONEQUILIBRIUM STATISTICAL PHYSICS AND SPECTRAL DECOMPOSITION OF THE LIOUVILLE OPERATOR

The importance of states with diagonal singularity for the approach to equilibrium has been emphasized repeatedly by Prigogine and co-workers [1–4]. These states correspond to a delocalized description because they are not reducible to pure states. Our formulation allows us to construct generalized spectral decompositions of the Liouville operator L for unstable systems with continuous spectrum. For such systems, the conventional spectral theory formulated in the Hilbert–Schmidt space has the following difficulties:

1. The eigenvalue problem is not computable [46,47]. For example, the Poincaré type divergences also appear in quantum mechanics [5,6,29,48–50].
2. The spectral resolutions as potentially existing mathematical entities do not reflect the dynamical processes involved, because the evolution is represented as a shift in the spectral representation space.

The natural spectral representations should include the lifetimes of the unstable states or the diffusion parameters in the spectrum. Such spectral decompositions can be constructed for unstable systems with a systematic method [16]. The key idea is the construction of an intermediate operator Θ which is intertwined with the Liouville operator L.

$$L\Omega = \Omega\Theta \quad \text{or} \quad L = \Omega\Theta\Omega^{-1} \tag{10}$$

The intermediate operator Θ generates irreversible probabilistic processes and has the same spectrum with L.

The intertwining relation (10) was obtained by Prigogine et al. [9]. Recently, Petrosky and Prigogine [15] pointed out that the intertwining relation can be used for the construction of the spectral decomposition of the Liouville operator. The method as reformulated by us [16] may also be considered as a generalization of the intertwining wave-operator method of scattering theory and of the partitioning technique of matrices.

In fact, the intermediate operator Θ is decomposable with respect to the correlation projectors $P_d, P_1, P_2, \ldots$, defined [16] as follows:

1. P_d is the projection onto the diagonal part of the states. The complement of P_d is the projection P_c onto the off-diagonal part. It is straightforward to see that

$$P_c^2 = P_c$$

$$P_0 + P_c = I$$

$$P_0 P_c = P_c P_0 = 0$$

2. P_n, $n = 1, 2, \ldots$ is the projection onto the space of states with degree of correlation n. This space consists of the states ρ with n-degree of correlation, that is, ρ satisfies the condition

$$(P_d L_1^{n'} \rho | = 0 \qquad \text{for } n' < n$$

$$(P_d L_1^{n'} \rho | \neq 0 \qquad \text{for } n' = n$$

where L_1, $L_1\rho = [V, \rho]$, is the interaction part of the Liouville operator $L = L_0 + L_1$. The degree of correlation n is the minimal positive integer power of the coupling operator L_1 that relates the diagonal with the off-diagonal operators.

The projectors P_n, $n = 1, 2, \ldots$ provide a further decomposition of P_c:

$$P_c = P_1 + P_2 + \cdots$$

The projectors P_d, P_1, P_2, ... form a discrete resolution of identity permuting with L_0:

$$\begin{aligned} P_d + P_1 + P_2 + \cdots &= I \\ P_d P_n &= 0 \\ P_n P_{n'} &= \delta_{nn'} P_n \\ L_0 P_d &= P_d L_0 \\ L_0 P_n &= P_n L_0 \end{aligned}$$

To write less formulas, we identify P_d with P_0. The spectral decomposition of L is obtained from the spectral decomposition of the intermediate operator Θ through the intertwining operator Ω. The intermediate operator Θ, as well as the similarity operator Ω, are obtained [16] from the auxiliary "creation" and "destruction" operators C_n and D_n

$$\Theta \equiv \sum_n (P_n L P_n + P_n L C_n P_n) = \sum_n \Theta_n \tag{11}$$

$$\Omega \equiv \sum_n (P_n + C_n) \tag{12a}$$

$$\Omega^{-1} \equiv \sum_n (P_n + D_n C_n)^{-1}(P_n + D_n) \tag{12b}$$

For any $n = 0, 1, 2, \ldots$, the operator C_n "creates" the complement of P_n acting on P_n, while the operator D_n "destroys" the complement of P_n, that is,

$$\begin{aligned} C_n &= Q_n C_n P_n \\ D_n &= P_n D_n Q_n \end{aligned}$$

with Q_n the ortho-complement of P_n: $Q_n = I - P_n$.

The creation and destruction operators C_n and D_n are obtained iteratively as solutions of the following nonlinear equations:

$$[L_0, P_m C_n] = (P_m C_n - P_m) L_1 (P_n + C_n) \tag{13a}$$

$$[L_0, D_n P_m] = (P_n + D_n) L_1 (P_m - D_n P_m) \tag{13b}$$

We remind the reader that the operator equation

$$[L_0, X] = Y$$

has the forward (+) and backward (−) solution [39]:

$$X^{\pm} = i \int_0^{\pm\infty} dt\, e^{-iL_0t} Y e^{+iL_0t}$$

We need therefore an additional rule that will serve as a selection principle or boundary condition. It is well known that for dynamical systems with continuous spectrum, the dynamical equations alone do not determine the solution. Additional boundary conditions have to be introduced based on physical considerations like the incoming/outgoing scattering conditions [26,51] or Sommerfeld's radiation conditions [52].

In our case, the additional rule is the time-ordering rule introduced by George [10], which is based on the standard convention that the components $P_m C_n$ and $D_n P_m$ signify transitions from n to m and from m to n correlation states, correspondingly. The solutions are chosen by ordering in time the processes represented by the components $P_m C_n$ and $D_n P_m$. The natural time ordering corresponds to an increase of correlations in the future. Therefore, for transitions corresponding to an increase of correlations, the forward (+) propagator is selected, while for transitions corresponding to decrease of correlations, the backward (−) propagator is selected. With this additional rule, Eqs. (9) are written

$$P_m C_n = i \int_0^{\pm\infty} dt\, e^{-iL_0t}(P_m C_n - P_m)L_1(P_n + C_n)e^{iL_0t} \qquad \begin{matrix} m > n \\ m < n \end{matrix} \tag{14a}$$

$$D_n P_m = i \int_0^{\pm\infty} dt\, e^{-iL_0t}(P_n + D_n)L_1(P_m - D_n P_m)e^{iL_0t} \qquad \begin{matrix} n > m \\ n < m \end{matrix} \tag{14b}$$

Using the dual pair $\{(\alpha|, (\alpha, \alpha'|; |\alpha), |\alpha\alpha')\}$, we obtain equations for the matrix elements of the operators C_n and D_n, which are solved iteratively. In the case of resonances, the time-ordering rule avoids the divergences because it serves as an analytic extension of the operators and gives rise to time-asymmetric spectral decompositions [16]. The equations (14) are a nonlinear generalization of the Lippmann–Schwinger equations for the Möller wave operators of scattering and may provide nonunitary intertwining operators even if the scattering asymptotic condition [26,51] between L_0 and L fails.

The algorithm for the construction of spectral decomposition of the operator L is the following:

1. Construct the creation and destruction operators C_n and D_n iteratively as solutions of the equations starting with $C_n^{[0]} = D_n^{[0]} = 0$.
2. Construct the intermediate operator Θ from Eq. (11) and find the spectral decomposition of Θ by solving the eigenvalue problem of the parts Θ_n in each P_n subspace.
3. Obtain the spectral decomposition of L from the spectral decomposition of Θ using the similarity Ω, Eqs. (10) and (12).

The resulting spectral decompositions are meaningless in the Hilbert–Schmidt space, but they acquire meaning in suitable dual pairs of states and observables. We consider these dual pairs as the natural generalization at the level of mixtures of the rigged Hilbert space extensions [15,16,27] which apply only at the level of wave functions. When the algebra of observables $\mathscr{A}$ is included in the dual $\mathscr{A}^\times$, we have the rigged Liouville space structure $\mathscr{A} \subset \mathscr{H} \otimes \mathscr{H}^\times \subset \mathscr{A}^\times$ [53].

VIII. INTRINSIC IRREVERSIBILITY OF SYSTEMS WITH A NONVANISHING COLLISION OPERATOR

The component Θ_0 of the intermediate operator Θ in the first approximation is the so-called $\lambda^2 t$-asymptotic approximation to the master equation [1–4,7–12,28]. Indeed from Eqs. (14a) and (11) we have

$$C_0^{[1]} = -i\lambda \int_0^{+\infty} dt\, e^{-iL_0 t} P_c L_1 P_0 = \lambda \frac{1}{i0 - L_0} P_c L_1 P_0$$

$$\Theta_0^{[2]} = P_0 L P_0 + P_0 L C_0^{[1]}$$

$$= -i\lambda^2 \int_0^{-\infty} dt\, P_0 L_1 P_c e^{-iL_0 t} P_c L_1 P_0$$

or

$$\Theta_0^{[2]} = \lambda^2 P_0 L_1 P_c \frac{1}{i0 - L_0} P_c L_1 P_0$$

The last expression is identical with the form of the asymptotic collision operator $\Psi^{[2]}(+i0)$ [4,9–12,28].

The operator $\Theta_0^{[2]}$ is the generator of the Pauli master equation. Indeed, for dynamical systems with a nonvanishing collision operator, the equation

$$\partial_t \rho = -i\Theta_0^{[2]} \rho$$

is the Pauli master equation [54]:

$$\partial_t \rho_\alpha = \int_\sigma d\beta [R_{\alpha\beta}\rho_\beta - R_{\beta\alpha}\rho_\alpha] \tag{15}$$

where $\rho_\alpha = (\rho \mid \alpha)$ and the transition rates $R_{\alpha\beta}$ are given by the formula

$$R_{\alpha\beta} = 2\pi\delta(\omega_\alpha - \omega_\beta) \mid V^d_{\alpha\beta} \mid^2$$

The operator Θ_0 allows, therefore, a classification of dynamical systems. If Θ_0 is a dissipative operator in the sense of Phillips [54], that is,

$$\text{Im}\ \Theta_0 \leq 0 \tag{16}$$

we have the irreversible representation of the evolution and the system is intrinsically irreversible. The dissipativity condition (16) was introduced by Prigogine and co-workers [4,9–12,28].

Dynamical systems that satisfy the dissipativity condition are the large Poincaré systems with persistent interactions, that is, systems for which the asymptotic condition of conventional scattering theory fails. Such systems include the matter–field interactions, electrons in lattices, interacting fields, and many-body systems.

IX. CONCLUSION

Intrinsic irreversibility cannot be formulated within the conventional quantum theory, as has been emphasized by Prigogine and co-workers [9,14,28–31], but requires an extension which we have formulated in terms of a suitable dual pair of observables and states with diagonal singularity. The extended quantum theory includes the states with diagonal singularity which express the nonlocality [31] and describe the approach to equilibrium for systems with a nonvanishing collision operator. For such unstable systems, irreversibility emerges naturally without any extraneous assumption like coarse graining or approximations.

The second-order contribution to the collision operator is Pauli's master equation, which we derived for transitions between states forming a continuum without any report to thermodynamic limits. This problem was one of the reasons why the derivation of the master equation was not rigorous [17]. The Pauli master equation (12) for a continuum of states is a generalization of the original equation [53], which described transitions between discrete states and defines a quantum Markov semigroup in the sense of Kossakowski [55] and Lindbland [56]. The quantum Markov semigroup which is the solution of Eq. (12) is not of the Lindbland–Kossakowski type because it acts on more general states than the tracial states. The master equations (12) define, therefore, a generalization of the known quantum Markov semigroups.

The Pauli master equation is mentioned also by Piron [57] as the prototype for irreversible quantum evolutions. However, Piron did not raise the problem of irreversibility, that is, how the irreversible master equation arises from the reversible Schrödinger evolution. We have shown [58] that the clarification of the problem of irreversibility in terms of systems with diagonal singularity leads to an extension of the conventional logic of Hilbert spaces, to a logic that allows only finite combinations of propositions-questions. The restriction to Hilbert space logic is, however, not a physical necessity but only a mathematical postulate [59] in order to force the poset of propositions to be essentially the lattice of Hilbert space projections. In fact, Varadarajan [60] and others have pointed out that "the result of von Neumann, Mackey Gleason and their successors has pretty much closed the door for the discovery of any non-trivial situation violating the canonical interpretations." The fact that only finite combinations of propositions are allowed for systems with diagonal singularity is a manifestation of the underlying instability of dynamics and shows clearly another aspect of our extended theory.

Classical systems with diagonal singularity due to Poincaré resonances lead also to an extension of Hamiltonian dynamics [9,28,61] and of the conventional Boole logic. However, these questions will be addressed in forthcoming publications [62,63]. We would only like to point out that chaotic systems also admit probabilistic extensions of the trajectory formulation through the generalized spectral decomposition of the evolution operators [16,64–67].

APPENDIX A. $\mathscr{A}^d$ AND INTEGRAL OPERATORS HAVE NO COMMON ELEMENTS

Proposition. *The multiplication operators on $\mathscr{L}^2_\sigma$, where σ is a subset of $\mathbb{R}$, cannot be bounded integral operators on $\mathscr{L}^2_\sigma$.*

Proof. The multiplication operators A^d are represented by the essentially bounded functions $A^d(\alpha) \in \mathscr{L}^\infty_\sigma$:

$$A^d\psi(\alpha) = A^d(\alpha)\psi(\alpha)$$

for every $\psi(\alpha) \in \mathscr{L}^2_\sigma$.

The integral operators A^c are represented by measurable kernels $A^c(\alpha, \alpha')$:

$$A^c\psi(\alpha) = \int_\sigma d\alpha'\, A^c(\alpha, \alpha')\psi(\alpha')$$

A^d is always different from A^c because of the following.

Lemma. *For any finite interval* $[a, b] \in \sigma$, *we have:*

$$\|A^d - A^c\| \geq \frac{\|A^d(\cdot)\|_2}{\sqrt{b-a}}$$

where $\|\cdot\|$ *is the operator norm and*

$$\|A^d\|_2 = \left\{\int_a^b d\alpha \,|\, A^d(\alpha)\,|^2\right\}^{1/2}$$

Proof. Because of the boundness of A^c, each function $\psi \in \mathscr{L}^2_\sigma$ belongs to its domain. In particular, taking $\psi = \mathbf{1}_{[a,b]}$, we see that $A^c(\alpha, \cdot)$ is integrable for almost all α on each finite interval. Therefore, taking the Fourier transform and applying the Riemann Lebesgue Lemma we have

$$\int_a^b d\alpha' \, K(\alpha, \alpha')e^{in\alpha'} \to 0 \qquad \text{as } n \to \infty$$

for almost all α.

Consider now the sequence of functions

$$e_n(\alpha) = \frac{1}{\sqrt{b-a}} e^{in\alpha}\mathbf{1}_{[a,b]} \qquad n = 0, 1, \ldots$$

Using the fact that $\|e_n\|_{L^2} = 1$ and applying Fatou Lemma we have

$$\begin{aligned}
\|A^d - A^c\|^2 &\geq \liminf_{n\to\infty} \|(A^d - A^c)e_n\|_2^2 \\
&= \liminf_{n\to\infty} \int_a^b d\alpha \,|\, A^d(\alpha)e_n(\alpha) - A^c e_n(\alpha)\,|^2 \\
&= \int_a^b d\alpha \, \frac{|\, A^d(\alpha)\,|^2}{b-a} = \frac{\|A^d\|_2^2}{b-a}
\end{aligned}$$

Here, if f_n, g_n are two complex sequences such that $f_n \to 0$ $(n \to \infty)$ and $|\, g_n\,| = |\, g\,|$, then $|\, f_n - g_n\,| \to |\, g\,|$ $(n \to \infty)$.

APPENDIX B. THE NORM OF THE DUAL SPACE $\mathscr{A}^\times$

Lemma. *Let* $\mathscr{A}_1$, $\mathscr{A}_2$ *be two Banach spaces with norms* $\|\cdot\|_1$, $\|\cdot\|_2$. *If the norm of the direct sum* $\mathscr{A}_1 \oplus \mathscr{A}_2$ *is* $\|\cdot\|_1 + \|\cdot\|_2$, *the norm* $\|\cdot\|^\times$ *of the dual space* $\mathscr{A}_1^\times \oplus \mathscr{A}_2^\times$ *is* $\max\{\|\cdot\|_1^\times, \|\cdot\|_2^\times\}$.

Proof. For $A \in \mathcal{A}$, $A = A_1 + A_2$, and any $\rho \in \mathcal{A}^\times$, $\rho = \rho_1 + \rho_2$, we have

$$
\begin{aligned}
|(\rho \,|\, A)| &= |(\rho_1 \,|\, A_1) + (\rho_2 \,|\, A_2)| \\
&\leq \|A_1\|_1 \, \|\rho_1\|_1^\times + \|A_2\|_2 \, \|\rho_2\|_2^\times \\
&\leq \max\{\|\rho_1\|_1^\times, \|\rho_2\|_2^\times\}(\|A_1\|_1 + \|A_2\|_2) \\
&= \max\{\|\rho_1\|_1^\times, \|\rho_2\|_2^\times\}\|A\|
\end{aligned}
$$

Therefore

$$
\|\rho\|^\times = \sup_{\|A\| \leq 1} |(\rho \,|\, A)| \leq \max\{\|\rho_1\|_1^\times, \|\rho_2\|_2^\times\}
$$

To end the proof, it is enough to show that for each $\epsilon > 0$ there is a $B \in \mathcal{A}$, with $\|B\| \leq 1$ such that

$$
(\rho \,|\, B) \geq \max\{\|\rho_1\|_1^\times, \|\rho_2\|_2^\times\} - \epsilon
$$

Indeed, suppose that

$$
\|\rho_1\|_1^\times \geq \|\rho_2\|_2^\times
$$

and let $\epsilon > 0$ be given. Since $\|\rho_1\|^\times = \sup_{\|A_1\| \leq 1} |(\rho_1 \,|\, A_1)|$, we can find an element $B_1 \in \mathcal{A}_1$, with $\|B_1\| \leq 1$, such that

$$
(\rho_1 \,|\, B_1) \geq \|\rho_1\|_1^\times - \epsilon
$$

Define

$$
B \equiv B_1 + B_2
$$

where B_1 satisfies the condition above and $B_2 = 0$. We have

$$
(\rho \,|\, B) = (\rho_1 \,|\, B_1) \geq \|\rho_1\|_1^\times - \epsilon = \max\{\|\rho_1\|_1^\times, \|\rho_2\|_2^\times\} - \epsilon
$$

APPENDIX C. KERNEL OPERATORS

Lemma. *Let ρ be a tracial operator on $\mathcal{L}_\sigma^2$ and $\{e_i\}$ an orthonormal basis. Define*

$$
\tilde{\rho}_{\alpha\alpha'} = \sum_i e_i(\alpha)(\rho^\dagger e_i)^*(\alpha')
$$

Then $\tilde{\rho}_{\alpha\alpha'}$ is the integral kernel of ρ, that is, for each $\varphi \in \mathscr{L}^2_\sigma$,

$$(\rho\varphi)(\alpha) = \int_\sigma d\alpha' \, \tilde{\rho}_{\alpha\alpha'} \, \varphi(\alpha')$$

Moreover, the diagonal values $\tilde{\rho}_{\alpha\alpha}$ are almost everywhere uniquely determined, independently of the choice of the orthonormal basis, and

$$\operatorname{tr}(\rho) = \int_\sigma d\alpha \, \tilde{\rho}_{\alpha\alpha}$$

Proof. If $\{e_i\}$ is an orthonormal basis in $\mathscr{L}^2_\sigma$, for each function φ and ψ from $\mathscr{L}^2_\sigma$ we have

$$\begin{aligned}
(\varphi \mid \rho\psi) &= \sum_i (\varphi \mid e_i)(e_i \mid \rho\psi) \\
&= \sum_i (\varphi \mid e_i)(\rho^\dagger e_i \mid \psi) \\
&= \sum_i \int_\sigma d\alpha \, \varphi^*(\alpha) e_i(\alpha) \int_\sigma d\alpha' (\rho^\dagger e_i)^*(\alpha')\psi(\alpha') \\
&= \int_{\sigma^2} d\alpha \, d\alpha' \left[\sum_i e_i(\alpha)(\rho^\dagger e_i)^*(\alpha') \right] \varphi^*(\alpha)\psi(\alpha') \\
&= \int_{\sigma^2} d\alpha \, d\alpha' \, \tilde{\rho}_{\alpha\alpha'} \, \varphi^*(\alpha)\psi(\alpha')
\end{aligned}$$

which implies that $\tilde{\rho}_{\alpha\alpha'}$ is the integral kernel of ρ. Because ρ is tracial, the series $\sum_i (e_i \mid \rho e_i)$ is absolutely convergent. So we have

$$\begin{aligned}
\operatorname{tr} &= \sum_i (e_i \mid \rho e_i) = \sum_i (\rho^\dagger e_i \mid e_i) = \sum_i \int_\sigma d\alpha (\rho^\dagger e_i)^*(\alpha) e_i(\alpha) \\
&= \int_\sigma d\alpha \sum_i e_i(\alpha)(\rho^\dagger e_i)^*(\alpha) = \int_\sigma d\alpha \, \tilde{\rho}_{\alpha\alpha}
\end{aligned}$$

Suppose now that we consider another orthonormal basis $\{e'_i\}$, to which we correspond, in the same way as shown above, the kernel $\tilde{\tilde{\rho}}_{\alpha\alpha'}$. Note first that for any bounded measurable function φ, the operator $\varphi\rho^\dagger$, defined by

$$f \mapsto \varphi\rho^\dagger f$$

is also tracial. Therefore,

$$\begin{aligned}
\operatorname{tr}(\varphi\rho^{\dagger}) &= \sum_i (e_i, \varphi\rho^{\dagger}e_i) \\
&= \sum_i \int_\sigma d\alpha\, e_i^*(\alpha)\varphi(x)(\rho^{\dagger}e_i)(\alpha) \\
&= \int_\sigma d\alpha \left[\sum_i e_i^*(\alpha)(\rho^{\dagger}e_i)(\alpha)\right]\varphi(\alpha) \\
&= \int_\sigma d\alpha \left[\sum_i e_i(\alpha)(\rho^{\dagger}e_i)^*(\alpha)\right]\varphi(\alpha) \\
&= \left[\int_\sigma d\alpha\, \tilde{\rho}_{\alpha\alpha}\, \varphi^*(\alpha)\right]^*
\end{aligned}$$

However, the trace does not depend on the choice of the basis, so repeating the chain of equalities above for $\{e_i'\}$, we obtain that for each bounded function φ,

$$\int_\sigma d\alpha\, \tilde{\rho}_{\alpha\alpha}\, \varphi^*(\alpha) = \int_\sigma d\alpha\, \tilde{\rho}_{\alpha\alpha}\, \varphi^*(\alpha)$$

which implies that $\tilde{\rho}_{\alpha\alpha} = \tilde{\tilde{\rho}}_{\alpha\alpha}$ almost everywhere.

ACKNOWLEDGMENTS

Our persistent and stimulating discussions with Professor I. Prigogine essentially shaped this work. We are also grateful to Professor J.-P. Antoine, A. Bohm, E. Brändas, M. Gadella, C. George, K. Gustafson, B. Misra, T. Petrosky, and A. Weron for several critical remarks. We thank Professors R. Laura and S. Tasaki for early discussions and remarks. This work received financial support from the European Commission ESPRIT P9282 ACTCS and the Belgian Government through the Interuniversity Attraction Poles.

REFERENCES

1. R. Brout and I. Prigogine, *Physica* **22**, 621 (1956).
2. I. Prigogine and R. Balescu, *Physica* **25**, 281 (1959).
3. J. Philippot, *Physica* **27**, 490 (1961).
4. I. Prigogine, *Non-equilibrium Statistical Mechanics* (Wiley, New York, 1962).
5. L. van Hove, *Physica* **21**, 901 (1955).
6. L. van Hove, *Physica* **22**, 343 (1956).
7. L. van Hove, *Physica* **23**, 441 (1957).

8. L. van Hove, *Physica* **25**, 268 (1959).
9. I. Prigogine, C. George, F. Henin, and L. Rosenfeld, *Chem. Scr.* **4,** 5 (1973).
10. C. George, *Physica* **65**, 277 (1973).
11. A. Grecos, T. Guo, and W. Guo, *Physica* **A80**, 421 (1975).
12. R. Balescu, *Equilibrium and Non-equilibrium Statistical Mechanics* (Wiley, New York, 1975).
13. T. Petrosky, I. Prigogine, and S. Tasaki, *Physica* **A173**, 175 (1991).
14. T. Petrosky and I. Prigogine, *Physica* **A175**, 146 (1991).
15. I. Antoniou and I. Prigogine, *Physica* **A192**, 443 (1993).
16. I. Antoniou and S. Tasaki, *Int. J. Quantum Chem.*, **46**, 425 (1993).
17. N. Hugenholtz "Leon van Hove and the Dutch Physics," in *Festschrift Leon van Hove* (World Scientific, Singapore, 1989).
18. M. Naimark, *Normed Algebra* (Wolfer-Noordhoff, Groningen, 1972).
19. J. Dixmier, *C*-algebras* (North-Holland, Amsterdam, 1977).
20. I. Segal, *Annals Math.* **48**, 930 (1947).
21. G. Emch, *Adv. Chem. Phys.* **22**, 315 (1972).
22. O. Bratteli and D. Robinson, *Operator Algebras and Quantum Statistical Mechanics*, Vols. I and II (Springer, New York, 1979, 1981).
23. R. Haag, *Local Quantum Physics* (Springer, Berlin, 1992).
24. P. A. M. Dirac, *The Principles of Quantum Mechanics* (Charendon Press, Oxford, 1958).
25. J. von Neumann, *Mathematical Foundations of Quantum Mechanics* (Princeton University Press, Princeton, NJ, 1955).
26. E. Prugovecki, *Quantum Mechanics in Hilbert Space* (Academic, New York, 1981).
27. A. Bohm, *Quantum Mechanics: Foundations and Applications* (Springer, Berlin, 1986).
28. I. Prigogine, *From Being to Becoming* (Freeman, San Francisco, 1980).
29. I. Prigogine, *Phys. Rep.* **219**, 93 (1992).
30. T. Petrosky and I. Prigogine, *Phys. Lett.* **A182**, 1 (1993).
31. T. Petrosky and I. Prigogine, "The Liouville Space Extension of Quantum Mechanics," in *Advances in Chemical Physics*, Vol. 99 (Wiley, New York, 1997), pp. 1–120.
32. K. Maurin, *Methods of Hilbert Spaces* (Polish Scientific Publishers, Warsaw, 1972).
33. K. Maurin, *General Eigenfunction Expansions and Unitary representations of Topological Groups* (Polish Scientific Publishers, Warsaw, 1968).
34. B. de Dormale and H.-F. Gautrin, *J. Math. Phys.* **16**, 2328 (1975).
35. J. M. Jauch and B. Misra, *Helv. Phys. Acta* **38**, 30 (1965).
36. J. M. Jauch, *Helv. Phys. Acta* **33**, 711 (1960).
37. J.-P. Antoine, G. Epifanio, and C. Trapani, *Helv. Phys. Acta* **56**, 1175 (1983).
38. R. Schatten, *Norm Ideals of Completely Continuous Operators* (Springer Verlag, Berlin, 1960).
39. T. Kato, *Perturbation Theory for Linear Operators*, second corrected printing of the second edition (Springer Verlag, Berlin, 1984).
40. N. Dunford and J. Schwartz, *Linear Operators I: General Theory* (Wiley, New York, 1988).
41. J.-P. Antoine and Karwowski, *Publ. R.I.M.S. Kyoto Univ.* **21**, 205 (1985).
42. G. Epifanio and C. Trapani, *J. Math. Phys.* **29**, 536 (1988); *ibid.* **32**, 1096 (1991).

43. P. Halmos and V. Sunder, *Bounded Integral Operators on L^2 Spaces* (Springer Verlag, Berlin, 1978).
44. S. Mikhlin and S. Prössdorf, *Singular Integral Operators* (Springer Verlag, Berlin, 1986).
45. J. Dixmier, *Von Neumann Algebras* (North-Holland, Amsterdam, 1981).
46. M. Pour-el and J. Richards, *Computability in Analysis and Physics* (Springer Verlag, Berlin, 1989).
47. I. Antoniou, pp. 109–110 of reference 29.
48. N. Hugenholtz, *Physica* **23**, 481 (1957).
49. G. Källen in *Proceedings of the XII Solvay Conference in Physics* (Interscience, New York, 1951).
50. T. Petrosky and I. Prigogine, *Physica* **A147**, 439 (1988).
51. W. Amrein, J. Jauch, and K. Sinha, *Scattering Theory in Quantum Mechanics* (Benjamin, Massachusetts, 1977).
52. A. Sommerfeld, *Partial Differential Equations in Physics* (Academic, New York, 1957).
53. I. Antoniou, M. Gadella, and Z. Suchanecki, "The Liouville Operator in Quantum Physics," in preparation.
54. W. Pauli, *Festschrift zum 60 Geburstage A. Sommerfields* (Hirzel, Leipzig, 1928), p. 30.
55. R. Phillips, *Trans. Am. Math. Soc.* **86**, 109 (1957); *Comm. Pure Appl. Math.* **12**, 249 (1959).
56. G. Lindbland, *Commun. Math. Phys.* **48**, 119 (1976).
57. C. Piron, *Foundations of Quantum Physics* (Benjamin, New York, 1976).
58. I. Antoniou and Z. Suchanecki, *Found. Phys.* **24**, 1439 1994).
59. G. W. Mackey, *The Mathematical Foundations of Quantum Mechanics* (Benjamin, New York, 1963).
60. V. S. Varadarajan, *Geometry of Quantum Theory* (Springer Verlag, Berlin, 1985).
61. T. Petrosky and I. Prigogine, *Proc. Natl. Acad. Sci. USA* **90**, 9393 (1993).
62. T. Petrosky and I. Prigogine, *Chaos Solitons and Fractals* (submitted).
63. I. Antoniou and Z. Suchanecki, *Classical Systems with Diagonal Singularity* (to be submitted).
64. I Antoniou and S. Tasaki, *J. Phys.* **A26**, 73 (1993).
65. I. Antoniou and S. Tasaki, *Physica* **A190**, 303 (1992).
66. S. Tasaki, Z. Suchanecki, and I. Antoniou, *Phys. Lett. A* **179**, 103 (1993).
67. S. Tasaki, I. Antoniou, and Z. Suchanecki, *Chaos Solitons and Fractals*, **4**, 227 (1994).

NONADIABATIC CROSSING OF DECAYING LEVELS

V. V. KOCHAROVSKY AND VL. V. KOCHAROVSKY

Institute of Applied Physics, Russian Academy of Science, Nizhny Novgorod, Russia, and International Solvay Institute for Physics and Chemistry, Brussels, Belgium

S. TASAKI

Institute for Fundamental Chemistry, Kyoto, Japan

CONTENTS

ABSTRACT

A closed set of equations for a nonadiabatic coupling of the amplitudes of decaying discrete states dressed by a continuum is derived by means of the complex spectral decomposition method developed by the Brussels–Austin group. The limitations of the previously used phenomenological equations are shown within the scope of the time-dependent N-level Friedrichs–Fano

---

Advances in Chemical Physics, Volume XCIX, Edited by I. Prigogine and Stuart A. Rice.
ISBN 0-471-16526-3

model. The following novel nonadiabatic effects due to the decay of discrete levels into a continuum are found: (1) nonunitary nonadiabaticity caused from unitary one, (2) nonadiabaticity via time-dependent coupling with a reservoir, and (3) additional cross-decay due to nonadiabaticity. Reasonable generalization of Berry's phase for decaying eigenstates and regeneration of driven decaying states via coherent mixing with a reservoir are discussed. The general results are applied to the quasi-energy state dynamics of a three-level molecule driven by dc and ac fields and to the nonsteady Born–Oppenheimer approach of molecules electronically excited to decaying states.

I. INTRODUCTION

As emphasized in the earlier works of the Brussels–Austin group (see [1]), there is an asymmetry in the quantum mechanical description of states or particles: Stable states or particles are defined as eigenstates of the total Hamiltonian of the system, but unstable states or particles cannot be treated in the same way. Indeed, as seen in the work, for example, of Friedrichs [2] and Fano [3], the total Hamiltonian of the system with discrete decaying states admits only the eigenstates corresponding to the real continuous spectrum and no counterparts corresponding to the unstable discrete states exist. Usually unstable (decaying) states or particles are assumed to be described enough by the eigenstates of an unperturbed Hamiltonian. This convention is satisfactory only in the lowest-order approximation with respect to the interaction responsible for the decay. There are, however, many phenomena where more accurate treatments of unstable states or particles and, thus, full consideration of the dressing effects, are required. Nonadiabatic coupling of unstable discrete states is one such phenomenon.

Recently, several theories [23–28] have been proposed where the decaying discrete states are described as the eigenstates of the total Hamiltonian corresponding to the complex eigenvalues. It is even possible to construct a complete set of basis vectors involving such decaying eigenstates [24,27]. These approaches open a new possibility of treating unstable states on the same basis as the stable ones. The complex spectral theory (see [29,30] and references therein) developed by the Brussels–Austin group was first formulated for the Liouville–von Neumann operator to deal with the irreversible phenomena described density matrices and also was applied to the Hamiltonian of the one-level Friedrichs–Lee model [27,28]. In this chapter, the complex spectral theory is used to study the nonadiabatic coupling of unstable (decaying) discrete levels for the most general N-level Friedrichs–Fano model.

The nonadiabatic coupling of discrete states has been analyzed extensively in quantum mechanics, starting from Landau [4] and Zener [5] (see [6–13]), and in the theory of wave propagation, plasma physics, acoustics, and inhomogeneous-wave-guide theory (see [13–19]). In these theories, the time evolution of the probability amplitudes of different levels or the spatial change of the amplitudes of different field modes (e.g., with different polarizations) are described by a set of linear ordinary differential equations with coefficients depending on time or on a coordinate along a propagation path, respectively. If the coefficients are constant, those equations possess a set of linearly independent harmonic solutions (in a nondegenerate case), that is, there are eigenstates with definite energies. When the coefficients vary in time or space, mixing of these solutions takes place. The mixing is exponentially small if the coefficients vary slowly enough, so that the adiabatic approximation is applicable [16–22]. Otherwise, nonadiabatic coupling of levels or eigenmodes causes new phenomena.

For decaying discrete states, the problem of nonadiabatic coupling is particularly interesting because it is crucial for solving various physical problems, for example, collisions of excited molecules [6–8], multiphoton ionization of atoms [20–22], dynamics of spins (or dipole momenta responsible for quantum transitions) driven by a coherent field [9–13], chemical reactions involving excited molecules, and so on. Usually, this problem is studied by phenomenological methods such as those using the nonhermitian Hamiltonian or nonadiabatic coupling equations for decaying partial states. However, the validity of different phenomenological equations and the limits of their applicability have not yet clearly been formulated. In the following, we derive consistently, for the first time, the fundamental equations of nonadiabatic coupling of decaying discrete levels for the N-level Friedrichs–Fano model and discuss the validity of the existing treatments as well as expected new features. As we shall see, this approach does not require any phenomenological nonhermitian Hamiltonians, but rather shows how they can be obtained as a specific approximation.

The Friedrichs–Fano model consists of discrete (partial) states $|\alpha\rangle$ ($\alpha = 1, \ldots, N$), and continuum reservoir states, $|\mathbf{k}\rangle$, with energy ω_k ($0 \leq \omega_k < +\infty$). The total Hamiltonian is then a sum of (1) the part $h(t)$ governing the evolution of the dynamical subsystem of the discrete states, $|\alpha\rangle$, (2) the interaction $v(t)$ between the discrete subsystem and the reservoir, and (3) the reservoir energy $\sum_{\mathbf{k}} \omega_k |\mathbf{k}\rangle\langle\mathbf{k}|$:

$$H(t) = h(t) + v(t) + \sum_{\mathbf{k}} \omega_k |\mathbf{k}\rangle\langle\mathbf{k}| \tag{1}$$

where

$$v(t) = \sum_{\alpha=1}^{N} \sum_{\mathbf{k}} [V_{\alpha\mathbf{k}}(t)|\mathbf{k}\rangle\langle\alpha| + V^*_{\alpha\mathbf{k}}(t)|\alpha\rangle\langle\mathbf{k}|] \tag{2}$$

$$h(t) = \sum_{\alpha,\alpha'=1}^{N} h_{\alpha\alpha'}(t)|\alpha\rangle\langle\alpha'| \equiv \sum_{\alpha=1}^{N} v_\alpha(t)|\phi_\alpha\rangle\langle\phi_\alpha| \tag{3}$$

The coupling parameters between two discrete levels, $h_{\alpha\alpha'}(t)$, and between a discrete level and the reservoir, $V_{\alpha\mathbf{k}}(t)$, depend on time explicitly. A $\mathbf{k}$-dependence of an interaction, $V_{\alpha\mathbf{k}}$, allows one to consider any type of reservoir energy spectrum (e.g., band-type spectrum) (cf. [31–34]). For simplicity, we use summations $\sum_{\mathbf{k}} \cdots$ instead of integrations $\int d\omega_{\mathbf{k}} \cdots$, bearing in mind the standard limit of the infinite quantization box [27,35], $L^3 \to \infty$. In this convention, the matrix element of the interaction between discrete states and the reservoir, $V_{\alpha\mathbf{k}}$, is of the order $L^{-3/2}$. In Eq. (3), $v_\alpha(t)$ and $|\phi_\alpha\rangle$ are, respectively, eigenvalues and eigenvectors for the partial Hamiltonian, $h(t)$, and both depend on time explicitly. If the reservoir is absent, the coupling among the discrete states is described by the subsystem Hamiltonian, $h(t)$, and the level coupling is described by the time evolution of the probability amplitudes corresponding to the instantaneous eigenfunctions, $|\phi_\alpha(t)\rangle$.

As discussed in Section II, the complex spectral theory [27] leads to the following decomposition of the total hermitian Hamiltonian, $H(t)$:

$$H(t) = \sum_{\alpha=1}^{N} \omega_\alpha(t)|\varphi_\alpha\rangle\langle\tilde{\varphi}_\alpha| + \sum_{\mathbf{k}} \omega_k|\varphi_{\mathbf{k}}\rangle\langle\tilde{\varphi}_{\mathbf{k}}| \tag{4}$$

where $|\varphi_\alpha\rangle$, $|\varphi_{\mathbf{k}}\rangle$ and $\langle\tilde{\varphi}_\alpha|$, $\langle\tilde{\varphi}_{\mathbf{k}}|$ are, respectively, right and left generalized eigenvectors of $H(t)$. The states $|\varphi_\alpha\rangle$ represent renormalized discrete states dressed by the definite superposition of continuum states, and the corresponding eigenfrequencies, $\omega_\alpha(t)$, are complex. The other states $|\varphi_{\mathbf{k}}\rangle$ of the reservoir with real eigenenergies, ω_k ($0 \le \omega_k < +\infty$), are not important if a reservoir memory time scale is less than a time scale of the variation of the Hamiltonian. Thus, the dynamics of the complex amplitudes of the generalized eigenstates, $|\varphi_\alpha\rangle$, yields a closed treatment of the nonunitary evolution of the discrete states in the presence of decay. It is the whole point of the present approach to nonadiabatic phenomena. In addition, it could provide a deeper insight into decaying states, because the nonsteady one-level Friedrichs model [2] or the steady N-level Fano model [3] cannot describe the interaction among decaying states.

This approach reveals novel nonadiabatic effects due to a time variation of the renormalization in generalized eigenstates, which were missed in previous phenomenological schemes [6–9,20–22]. For the N-level Friedrichs–Fano model, we can explicitly treat the discrete-state redressing due to the time dependence of the subsystem Hamiltonian (3), $h(t)$, as well as the interaction (2), $v(t)$, between the subsystem and the reservoir. Note that the Friedrichs–Fano model does not include the so-called counter-rotating terms [58] in the subsystem–reservoir interaction. However, we expect that the main results of the present work are unaffected even if these effects are included.

In Section II we construct the complex spectral decomposition of the Hamiltonian for the N-level Friedrichs–Fano model (1)–(3). The fundamental equations of nonadiabatic crossing of decaying levels are derived in Section III. Section IV is devoted to a discussion of the novel nonadiabatic effects arising from a decay of discrete levels and to a "resonance" approximation leading to the nonhermitian Hamiltonian similar to phenomenological ones. A generalization of Berry's phase for decaying states is given in Section V. The time evolution of Lyapunov-type quantity and regeneration of driven decaying states is discussed in Section VI. In Sections VII and VIII, two basic problems in quantum optics and in quantum chemistry are treated as examples of the present theory. In Section VII, novel nonadiabatic effects are discussed in quasi-energy-state dynamics of a three-level molecule driven by nonsteady dc and ac electric fields and, in Section VIII, we formulate a generalization of the Born–Oppenheimer approach to the case with decaying electronic states and briefly discussed expected new features. The main conclusions are summarized in Section IX.

II. GENERALIZED EIGENSTATES OF THE N-LEVEL FRIEDRICHS–FANO MODEL

The problem of the diagonalization of the total Hamiltonian (1) was put forward in connection with the problem of atomic autoionization [36,37] and was solved by Friedrichs [2,38] for $N = 1$ and by Fano [3] for an arbitrary N with several continua. (A similar model was considered [39] in the context of neutral kaon decay.) However, they found only the real spectral decomposition, $H(t) = \sum_{\mathbf{k}} \omega_k |\phi_{\mathbf{k}}^{\mathrm{F}}\rangle\langle\phi_{\mathbf{k}}^{\mathrm{F}}|$, which "dilutes" discrete states throughout the continuum of new eigenstates, $|\phi_{\mathbf{k}}^{\mathrm{F}}\rangle$, with the unperturbed spectrum $0 \le \omega_k < +\infty$. Taking into account the orthonormalization, $\langle\phi_{\mathbf{k}}^{\mathrm{F}}|\phi_{\mathbf{k}'}^{\mathrm{F}}\rangle = \delta_{\mathbf{k}\mathbf{k}'}$, we find the following state (cf. [3,40]):

$$|\phi_{\mathbf{k}}^{\mathrm{F}}\rangle = |\mathbf{k}\rangle + \sum_{\alpha=1}^{N} a_{\mathbf{k}\alpha}(t)\left[|\alpha\rangle + \sum_{\mathbf{k}'} \frac{V_{\alpha\mathbf{k}'}(t)}{\omega_k - \omega_{k'} + i\epsilon}|\mathbf{k}'\rangle\right] \tag{5}$$

and the mixing coefficient, $a_{\mathbf{k}\alpha}$:

$$a_{\mathbf{k}\alpha}(t) = \sum_{\beta=1}^{N} \{[\omega_k I - K(\omega_k)]^{-1}\}_{\alpha\beta} V^*_{\beta\mathbf{k}} \equiv \sum_{\beta\beta'} \frac{\xi_{\alpha\beta}(\omega_k)\xi^{-1}_{\beta\beta'}(\omega_k)}{\omega_k - \mathscr{H}_\beta(\omega_k)} V^*_{\beta'\mathbf{k}} \tag{6}$$

where the symbol $\epsilon(\to +0)$ indicates the usual Landau's rule of evaluating the pole contribution [12,13,27,28,35], I is the unit $N \times N$ matrix, and $K(\omega)$ stands for the time- and frequency-dependent $N \times N$ matrix:

$$K(\omega)_{\alpha\alpha'} = h_{\alpha\alpha'} + \sum_{\mathbf{k}'} \frac{V^*_{\alpha\mathbf{k}'} V_{\alpha'\mathbf{k}'}}{\omega - \omega_{k'} + i\epsilon} \tag{7}$$

In Eq. (6), $\mathscr{H}_\alpha(\omega)$ (with $\alpha = \beta$) is the eigenvalue of the matrix $K(\omega)$, and $\xi_{\alpha'\alpha}(\omega)$ and $\xi^{-1}_{\alpha\alpha'}(\omega)$ are components of its right and left eigenvectors:

$$K(\omega)\,|\,u_\alpha(\omega)\rangle = \mathscr{H}_\alpha(\omega)\,|\,u_\alpha(\omega)\rangle; \qquad \langle v_\alpha(\omega)\,|\,K(\omega) = \mathscr{H}_\alpha(\omega)\langle v_\alpha(\omega)\,| \tag{8}$$

$$|\,u_\alpha(\omega)\rangle = \sum_{\alpha'=1}^{N} \xi_{\alpha'\alpha}(\omega)\,|\,\alpha'\rangle; \qquad \langle v_\alpha(\omega)\,| = \sum_{\alpha'=1}^{N} \xi^{-1}_{\alpha\alpha'}(\omega)\langle\alpha'\,| \tag{9}$$

Here we assume that, for any value of ω, the matrix (7) has N different eigenvalues. Then the right and left eigenvectors $\{|\,u_\alpha(\omega)\rangle, \langle v_\alpha(\omega)\,|\}$ form a complete biorthogonal basis in the dynamical subspace spanned by $\{|\,\alpha\rangle\}$.

When the transition-matrix element, $V_{\alpha\mathbf{k}}$, as well as the unperturbed continuum state, $|\,\mathbf{k}\rangle = |\,\omega\rangle$, depend only on the energy ω_k, that is, $|\,\mathbf{k}\rangle = |\,\omega_k\rangle$, and Eq. (5) takes a familiar form:

$$|\,\phi^{\mathrm{F}}_\omega\rangle = |\,\omega\rangle + \sum_{\mu=1}^{N} a_{\omega\mu}(t)\left[\,|\,\mu\rangle + \sum_{\omega'} \frac{\tilde{V}_{\mu\omega'}(t)}{\omega - \omega' + i\epsilon}\,|\,\omega'\rangle\right]$$

with

$$a_{\omega\mu} = \frac{\tilde{V}^*_{\mu\omega}}{(1 - i\pi/\zeta)(\omega - \bar{\omega}_{\mu\omega})}, \qquad \frac{1}{\zeta} = \sum_{\mu=1}^{N} \frac{|\,\tilde{V}_{\mu\omega}\,|^2}{\omega - \bar{\omega}_{\mu\omega}}$$

where $|\,\mu\rangle$ is the eigenvector of the Hermitian matrix:

$$\left(h_{\alpha\alpha'} + \text{V.p.}\sum_{\omega'} \frac{V^*_{\alpha\omega'} V_{\alpha'\omega'}}{\omega - \omega'}\right)$$

$\bar{\omega}_{\mu\omega}$ is the corresponding eigenvalue, and $\tilde{V}_{\mu\omega}(t) \equiv \langle\omega\,|\,v(t)\,|\,\mu\rangle$ is the transition amplitude between the new discrete state $|\,\mu\rangle$ and the continuum state $|\,\omega\rangle$. In this nondegenerate case, the eigenstates $|\,\phi^{\mathrm{F}}_\omega\rangle$ differ from Fano's [3] only by phase factors.

Note that, because the matrix $K(\omega)$ is not hermitian, its right and left eigenvectors are not hermitian conjugations:

$$\xi^{-1}_{\alpha\alpha'}(\omega) \neq \xi^{*}_{\alpha'\alpha}(\omega) \tag{10}$$

As mentioned in the Introduction, this real spectral decomposition does not contain the parts corresponding to the unstable states and there is no well-defined manner to describe the level crossing phenomena involving unstable states. For the one-level Friedrichs–Lee model [2,38], Nakanishi [41] introduced a generalized eigenstate corresponding to a complex eigenvalue of the total Hamiltonian. Sudarshan, Chiu, and Gorini [24] then showed that there exist a complete set of left and right generalized eigenstates of the *total* Hamiltonian (involving Nakanishi's state), which provides a complex spectral decomposition of the total Hamiltonian. The equivalent complete set was constructed perturbatively by the complex spectral theory developed by the Brussels–Austin group [27,28]. For the N-level Friedrichs–Fano model, the complex spectral decomposition can be obtained as in the one-level Friedrichs model. One can derive it by a perturbation method adopting a time-ordered boundary condition [27,28]: small denominators relating to the transition from discrete to continuum states and regularized as retarded propagators and those relating to the transition from continuum to discrete states as advanced propagators (of course, another method, for example, the resolvent method, will lead to the same result).

The calculations are similar to the case of $N = 1$ [27] and here we only list the results:

$$H(t)|\varphi_\alpha\rangle = \omega_\alpha|\varphi_\alpha\rangle \qquad \langle\tilde{\varphi}_\alpha|H(t) = \omega_\alpha\langle\tilde{\varphi}_\alpha| \tag{11}$$

$$H(t)|\varphi_{\mathbf{k}}\rangle = \omega_k|\varphi_{\mathbf{k}}\rangle \qquad \langle\tilde{\varphi}_{\mathbf{k}}|H(t) = \omega_k\langle\tilde{\varphi}_{\mathbf{k}}| \tag{12}$$

where

$$|\varphi_\alpha\rangle = g_\alpha[\eta'_\alpha(\omega_\alpha)]^{-1/2}\sum_{\alpha'=1}^{N}\xi_{\alpha'\alpha}(\omega_\alpha)\left\{|\alpha'\rangle + \sum_{\mathbf{k}'}\frac{V_{\alpha'\mathbf{k}'}}{[\omega_\alpha - \omega_{\mathbf{k}'}]_+}|\mathbf{k}'\rangle\right\} \tag{13}$$

$$\langle\tilde{\varphi}_\alpha| = g_\alpha^{-1}[\eta'_\alpha(\omega_\alpha)]^{-1/2}\sum_{\alpha'=1}^{N}\xi^{-1}_{\alpha\alpha'}(\omega_\alpha)\left\{\langle\alpha'| + \sum_{\mathbf{k}'}\langle\mathbf{k}'|\frac{V^{*}_{\alpha'\mathbf{k}'}}{[\omega_\alpha - \omega_{\mathbf{k}'}]_+}\right\} \tag{14}$$

$$\begin{aligned}|\varphi_{\mathbf{k}}\rangle &= |\phi^{\mathrm{F}}_{\mathbf{k}}\rangle - \sum_{\alpha'=1}^{N}|\varphi_{\alpha'}\rangle\langle\tilde{\varphi}_{\alpha'}|\phi^{\mathrm{F}}_{\mathbf{k}}\rangle \\ &= |\mathbf{k}\rangle + \sum_{\alpha=1}^{N}\tilde{a}_{\mathbf{k}\alpha}(t)\left[|\alpha\rangle + \sum_{\mathbf{k}'}\frac{V_{\alpha\mathbf{k}'}(t)}{\omega_k - \omega_{k'} + i\epsilon}|\mathbf{k}'\rangle\right]\end{aligned} \tag{15}$$

$$\langle \tilde{\varphi}_{\mathbf{k}} | = \langle \phi_{\mathbf{k}}^{\mathrm{F}} | \tag{16}$$

with

$$\tilde{a}_{\mathbf{k}\alpha} = \sum_{\beta\beta'} \frac{\xi_{\alpha\beta}(\omega_k)\xi_{\beta\beta'}^{-1}(\omega_k)}{\tilde{\eta}_{\beta}(\omega_k)} V_{\beta'\mathbf{k}}^{*} \tag{17}$$

The complex eigenvalues, ω_α, are obtained as the solutions of the dispersion relations:

$$\omega_\alpha = \mathscr{H}_\alpha(\omega_\alpha) \qquad (\alpha = 1, \ldots, N) \tag{18}$$

which are assumed to have unique solutions (in the lower half-plane for decaying levels). The factor g_α is a normalization constant to be discussed in Sections IV and V, and $\eta'_\alpha(\omega)$ is the derivative given by

$$\eta'_\alpha(\omega) \equiv \frac{d}{d\omega} \{\omega - \mathscr{H}_\alpha(\omega)\} \tag{19}$$

The expression $1/[\omega_\alpha - \omega_{\mathbf{k}'}]_+$ stands for the analytic extension from the upper half-plane with respect to ω_α and, for $\mathrm{Im}\ \omega_\alpha < 0$, can be written in terms of the delta function with complex arguments [24–28,41,42]:

$$\frac{1}{[\omega_\alpha - \omega_{\mathbf{k}'}]_+} = \frac{1}{\omega_\alpha - \omega_{\mathbf{k}'}} - 2\pi i \delta(\omega_{k'} - \omega_\alpha) \tag{20}$$

The distributions, $1/\tilde{\eta}_\alpha(\omega)$, in the formula (17) can also be expressed in terms of the delta functions with complex arguments:

$$\frac{1}{\tilde{\eta}_\alpha(\omega)} \equiv \frac{1}{\omega - \mathscr{H}_\alpha(\omega)} + 2\pi i \frac{\delta(\omega - \omega_\alpha)}{\eta'_\alpha(\omega_\alpha)} \tag{21}$$

As verified by direct calculations, the new eigenstates form a complete biorthonormal system:

$$\langle \tilde{\varphi}_\alpha | \varphi_{\alpha'} \rangle = \delta_{\alpha\alpha'}, \qquad \langle \tilde{\varphi}_\alpha | \varphi_{\mathbf{k}} \rangle = 0, \qquad \langle \tilde{\varphi}_{\mathbf{k}} | \varphi_{\mathbf{k}'} \rangle = \delta_{\mathbf{k}\mathbf{k}'}, \qquad \langle \tilde{\varphi}_{\mathbf{k}} | \varphi_\alpha \rangle = 0 \tag{22}$$

$$\sum_{\alpha=1}^{N} | \varphi_\alpha \rangle \langle \tilde{\varphi}_\alpha | + \sum_{\mathbf{k}} | \varphi_{\mathbf{k}} \rangle \langle \tilde{\varphi}_{\mathbf{k}} | = I \tag{23}$$

According to Eqs. (8), (9), and (18), the complex eigenvalue problem of the total Hamiltonian is related to the eigenvalue problem of an "eigenvalue-dependent" operator $K(\omega)$ of Eq. (7). Operators like $K(\omega)$ arise naturally in the projection operator or partitioning techniques and are known as Livsic operators [43] or collision operators [29,44,45,46]. The eigenvalue problem of such eigenvalue-dependent operators was first discussed by Lowdin [44] and Bartlett and Brändas [45]. Grecos, Guo, and Guo [46] and Petrosky and Prigogine [29] used this method in the context of the subdynamics theory to define generalized eigenvectors. Also, Chiu and Sudarshan [39] used it in the context of the kaon decay problem to extract the evolution of the subsystem consisting of the neutral kaon and its antiparticle.

Because the generalized eigenstates (13)–(15) contains the delta functions with complex arguments, they cannot live in the usual Hilbert space. Such objects acquire mathematical meaning as linear functionals over a restricted class of Hilbert-space vectors (test vectors) [24–28,47], that is, through the use of generalized function theory (e.g., [42]). For the N-level Friedrichs–Fano model, as in the one-level Friedrichs model [28], the states $\langle\varphi_\alpha|$ and $\langle\tilde{\varphi}_\alpha|$ can be defined as linear functionals over the spaces Φ_+ and Φ_- respectively, where $\Phi_\pm$ is the space of all wave functions $|\psi\rangle$ such that the component $\langle\mathbf{k}|\psi\rangle$ belongs, as a function of ω_k, to the intersection between the Schwartz class and the upper or lower Hardy class, $\mathscr{S} \cap \mathscr{H}_\pm$, respectively. In other words, they belong to the topological duals of the test spaces: $\langle\varphi_\alpha| \in \Phi_+^\dagger$ and $\langle\tilde{\varphi}_\alpha| \in \Phi_-^\dagger$. Mathematically speaking, this extension to the non-Hilbert generalized space enables the hermitian operator in the Hilbert space to admit complex eigenvalues (the notion of hermiticity is well-defined *only* for the Hilbert space operators).

Because a one-to-one correspondence exists between the complex eigenstates and the unperturbed states, one can introduce a transformation connecting the two [27]: $\Lambda^{-1}|\beta\rangle = |\varphi_\beta\rangle$, $(\beta = 1, \ldots, N, \mathbf{k})$, where

$$\Lambda \equiv \sum_{\alpha=1}^{N} |\alpha\rangle\langle\tilde{\varphi}_\alpha| + \sum_{\mathbf{k}} |\mathbf{k}\rangle\langle\tilde{\varphi}_\mathbf{k}|, \qquad \Lambda^{-1} \equiv \sum_{\alpha=1}^{N} |\varphi_\alpha\rangle\langle\alpha| + \sum_{\mathbf{k}} |\varphi_\mathbf{k}\rangle\langle\mathbf{k}| \tag{24}$$

If the system is integrable, such transformation is always unitary. However, as clearly seen from Eqs. (24), the transformation Λ is not unitary as a result of the nonintegrability of the system. In case of time-independent systems, the nonunitary transformation Λ intertwines the total Hamiltonian with a diagonal dissipative operator [27]. In this respect, the use of such a transformation can be considered an extension of the generalized scattering theory of Sudarshan [48] (see also [47]) where the total Hamiltonian is intertwined with a comparison Hamiltonian, instead of the unperturbed one.

The $N \times N$-matrix (7), $K(\omega)$, which provides complex eigenvalues, is not Hermitian. Thus, the matrix may not be diagonalizable and may admit the Jordan block structure. This structure will be transferred to the complex spectral decomposition of the Hamiltonian. Clearly, the appearance of the Jordan blocks corresponds to the degeneracy of the complex eigenenergies corresponding to several decaying states. As is well known, the appearance of the Jordan blocks in the generator of motion implies a deviation from the purely exponential and/or oscillatory behaviors, namely, there appears a polynomial factor: $t^n \exp(i\omega_\alpha t)$, where n is an integer and ω_α may be complex. In the time-independent case, Arecchi and Courtens [49] have considered a simple model which exhibits nonexponential decay of the form $t \exp(-\gamma t)$. For our knowledge, the physical examples of such a nonexponential decay have not yet been considered for explicitly time-dependent Hamiltonians, $H = H(t)$.

III. FUNDAMENTAL EQUATIONS

We now have *dressed* unstable (decaying) states which are expected to provide a consistent description of the level coupling involving decay into a continuum. We shall derive the coupling equations for these states in the most interesting case, where the total Hamiltonian varies slowly over the oscillation periods of all discrete eigenstates, $2\pi/\nu_\alpha$ for $|\phi_\alpha\rangle$ in Eq. (3) or $2\pi/\omega'_\alpha$ for $|\varphi_\alpha\rangle$ in Eq. (4); nevertheless nonadiabatic effects appear due to a crossing, or even a transitory approach, of some levels:

$$|\omega'_\alpha(t) - \omega'_{\alpha'}(t)| \ll (\omega'_\alpha + \omega'_{\alpha'})/2; \qquad \omega'_\alpha \equiv \operatorname{Re} \omega_\alpha(t) \tag{25}$$

A decay yields nontrivial effects upon a nonadiabatic coupling if a crossing lasts longer than inverse decay rates. Otherwise, all decay rates can be neglected in the coupling equations. Therefore, we only deal with the case where the change of the Hamiltonian is relatively slow, that is, where the following inequality holds during some period, at least for some levels:

$$\tau^{-1} \equiv |(dH/dt)H^{-1}| \lesssim -\omega''_\alpha; \qquad \omega''_\alpha \equiv \operatorname{Im} \omega_\alpha(t) \tag{26}$$

Then, a decay of generalized eigenstates $|\varphi_\alpha\rangle$ can be treated as exponential with current decay rates $\omega''_\alpha(t)$. Of course, we assume that $-\omega''_\alpha \ll \omega'_\alpha$, otherwise the concept of discrete levels is physically meaningless.

In the situation stated above, dressing of generalized eigenstates can set up adiabatically in spite of nonadiabatic level crossing. (An analogous situation is well known for a transient radiation of moving charges in electrodynamics of continuous inhomogeneous media [50].) As a result, the complex spectral decomposition (4) allows one to take into account consistently time dependence both of a dynamical subsystem and of its interaction with a reservoir, as well as to keep a closed form of equations of a nonadiabatic discrete-level coupling. The decisive step is the introduction of the complex amplitudes, $f_\alpha(t)$ and $f_{\mathbf{k}}(t)$, of time-dependent generalized eigenstates, $|\varphi_\alpha\rangle$ and $|\varphi_{\mathbf{k}}\rangle$, respectively, in a rigged Hilbert space (cf. Eq. (23) and also [24–28,35]):

$$|\psi(t)\rangle \equiv \sum_{\alpha=1}^{N} b_\alpha(t)|\alpha\rangle + \sum_{\mathbf{k}} b_{\mathbf{k}}(t)|\mathbf{k}\rangle = \sum_{\alpha=1}^{N} f_\alpha(t)|\varphi_\alpha\rangle + \sum_{\mathbf{k}} f_{\mathbf{k}}(t)|\varphi_{\mathbf{k}}\rangle \tag{27}$$

$$f_\alpha(t) = \sum_{\alpha'=1}^{N} \langle\tilde{\varphi}_\alpha|\alpha'\rangle b_{\alpha'} + \sum_{\mathbf{k}'} \langle\tilde{\varphi}_\alpha|\mathbf{k}'\rangle b_{\mathbf{k}'} \tag{28}$$

The Schrödinger equation, $d|\psi\rangle/dt = -iH|\psi\rangle$, yields the exact equations for the amplitudes:

$$\dot{f}_\alpha + i\omega_\alpha(t) f_\alpha = -\sum_{\alpha'=1}^{N} \langle\tilde{\varphi}_\alpha|\dot{\varphi}_{\alpha'}\rangle f_{\alpha'} - \sum_{\mathbf{k}'} \langle\tilde{\varphi}_\alpha|\dot{\varphi}_{\mathbf{k}'}\rangle f_{\mathbf{k}'} \tag{29}$$

$$\dot{f}_{\mathbf{k}} + i\omega_k f_{\mathbf{k}} = -\sum_{\alpha'=1}^{N} \langle\tilde{\varphi}_{\mathbf{k}}|\dot{\varphi}_{\alpha'}\rangle f_{\alpha'} - \sum_{\mathbf{k}'} \langle\tilde{\varphi}_{\mathbf{k}}|\dot{\varphi}_{\mathbf{k}'}\rangle f_{\mathbf{k}'} \tag{30}$$

where the dot stands for the time derivative, d/dt.

For the general situations stated above, it is possible, as a natural approximation, to neglect the last continuous sum of Eq. (29) as compared to the first discrete sum; even if nonadiabatic effects are small, for example, $|\langle\tilde{\varphi}_\alpha|\dot{\varphi}_{\alpha'}\rangle|, |\langle\tilde{\varphi}_{\alpha'}|\dot{\varphi}_\alpha\rangle| \ll |\omega_\alpha - \omega_{\alpha'}|$, (see [19]) the last term of Eq. (29) may be negligible as compared to the left-hand-side, $i\omega_\alpha f_\alpha$. In this case, the time evolution of the discrete amplitudes, f_α, is independent of the amplitudes of the (generalized) reservoir eigenstates, $f_{\mathbf{k}}(t)$, and thus it is not necessary to consider their evolution (30). Under this approximation, the bare amplitudes, b_α and $b_{\mathbf{k}}$, are related to the generalized amplitudes as

$$b_\alpha = \sum_{\alpha'=1}^{N} \langle\alpha|\varphi_{\alpha'}\rangle f_{\alpha'}, \qquad b_{\mathbf{k}} = \sum_{\alpha'=1}^{N} \langle\mathbf{k}|\varphi_{\alpha'}\rangle f_{\alpha'} \tag{31}$$

Physically speaking, the reduction to the closed description in the dynamical subspace, $\bar{h}$, spanned by the generalized eigenstates $|\varphi_\alpha\rangle$ results from a random, incoherent distribution of reservoir-eigenstate amplitudes, $f_{\mathbf{k}}$, when they have no time for mutual phasing during a nonadiabatic crossing (such a case may take place, for example, if the reservoir state is close to the vacuum). In other words, the present approximation takes into account the reservoir effects arising from a discrete-state decay into a continuum, but excludes an effect of macroscopic dynamics of the reservoir (e.g., coherent wave packet propagation) as well as the non-Markovian effects arising from the reservoir dynamics in the initial Zeno and final polynomial-decay stages [31,51–56]. Moreover, as the generalized discrete states, $|\varphi_\alpha\rangle$, correspond to the Gamow vectors (e.g., [26]), the coordinate representation of their field parts has exponentially growing terms with respect to the coordinates and, as a result, the present approximation, where only the generalized discrete states are retained, cannot correctly treat the field propagation effects. In this Chapter, we focus only on the nonadiabatic decaying-level coupling, where this approximation is expected to be valid. We are not going to discuss the precise validity conditions of this approximation or more general situations where the full set of Eqs. (29)–(30) is necessary.

In conclusion, we find the fundamental set of equations taking into account an exponential decay of generalized discrete states:

$$\dot{f}_\alpha + i\omega_\alpha(t) f_\alpha = -\sum_{\alpha'=1}^{N} \langle \tilde{\varphi}_\alpha | \dot{\varphi}_{\alpha'} \rangle f_{\alpha'} \tag{32}$$

Then, through Eq. (31), one can find the amplitudes, b_α, of the initial partial states, $|\alpha\rangle$, which we are often interested in. This corresponds to the analogous set of equations for linear mode coupling in a wave-propagation theory [19]. (An introduction of bras different from kets corresponds to an introduction of so-called "transfer" modes which possess polarizations different from those of primary modes [14,15].) The matrix of coupling coefficients, $i\langle \tilde{\varphi}_\alpha | \dot{\varphi}_{\alpha'} \rangle$, is not hermitian in general and is evaluated in the following sections. Indeed, by differentiating the biorthogonality condition for the generalized basis [see Eqs. (22)], we obtain

$$[i\langle \tilde{\varphi}_{\alpha'} | \dot{\varphi}_\alpha \rangle]^* = -i\langle \dot{\varphi}_\alpha | \tilde{\varphi}_{\alpha'} \rangle = i\langle \varphi_\alpha | \dot{\tilde{\varphi}}_{\alpha'} \rangle \neq i\langle \tilde{\varphi}_\alpha | \dot{\varphi}_{\alpha'} \rangle \tag{33}$$

where the star is the complex conjugation. One can always choose a basis with zero diagonal elements [19,20], $\langle \dot{\tilde{\varphi}}_\alpha | \varphi_\alpha \rangle \equiv -\langle \tilde{\varphi}_\alpha | \dot{\varphi}_\alpha \rangle = 0$, that is, one can always normalize the time-dependent amplitudes so that the snapshot eigenfrequencies do not depend explicitly on the time derivatives of the Hamiltonian parameters (1). As shown in Section V, this choice takes into account complex Berry's phase.

In a simple phenomenological approach [6–9,20–22], one starts from the equations for amplitudes, $\bar{f}_\alpha$, corresponding to the partial states, $|\phi_\alpha\rangle$, which are analogous to Eq. (32),

$$\dot{\bar{f}}_\alpha + i\nu_\alpha(t)\bar{f}_\alpha = -\sum_{\alpha'=1}^{N} \langle \phi_\alpha | \dot{\phi}_{\alpha'} \rangle \bar{f}_{\alpha'} \tag{34}$$

and adds an imaginary part to the normal frequencies, $\nu_\alpha(t)$, or to the partial frequencies, $h_{\alpha\alpha}(t)$ [i.e., the diagonal components of the Hamiltonian (3)], in order to take into account the decay. In this approach, the coupling-coefficient matrix, $(i\langle \phi_\alpha | \dot{\phi}_{\alpha'} \rangle)$, does not include the reservoir effects and thus the interplay between the nonadiabatic coupling and reservoir effects is treated inconsistently. As discussed in the next section, the reservoir dependence of the coupling-coefficient matrix, $(i\langle \tilde{\varphi}_\alpha | \dot{\varphi}_{\alpha'} \rangle)$, is the very origin of the novel nonadiabatic effects of decay-level coupling.

Moreover, the relation (28) indicates the necessity of going out of the usual dynamical subspace spanned by discrete partial states, $|\alpha\rangle$, and introducing their nontrivial superpositions with the continuous partial states, $|\mathbf{k}\rangle$, [see Eq. (31)] to obtain the closed set of equations (32) (i.e., to construct the correct nonhermitian time-dependent Schrödinger equation). This result is very difficult to guess phenomenologically, that is, without using the complex spectral decomposition (4) and introducing explicitly dressed discrete decaying levels (cf. [9,20]).

The presence of a nonadiabatic behavior is shown in Eq. (32) explicitly through the time derivatives, $|\dot{\varphi}_{\alpha'}\rangle$. Once we know their time dependence, as well as that of the eigenfrequencies, several methods are available for solving the fundamental equation (32). Usually, two different cases have been studied: (1) where the Hamiltonian is a periodic function of time and (2) where the Hamiltonian changes only during some finite time interval, $[0, t_s]$. In the former, the problem is to find a monodromy matrix and quasi-energies of the set of equations (32) [10–12,57]. In the latter, one usually considers the so-called depletion problem, that is, the calculation of final amplitudes, $f_\alpha(t_s)$, of all discrete states starting from any given state, $|\varphi_{\alpha_0}\rangle : f_\alpha(0) = \delta_{\alpha\alpha_0}$. The simplest method of solving the depletion problem is based on a perturbation theory in the case of a weak (but nonadiabatic) coupling, when $|f_{\alpha_0}(t)| \gg |f_\alpha(t)|$ for all $\alpha \neq \alpha_0$ [8,9,12,17–19]. Concerning this and other methods, see reviews [16–19] and references therein.

IV. NOVEL NONADIABATIC EFFECTS DUE TO DECAY

First, we remark that the nonadiabatic terms in Eq. (32) (right-hand side) exist even for noncrossing levels whose complex energies are constant or almost constant in time, $\omega_\alpha(t) \simeq$ constant, because of the redressing of the time-dependent reservoir states. To extract systematically all the effects missed in the previous phenomenological approaches, we calculate the coupling-coefficient matrix, $(i\langle\tilde{\varphi}_\alpha|\dot{\varphi}_{\alpha'}\rangle)$, explicitly from the formula for the complex eigenstates (13) and (14). Tedious but straightforward calculation for $\alpha \neq \beta$ gives

$$\langle\tilde{\varphi}_\alpha|\dot{\varphi}_\beta\rangle = C_{\alpha\beta} \sum_{\alpha', \beta'=1}^{N} \xi_{\alpha\alpha'}^{-1}(\omega_\alpha)\xi_{\beta'\beta}(\omega_\beta) \times \left\{\dot{h}_{\alpha'\beta'} + \sum_{\mathbf{k}'} \frac{\dot{V}_{\alpha'\mathbf{k}'}^* V_{\beta'\mathbf{k}'}}{[\omega_\beta - \omega_{k'}]_+} + \sum_{\mathbf{k}'} \frac{V_{\alpha'\mathbf{k}'}^* \dot{V}_{\beta'\mathbf{k}'}}{[\omega_\alpha - \omega_{k'}]_+}\right\} \tag{35}$$

where the dots stand for the derivative with respect to the time and the constant factor, $C_{\alpha\beta}$, is given by

$$C_{\alpha\beta} = \frac{1}{\omega_\beta - \omega_\alpha}\left(\frac{g_\beta}{g_\alpha}\right)\frac{1}{\sqrt{\eta_\alpha'(\omega_\alpha)\eta_\beta'(\omega_\beta)}} \tag{36}$$

The matrix (35) depends on the reservoir not only explicitly, but also implicitly, through the normalization constants, g_α, $\eta_\alpha'(\omega_\alpha)$, and the matrix elements, $\xi_{\alpha\alpha'}^{-1}(\omega_\alpha)$, $\xi_{\beta\beta'}(\omega_\beta)$ [cf. Eqs. (8), (9), (18), and (19)]. Note that when there is no interaction with the reservoir, $V_{\alpha'\mathbf{k}'} = 0$, only the first term of Eq. (35) remains and the matrix $\xi_{\alpha\alpha'}$ is determined by the dynamical (partial) Hamiltonian, $h_{\alpha\alpha'}$, being unitary and frequency independent. In this case, according to the definitions (7)–(9), we have

$$|\varphi_\alpha\rangle = g_\alpha|\phi_\alpha\rangle \equiv g_\alpha|u_\alpha\rangle, \qquad \langle\tilde{\varphi}_{\alpha'}| = g_{\alpha'}^{-1}\langle\phi_{\alpha'}| \equiv g_{\alpha'}^{-1}\langle v_{\alpha'}| \tag{37}$$

so that Eq. (32) coincides with Eq. (34), where $\bar{f}_\alpha = g_\alpha f_\alpha$, and g_α can be so chosen that no diagonal coupling coefficients there appear, for example, $g_\alpha = 1$ or $g_\alpha = \bar{g}_\alpha$ with

$$\dot{\bar{g}}_\alpha/\bar{g}_\alpha = -\langle\phi_\alpha|\dot{\phi}_\alpha\rangle \equiv -\sum_{\alpha'=1}^{N} \xi_{\alpha\alpha'}^{-1}\,\dot{\xi}_{\alpha'\alpha} \tag{38}$$

As explained in the previous section, in the phenomenological approaches, decay is taken into account by adding imaginary parts to the partial energies, $h_{\alpha\alpha}(t)$, or normal energies, $v_\alpha(t)$. But the unperturbed coupling coefficients, $\langle\phi_{\alpha'}|\dot{\phi}_\alpha\rangle$, are still used. On the contrary, formula (35) takes into account, consistently, the effects of decay even on the coupling coefficients and yields the following possibilities:

1. Let us assume that the interaction is such that the second and third terms of Eq. (35) are much smaller than the first. Usually, only this case is

considered, and the interaction with the reservoir is supposed to be time independent. In spite of this assumption, the decay rates acquire time dependence induced by the time dependence of the subsystem Hamiltonian (3), $h(t)$. And this time dependence, $-\omega''_\alpha(t)$, can yield the nonadiabatic effects by itself. On the other hand, since the matrix elements $\xi_{\alpha\alpha'}$ depend on the interaction between the subsystem and the reservoir, even the first term of Eq. (35) produces nonadiabatic effects, the order of which is $O(V^2)f_\alpha$, or the same order of the decay term, $-\omega''_\alpha f_\alpha$. As a result, the nonadiabatic effects of decay cannot be discussed separately from the unitary nonadiabatic effects, even if only the subsystem changes in time, $h_{\alpha\alpha'} = h_{\alpha\alpha'}(t)$.

2. Let the dynamical partial coupling, $h_{\alpha\alpha'}$, and normal frequencies, ν_α, be constant or vary adiabatically slowly, so that the unitary nonadiabatic coupling between eigenstates is absent in the usual approximation. In fact, nonadiabatic effects are still possible, owing to the time dependence of the interaction with reservoir, $V_{\alpha\mathbf{k}}(t)$, that is, to the second and third terms of Eq. (35). In other words, the time-dependent reservoir-subsystem coupling induces time-dependent redressing of the generalized eigenstates and time-dependent decay rates, $-\omega''_\alpha(t)$. As a result, nonadiabatic effects will come out. [If the redressing is neglected, one has $\langle\bar{\varphi}_\alpha|\dot{\varphi}_\beta\rangle = 0$ and, according to Eq. (32), independent eigenstates, even if the decay rates strongly depend on time. Such situation would be unphysical and, thus, is impossible.]

3. Let the eigenfrequencies, ω'_α, decay rates, ω''_α, and the matrix of coupling coefficients, $i\langle\tilde{\varphi}_\alpha|\dot{\varphi}_\beta\rangle$, be constant or vary slowly enough in a certain time interval. Then the coupling equation (32) possess N linearly independent solutions characterized by new complex eigenfrequencies, $\tilde{\Omega}_\alpha = \tilde{\Omega}'_\alpha + i\tilde{\Omega}''_\alpha$. Ignoring the reservoir, that is, assuming the matrix of coupling coefficients to be Hermitian, one finds only corrections to the real parts, $\tilde{\Omega}'_\alpha - \omega_\alpha \neq 0$, of the initial eigenfrequencies, $\omega_\alpha = \nu_\alpha$, and, hence, the nonadiabatic effects appear as a harmonic beating, that is, the periodic exchange of populations (and energies) among generalized eigenstates, $|\varphi_\alpha\rangle$. Besides, in the phenomenological approach, the decay term, $-\omega''_\alpha f_\alpha$, is introduced to the left-hand side of Eq. (34) independently. On the other hand, the consistent approach based on Eq. (32) reveals, in addition, the existence of the corrections to the imaginary parts of the eigenfrequencies, $\tilde{\Omega}''_\alpha - \omega''_\alpha \neq 0$, due to the nonhermiticity of the coupling-coefficient matrix, $i\langle\tilde{\varphi}_\alpha|\dot{\varphi}_\beta\rangle$ [cf. Eq. (33)]. Thus, side by side with beating, an additional cross-relaxation of generalized eigenstates appears. When a harmonic beating is suppressed, simple acceleration or slowing down of the decay of a certain superposition of generalized eigenstates is possible.

For concrete calculations, we need further auxiliary variables—the eigenvalues of the $N \times N$-matrix $K(\omega)$, $\mathscr{H}_\alpha(\omega)$, the matrix elements diagonalizing

$K(\omega)$, $\xi_{\alpha\alpha'}$, and the complex eigenvalues, ω_α. We obtain all these in Section VII for the simplest example in quantum optics, namely, for a system with two crossing levels ($N = 2$), which is already suitable for most of the applications. (The phenomenological approach in this case had been improved slightly in [7], but it is not sufficient to describe rigorously the nonadiabatic effects due to level mixing via reservoir.)

Finally, we discuss a "resonance" approximation implicitly used in the previous works on level crossing, where relaxation is taken into account via a perturbation method; (cf., [6–8,13,58]). This approximation does not hold unless the condition explained in the first paragraph of Section III is satisfied, that is, unless all coupled levels are relatively close to each other and their decay is not too strong:

$$-\omega''_\alpha(t), \qquad |\omega'_\alpha(t) - \omega'_{\alpha'}(t)| \ll \omega_0(t) = \frac{1}{N}\sum_{\beta=1}^{N} \omega'_\beta(t) \tag{39}$$

Now we show that, under an appropriate approximation, the consistent formulas (35) and (7)–(9) lead to the "resonance" approximation with a new explicit form of a true nonhermitian Hamiltonian. Bearing in mind a smooth dependence of the molecule-reservoir coupling, $V_{\alpha\mathbf{k}}$, on a mode frequency, $\omega_{k'}$, the frequency in the matrix (7), $K(\omega)$, can be fixed to the average energy of coupling levels, $\omega = \omega_0(t)$, and a small difference between the dispersive parameters (19) and unity can be ignored: $\eta'_\alpha(\omega_\alpha) \simeq 1$. Then, using the tilde to denote the solutions of the eigenvalue problem (8) and (9) for the fixed nonhermitian matrix $\tilde{K}_{\alpha\alpha'} = K_{\alpha\alpha'}(\omega_0)$,

$$\begin{gathered}\tilde{\omega}_\alpha = \mathscr{H}_\alpha(\omega_0), \qquad |\tilde{u}_\alpha\rangle = |u_\alpha(\omega_0)\rangle, \qquad \langle\tilde{v}_\alpha| = \langle v_\alpha(\omega_0)|, \\ \tilde{\xi}_{\alpha'\alpha} = \xi_{\alpha'\alpha}(\omega_0) \neq \hat{\xi}^{-1*}_{\alpha\alpha'}\end{gathered} \tag{40}$$

we can write the approximate formula for the matrix of coupling coefficients (35) for $\alpha \neq \beta$:

$$\langle\tilde{\varphi}_\alpha|\dot{\varphi}_\beta\rangle \simeq g_\beta g_\alpha^{-1}\langle\tilde{v}_\alpha|\dot{\tilde{u}}_\beta\rangle = \frac{1}{\tilde{\omega}_\beta - \tilde{\omega}_\alpha}\left(\frac{g_\beta}{g_\alpha}\right)\sum_{\alpha',\beta'=1}^{N} \tilde{\xi}^{-1}_{\alpha\alpha'}\dot{\tilde{K}}_{\alpha'\beta'}\tilde{\xi}_{\beta'\beta} \tag{41}$$

This result corresponds to the proper neglection of the reservoir degrees of freedom and the approximate substitution

$$|\varphi_{\alpha'}\rangle \to g_{\alpha'}|\tilde{u}_{\alpha'}\rangle, \qquad \langle\tilde{\varphi}_\alpha| \to g_\alpha^{-1}\langle\tilde{v}_\alpha|, \qquad h_{\alpha\alpha'} \to \tilde{K}_{\alpha\alpha'}(t) \tag{42}$$

[cf. Eq. (37)], and thus, to the replacement of the total hermitian Hamiltonian (1) ≡ (4) with the nonhermitian Hamiltonian:

$$\tilde{K}(t) = K(\omega_0, t) \equiv \sum_{\alpha=1}^{N} \tilde{\omega}_\alpha(t) \,|\, \tilde{u}_\alpha(t)\rangle\langle\tilde{v}_\alpha(t)\,| \tag{43}$$

which acts on the N-dimensional Hilbert space spanned by a complete biorthogonal basis $\{|\,\tilde{u}_\alpha\rangle, \langle\tilde{v}_\alpha\,|\}$. Note that, in general, there appear nonhermitian corrections in all $N \times N$ elements of the dynamical (partial) hermitian Hamiltonian (3), $h_{\alpha\alpha'}$. So, even in the "resonance" approximation (39)–(41), the coupling equations (32) for decaying eigenstates are different from the phenomenological equations (34) with improved normal ($v_\alpha \to \tilde{\omega}_\alpha$) or partial ($h_{\alpha\alpha} \to \tilde{K}_{\alpha\alpha}$) frequencies, because such improvements cannot take the proper coefficients of a nonadiabatic coupling (41) into account. The normalization factors, g_α, in Eqs. (41) and (42) may be chosen arbitrarily, for example, $g_\alpha = 1$, but, as in the case without reservoir (38), there is a choice, $g_\alpha = \tilde{g}_\alpha$, which reduces the diagonal coupling coefficients in Eq. (32) to zero:

$$\dot{\tilde{g}}_\alpha/\tilde{g}_\alpha = -\langle\tilde{v}_\alpha\,|\,\dot{\tilde{u}}_\alpha\rangle \equiv -\sum_{\alpha'=1}^{N} \tilde{\xi}^{-1}_{\alpha\alpha'}\,\dot{\tilde{\xi}}_{\alpha'\alpha} \simeq -\sum_{\alpha'=1}^{N} \xi^{-1}_{\alpha\alpha'}(\tilde{\omega}_0)\dot{\xi}_{\alpha'\alpha}(\tilde{\omega}_0) \tag{44}$$

The accuracy of the "resonance" approximation relies mainly on the smallness of the parameters in Eq. (39), but its detail depends on the concrete oscillograms of the functions (35) and (41), so that the accuracy should be examined case by case.

V. GENERALIZED COMPLEX BERRY'S PHASE

Let us show that the present method provides a generalization of the Berry's phase. As mentioned before, the generalized eigenfunctions (13)–(14), $\{|\,\varphi_\alpha\rangle, \langle\tilde{\varphi}_\alpha\,|\}$, contain an arbitrary normalization constant g_α, which can always be determined so that $\langle\tilde{\varphi}_\alpha\,|\,\dot{\varphi}_\alpha\rangle = 0$. This relation means, according to Eqs. (27)–(29), that for each instantaneous eigenstate, $|\,\psi\rangle = f_\alpha\,|\,\varphi_\alpha\rangle$, the local phase shift is determined by the solution of the dispersion relation (18) via $\dot{f}_\alpha/f_\alpha = -i\omega_\alpha(t)$ and does not depend explicitly on any time derivative. Now we show that $\langle\tilde{\varphi}_\alpha\,|\,\dot{\varphi}_\alpha\rangle = 0$. Straightforward calculation gives

$$\langle\tilde{\varphi}_\alpha\,|\,\dot{\varphi}_\alpha\rangle = \frac{d}{dt}\ln g_\alpha + \langle\tilde{\varphi}_\alpha\,|\,\dot{\varphi}_\alpha\rangle_1 \tag{45}$$

where

$$\langle\tilde{\varphi}_\alpha|\dot{\varphi}_\alpha\rangle_1 \equiv \langle\tilde{\varphi}_\alpha|\dot{\varphi}_\alpha\rangle\Big|_{g_\alpha=1} = \sum_{\alpha'=1}^{N} \frac{\xi_{\alpha\alpha'}^{-1}(\omega_\alpha)}{[\eta'_\alpha(\omega_\alpha)]^{1/2}}$$
$$\times \left\{\frac{d}{dt}\frac{\xi_{\alpha'\alpha}(\omega_\alpha)}{[\eta'_\alpha(\omega_\alpha)]^{1/2}} - \sum_{\mathbf{k}} \frac{V^*_{\alpha'\mathbf{k}}}{[\omega_\alpha-\omega_k]^3_+} \sum_{\beta'=1}^{N} \frac{V_{\beta'\mathbf{k}}\xi_{\beta'\alpha}(\omega_\alpha)}{[\eta'_\alpha(\omega_\alpha)]^{1/2}}\frac{d\omega_\alpha}{dt}\right.$$
$$\left. + \sum_{\mathbf{k}} \frac{V^*_{\alpha'\mathbf{k}}}{[\omega_\alpha-\omega_k]^2_+}\frac{d}{dt}\sum_{\beta'=1}^{N} \frac{V_{\beta'\mathbf{k}}\xi_{\beta'\alpha}(\omega_\alpha)}{[\eta'_\alpha(\omega_\alpha)]^{1/2}}\right\} \tag{46}$$

Therefore, the condition $\langle\tilde{\varphi}_\alpha|\dot{\varphi}_\alpha\rangle = 0$ leads to the equation for g_α [19]:

$$\frac{d}{dt}\ln g_\alpha = -\langle\tilde{\varphi}_\alpha|\dot{\varphi}_\alpha\rangle_1 \tag{47}$$

which admits an exponential function of some complex phase as a solution:

$$g_\alpha = \exp i\Delta_\alpha(t), \qquad \Delta_\alpha = i\int_0^t \langle\tilde{\varphi}_\alpha|\dot{\varphi}_\alpha\rangle_1\, dt' \tag{48}$$

The quantity $\langle\tilde{\varphi}_\alpha|\dot{\varphi}_\alpha\rangle_1$ can also be expressed as

$$\langle\tilde{\varphi}_\alpha|\dot{\varphi}_\alpha\rangle_1 = \frac{1}{2\dot{\omega}_\alpha}\{\langle\tilde{\varphi}_\alpha|\dot{H}(t)|\dot{\varphi}_\alpha\rangle - \langle\dot{\tilde{\varphi}}_\alpha|\dot{H}(t)|\varphi_\alpha\rangle\}_{g_\alpha=1} \tag{49}$$

Therefore, if the eigenfrequency, ω_α, is real and the right and left eigenstates coincide, $|\varphi_\alpha\rangle = |\tilde{\varphi}_\alpha\rangle$, the quantity, $\langle\tilde{\varphi}_\alpha|\dot{\varphi}_\alpha\rangle_1$, becomes purely imaginary and thus the phase, Δ_α, becomes real (as discussed in [11] and [59]). However, when decay exists, the phase Δ_α becomes complex. [This result may be understood by comparing Eqs. (38) and (44) which are the approximate versions of Eq. (47).]

The phase Δ_α provides a generalization of the Berry's phase, $\Delta_\alpha(t_s)$, if one supposes that an adiabatic approximation is valid and a Hamiltonian returns to the initial value, $H(t_s) = H(0)$, at some moment, t_s [59] (see also [60]). In this case, the Schrödinger equation

$$i\frac{\partial}{\partial t}|\psi(t)\rangle = H(t)|\psi(t)\rangle, \qquad |\psi(0)\rangle = |\varphi_\alpha(0)\rangle_1 \qquad g_\alpha \equiv 1 \tag{50}$$

together with the adiabatic assumption $|\psi(t)\rangle = a_\alpha(t)|\varphi_\alpha(t)\rangle_1$ gives a closed equation for the amplitude a_α:

$$\dot{a}_\alpha(t) + i\omega_\alpha(t)a_\alpha(t) = -\langle\tilde{\varphi}_\alpha|\dot{\varphi}_\alpha\rangle_1 a_\alpha(t), \qquad a_\alpha(0) = 1 \tag{51}$$

which leads to

$$a_\alpha(t_s) = \exp\left[-i\int_0^{t_s}\omega_\alpha(t')\,dt'\right]\exp\left[-\int_0^{t_s}\langle\tilde{\varphi}_\alpha|\dot{\varphi}_\alpha\rangle_1\,dt'\right] \tag{52}$$

Thus, the factor (48) characterizes that variance of a complex amplitude of the generalized eigenvector, $|\varphi_\alpha\rangle_1$, which takes place in addition to the usual time dependence, $f_\alpha(0)\exp[-i\int_0^{t_s}\omega_\alpha(t')\,dt']$, that is, in addition to a dynamical rotation of the wave-function phase and an exponential decay of the amplitude determined by Eq. (32). Thus, we find a rigorous justification for the phenomenological approach to the Berry's phase in the system with a nonhermitian Hamiltonian [61].

The present analysis [cf. Eq. (46)] shows that a decay yields a nontrivial dependence of a generalized complex Berry's phase on a trajectory of a generalized eigenstate, $|\varphi_\alpha\rangle_1$, in the extended Hilbert space. It can be shown that the Berry-phase construction for nonadiabatic unclosed trajectories suggested in [62,63] is still valid for decaying generalized states. We are not going to discuss a similar generalization for the formula (48), because for a nonadiabatic crossing, a total complex amplitude of the discrete-level-occupation probability is of physical interest, but not its separate factor, g_α. Exhaustive information is contained in a solution of coupling equations (32), where a generalized-eigenstate normalization is so chosen that the normalization factor, g_α, takes into account the generalized complex Berry's phase (48).

VI. TIME EVOLUTION OF LYAPUNOV-TYPE QUANTITY

Through complex spectral decomposition, one can introduce a representation for the time-independent one-level Friedrichs model [27], where the dynamics is manifestly irreversible, and a Lyapunov-type quantity which decreases monotonically. Here we disuss changes in these aspects caused by the explicit time dependence of the Hamiltonian.

As in the time-independent Friedrichs model [27], we introduce a new representation in the N-level case through

$$|\Psi_\Lambda\rangle \equiv \Lambda|\Psi\rangle \tag{53}$$

where the nonunitary transformation, Λ, is defined in Eq. (24). Then the Schrödinger equation becomes

$$i\frac{\partial}{\partial t}|\Psi_\Lambda\rangle \equiv i\frac{\partial}{\partial t}\Lambda|\Psi\rangle = \Lambda H(t)\Lambda^{-1}|\Psi_\Lambda\rangle + i\left(\frac{\partial\Lambda}{\partial t}\right)\Lambda^{-1}|\Psi_\Lambda\rangle \equiv H_\Lambda(t)|\Psi_\Lambda\rangle \tag{54}$$

where the transformed Hamiltonian, $H_\Lambda(t)$, is given by

$$H_\Lambda(t) = \sum_{\alpha=1}^{N}\omega_\alpha(t)|\alpha\rangle\langle\alpha| + \sum_{\mathbf{k}}\omega_k|\mathbf{k}\rangle\langle\mathbf{k}| + i\left(\frac{\partial\Lambda}{\partial t}\right)\Lambda^{-1} \tag{55}$$

Note that the transformed Hamiltonian is not diagonal because of the third term, which arises from the nonadiabatic behavior. Indeed, the third term, restricted to the dynamical subspace spanned by $\{|\alpha\rangle\}$, precisely corresponds to the nonadiabatic coupling term of Eq. (32):

$$\begin{aligned} i\frac{\partial\Lambda}{\partial t}\Lambda^{-1}\bigg|_{\text{dis}} &\equiv P_d\left(i\frac{\partial\Lambda}{\partial t}\Lambda^{-1}\right)P_d = \sum_{\alpha,\alpha'=1}^{N} i\langle\dot{\tilde{\varphi}}_\alpha|\varphi_{\alpha'}\rangle|\alpha\rangle\langle\alpha'| \\ &\equiv -\sum_{\alpha,\alpha'=1}^{N} i\langle\tilde{\varphi}_\alpha|\dot{\varphi}_{\alpha'}\rangle|\alpha\rangle\langle\alpha'| \end{aligned} \tag{56}$$

where $P_d \equiv \sum_{\alpha=1}^{N}|\alpha\rangle\langle\alpha|$ is the projection operator to the space of partial discrete states.

The appearance of the nondiagonal part to the transformed Hamiltonian due to the nonadiabatic effects suggests the qualitative change of the dynamics as compared with the time-independent case. Thus it is interesting to see the behavior of the Lyapunov-type variable defined similarly to that for the one-level Friedrichs model [27]:

$$\mathcal{Y} \equiv \Lambda^\dagger\Lambda - I = \sum_{\alpha=1}^{N}|\tilde{\varphi}_\alpha\rangle\langle\tilde{\varphi}_\alpha| \tag{57}$$

Its expectation value for a given state, $|\psi(t)\rangle$, is

$$\mathcal{Y}_\psi \equiv \langle\psi(t)|\mathcal{Y}|\psi(t)\rangle = \sum_{\alpha=1}^{N}|f_\alpha(t)|^2 \tag{58}$$

which also represents the total population of the subspace of the generalized discrete states. Within the range of their validity, the level-coupling

Eq. (32) leads to

$$\frac{d\mathscr{Y}_\psi}{dt} = 2\sum_{\alpha=1}^{N} \text{Im}\ \omega_\alpha(t)\,|f_\alpha|^2 - \sum_{\alpha,\alpha'=1}^{N} \{\langle\dot{\varphi}_{\alpha'}|\tilde{\varphi}_\alpha\rangle + \langle\tilde{\varphi}_{\alpha'}|\dot{\varphi}_\alpha\rangle\} f_\alpha f^*_{\alpha'} \tag{59}$$

As Im $\omega_\alpha(t) < 0$ for decaying levels, the first term always drives $\mathscr{Y}_\psi$ to monotonic decrease. The second term resulting from the nonadiabatic effects may cause a qualitative change of the evolution of the Lyapunov-type quantity $\mathscr{Y}_\psi$.

Hence, the nonadiabatic interaction between the discrete subsystem and the reservoir can alter the qualitative behavior of the net decay property of the discrete subsystem. In particular, the population of some levels may even be regenerated via faster and deeper depletion of other discrete levels and their redressing with the reservoir states. So, the nonadiabatic coupling among discrete levels can change their lifetimes and provide a way to investigate the very nature of unstable states (or particles). In a wave-propagation theory, an analogous formula gives a variation law of the wave-field energy [13–19]. It cannot be seen in the nonsteady (driven) one-level Friedrichs model, although in this case the monotonicity of the evolution of the Lyapunov-type quantity may be also destroyed by a time-dependent (e.g., periodic) external force, as was shown by Rosenberg and Petrosky [64].

VII. LINEAR COUPLING OF EXCITED QUASI-ENERGY STATES IN A THREE-LEVEL MOLECULE DRIVEN BY NONSTEADY dc AND ac FIELDS

To illustrate the general theory of Sections III and IV, let us study a nonadiabatic effect of spontaneous decay in the coherent dynamics of a three-level molecule with nondegenerate energy levels, E_0, E_1, and E_2 ($E_0 < E_1 < E_2$) under external electric fields. The external fields which we consider consist of nonstationary dc and ac components: the former affects the dipole momenta and energy levels via the usual Stark effect and the latter is nearly in resonance with the transition between two excited levels, $E_2 - E_1$ (optical Stark effect). A three-level scheme is the basic one in quantum optics and proves its value in treating the resonance interaction of a molecule with quasi-monochromatic electromagnetic fields which interact mainly with three transitions: $E_2 - E_1$, $E_1 - E_0$, and $E_2 - E_0$.

A. Reduction to the Friedrichs–Fano Model

A nonstationary (pulse-like) dc field, $\vec{E}_d(t)$, causes temporal changes of three dipole momenta, $\vec{\mu}(t)$, $\vec{\mu}_1(t)$, $\vec{\mu}_2(t)$, corresponding to the transitions $E_2 - E_1$,

$E_1 - E_0$, and $E_2 - E_0$, respectively, and of three energy levels, $E_{0,1,2}(t)$, via the Stark effect. For simplicity, we assume that the Stark effect for the excited levels 1 and 2 is quadratic. It is realized if the energy states, $|1\rangle$ and $|2\rangle$, are in opposite parity eigenstates and, hence, have zero constant dipole momenta. Furthermore, if the ground state, $|0\rangle$, is in a parity eigenstate, one optical transition, $1 \to 0$ or $2 \to 0$, is forbidden in the dipole approximation (as $\vec{\mu}_1$ or $\vec{\mu}_2$ is zero) and a spontaneous decay of the mixed excited states takes place through another optical transition (e.g., $2 \to 0$).

A nonsteady external ac field,

$$\vec{E}_a(t) = \frac{1}{2}\vec{\mathscr{E}}(t)\exp\left[-i\int_0^t \Omega(t')\,dt'\right] + \text{c.c.} \qquad |\dot{\Omega}/\Omega| \ll \Omega \tag{60}$$

is assumed to mix two excited levels, that is, to be nearly in resonance with the transition $1 \leftrightarrow 2$:

$$|E_2 - E_1 - \Omega| \ll \Omega \sim E_2 - E_1 \ll E_{2,1} - E_0 \tag{61}$$

In Eq. (61), we further assume that the transition frequency between excited levels is much lower than the optical frequencies of the transitions to the ground level, E_0. Then, as is well-known, one may ignore the spontaneous transition between excited levels, $1 \leftrightarrow 2$, because the probability of the transition (i.e., Einstein's coefficient), is proportional to the cube of the transition frequency, $A(\omega) \propto \omega^3$.

In addition, the spontaneous optical transitions to the ground level ($2 \to 0$, $1 \to 0$) are taken into account because of their principal role. This process arises from the interaction between excited molecular states and the vacuum of optical modes, which is described by the second quantized electric field:

$$\vec{E}_V = i\sum_{p=1}^{2}\sum_{\mathbf{k}}\sqrt{\frac{2\pi\omega_k}{L^3}}\,\vec{e}_{\mathbf{k}}^{(p)}(\hat{b}_{\mathbf{k}}^{(p)}A^{\mathbf{k}} - \hat{b}_{\mathbf{k}}^{(p)\dagger}A_{\mathbf{k}}^*) \tag{62}$$

where $\hat{b}_{\mathbf{k}}^{(p)}$ and $\hat{b}_{\mathbf{k}}^{(p)\dagger}$ ($p = 1, 2$) are annihilation and creation operators of a mode, $A_{\mathbf{k}} = \exp(i\mathbf{k}\cdot\mathbf{r})$ (in a free space), with a frequency, ω_k, and a (unit) polarization vector, $\vec{e}_{\mathbf{k}}^{(p)} \perp \mathbf{k}$, $\vec{e}_{\mathbf{k}}^{(1)} \perp \vec{e}_{\mathbf{k}}^{(2)}$. We show that, under the influence of the nonsteady Stark effect, the spontaneous decay can cause a nontrivial effect on the evolution of level populations. In particular, there is a possibility of speeding up (or slowing down) the decay rate of some level through the nonadiabatic interaction with other levels which decay faster (or slower).

Now we consider the Hamiltonian of the system. Because we are not interested in the induced molecular emission and, hence, in many-optical-photon states, the total Hilbert space may be restricted to the direct sum of one-photon states with ground molecular level 0, $|\mathbf{k}p\rangle$, and two different molecular excited states without photons, $|\bar{1}\rangle$ and $|\bar{2}\rangle$. Under the well-known rotating wave approximation [12,35,58], we obtain a time-dependent Hamiltonian describing the interaction between two excited levels ($N = 2$):

$$\begin{aligned}\bar{H} &= \bar{E}_1(t)|\bar{1}\rangle\langle\bar{1}| + \bar{E}_2(t)|\bar{2}\rangle\langle\bar{2}| \\ &\quad - \gamma(t)\exp\left(-i\int_0^t \Omega\, dt'\right)|\bar{2}\rangle\langle\bar{1}| - \gamma^*(t)\exp\left(i\int_0^t \Omega\, dt'\right)|\bar{1}\rangle\langle\bar{2}| \\ &\quad + \sum_{\beta=1}^{2}\sum_{p=1}^{2}\sum_{\mathbf{k}} i\sqrt{\frac{2\pi\omega_k}{L^3}}\{(\vec{\mu}_\beta^*(t)\cdot\vec{e}_{\mathbf{k}}^{(p)})\exp[-i\mathbf{k}\cdot\mathbf{r}(t)]|\mathbf{k}p\rangle\langle\bar{\beta}| \\ &\quad - (\vec{\mu}_\beta(t)\cdot\vec{e}_{\mathbf{k}}^{(p)})\exp[i\mathbf{k}\cdot \mathrm{r}(t)]|\bar{\beta}\rangle\langle\mathbf{k}p|\} \\ &\quad + \sum_{p=1}^{2}\sum_{\mathbf{k}}\omega_k|\mathbf{k}p\rangle\langle\mathbf{k}p|\end{aligned} \tag{63}$$

Here $\bar{E}_{1,2} = E_{1,2} - E_0$ denote energy differences between excited and ground levels and we introduce the Rabi frequency, $\gamma(t) = \vec{\mathscr{E}}(t)\cdot\vec{\mu}(t)/2$, which corresponds to the energy of the interaction between the molecule and the ac field, $\vec{\mathscr{E}}$. The Rabi frequency can always be made real by properly redefining the time-dependent frequency $\Omega(t)$ in Eq. (60). The interaction between the molecule and the nonstationary dc field, $\vec{E}_d(t)$, is taken into account implicitly in the time-dependent parameters $E_{1,2}(t)$ and $\vec{\mu}_{1,2}(t)$, $\vec{\mu}(t)$. Note that the interaction between the molecule and optical modes depends on the position of the molecule, $\mathbf{r}(t)$.

The rapid oscillations of the dynamical (partial) molecular Hamiltonian can be removed by rotating a phase of one excited state with frequency Ω, that is, by introducing new states, for example, $|1\rangle = \exp(i\int_0^t \Omega\, dt')|\bar{1}\rangle$ and $|2\rangle = |\bar{2}\rangle$. Then, the Hamiltonian, $\bar{H}(t)$, reduces to that, $H(t)$, of the Friedrichs–Fano model, (1)–(3), with the parameters

$$h_{11} = \bar{E}_1 + \Omega \equiv \Omega_1'(t), \qquad h_{22} = \bar{E}_2 \equiv \Omega_2'(t), \qquad h_{21} \equiv h_{12}^* = -\gamma(t) \tag{64}$$

$$V_{\alpha\mathbf{k}p}(t) = i\sqrt{\frac{2\pi\omega_k}{L^2}}(\vec{\mu}_\alpha^*(t)\cdot\vec{e}_{\mathbf{k}}^{(p)})\exp\left[i\delta_{1\alpha}\int_0^t \Omega\, dt' - i\mathbf{k}\cdot\mathbf{r}(t)\right] \tag{65}$$

In the absence of an interaction with the reservoir, the dynamical (partial) Hamiltonian (3), $h(t)$, admits real eigenvalues known as quasienergies:

$$\nu_{1,2}(t) = \tfrac{1}{2}[\Omega_1' + \Omega_2' \mp \sqrt{(\Omega_2' - \Omega_1')^2 + 4|\gamma|^2}] \tag{66}$$

B. Coupling Equations for Decaying Discrete States

Because the system under consideration is equivalent to the two-level Friedrichs–Fano model, the evolution of amplitudes of the decaying discrete states is described by Eq. (32), derived in Section III. To write down the equations, we need instantaneous discrete eigenvalues of the total Hamiltonian, ω_α, and mixing matrix elements, $\{\langle\tilde{\varphi}_\alpha|\dot{\varphi}_\beta\rangle\}$.

As discussed in Section II [cf. Eqs. (7–9) and (18)], the discrete eigenvalues are determined from the matrix, $\{K(\omega)_{\alpha\alpha'}\}$, defined in Eq. (7). To avoid the QED ultraviolet divergences in the integrals involved in the matrix, K, we assume that the dipole momenta $\vec{\mu}_\beta$ have additional dependence on the problem frequency ω_k in such a way that they are constant in relatively narrow frequency intervals, $2\bar{\omega}$, and zero otherwise: $\vec{\mu}_{1,2}(\omega) =$ const for $\omega \in (\bar{E}_{1,2} - \bar{\omega}, \bar{E}_{1,2} + \bar{\omega})$ and $\vec{\mu}_{1,2}(\omega) = 0$ for $\omega \notin (\bar{E}_{1,2} - \bar{\omega}, \bar{E}_{1,2} + \bar{\omega})$. Note that the auxiliary band width, $2\bar{\omega}$, does not enter the final results when the condition $\bar{E}_2 - \bar{E}_1 \ll \bar{\omega} \ll \bar{E}_1 + \bar{E}_2$ is satisfied.

Then, integrating over orientations of a wave vector, $\mathbf{k}$, and over frequency, $\omega = c|\mathbf{k}|$, we have

$$K(\omega, t)_{11,22} \simeq \Omega_{1,2}'(t) - i\frac{2}{3}\left(\frac{\omega}{c}\right)^3 |\vec{\mu}_{1,2}(t)|^2 \equiv \Omega_{1,2}'(t) - i\Gamma_{1,2}(\omega, t) \tag{67}$$

$$\begin{aligned} K(\omega, t)_{12}^* + \gamma(t) &= -[K(\omega, t)_{21} + \gamma(t)] \\ &\simeq i\frac{2}{3}\left(\frac{\omega}{c}\right)^3 (\vec{\mu}_1^*(t)\cdot\vec{\mu}_2(t)) \exp\left(+i\int_0^t \Omega\, dt'\right) \end{aligned} \tag{68}$$

Hence, we have

$$[K(\omega, t)_{\alpha\alpha'}] = \begin{bmatrix} \Omega_1'(t) - i\Gamma_1(\omega, t), & -\gamma^*(t) \\ -\gamma(t), & \Omega_2'(t) - i\Gamma_2(\omega, t) \end{bmatrix} \tag{69}$$

and, assuming further, for simplicity, that the dipole momenta of optical transitions are almost orthogonal, $\vec{\mu}_1^* \cdot \vec{\mu}_2 \simeq 0$, or

$$|\vec{\mu}_1^* \cdot \vec{\mu}_2| \ll ||\vec{\mu}_1|^2 - |\vec{\mu}_2|^2| \tag{70}$$

we can diagonalize the matrix (69) by the nonhermitian matrix of the frequency-dependent transformation (9), $(\xi_{\alpha\alpha'})$,

$$(\xi_{\alpha\alpha'}) = \frac{1}{\sqrt{\theta^2 + |\gamma|^2}} \begin{pmatrix} \gamma^* & -\theta \\ \theta & \gamma \end{pmatrix}, \qquad (\xi^{-1}_{\alpha\alpha'}) = \frac{1}{\sqrt{\theta^2 + |\gamma|^2}} \begin{pmatrix} \gamma & \theta \\ -\theta & \gamma^* \end{pmatrix} \tag{71}$$

$$\theta(\omega) \equiv \tfrac{1}{2}\{\Omega_1' - i\Gamma_1(\omega) - \Omega_2' + i\Gamma_2(\omega) + \sqrt{[\Omega_2' - i\Gamma_2(\omega) - \Omega_1' + i\Gamma_1(\omega)]^2 + 4|\gamma|^2}\} \tag{72}$$

The eigenvalues of the matrix (69) are complex: $\mathscr{H}_1(\omega) = \Omega_1' - i\Gamma_1(\omega) - \theta(\omega)$ and $\mathscr{H}_2(\omega) = \Omega_2' - i\Gamma_2(\omega) + \theta(\omega)$. They enter the characteristic equation (18), which we have to solve.

Under the "resonance" approximation explained at the end of Section IV, which is valid here because the condition (61), the discrete eigenvalues of the total Hamiltonian (i.e., generalized quasi-energies; cf. [10–12]), are given by

$$\begin{aligned} \omega_{1,2}(t) \simeq \tilde{\omega}_{1,2}(t) &= \tfrac{1}{2}[\Omega_1 + \Omega_2 \mp \sqrt{(\Omega_2 - \Omega_1)^2 + 4|\gamma|^2}] \\ \Omega_{1,2}(t) &\equiv \Omega_{1,2}'(t) - i\bar{\Gamma}_{1,2}(t) \\ \bar{\Gamma}_{1,2} &= \Gamma_{1,2}((\Omega_1' + \Omega_2')/2) \simeq \Gamma_{1,2}(\Omega_{1,2}') \end{aligned} \tag{73}$$

The relaxation rates, $\bar{\Gamma}_{1,2}$, are equal to one half of the Einstein's coefficients. $A(\Omega_{1,2}')$, at the partial (optical) transition frequencies, $\Omega_{1,2}'$. In this case, the corresponding generalized eigenstates (i.e., quasi-energy states), are given in terms of Eqs. (13) and (14) of Section II by changing $\eta_\alpha(\omega_\alpha)'$ to 1, $\xi_{\alpha\alpha'}$ to $\tilde{\xi}_{\alpha\alpha'}$ and $\xi^{-1}_{\alpha\alpha'}$ to $\tilde{\xi}^{-1}_{\alpha\alpha'}$, where $\tilde{\xi}_{\alpha\alpha'}$ and $\tilde{\xi}^{-1}_{\alpha\alpha'}$ are obtained from Eq. (71) by changing $\theta(\omega)$ to

$$\theta_0 = \tfrac{1}{2}[\Omega_1 - \Omega_2 + \sqrt{(\Omega_2 - \Omega_1)^2 + 4|\gamma|^2}] \tag{74}$$

We now consider the mixing matrix elements, $\langle\tilde{\varphi}_\alpha|\dot{\varphi}_\beta\rangle$. Using the regularization explained above, the integrals appearing in the mixing matrix elements are calculated as

$$\sum_{p=1}^{2} \sum_{\mathbf{k}} \frac{V^*_{\alpha\mathbf{k}p} \dot{V}_{\beta\mathbf{k}p}}{[\omega_\alpha - \omega_k]_+} \simeq -i\frac{2}{3}\left(\frac{\omega_\alpha}{c}\right)^3 (\vec{\mu}_\alpha \cdot (\dot{\vec{\mu}}^*_\beta + i\delta_{1\beta}\Omega\dot{\vec{\mu}}^*_\beta)) \times \exp\left\{i(\delta_{1\beta} - \delta_{1\alpha})\int_0^t \Omega\, dt'\right\} \tag{75}$$

Assuming the validity of the "resonance" approximation, the mutual orthogonality of excited dipole momenta as before, and a slow enough change of the parameter such that

$$|\dot{\vec{\mu}}_{1,2}|/|\vec{\mu}_{1,2}|, \qquad |\dot{\tilde{\omega}}_{1,2}|/|\tilde{\omega}_2 - \tilde{\omega}_1| \ll \Omega \tag{76}$$

we have

$$\langle\tilde{\varphi}_1|\dot{\varphi}_2\rangle = \frac{g_2}{g_1}\left[\frac{\theta_0^2\dot{\gamma} - \gamma^2\dot{\gamma}^* + \theta_0\gamma(\dot{\Omega}_2 - \dot{\Omega}_1)}{(\theta_0^2 + |\gamma|^2)\sqrt{(\Omega_2 - \Omega_1)^2 + 4|\gamma|^2}} + \frac{3\Omega\theta_0\gamma\bar{\Gamma}_1}{(\theta_0^2 + |\gamma|^2)\Omega_1'}\right] \tag{77}$$

$$\langle\tilde{\varphi}_2|\dot{\varphi}_1\rangle = -\frac{g_1}{g_2}\left[\frac{\theta_0^2\dot{\gamma}^* - \gamma^{*2}\dot{\gamma} + \theta_0\gamma^*(\dot{\Omega}_2 - \dot{\Omega}_1)}{(\theta_0^2 + |\gamma|^2)\sqrt{(\Omega_2 - \Omega_1)^2 + 4|\gamma|^2}} - \frac{3\Omega\theta_0\gamma^*\bar{\Gamma}_1}{(\theta_0^2 + |\gamma|^2)\Omega_1'}\right] \tag{78}$$

When the Rabi frequency, γ, is real, the mixing matrix elements can be rewritten in a compact form:

$$\langle\tilde{\varphi}_1|\dot{\varphi}_2\rangle = -\frac{g_2}{g_1}\left[\frac{\dot{q}}{2(1+q^2)} - \frac{3\Omega\gamma\bar{\Gamma}_1}{2\Omega_1'|\gamma|\sqrt{1+q^2}}\right] \tag{77'}$$

$$\langle\tilde{\varphi}_2|\dot{\varphi}_1\rangle = \frac{g_1}{g_2}\left[\frac{\dot{q}}{2(1+q^2)} + \frac{3\Omega\gamma\bar{\Gamma}_1}{2\Omega_1'|\gamma|\sqrt{1+q^2}}\right] \tag{78'}$$

where $q \equiv (\Omega_1 - \Omega_2)/(2\gamma)$. The coupling equations of the decaying levels are then given by Eq. (32):

$$\dot{f}_1 + i\tilde{\omega}_1 f_1 = -\langle\tilde{\varphi}_1|\dot{\varphi}_2\rangle f_2, \qquad \dot{f}_2 + i\tilde{\omega}_2 f_2 = -\langle\tilde{\varphi}_2|\dot{\varphi}_1\rangle f_1 \tag{79}$$

The conditions assumed above are equivalent to those assumed in the phenomenological approaches leading to Eq. (34). However, the latter treats the nonadiabatic level mixing inconsistently, and we discuss the possible differences between the phenomenological and present approaches in the next section.

C. Novel Nonadiabatic Effects—Comparison with the Phenomenological Approach

The novel nonadiabatic effects discussed in Section IV are not taken into account by the phenomenological approach based on Eq. (34), where the mixing matrix elements are calculated in terms of the unperturbed eigenvectors. When the Rabi frequency, γ, is real (which is assumed hereafter), they are given by

$$\langle \phi_2 | \dot{\phi}_1 \rangle = -\langle \phi_1 | \dot{\phi}_2 \rangle = \frac{\dot{q}'}{2(1 + q'^2)}, \qquad q'(t) \equiv \frac{\Omega'_1 - \Omega'_2}{2\gamma} \tag{80}$$

Note that q' coincides with the real part of q defined just after Eq. (78′). Therefore, these effects can clearly be understood either by comparing the numerical solutions of the phenomenological equation (34) and our equation (32), or as the deviation of the experiments from the prediction based on the phenomenological equation (34). However, we argue that the experiments will confirm the predictions based on the correct equation (32). In the rest of this section, we investigate the observability of the expected difference between the phenomenological and our approach. Note that, because of the explicit lifetime dependence on the mixing matrix elements seen in Eqs. (77) and (78), all cases explained in Section IV are possible.

The new effects would be observable if the difference between the mixing matrix elements of the phenomenological and the present approach is greater than or of the same order as the phenomenological matrix elements:

$$|\langle \tilde{\varphi}_2 | \dot{\varphi}_1 \rangle - \langle \phi_2 | \dot{\phi}_1 \rangle| \gtrsim |\langle \phi_2 | \dot{\phi}_1 \rangle| \tag{81}$$

First, to observe the populations of and the mixing between the two excited levels, their linewidths should be smaller than both the Rabi frequency and the detuning. Because $\bar{\Gamma}_{1,2} \sim |\bar{\Gamma}_2 - \bar{\Gamma}_1|$, this implies

$$2|\gamma| \sim |\Omega'_2 - \Omega'_1| \gtrsim |\bar{\Gamma}_2 - \bar{\Gamma}_1|, \qquad \text{that is, } |q'| \sim 1 \gtrsim |q''| \tag{82}$$

where $q'' \equiv (\bar{\Gamma}_2 - \bar{\Gamma}_1)/(2\gamma)$ is the imaginary part of q defined in Eq. (78′). Therefore, by approximating $g_2/g_1 \simeq 1$ and by keeping the terms up to the first order in q'', we have

$$\langle \tilde{\varphi}_2 | \dot{\varphi}_1 \rangle - \langle \phi_2 | \dot{\phi}_1 \rangle \simeq i \frac{\dot{q}''}{2(1 + q'^2)} - iq'' \frac{q'\dot{q}'}{(1 + q'^2)^2} \tag{83}$$

which reduces the observability condition (81) to the following in the lowest order in q'':

$$|\dot{q}''| \gtrsim |\dot{q}'| \tag{84}$$

The condition (84) is satisfied in the following cases:

$$|\dot{\bar{\Gamma}}_1 - \dot{\bar{\Gamma}}_2| \gtrsim |\dot{\Omega}'_1 - \dot{\Omega}'_2| \qquad \text{for} \left|\frac{\dot{\Omega}'_1 - \dot{\Omega}'_2}{\Omega'_1 - \Omega'_2}\right| > \left|\frac{\dot{\gamma}}{\gamma}\right| \tag{85}$$

$$|\dot{\bar{\Gamma}}_1 - \dot{\bar{\Gamma}}_2| \gtrsim |2q'\dot{\gamma}| \qquad \text{for} \left|\frac{\dot{\Omega}'_1 - \dot{\Omega}'_2}{\Omega'_1 - \Omega'_2}\right| < \left|\frac{\dot{\gamma}}{\gamma}\right| \tag{86}$$

$$\left|\frac{\bar{\Gamma}_1 - \bar{\Gamma}_2}{\Omega'_1 - \Omega'_2}\right| \gtrsim \left|1 - \frac{(\dot{\Omega}'_2 - \dot{\Omega}'_1)\gamma}{(\Omega'_2 - \Omega'_1)\dot{\gamma}}\right| \qquad \text{for } \dot{\bar{\Gamma}}_1 - \dot{\bar{\Gamma}}_2 \simeq 0 \tag{87}$$

In a separate paper, by comparing the solutions to the phenomenological and the present equations (79), we show that the nonadiabatic effects of decay may be of the relative order of unity. Also, further possibilities for realization and optimization of the observability condition (84) are discussed elsewhere [70]. Here we mention just one example: if the time evolution of the resonance detuning, $\Omega'_2 - \Omega'_1 = \bar{E}_2 - \bar{E}_1 - \Omega$, is synchronized with that of the Rabi frequency, $\gamma = \vec{\mathscr{E}} \cdot \vec{\mu}/2$, such that their ratio, q', is (nearly) constant in time, the phenomenological mixing matrix (nearly) vanishes and the novel nonadiabatic effects are observable under a weaker condition, that is, whenever the time evolution of the difference of the lifetime, $\bar{\Gamma}_2 - \bar{\Gamma}_1$, is not synchronized with that of the Rabi frequency. Note that the conditions discussed above are derived under the assumption of (approximate) orthogonality (69) of the dipole momenta of the two excited levels. If this assumption is violated, the expressions of the mixing matrix elements become more complicated than Eqs. (77) and (78), but the results discussed above are still expected to be valid.

To realize the situations (85)–(87) experimentally, one needs a tuning among the time evolutions of: (1) the dc field, $\vec{E}_d(t)$, which controls the decay rates, $\bar{\Gamma}_{1,2} \propto |\vec{\mu}_{1,2}|^2$, the Rabi frequency, $\gamma \propto \mu$, and the discrete level energies, $\Omega'_{1,2} \sim \bar{E}_{1,2}$; (2) the ac field, $\vec{\mathscr{E}}(t)$, which controls the mixing of the excited levels, that is, the Rabi frequency, $\gamma \propto \mathscr{E}$; and (3) the frequency, Ω, of the ac field, which controls the partial energy, $\Omega'_1 = \bar{E}_1 + \Omega$, and, thus, the detuning frequency, $\Omega'_1 - \Omega'_2$, between the two excited levels. For example, an appropriate synchronization between $\vec{\mathscr{E}}(t)$ and $\vec{E}_d(t)$ would lead to a constant Rabi frequency, $\gamma \simeq$ const, and that between $\Omega(t)$ and $\vec{E}_d(t)$ to a constant detuning frequency, $\Omega'_1 - \Omega'_2 \simeq$ const.

Now let us give typical figures of the parameters. The typical values of the optical dipole momenta are μ, $\mu_{1,2} \simeq 1\ \text{D} = 10^{-18}$ cgsu, and the electric static field in the range of $|\vec{E}_d| \simeq 1 - 100$ MV/cm can mix molecular (at least vibrational–rotational) eigenstates and change optical dipole momenta. The typical frequencies of optical transitions, natural bandwidths, and vibrational–rotational splitting of electronic levels are respectively,

$$\Omega'_{1,2} \simeq 2 \cdot 10^{15}\ \mathrm{sec}^{-1}, \qquad \bar{\Gamma}_{1,2} \simeq 2 \cdot 10^{5}\ \mathrm{sec}^{-1}, \qquad \Omega \simeq E_2 - E_1 \simeq 10^{13}\ \mathrm{sec}^{-1} \tag{88}$$

Then, according to the observability (82) of mixing of the two excited levels, the Rabi frequency should have a value of $\gamma \gtrsim 10^5\ \mathrm{sec}^{-1}$, which corresponds to the field strength of $\mathscr{E} \gtrsim 0.06$ V/cm. All of these values are experimentally accessible. Also, it is not difficult to create a nonsteady electrostatic field, $\vec{E}_d(t)$, which varies in a few microseconds. For example, it can be realized for a molecule moving with thermal velocity, $v_T \sim 10^5$ cm/sec, through an inhomogeneous static field with millimeter-length scale.

The main experimental difficulties are preparing identically oriented and excited molecules in a small given region with nonsteady dc and ac field, and performing accurate spectroscopic measurements of populations of decaying quasi-energy levels before and after the crossing, which is separated by a few microseconds. However, we expect that recent rapid progress in the quantum optical experiments of a single molecule would remove such difficulties and allow us to detect the nonadiabatic effects experimentally.

VIII. NONADIABATIC EFFECTS IN MOLECULES—BORN–OPPENHEIMER APPROACH

Nonadiabatic crossing of discrete levels is also important in the Born–Oppenheimer approach, where the effective dynamics for the slow nuclear motion is derived by separating it from the fast electronic motion [65] and the former plays the role of a slowly varying external field on the latter. The original Born–Oppenheimer approach, which takes into account only the stable electronic states, has been generalized to deal with decaying electronic states by several authors (see [66–68] and references therein). In this section, which emphasizes a geometrical-phase contribution [69], we deal with the system described by the following Hamiltonian:

$$H_{\mathrm{tot}} = \sum_j \frac{-1}{2M_j} \nabla_j^2 + W(R) + H_{\mathrm{FF}}(R) \tag{89}$$

where M_j is the mass of jth nuclear; ∇_j is the derivative with respect to its coordinate, R_j; $W(R)$ is the interaction energy among nuclei, with R the abbreviation of a set of nuclear coordinates, $R = \{R_j\}$; and $H_{\mathrm{FF}}(R)$ is the electronic Hamiltonian of Friedrichs–Fano type (1) with R-dependent parameters [i.e., $h_{\alpha\alpha'} = h_{\alpha\alpha'}(R)$, $\omega_k = \omega_k(R)$, and $V_{\alpha\mathbf{k}} = V_{\alpha\mathbf{k}}(R)$]. According to Section II, for each parameter R, the electronic Hamiltonian, $H_{\mathrm{FF}}(R)$, admits a complex spectral decomposition,

$$H_{\mathrm{FF}}(R) = \sum_{\alpha=1}^{N} \omega_\alpha(R) | \varphi_\alpha(R)\rangle\langle\tilde{\varphi}_\alpha(R)| + \sum_{\mathbf{k}} \omega_k(R) | \varphi_{\mathbf{k}}(R)\rangle\langle\tilde{\varphi}_{\mathbf{k}}(R)| \tag{90}$$

where the right and left eigenvectors, $\{|\varphi_\alpha(R)\rangle, |\varphi_{\mathbf{k}}(R)\rangle$ and $\langle\tilde{\varphi}_\alpha(R)|, \langle\tilde{\varphi}_{\mathbf{k}}(R)|\}$, form a complete biorthonormal basis [cf. Eqs. (22) and (23)] and the discrete eigenvalues, $\omega_\alpha(R)$, can have a nonzero imaginary part, $\operatorname{Im}\omega_\alpha \neq 0$.

Using the Born–Oppenheimer approach, we look for a solution to the Schrödinger equation,

$$i\frac{\partial}{\partial t}\Psi = H_{\mathrm{tot}}\Psi \tag{91}$$

of the form

$$\Psi = \sum_{\alpha=1}^{N} \Phi_\alpha(R, t) | \varphi_\alpha(R)\rangle + \sum_{\mathbf{k}} \Phi_{\mathbf{k}}(R, t) | \varphi_{\mathbf{k}}(R)\rangle \tag{92}$$

Dirac's notation is used above to express the electronic wave functions and the coordinate representation is used for the nuclear parts. By substituting Eqs. (89) and (92) into Eq. (91) and using Eq. (90), one obtains the equation of motion for the nuclear part, Φ_α:

$$\begin{aligned} i\frac{\partial}{\partial t}\Phi_\alpha(R, t) &= \left[\sum_j \frac{-1}{2M_j}\nabla_j^2 + W(R) + \omega_\alpha(R)\right]\Phi_\alpha(R, t) \\ &+ \sum_j \frac{-1}{M_j}\left[\sum_{\beta=1}^{N} \langle\tilde{\varphi}_\alpha(R)|\nabla_j|\varphi_\beta(R)\rangle\nabla_j\Phi_\beta(R, t)\right. \\ &\left.+ \sum_{\mathbf{k}} \langle\tilde{\varphi}_\alpha(R)|\nabla_j|\varphi_{\mathbf{k}}(R)\rangle\nabla_j\Phi_{\mathbf{k}}(R, t)\right] \\ &+ \sum_j \frac{-1}{2M_j}\left[\sum_{\beta=1}^{N} \langle\tilde{\varphi}_\alpha(R)|\nabla_j^2|\varphi_\beta(R)\rangle\Phi_\beta(R, t)\right. \\ &\left.+ \sum_{\mathbf{k}} \langle\tilde{\varphi}_\alpha(R)|\nabla_j^2|\varphi_{\mathbf{k}}(R)\rangle\Phi_{\mathbf{k}}(R, t)\right] \end{aligned} \tag{93}$$

The function $\Phi_{\mathbf{k}}$ satisfies the equation similar to Eq. (93). As before, we neglect the contributions from the continuum parts $\Phi_{\mathbf{k}}$ as well as the transitions between discrete and continuum parts. Then, because

$$\langle \tilde{\varphi}_\alpha(R) | \nabla_j^2 | \varphi_\beta(R) \rangle \simeq \nabla_j(\langle \tilde{\varphi}_\alpha(R) | \nabla_j | \varphi_\beta(R) \rangle) + \sum_{\gamma=1}^{N} \langle \tilde{\varphi}_\alpha(R) | \nabla_j | \varphi_\gamma(R) \rangle \langle \tilde{\varphi}_\gamma(R) | \nabla_j | \varphi_\beta(R) \rangle$$

Eq. (93) leads to

$$i \frac{\partial}{\partial t} \Phi_\alpha(R, t) = \sum_j \frac{1}{2M_j} \sum_{\beta, \gamma=1}^{N} \left[\delta_{\alpha\beta} \frac{1}{i} \nabla_j - A_{\alpha\beta}^j(R) \right] \times \left[\delta_{\beta\gamma} \frac{1}{i} \nabla_j - A_{\beta\gamma}^j(R) \right] \Phi_\gamma(R, t) + [W(R) + \omega_\alpha(R)] \Phi_\alpha(R, t) \tag{94}$$

where the "vector potential" $A_{\alpha\beta}^j$ is given by

$$A_{\alpha\beta}^j(R) \equiv i \langle \tilde{\varphi}_\alpha(R) | \nabla_j | \varphi_\beta(R) \rangle \tag{95}$$

Equation (94) is a fundamental equation which governs the molecular motion, including the effects of the decay of discrete levels, and provides a generalization of the Born–Oppenheimer equation. As in the case of stable electronic states [69], the "vector potentials" $A_{\alpha\beta}^j(R)$ correspond to (complex) Berry's phase, which is discussed in Section V.

Some remarks are in order about the boundary conditions posed on Φ_α. The total wave function Ψ should be single-valued with respect to the nuclear coordinates, R, and be symmetric (antisymmetric) with respect to any exchange of a pair of identical Bosonic (Fermionic) nuclei. The same condition should be satisfied by the product $\Phi_\alpha(R, t) | \varphi_\alpha(R) \rangle$. This requirement and the symmetry properties of the electronic wave function, $| \varphi_\alpha(R) \rangle$, determine boundary conditions for the nuclear part, $\Phi_\alpha(R, t)$, under which the equation of motion (94) should be solved. For example, if the electronic wave function $| \varphi_\alpha(R) \rangle$ is double-valued with respect to R, one should find a solution to Eq. (94) which is also double-valued so that the product $\Phi_\alpha(R, t) | \varphi_\alpha(R) \rangle$ becomes single-valued. It is then convenient to redefine electronic wave functions, $| \varphi_\alpha(R) \rangle$, such that they are single-valued in the nuclear coordinates, R, and are symmetric with respect to the exchange of any pair of identical nuclei. Using this convention, one can solve the equation of motion (94) under the usual boundary conditions: $\Phi_\alpha(R)$ is single-valued in R and is symmetric (antisymmetric) with respect to any exchange of a pair of identical Bosonic (Fermionic) nuclei. For stable electronic states, this requirement is essential to understand the appearance of a "vector potential" in the presence of conical intersections [69].

The potential $\omega_\alpha(R)$ induced by the electronic motion is, in general, complex and produces the decay of nuclear motion, which has been attracting many researchers' attention (e.g., [66,68]). Here we remark that the molecular-configuration dependence of the imaginary part of $\omega_\alpha(R)$ would play a significant role in the wave-packet propagation: Suppose that there are two different wave-packet motions, and one passes a molecular configuration with large $|\mathrm{Im}\,\omega_\alpha(R)|$, but the other does not. Then the latter motion is more stable than the former. Because wave-packet motions can be controlled by lasers, such differences may be measurable experimentally.

Corresponding to the novel nonadiabatic effects discussed in Section IV, the effects of decay also appear in the "vector potentials," $A^j_{\alpha\beta}(R)$, with different electronic indices ($\alpha \neq \beta$), and are responsible for nonadiabatic transitions between different electronic states.

Moreover, the decay would cause a new feature even in the adiabatic approximation, where Eq. (94) reduces to

$$i\frac{\partial}{\partial t}\Phi_\alpha(R,t) = \left\{\sum_j \frac{1}{2M_j}\left[\frac{1}{i}\nabla_j - A^j_{\alpha\alpha}(R)\right]^2 + W(R) + \omega_\alpha(R)\right\}\Phi_\alpha(R,t) \quad (96)$$

For stable electronic states, one can make electronic wave functions real-valued (except with such complications as conical intersections and magnetic fields) [69]. In these cases, the "vector potentials," $A^j_{\alpha\alpha}(R)$, vanish. In the presence of decay, however, the electronic wave functions are not real-valued and nonvanishing complex "vector potentials," $A^j_{\alpha\alpha}(R)$, appear generically, even without complications such as conical intersections and magnetic fields.

Applications to concrete systems will be reported elsewhere.

IX. CONCLUSIONS

We demonstrate a modification of nonadiabatic effects in a time-dependent Friedrichs–Fano model due to the presence of the decay of discrete states into a continuum. The phenomena described here arise from the nonunitary subsystem dynamics and are connected with a renormalization and redressing of unstable (decaying) discrete states. We argue that this nonunitary evolution can be accounted for using the closed set of nonadiabatic equations (32), provided that a reservoir does not contain any artificially induced macroscopic structures like coherent wave packets. It is remarkable that, in this model, all nonadiabatic effects can be treated via the evolution of pure decaying states.

A consistent analysis of the problem is carried out on the basis of the complex spectral decomposition of the whole Hamiltonian. We find a series of novel nonadiabatic effects missed in the previous phenomenological decay theories: (1) a time dependence of a hermitian Hamiltonian of a discrete subsystem can cause a nonadiabaticity not only of a unitary evolution, but also of a non-unitary evolution; (2) the nonadiabatic effects can arise from a time dependence of an interaction with a reservoir, even if the Hamiltonian of the discrete subsystem is constant in time; (3) a cross-relaxation of dressed states is possible owing to their nonadiabatic coupling.

A nonadiabatic crossing of decaying levels may play an important role in quantum optics and in quantum chemistry of excited molecules. In this chapter, we have shown the nontrivial role spontaneous emission in the course of the coherent evolution of quasi-energy states for a three-level molecule driven by nonsteady dc and ac fields and have investigated the condition of its experimental observability. Also, we have discussed a modification of the Born–Oppenheimer approach and its outcomes in the presence of the decay of electronic states.

As mentioned in Section II, the complex spectral decomposition of the Hamiltonian, contrary to the real (Hilbert-space) spectral decomposition, may admit Jordan blocks, reflecting the degeneracy of decaying eigenstates due to a symmetry of the Hamiltonian, $H(t)$. If the Jordan-block structure appears during a finite time interval in the course of the level crossing, the nonadiabatic dynamics (32) will change significantly. In such a case, one more new nonadiabatic effect can be expected, and the basic equations (32), (79), and the Born–Oppenheimer equation (94), have to be reconsidered. This possibility will be discussed elsewhere.

ACKNOWLEDGMENTS

The authors thank Prof. I. Prigogine for his continuous interest, valuable comments, critical reading of the manuscript, and hospitality during their stay in Brussels. They also thank Profs. I. Antoniou, A. Bohm, and E. C. G. Sudarshan for several fruitful discussions and useful comments. Particularly, Prof. I. Antoniou kindly provided the historical remarks on Livsic's method [after Eq. (23)]. S.T. is grateful to Prof. K. Fukai and Dr. M. Nagaoka for their encouragement and valuable comments and for turning the authors' attention to [66–68]. This work is partly supported by the Belgian Government (under the contract "Pole d'attraction interuniversitaire"), the European Communities Commission DG III/ESPRIT—Project ACTCS 9282 and the contract n° 27155.1/BAS, the U.S. Department of Energy, Grant N° FG05-88ER13897, the Robert A. Welch Foundation, and the Ministry of Education, Science, and Culture of Japan (a Grant-in-Aid for Scientific Research and a grant under the International Scientific Research Program).

REFERENCES

1. I. Prigogine, *From Being to Becoming* (Freeman, New York, 1980).
2. K. Friedrichs, *Comm. Pure Appl. Math.* **1**, 361 (1948).
3. U. Fano, *Phys. Rev.* **124**, 1866 (1961).
4. L. D. Landau, *Phys. Z. Sowjetunion* **1**, 88 (1932).
5. C. Zener, *Proc. Roy. Soc. London A* **137**, 696 (1932).
6. V. M. Akulin and W. P. Schleich, *Phys. Rev. A* **46**, 4110 (1992); V. A. Bazylev, N. K. Zhevago, and M. I. Chibisov, *Sov. Phys. JETP* **42**, 436 (1975); *Zh. Eksp. Teor. Fiz.* **71**, 1285 (1976); V. P. Krainov and A. V. Kruglikov, *Opt. Spectrosk. (USSR)* **50**, 825 (n. 5) (1981).
7. A. Z. Devdariani, V. N. Ostrovskii, and U. N. Sebyakin, *Sov. Phys. JETP* **44**, 480 (1976); *Zh. Eksp. Teor. Fiz.* **71**, 909 (1976).
8. E. E. Nikitin and S. Ya. Umanskii, *Theory of Slow Atomic Collisions* (Springer, Berlin, 1984); F. H. M. Faisel, *Electron-Atom and Electron-Molecule Collisions*, J. Hinze (ed.) (Plenum, New York, 1983).
9. E. Shimshoni and Y. Gefen, *Ann. Phys.* **210**, 16 (1991); P. Ao and J. Rammer, *Phys. Rev. Lett.* **62**, 3004 (1989).
10. K.-A. Suominan and B. M. Garraway, *Phys. Rev. A* **45**, 374, 3060 (1992); *Opt. Comm.* **82**, 260 (1991); G. Hermann, G. Lasnitschka, H. Richter, and A. Scharmann, *Z. Phys. D* **18**, 11 (1991); J.-N. Lopez-Castillo, A. Filali-Mouhim, and J.-P. Jay-Gerin, *J. Chem. Phys.* **97**, 1905 (1992); F. Grobmann and P. Hanggi, *Europhys. Lett.* **18**, 571 (1992); *Chem. Phys.* **170**, 295 (1993); Y. Dakhnovskii, *J. Chem. Phys.* **100**, 6492 (1994).
11. H. P. Breuer and M. Holthaus, *Phys. Lett. A* **140**, 507 (1989); A. N. Seleznyava, *J. Phys. A* **26**, 981 (1993).
12. V. A. Kovarskii, N. F. Perel'man, I. Sh. Averbukh, S. A. Baranov, and S. S. Todirashku, *Nonadiabatic Transitions in a Strong Electromagnetic Field* (Shtiintsa, Kishinev, 1980) in Russian.
13. I. S. Erokhin and S. S. Moiseev, *Problems in Plasma Theory* (Atomizdat, Moscow, 1993) p. 146; V. I. Karas', S. S. Moiseev, and V. E. Novikov, *Zh. Eksp. Teor. Fiz.* **67**, 1702 (1974).
14. V. V. Zheleznyakov, E. V. Suvorov, and V. E. Shaposhnikov, *Sov. Astron.* **18**, 142 (1975).
15. N. E. Sazonov, *Astrofiz.* **10**, 405 (1974).
16. L. M. Brekhovskikh, *Waves in Layered Media* (Academic Press, New York, 1980).
17. G. M. Zaslavskii, V. P. Meitlis, and N. N. Filonenko, *Coupled Waves in Inhomogeneous Media* (Nauka, Novosibirsk, 1982) in Russian.
18. V. V. Shevchenko, *Continuous Transitions in Open Waveguides* (Nauka, Moscow, 1969) in Russian.
19. V. V. Zheleznyakov, V. V. Kocharovsky, and Vl. V. Kocharovsky, *Sov. Phys. Usp.* **26**, 877 (1983).
20. H. C. Baker, *Phys. Rev. A* **30**, 773 (1984).
21. H. C. Baker and R. L. Singleton, Jr., *Phys. Rev. A* **42**, 10 (1990).
22. G. Dattoli, A. Torre, and R. Mignani, *Phys. Rev. A* **42**, 1467 (1990); M. Poirier, J. Reit, D. Normand, and J. Morellec, *J. Phys. B* **17**, 4135 (1984).
23. J. Aguilar and J. M. Combes, *Commun. Math. Phys.* **22**, 269 (1971); E. Balslev and J. M. Combes, *Commun. Math. Phys.* **22**, 280 (1971); for other references, see 68 and *Resonances*,

E. Brändas and N. Elander (eds.), Springer Lecture Notes in Physics, Vol. 325 (Springer, Berlin, 1989).

24. E. C. G. Sudarshan, C. Chiu, and V. Gorini, *Phys. Rev. D* **18**, 2914 (1978); G. Parravicini, V. Gorini, and E. C. G. Sudarshan, *J. Math. Phys.* **21**, 2208 (1980).
25. T. K. Bailey and W. C. Schieve, *Nuovo Cimento* **A47**, 231 (1978).
26. A. Bohm, *J. Math. Phys.* **22**, 2813 (1981); A. Bohm, M. Gadella, and G. B. Mainland, *Am. J. Phys.* **57**, 1103 (1989); A. Bohm and M. Gadella, *Dirac kets, Gamowvectors and Gelfand Triplets*, Springer Lecture Notes on Physics, Vol. 348 (Springer, Berlin, 1989).
27. T. Petrosky, I. Prigogine, and S. Tasaki, *Physica* **A173**, 175 (1991).
28. I. E. Antoniou and I. Prigogine, *Physica* **A192**, 443 (1993).
29. T. Petrosky and I. Prigogine, *Physica* **A175**, 146 (1991).
30. I. Prigogine, *Phys. Rep.* **219**, 93 (1992).
31. R. F. Sawyer, *Phys. Rev. Lett.* **69**, 2457 (1992).
32. S. Ya. Kilin and D. S. Mogilevtsev, *Laser Phys.* **2**, 153 (1992).
33. F. Capasso, C. Sirtori, J. Faist, D. L. Sivco, S.-N. G. Chu, and A. Y. Cho, *Nature* **358**, 565 (1992).
34. G. Duerinckx, *J. Phys. A* **17**, 385 (1984).
35. A. Bohm, *Quantum Mechanics* (Springer, New York, 1979).
36. O. K. Rice, *J. Chem. Phys.* **1**, 375 (1933).
37. U. Fano, *Nuovo Cimento* **12**, 156 (1935).
38. T. D. Lee, *Phys. Rev.* **95**, 1329 (1954).
39. C. B. Chiu and E. C. G. Sudarshan, *Phys. Rev. D* **42**, 3712 (1990); L. A. Khalfin, Usp. Fiz. Nauk **162**, 179 (1992).
40. N. G. Van Kampen, *Physica* **A147**, 165 (1987).
41. N. Nakanishi, *Prog. Theor. Phys.* **19**, 607 (1958).
42. I. Gel'fand and G. Shilov, *Generalized Functions*, Vol. 2 (Academic Press, New York, 1968); *ibid.*, Vol. 3 (Academic Press, New York, 1967); I. Gel'fand and N. Vilenkin, *Generalized Functions*, Vol. 4 (Academic Press, New York (1964).
43. M. Livsic, *Am. Math. Soc. Trans. Ser. 2*, **16**, 427 (1936); J. Howland, *J. Math. Anal. Appl.* **50**, 415 (1975).
44. P.-O. Lowdin, *J. Math. Phys.* **3**, 969 (1962).
45. R. Bartlett and E. Brändas, *J. Chem. Phys.* **59**, 2032 (1973).
46. A. Grecos, T. Guo, and W. Guo, *Physica* **80A**, 421 (1975).
47. E. C. G. Sudarshan and C. B. Chiu, *Phys. Rev. D* **47**, 2602 (1993).
48. E. C. G. Sudarshan, *Relativistic Particle Interactions* (Proceedings of the 1961 Brandies Summer Institute (Benjamin, New York, 1962).
49. F. T. Arecchi and E. Courtens, *Phys. Rev. A* **2**, 1730 (1970).
50. V. L. Ginzburg and V. N. Tsytovich, *Transient Radiation and Transient Scattering* (Nauka, Moscow, 1985) in Russian; V. A. Davydov, *Izv. Vyssh. Uchebn. Zaved.* **23**, 982 (1980); *ibid.*, **25**, 1429 (1982); *ibid.*, **26**, 1134 (1983).
51. L. A. Khalfin, *Zh. Eksp. Teor. Fiz.* **33**, 1371 (1957); [*Sov. Phys.-JETP* **6**, 1053 (1958)]; Usp. Fiz. Nauk **160**, 185 (1990).
52. L. Fonda, G. C. Ghirardi, and A. Rimini, *Rep. Prog. Phys.* **41**, 587 (1978).
53. B. Misra and E. C. G. Sudarshan, *J. Math. Phys.* **18**, 756 (1977).

54. A. Peres, *Ann. Phys.* **129**, 33 (1980).
55. K. J. F. Gaemers and T. D. Visser, *Physica* **A153**, 234 (1988).
56. T. Petrosky, S. Tasaki, and I. Prigogine, *Phys. Lett. A* **151**, 109 (1990); *Physica* **A170**, 306 (1991).
57. V. A. Yakubovich and V. M. Starjinskii, *Linear Differential Equations with Periodic Coefficients and Their Applications* (Nauka, Moscow, 1972) in Russian.
58. C. H. Townes and A. L. Schawlow, *Microwave Spectroscopy* (McGraw-Hill, New York, 1955); R. H. Pantell and H. E. Puthoff, *Fundamentals of Quantum Electronics* (Wiley, New York, 1969); L. Allen, and J. H. Eberly, *Optical Resonance and Two-Level Atoms* (Wiley, New York, 1975).
59. M. V. Berry, *Proc. Roy. Soc. London* **392**, 45 (1984).
60. A. Bohm, "The Geometric Phase in Quantum Physics," in *Recent Problems in Mathematical Physics*, I. Ibort and M. A. Rodriguez (eds.) (Kluwer, The Netherlands, 1993).
61. J. C. Garrison and E. M. Wright, *Phys. Lett. A* **128**, 177 (1988); A. Kvitsinsky and S. Puttermann, *J. Math. Phys.* **32**, 1403 (1991); G. Nenciu and G. Rasche, *J. Phys. A* **25**, 5741 (1992).
62. B. Simon, *Phys. Rev. Lett.* **51**, 2167 (1983); Y. Aharonov and J. Anandan, *Phys. Rev. Lett.* **58**, 1593 (1987).
63. J. Samuel and R. Bhandari, *Phys. Rev. Lett.* **60**, 2339 (1988).
64. M. Rosenberg and T. Petrosky, 1996, unpublished.
65. For example, see C. J. Ballhausen and A. E. Hansen, *Annu. Rev. Phys. Chem.* **23**, 15 (1972); C. A. Mead, in *Mathematical Frontiers in Computational Chemical Physics*, D. G. Truhlar (ed.) (Springer, New York, 1988).
66. J. C. Y. Chen, *Adv. Radiat. Chem.* **1** 245 (1969).
67. D. A. Micha, *Adv. Quantum Chem.* **8**, 231 (1974).
68. T. Yamabe, A. Tachibana, and K. Fukui, *Adv. Quantum Chem.* **11**, 195 (1978); A. Tachibana, T. Yamabe, and K. Fukui, *Mol. Phys.* **37**, 1045 (1979).
69. For example, see C. A. Mead, *Rev. Mod. Phys.* **64**, 51 (1992).
70. V. V. Kocharovsky, Vl. V. Kocharovsky, E. V. Derishev, S. A. Litvak, and I. A. Shereshevsky, Technical Digest European Quantum Electronics Conf. (Hamburg, September 8–13, 1996).

non of scattering of a point particle in an open configuration of disk scatterers. The dynamics of this billiard is chaotic on the set of trapped trajectories which forms a fractal repeller [8]. In this regard, the diffusion coefficient is given as the escape rate out of the disk scatterer. The escape rate appears to be an exact property of the Liouvillian dynamics of an ensemble of particles in the system. The surprise has certainly been that the exactness of this property is not incompatible with the fact that the escape rate characterizes an exponential decay in time and is, therefore, typically irreversible. On these results, Gaspard and Nicolis [9] derived a formula which gives the diffusion coefficient as the large-scale limit of the difference between the Lyapunov exponent and the Kolmogorov–Sinai entropy per unit time characterizing the fractal repeller. A fundamental connection was so established between a transport and an irreversible property like diffusion and the characteristic properties of chaos and fractality of the microscopic dynamics.[1] More recently, Gaspard and Baras [10] studied these fractal properties in the Lorentz gas and showed that there also exists a direct formula between the diffusion coefficient and the Hausdorff codimension of the fractal repeller. Furthermore, another formula has been derived by Cvitanović, Eckmann, and Gaspard [11,12], which gives the diffusion coefficient directly in terms of the periodic orbits of the system.

To show that diffusion can be understood as an exact property of the Liouvillian dynamics, the author [13] invented a simple and exactly solvable area-preserving map called the multibaker, which sustains deterministic diffusion, and showed that the spectrum of Ruelle resonances (i.e., the spectrum of the Perron–Frobenius operator of the Liouvillian dynamics), contains the spectrum of eigenvalues of the phenomenological diffusion equation in the large-system limit. This other fundamental result has been instrumental in proving the relevance of Ruelle resonances (which are exact dynamical properties) in diffusion. It was shown that the relaxation properties of the eigenmodes of diffusion can be understood as strict properties of the classical dynamics of statistical ensembles, rather than as approximate properties obtained by invoking a Markovian assumption like in the Boltzmann equation [13]. Moreover, Gaspard and Alonso [14] showed, for the disk scatterers, that the Ruelle resonances can be calculated numerically with high precision and are identical with the decay rates obtained in numerical simulations of escape processes.

Based on these results, the eigenstates associated with the Ruelle resonances and corresponding spectral decompositions have been derived in several piecewise-linear mappings [13,15–17] and, finally, in the multibaker

[1] Very recently, these results have been extended to general transport and reaction-rate properties by Dorfman and Gaspard [52].

mapping as well as in hyperbolic flows with spatial periodicity by the author [18] (see Fig. 1). The main particularity of these spectral decompositions is that they involve distributions rather than functions. In one-dimensional mappings, the right eigenvectors are functions and the left eigenvectors are distributions [15]; in volume-preserving maps and flows, both the right and the left eigenvectors are distributions [13,15,17,18]. As a consequence, the spectral decomposition of the Perron–Frobenius operator should be understood as a bidistribution which maps the initial density and the final observable directly onto a number [18]. Another particularity comes from the multiplicity of the Ruelle resonances in symplectic systems. This is a result of the periodic orbit theory of the Perron–Frobenius operator [6,19–21], which implies that this operator can only be reduced into Jordan blocks. With these results, fundamental aspects of irreversibility, proposed long ago by Prigogine, are finding striking theoretical confirmations.[2]

More recently, Tasaki and Gaspard [28] constructed nonequilibrium steady-states corresponding to concentration gradients in the dyadic multibaker (see Fig. 2). Here also, distributions appear whose cumulative functions are given in terms of the highly irregular Takagi functions. It is natural to find distributions here also because constant gradients result from the oscillatory diffusive eigenmodes (which are themselves distributions) in the limit where the wavenumber grows with the amplitude [28].

Such steady states have never before been exactly constructed even in a simple model, so that it was not realized that we need the concepts of distribution and of fractal to really understand the deep nature of such nonequilibrium states. For these reasons, this fundamental advance had to await a century of efforts in nonequilibrium statistical mechanics where the solutions of the Liouvillian dynamics have always been assumed to be regular functions. This assumption turns out to be too restrictive and has been misleading us toward the apparent necessity of approximating the Liouvillian dynamics through stochastic dynamics, as in the many known

[2] We emphasize that the approach we followed strictly preserves the microscopic equations of motion and a splitting of the time evolution into forward and backward semigroups [18] is only conceived at the level of the Liouvillian dynamics of statistical ensembles, a point of view advocated by Prigogine [22]. Another scheme exists originally proposed by Gauss [23], revived in a modern context by Smale [24], and then worked out by Hoover and Nosé [25], in which the microscopic equations of motion are altered to achieve thermalization. In this Gaussian or Nosé–Hoover scheme, the forward and backward time evolutions admit two distinct invariant measures and phase-space volumes are not preserved. In this context, relationships between volume contractivity and transport coefficients have been derived [26,27] which are similar to the Gaspard–Nicolis formulas [9] but which result from the alteration of the microscopic dynamics rather than from appropriate boundary conditions, as in the author's approach.

master equations. Of course, the value of such approximations is in their comparison with experiments. The value of such approximations lies in the extremely large number of degrees of freedom (10^{23}) of macroscopic systems, such that the fractal and chaotic properties are hidden on the smallest scales of phase space. Nevertheless, we may wonder if some of the aforementioned fundamental properties will not turn out to be experimentally observable and provide hints for the validity of the newly obtained theoretical results.

It is the purpose of this chapter to argue that such experimental evidence is available in the well-known phenomena of Brownian motion or other noises, such as the Nyquist noise in electric circuits. Indeed, the dynamical randomness of the Brownian motion of a colloidal particle in a fluid appears to be the macroscopically observational part of the microscopic chaos, which is visible as the emerged part of an iceberg is. At the basis of the argument is a recent work by Gaspard and Wang [30], who classified a series of different random processes (including the stochastic processes describing Brownian motion) in terms of ϵ-entropy per unit time. This clas-

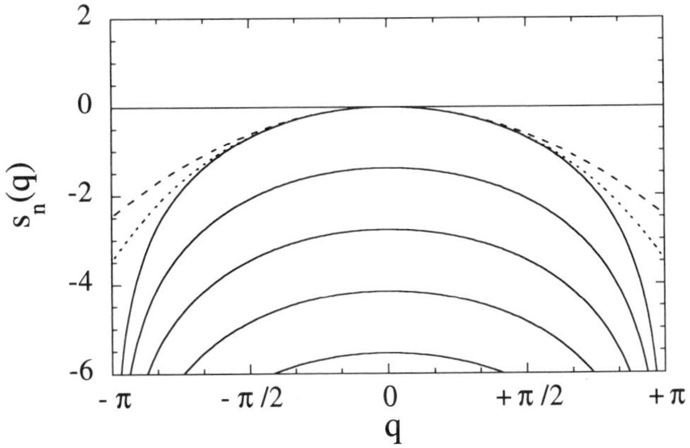

(*a*)

Figure 1. (*a*) Eigenvalues of the Liouvillian dynamics of the 4-adic multibaker map versus the wavenumber q. The eigenvalue problem of the Perron–Frobenius operator $\hat{P}^t$ is written as $\hat{P}^t\psi_n(q) = \exp[s_n(q)t]\psi_n(q)$ where $\psi_n(q)$ is a distribution. The eigenvalues of the 4-adic multibaker map are given by $s_n(q) = 2\ \ln[\cos(q/2)/2^n]$ with $n = 0, 1, 2, \ldots$ and are drawn as solid lines. The long dashed line is the diffusion approximation $s_D(q) = -q^2/4$ while the short dashed line is the Burnett approximation $s_B(q) = -q^2/4 - q^4/96$ of the leading eigenvalue $n = 0$ (see [12,13]). (*b*) The real parts Re $G^{(0)}(x)$ of the cumulative functions of the right eigendistributions integrated along the x axis for different values of the wavenumber q. The variations increase at large q values. Beyond $|q| \leq 2\pi/3$, the integrated eigendistributions do not exist as functions (see [17]). (*c*) The real parts Re $G^{(1)}(x)$ of the doubly integrated eigendistributions which exist until $|q| = 2\ \arccos(1/4)$ (see [18]).

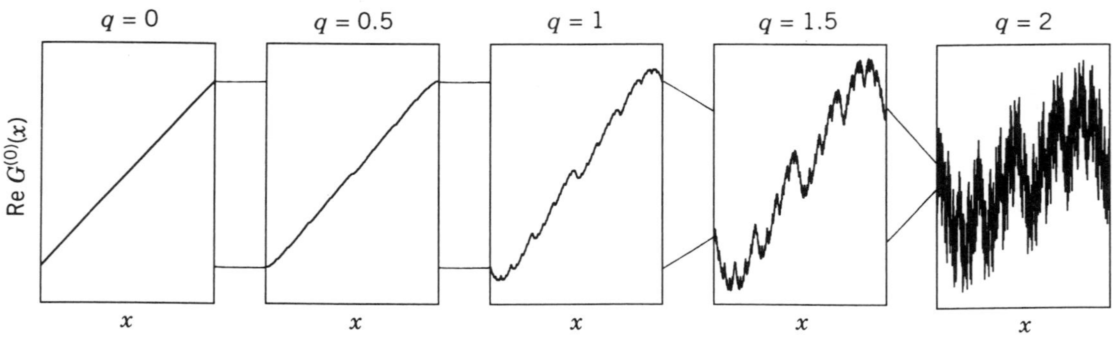

(*b*)

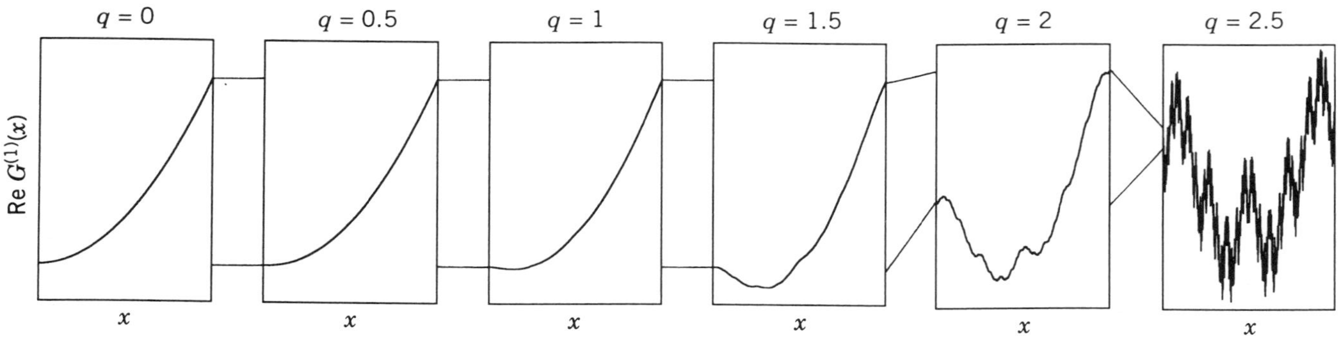

(*c*)

Figure 1. (*Continued*)

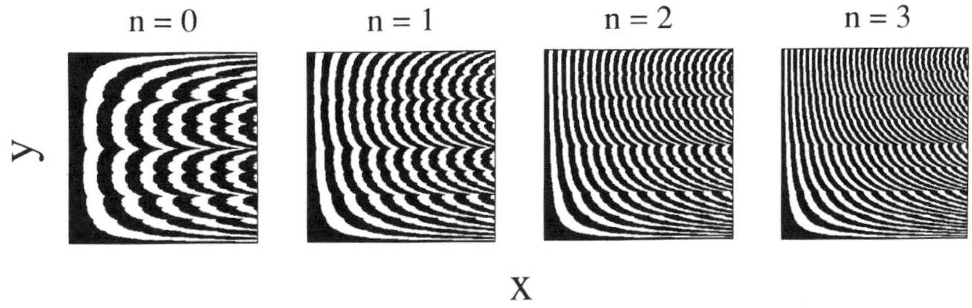

Figure 2. Contour plots of the cumulative functions $G(n, x, y) = gx[ny + T(y)]$ of the eigendistribution corresponding to a concentration gradient g in the dyadic multibaker map (see [28]). $T(y)$ is the Takagi function (see [29]). The phase space is composed of successive squares (n, x, y) with $n \in \mathbb{Z}$, but only four are drawn, namely $n = 0, 1, 2, 3$. The contour lines are separated by 0.05 and the interline spaces are successively filled in black and white. The region where $0 < G < 0.05$ is in black.

sification allows us to compare the dynamical randomness of different systems or different levels in the same system.

The chapter is organized as follows. In Section II, we summarize the results of Lyapunov exponents in microscopic systems. Section III is devoted to the Kolmogorov–Sinai entropy per unit time and to the ϵ-entropy per unit time. In Section IV, devoted to Brownian motion, we show the basic result of this chapter, that the ϵ-entropy per unit time provides a lower bound on the sum of positive Lyapunov exponents. Accordingly, a measure of the ϵ-entropy can give evidence of the positivity of Lyapunov exponents and, in this way, of microscopic chaos. Conclusions are drawn in Section V.

II. LYAPUNOV SPECTRUM

A. Linear Stability and Lyapunov Exponent

The microscopic dynamics of atoms and molecules in gases, liquids, and even solids is well described in statistical mechanics by classical Hamiltonian systems:

$$\dot{\mathbf{q}} = \frac{\partial H}{\partial \mathbf{p}}, \qquad \dot{\mathbf{p}} = -\frac{\partial H}{\partial \mathbf{q}} \tag{2.1}$$

If boundary conditions are added to Eq. (2.1), elastic collisions on hard walls or between hard spheres can be described. All these mechanical systems have a symplectic phase space. Linear stability of the solutions of

Eq. (2.1) is obtained from the time evolution of infinitesimal perturbations ruled by

$$\delta\dot{\mathbf{q}} = \frac{\partial^2 H}{\partial \mathbf{p}\, \partial \mathbf{q}} \cdot \delta\mathbf{q} + \frac{\partial^2 H}{\partial \mathbf{p}^2} \cdot \delta\mathbf{p}$$
$$\delta\dot{\mathbf{p}} = -\frac{\partial^2 H}{\partial \mathbf{q}^2} \cdot \delta\mathbf{q} - \frac{\partial^2 H}{\partial \mathbf{q}\, \partial \mathbf{p}} \cdot \delta\mathbf{p} \tag{2.2}$$

or equivalent equations for hard-sphere systems. The Lyapunov exponents are then defined by

$$\lambda(\mathbf{q}_0, \mathbf{p}_0; \delta\mathbf{q}_0, \delta\mathbf{p}_0) = \lim_{t\to\infty} \frac{1}{t} \log \frac{(\delta\mathbf{q}_t^2 + \delta\mathbf{p}_t^2)^{1/2}}{(\delta\mathbf{q}_0^2 + \delta\mathbf{p}_0^2)^{1/2}} \tag{2.3}$$

where the choice of the basis of the logarithms determines the units (bit, nat, or digit per second). In ergodic systems, the limit $t \to \infty$ performs a time average which erases the dependency of the Lyapunov exponents on the particular initial conditions $(\mathbf{q}_0, \mathbf{p}_0)$. On the other hand, a discrete dependency on the direction of the initial perturbation $(\delta\mathbf{q}_0, \delta\mathbf{p}_0)$ remains. A discrete set of values are selected which form the spectrum of Lyapunov exponents [1]

$$\lambda_{\max} = \lambda_{n_{\max}} \geq \lambda_{n_{\max}-1} \geq \cdots \geq 0 \geq \cdots \geq -\lambda_{n_{\max}-1} \geq -\lambda_{n_{\max}} = -\lambda_{\max} \tag{2.4}$$

corresponding to the different directions of instability (or stability). In symplectic systems, the Lyapunov exponents are known to appear in pairs $(\lambda_n, -\lambda_n)$. For N particles in a box, the energy is the only general constant of motion and there are at most $n_{\max} = 3N - 1$ positive Lyapunov exponents. In the hard-sphere gas, it can be shown that these $3N - 1$ Lyapunov exponents are indeed positive [4].

B. Lyapunov Spectrum in Many-Body Classical Systems

The Lyapunov spectrum has been numerically calculated for several many-particle systems and it was empirically observed that the spectrum is often of the form [7]

$$\lambda_n = \lambda_{\max}\left(\frac{n}{n_{\max}}\right)^{\beta} \tag{2.5}$$

with $\beta \simeq 1$ for the one-dimensional Fermi–Pasta–Ulam chain at high temperature as well as for three-dimensional solids, but $\beta \simeq 1/3$ for a dense

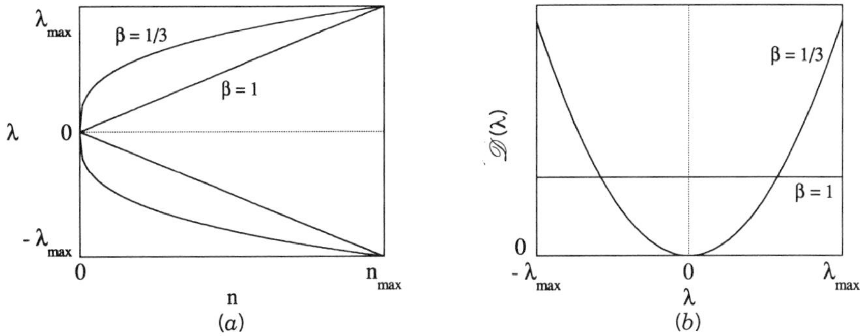

Figure 3. (*a*) Schematic behavior of the spectrum of Lyapunov exponents in three-dimensional solids ($\beta \simeq 1$) and dense fluids ($\beta \simeq 1/3$). (*b*) Corresponding density of Lyapunov exponents.

three-dimensional fluid with repulsive Lennard–Jones interactions [7]. We can also introduce the density of Lyapunov exponents as $\mathscr{D}(\lambda) = dn/d\lambda$ (see Fig. 3).

In a dilute gas, the maximum Lyapunov exponent can be evaluated from the geometry of a collision according to Krylov's argument [4,5]. If d is the effective diameter of the gas particles, l the mean free path, and v the mean velocity, we have [5]

$$\lambda_{\max} \sim \frac{v}{l} \log \frac{2l}{d} \tag{2.6}$$

The maximum Lyapunov exponent can be expressed in terms of the thermodynamic quantities of the gas, such as the density ρ and the temperature T, according to $l \simeq 1/\rho\pi d^2$ and $v \sim (k_B T/m)^{1/2}$ where m is the mass of the particles. If we assume that the effective diameter is constant, we obtain [6]

$$\lambda_{\max} \sim \rho\pi d^2 \left(\frac{k_B T}{m}\right)^{1/2} \log \frac{2}{\rho\pi d^3} \sim 10^{10} \text{ digits/sec} \tag{2.7}$$

for a gas like argon at room temperature and pressure.

The Lyapunov exponents in microscopic chaos reach enormous values with respect to macroscopic chaos. Therefore, a measurement of the Lyapunov quantities is not practically possible because of our inability to follow the motion of each individual particle in a fluid. In the following section, we describe the concept of entropy per unit time, which is of a much broader generality.

III. ENTROPY PER UNIT TIME

A. Kolmogorov–Sinai Entropy per Unit Time

1. *Generalities*

If dynamical instability is quantitatively measured by the Lyapunov exponents, dynamical randomness is characterized by the entropy per unit time [1]. The entropy per unit time is a transposition of the concept of thermodynamic entropy per unit volume from space translations to time translations. As Boltzmann showed, the entropy is the logarithm of the number of complexions, that is, the number of states that are possible in a certain volume and under certain constraints. In the time domain, the number of complexions becomes the number of possible trajectories in a given time interval. The entropy per unit time is therefore an estimation of the rate at which the number of possible trajectories grows with the length of the time interval. In a chaotic system, the set of trajectories appears as a tree with a growing number of branches and the entropy per unit time is the branching rate.

This scheme does not contradict the famous Cauchy theorem, which asserts the unicity of the trajectory issued from given initial conditions. Indeed, as in statistical mechanics, the counting proceeds with the constraint that the trajectories belong to cells of phase space. Since each cell is a continuum, the counting becomes nontrivial. The existence of a positive branching rate is induced by a stretching and folding mechanism in phase space. Indeed, an initial cell may be stretched into a long, thin cell that will overlap several cells at the next time unit.

The counting may be purely topological, in which case the topological entropy per unit time is defined and the tree can be represented as an oriented graph. However, the tree may be weighted by transition probabilities, which leads to the definition of the Kolmogorov–Sinai entropy per unit time.

The concept of entropy per unit time was introduced by Shannon in his famous information theory [31]. The purpose was to characterize random processes which emit or produce symbols per unit time. Physicochemical systems are therefore considered like sources of information. It is important to critically review the evolution of the idea of information since 1948. Entropy is considered as a rate property in Shannon's text (actually as an entropy per symbol), that is, as a dynamical property of a physical communication device. Later, Kolmogorov and Sinai [32] proposed an operational and rigorous definition in the context of dynamical systems and ergodic theory. More recently, the works of Chaitin and others [33] have shown that the entropy per unit time plays a fundamental role in our logical understanding of randomness. Consequently, this concept is one of the best documented: it permeates several fields, from dynamical system theory—and its physicochemical applications—to mathematics and logics, and allows us to establish well-founded connections. Here we adopt the following operational interpretation. The entropy per unit time is the minimum number of symbols which needs to be recorded per unit time to reconstruct the unique trajectory followed by the system during the observation period [30].

2. *Definition and Some Properties*

We now summarize and comment on the definition of the Kolmogorov and Sinai (KS) entropy per unit time [1,34].

We denote by $\mathbf{X} = (\mathbf{q}, \mathbf{p})$ the phase-space variables of our dynamical system. The flow resulting from the integration of Hamilton's equations (2.1) is denoted by $\mathbf{X}_t = \Phi^t(\mathbf{X}_0)$. The phase space Γ of the dynamical system is partitioned into mutually disjoint cells: $\mathscr{P} = \{C_1, C_2, \ldots, C_M\}$. A probability measure μ is defined which is invariant under time evolution. The probabilities $\mu(\omega_0 \omega_1 \cdots \omega_{n-1})$ to visit successively the cells $C_{\omega_0}, C_{\omega_1}, \ldots, C_{\omega_{n-1}}$ at times $t = 0, \Delta t, \ldots, (n-1)\,\Delta t$ are then calculated.

These probabilities are actually n-time correlation functions because

$$\mu(\omega_0 \omega_1 \cdots \omega_{n-1}) = \int_\Gamma \mu(d\mathbf{X}) I_{\omega_0}(\mathbf{X}) I_{\omega_1}(\Phi^{\Delta t}\mathbf{X}) \cdots I_{\omega_{n-1}}(\Phi^{(n-1)\Delta t}\mathbf{X})$$
$$= \langle \hat{U}^{t_0} I_{\omega_0} \hat{U}^{t_1} I_{\omega_1} \cdots \hat{U}^{t_{n-1}} I_{\omega_{n-1}} \rangle_\mu \quad \text{with } t_k = k\,\Delta t \qquad (3.1)$$

where $I_{\omega_i}(\mathbf{X})$ are the characteristic functions of the cells C_{ω_i} and where $\hat{U}^t A(\mathbf{X}) = A(\Phi^t \mathbf{X})$ is the Koopman classical evolution operator [34]. This remark is important in the context of nonequilibrium statistical mechanics where the hierarchy of the n-time correlation functions is at the basis of our knowledge of the dynamics in the thermodynamic system. In particular, autocorrelation functions defining the transport properties by the Green–Kubo formulas are two-time correlation functions like $\langle A \hat{U}^t A \rangle_\mu - \langle A \rangle_\mu^2$ [35]. At the top of the hierarchy, we find thermodynamic averages like $\langle A \rangle_\mu$ which are time independent. Because the KS entropy per unit time is based on these n-time correlation functions, we see that it characterizes extremely fine correlations between the observables at different successive times.

The entropy per unit time of the partition is then [32,34]

$$h(\mathscr{P}) = \lim_{n\to\infty} -\frac{1}{n\,\Delta t} \sum_{\omega_0 \cdots \omega_{n-1}} \mu(\omega_0 \omega_1 \cdots \omega_{n-1}) \log \mu(\omega_0 \omega_1 \cdots \omega_{n-1}) \qquad (3.2)$$

and the KS entropy is defined by [32,34]

$$h_{\mathrm{KS}} = \mathrm{Sup}_{\mathscr{P}}\, h(\mathscr{P}) \tag{3.3}$$

Taking the supremum is a way to define a quantity that does not refer any longer to an arbitrary partition but which is intrinsic to the dynamical system Φ^t and the invariant measure μ. Nevertheless, positivity of $h(\mathscr{P})$ for an arbitrary partition $\mathscr{P}$ is already sufficient to have dynamical randomness and this is a main step in our argument.

The interpretation of $h(\mathscr{P})$ is provided by the Shannon–McMillan–Breiman theorem [34], which states that if the system is ergodic and if a trajectory is successively visiting the cells $C_{\omega k}$ at times $t_k = k\,\Delta t$, we have

$$\mu(\omega_0\,\omega_1 \cdots \omega_{n-1}) \sim \exp[-n\,\Delta t\, h(\mathscr{P})] \tag{3.4}$$

for almost all trajectories of the system. In this sense, the entropy per unit time is the decay rate of the n-time correlation functions. We emphasize that the rate is taken by increasing the number of time intervals, but not by increasing the period of any time interval, as is done to define the property of mixing.

To illustrate the preceding discussion, we consider a system like a protein or an ion channel which may be in two distinct conformations, 0 and 1, and which is coupled to an environment like a liquid. We may choose the partition $\mathscr{P}$ to distinguish the two conformations, 0 and 1, without distinguishing among the various states of the environment. The positivity of $h(\mathscr{P})$ for this partition would imply that the switching process of the protein or of the ion channel between both conformations is random because the different sequences $\omega_0\,\omega_1 \cdots \omega_{n-1}$ are then equally probable according to the Shannon–McMillan–Breiman theorem. Another example is provided by the process of coin tossing, where the n-time correlations functions (3.4) decay like 2^{-n}.

In bounded classical systems with a finite number of degrees of freedom, dynamical randomness is produced by the sensitivity to initial conditions, which is characterized by positive Lyapunov exponents. Pesin's theorem expresses this result by the statement that the KS entropy is the sum of positive average Lyapunov exponents,

$$h_{\mathrm{KS}} = \sum_{\lambda_n > 0} \lambda_n \tag{3.5}$$

in ergodic and hyperbolic systems [1,36].

3. *The KS Entropy in Many-Body Classical Systems*

Using the results of Section II, we can evaluate the value of the KS entropy in a thermodynamic system. For systems with many Lyapunov exponents, the sum (3.5) can be approximated by an integral so that [6]

$$h_{\mathrm{KS}} \simeq \int_0^{n_{\max}} \lambda_n \, dn = \int_0^{\lambda_{\max}} \lambda \mathscr{D}(\lambda) \, d\lambda$$
$$\simeq \frac{1}{\beta + 1} n_{\max} \lambda_{\max} \tag{3.6}$$

where we used Eq. (2.5). An entropy per unit time and volume can also be defined as the thermodynamic limit of the ratio of the KS entropy to the volume [6],

$$h^{(\text{time, volume})} = \lim_{\substack{V, N \to \infty \\ N/V = \rho}} \frac{h_{\mathrm{KS}}}{V} \simeq \frac{3}{\beta + 1} \rho \lambda_{\max} \tag{3.7}$$

where ρ is the density of particles and the maximum Lyapunov exponent is given by Eq. (2.7) for a dilute gas. We may therefore obtain the entropy per unit time and volume as thermodynamic quantities characterizing the dilute gas,

$$h^{(\text{time, volume})} \sim \rho^2 \pi d^2 \left(\frac{k_{\mathrm{B}} T}{m} \right)^{1/2} \log \frac{2}{\rho \pi d^3} \tag{3.8}$$

as compared with the standard entropy per unit volume which is

$$S^{(\text{volume})} \simeq \rho \ln \frac{(2\pi k_{\mathrm{B}} T/m)^{3/2}}{\rho \, \Delta^3 x \, \Delta^3 v} + \frac{5}{2} \rho \tag{3.9}$$

In classical statistical mechanics, the entropy per unit volume is an ϵ-entropy where $\epsilon = \Delta^3 x \, \Delta^3 v$ is the volume of the phase-space cells of the one-body system so that the entropy constant is not fixed outside quantum mechanics. The main difference between both entropies is that the first is clearly of dynamical origin because Eq. (3.8) vanishes with the diameter of the particles ($d \to 0$). However, because the standard entropy varies so slowly with this diameter, it is well approximated by its value for the ideal gas, namely Eq. (3.9).

For a gas like argon at room temperature (300 K) and pressure (1 atm), the space–time entropy (3.8) already reaches extremely high values like 10^{29}–10^{30} digits/sec·cm^3. Accordingly, it is practically impossible to follow in detail the individual trajectories in a gas requiring such a high data accumulation rate, although the fastest data-acquisition systems technologically available have rates below or around 10^9 digits/sec.

Nevertheless, combining Eqs. (3.3) and (3.5), we obtain a lower bound on the sum of positive Lyapunov exponents if we measure a positive entropy per unit time for a particular partition $\mathscr{P}$:

$$0 \leq h(\mathscr{P}) \leq \sum_{\lambda_n > 0} \lambda_n \tag{3.10}$$

Even if the lower bound is much smaller than the value expected for the sum of positive Lyapunov exponents in classical statistical mechanics, we would have evidence of the positivity of some exponents; this is the crucial step of our argument.

4. *The KS Entropy and Transport Properties*

Before explaining the method and determining where we can measure $h(\mathscr{P})$, we complete this section by showing how the preceding quantities characterizing microscopic chaos are related to the transport properties with the Gaspard–Nicolis formula [9].

In systems that are no longer bounded but which allow the escape of particles or of representative points of a phase-space configuration, the KS entropy is not equal to the sum of positive Lyapunov exponents but is diminished by the escape rate γ according to [1,37,38]

$$h_{\text{KS}} = \sum_{\lambda_n > 0} \lambda_n - \gamma \tag{3.11}$$

This is the case in classical scattering systems where γ is the leading Ruelle resonance of the Perron–Frobenius operator which controls the time evolution of the two-time correlation functions [8,9,13,18]. In open systems, there is a cascade mechanism over the hierarchy of n-time correlation functions. Open systems are most natural systems common in many fields of physics and chemistry where collisions occur between particles, in particular, in unimolecular and bimolecular reactions. In classical models of such processes, the time evolution is assumed to obey Hamilton's equations (2.1) and statistical ensembles are introduced to define the different reaction rates and cross sections.

Coming back to Eq. (3.11), we see that the Pesin formula (3.5) is recovered if the escape rate vanishes, which is the case in bounded systems. On the other hand, we may open a previously bounded system by selecting a subset of its trajectories in phase space. The trajectories continue to obey to

Hamilton's equations. The selection is carried out by requiring that the trajectories satisfy particular constraints like boundary conditions on a fictitious or real surface in physical or phase spaces. Examples of such a procedure are described for the Lorentz gas and the multibaker map elsewhere [9,13]. In general, the subset of trajectories is a fractal. In hyperbolic systems like the Lorentz gas and the multibaker map, the escape rate out of this fractal is given by Eq. (3.11). However, if the boundary condition corresponds to the escape of a diffusive particle out of an open billiard like a slab of thickness L cut out of the Lorentz gas, the escape rate is given in terms of the diffusion coefficient according to $\gamma = D(\pi/L)^2$ [8]. The fractal $\mathcal{F}_L$ is then composed of all the periodic and nonperiodic trajectories that are indefinitely trapped in the open billiard. Using Eq. (3.11), Gaspard and Nicolis [9] obtained the diffusion coefficient by

$$D = \lim_{L\to\infty} \left(\frac{L}{\pi}\right)^2 \left[\sum_{\lambda_n > 0} \lambda_n(\mathcal{F}_L) - h_{\mathrm{KS}}(\mathcal{F}_L) \right] \tag{3.12}$$

where the quantities are evaluated for the natural ergodic measure on the fractal repeller $\mathcal{F}_L$, that is, the invariant measure where the probability weight of each orbit is inversely proportional to its stretching or instability factor. This measure is natural in the sense that it is selected in numerical simulations of the escape process and reduces to the Liouville invariant measure in the large-system limit where the escape rate vanishes.[3]

Equation (3.12) gives us an insight into the connection between the chaotic and the transport properties. Microscopic dynamics develops an enormous dynamical randomness, as shown with Eq. (3.8). The transport processes make use of only a tiny portion of this randomness. Indeed, for a macroscopic distance $L \sim 1$ cm and a typical diffusion coefficient in a dilute gas given by $D \sim lv \sim 1$ cm^2/sec, we have an escape rate of the order of $\gamma \sim 5$ digits/sec, which should be compared with the enormous values (10^{29}–10^{30} digits/sec) for the KS entropy or the sum of positive Lyapunov exponents.

In the next subsection, we describe the concept of ϵ-entropy per unit time, which allows us to characterize stochastic processes.

[3] Variational principles may select other seemingly natural invariant measures, but they depend on the quantity that is varied. For this reason, we adopt the functional point of view in which the natural invariant measure is the one selected by direct numerical simulation of statistical ensembles according to Hamilton's equations on a typical machine. It is equivalent to the criterion adopted by Ruelle who proposed, based on Kifer's work, selecting the invariant measure that is the noiseless limit of the dynamical system forced by a white noise [1,39].

B. ϵ-Entropy per Unit Time

1. *Langevin Processes and* ϵ*-Entropy*

Many physicochemical processes are described by stochastic processes given by the Langevin equations

$$\dot{\mathbf{X}} = \mathbf{F}(\mathbf{X}) + \mathbf{W}(t) \tag{3.13}$$

where $\mathbf{W}$ is a vector of white noises:

$$\begin{aligned} \langle W_i(t) \rangle &= 0 \\ \langle W_i(t) W_j(t') \rangle &= 2D_{ij}\,\delta(t - t') \end{aligned} \tag{3.14}$$

The probability density representative of a statistical ensemble is known to obey a Fokker–Planck equation. Figure 4 shows a typical trajectory of a Langevin process.

Such stochastic processes certainly have a dynamical randomness which is much larger than in deterministic dynamical systems where $\mathbf{W}(t) = 0$. However, the Lyapunov exponents are of little use in the characterization of randomness. On the other hand, the concept of entropy remains operative. If μ is the equilibrium invariant measure of the Fokker–Planck equation, the definition (3.2) of the entropy per unit time of a partition $\mathscr{P}$ remains applicable. However, the supremum (3.3) over all the partitions is infinite; thus the KS entropy of stochastic processes is infinite and also of little use. We can therefore define an intermediate quantity which is the entropy per unit time of partitions $\mathscr{P}_\epsilon$ into cells of diameter ϵ:

$$h(\epsilon) \equiv h(\mathscr{P}_\epsilon) \tag{3.15}$$

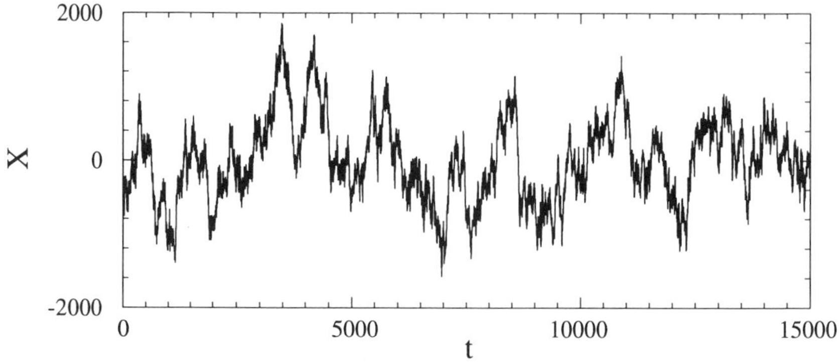

Figure 4. Typical trajectory of the Ornstein–Uhlenbeck process for which $\dot{X} = -aX + W(t)$.

This so-called ϵ-entropy is a function of the cell diameter ϵ. The standard entropy given by Eq. (3.9) is an example of such an entropy. Different definitions are possible, based on the way the cell diameters are defined, but the different definitions are equivalent as far as the ϵ-entropies depends on ϵ. In this sense, each stochastic process is characterized by the functional dependency of its ϵ-entropy on ϵ. We come back to this point later.

In the context of random processes, ϵ-entropies were first considered under a different name by Shannon [31]. Later, Kolmogorov and coworkers clarified the concept and named it [40]. In the definition of Kolmogorov [40], the cell diameter is a distance in the functional space of the trajectories of the process (3.13). Here, we adopt the definition shown in Eq. (3.15) to emphasize its connection with the entropy per unit time of Section III.A [30]. With this definition, the KS entropy is the limit of the $\epsilon \to 0$.

Gaspard and Wang [30] recently classified a series of random and stochastic processes in terms of the ϵ-entropy per unit time. In particular, for Mandelbrot's fractional Brownian motions [41], we have [30]

$$h(\epsilon) \sim \left(\frac{1}{\epsilon}\right)^{1/H} \qquad 0 < H < 1 \tag{3.16}$$

The ϵ-entropy allows us to compare the degrees of randomness of different stochastic processes according to the value and the rapidity of increase of the entropy as $\epsilon \to 0$.

Several numerical methods have been proposed to evaluate the ϵ-entropy per unit time of a signal, for example, the Cohen and Procaccia method [42] or the original Kolmogorov–Tikhomirov method [40]. The paper by Gaspard and Wang [30] gives a review of these methods and examples of applications.

2. *Lorentz–Boltzmann Process*

Before going to the next section, which is devoted to Brownian motion and the practical evaluation of the lower bound on the Lyapunov exponents, we would like to give a further theoretical example of ϵ-entropy and some of its consequences in the context of nonequilibrium statistical mechanics. The Boltzmann equation gives a very good description of the kinetic processes taking place in dilute gases. It is interesting to evaluate the randomness assumed at this level of description in order to compare it with the actual randomness of a gas as evaluated in Section III.A.

The first point is that the nonlinear Boltzmann equation is not the master equation of a random process. Therefore, several authors [43] have proposed a master equation for the fluctuations in the variables ruled by Boltzmann's equation. Here, we consider for simplicity the linearized Boltzmann equation, which is itself the master equation of the random process ruling the velocity of a particle of the fluid [44]. We assume that the process is uniform in space. The linearized Boltzmann equation is then

$$\frac{\partial f(\mathbf{v}_1)}{\partial t} = \rho \int d^3v_2\, d^2\Omega\, |\mathbf{v}_1 - \mathbf{v}_2|\, \sigma(\theta, \varphi, \mathbf{v}_1, \mathbf{v}_2) f_{\text{eq}}(v_2)[f(\mathbf{v}_1') - f(\mathbf{v}_1)] \quad (3.17)$$

where ρ is the particle density, σ is the differential cross section of the binary collision, and $f_{\text{eq}}(v)$ the equilibrium velocity distribution of the fluid particles. Equation (3.17) describes the time evolution of the probability density of the velocity of a test particle 1 undergoing multiple collisions with other particles, 2, in the gas. The outcoming velocities $\mathbf{v}_1'$ and $\mathbf{v}_2'$ after each binary collision are uniquely determined by the velocities $\mathbf{v}_1$ and $\mathbf{v}_2$ of the two particles entering the collision together with the impact unit vector locating the relative positions of particles 1 and 2 at the point of closest approach.

The Boltzmann–Lorentz equation (3.17) describes the successive collisions as random events where the velocity $\mathbf{v}_2$ of the bath particle as well as the solid angle $\Omega = (\theta, \varphi)$ of the aforementioned impact unit vector are random variables. The Boltzmann–Lorentz equation has the form of a birth-and-death process in the continuous-velocity and solid-angle variables.

Gaspard and Wang [30] evaluated the ϵ-entropy per unit time of this single-particle process for hard spheres of mass m and diameter d and obtained

$$h_{\text{one-body}}(\Delta t\, \Delta^3 v\, \Delta^2\Omega) \simeq \rho d^2 \left(\frac{\pi k_B T}{m}\right)^{1/2} \ln\left(\frac{1595.2 \times k_B T}{\rho d^2 m\, \Delta t\, \Delta^3 v\, \Delta^2 \Omega}\right) \quad (3.18)$$

We see that this ϵ-entropy per unit time corresponds to the random choice of six continuous random variables at time intervals separated by the mean intercollisional time. The six continuous random variables are the three velocity components $\mathbf{v}_2$ of the particles of the bath, the two angles $\Omega = (\theta, \varphi)$ of the impact unit vector, and the random intercollisional time t. As a consequence, ϵ is the product of the ϵ's for each random variables: $\epsilon = \Delta t\, \Delta^3 v\, \Delta^2\Omega$.

This entropy can be compared with the maximum Lyapunov exponent (2.7). The entropy of the full process is obtained by multiplication with the number of particles in the fluid. In this case, comparison can be made with the entropy (3.8). We see that randomness of a stochastic process like the Lorentz–Boltzmann may exceed the upper bound given by the KS entropy if the ϵ quantities become too small. We can estimate the minimum value of ϵ allowed by the microscopic chaos of Newton's equations by comparing

Eq. (3.18) with Eq. (2.7). We find:

$$\Delta t\, \Delta^3 v\, \Delta^2 \Omega \gtrsim d\, \frac{k_B T}{m} \tag{3.19}$$

which gives the limit of applicability of the stochastic process. Crossing this limit would lead us to suppose that the process is more random than it actually is according to its deterministic evolution.

IV. BROWNIAN MOTION AND NOISES

Brownian motion is the irregular motion of a colloidal particle submitted to incessant collisions with the atoms or molecules of the gas or liquid. At the microscopic level, it is described by the classical Hamiltonian [44],

$$H = \frac{\mathbf{P}^2}{2M} + \sum_{i=1}^{N} \frac{\mathbf{p}_i^2}{2m} + \sum_{1 \le i < j \le N} V(|\mathbf{r}_i - \mathbf{r}_j|) + \sum_{i=1}^{N} \tilde{V}(|\mathbf{R} - \mathbf{r}_i|) \tag{4.1}$$

where M is the mass of the colloidal particle, m the mass of the atoms or molecules, V is the interparticle potential (e.g., the Lennard–Jones interaction in rare gases), and $\tilde{V}$ is the interaction potential between the atoms or molecules and the colloidal particle. A Fokker–Planck equation for the position and velocity of the colloidal particle, $\mathbf{X} = (\mathbf{R}, \mathbf{V} = \mathbf{P}/M)$, has been derived from the Hamiltonian (4.1) in the limit of large M/m ratios [44,45]. In the hydrodynamic limit of large spatial distances, we obtain the simplest Langevin equation for the lone position of the colloidal particle, $\mathbf{X} = \mathbf{R}$, with an isotropic diffusion tensor $D_{ij} = D\delta_{ij}$ in Eqs. (3.13) and (3.14). The diffusion coefficient of a colloidal particle of radius a is given by [46]

$$D = \frac{k_B T}{6\pi a \eta} \tag{4.2}$$

where η is the fluid viscosity (for water, $\eta = 0.01$ g/cm · sec; for air or argon, $\eta \simeq 2 \times 10^{-4}$ g/cm · sec; at room temperature and pressure). In the classical experiments on Brownian motion by Perrin [47], the radii of the colloidal particles were in the range 0.1–0.5 μm.

For Brownian motion or Langevin processes (3.13) like the Ornstein–Uhlenbeck process, the exponent of the ϵ-entropy (3.16) takes the value $H = 1/2$. More precisely, we obtain [30]

$$h(\epsilon) \simeq \frac{D}{\epsilon^2} \tag{4.3}$$

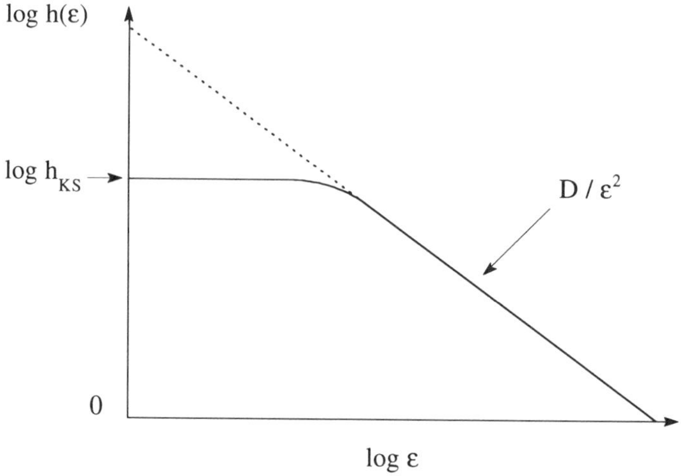

Figure 5. Schematic behavior of the ϵ-entropy in a system like the Lorentz gas. At large values of ϵ, the ϵ-entropy scales like $h \sim D/\epsilon^2$, although it saturates at the KS entropy at small values of ϵ (see [13,30]).

in digits per unit time where D is the diffusion coefficient and ϵ is the diameter of cells in the physical position space. Equation (4.3) can be obtained by a simple scaling argument from the observation that the entropy has the units of the inverse of time while ϵ is a distance in space.[4]

The increase of the entropy with $\epsilon \to 0$ shows that randomness exists on arbitrarily small scales in Brownian motion, as we see in Fig. 4. Nevertheless, on very small scales below the mean free path, the deterministic dynamics of the fluid particles becomes apparent and the ϵ-entropy (4.3) cannot grow indefinitely because the scaling property of the trajectory is only valid on scales larger than the mean free path according to kinetic theory. Therefore, the ϵ-entropy saturates at a value given by the KS entropy because the KS entropy is the supremum of all the partition entropies according to Eq. (3.3), as shown in Fig. 5 for a system like the Lorentz gas.

We now reach the conclusion of our argument. The ϵ-entropy (4.3) is an estimation of the entropy per unit time of a partition $\mathscr{P}_\epsilon$ where the colloidal particle belongs to cells of radius ϵ in the space $\mathbf{R}$ while all the other variables, $\mathbf{P}$, $\mathbf{r}_1, \ldots, \mathbf{r}_N, \mathbf{p}_1, \ldots, \mathbf{p}_N$, remain arbitrary. In view of Eq. (3.10), this partition entropy $h(\mathscr{P}_\epsilon) = h(\epsilon)$ gives us a lower bound on the sum of positive

[4] We note that, using the definition (3.2) for the ϵ-entropy, there is a further dependency on the time Δt between the samplings, as discussed by Gaspard and Wang [30], and the result (4.3) holds for small enough Δt.

TABLE I

ϵ-Entropy Per Unit Time and Lyapunov Exponents of Typical Brownian Motions and Fluids[a]

Fluid	a (μm)	D (cm^2/sec)	$h(\epsilon)$ (digits/sec)	$\lambda_{\max}$ (digits/sec)	$\sum_{\lambda_n>0} \lambda_n$ (digits/sec · cm^3)
Water	0.5	4.4×10^{-9}	1.8	10^{12}	10^{34}–10^{35}
	0.1	2.2×10^{-8}	8.8	10^{12}	10^{34}–10^{35}
Air or Ar	0.5	2.2×10^{-7}	88	10^{10}	10^{29}–10^{30}
	0.1	1.1×10^{-6}	440	10^{10}	10^{29}–10^{30}

[a] For colloidal particles of radius a in suspension in the fluid at room temperature (300 K) and pressure (1 atm), values of its diffusion coefficient D given by Eq. (4.2), of the ϵ-entropy per unit time $h(\epsilon) \simeq D/\epsilon^2$ for a resolution $\epsilon = 0.5\ \mu$m, which is the lower bound on the sum of positive Lyapunov exponents. For comparison, we give an estimation of the maximum Lyapunov exponent as well as of the sum of positive Lyapunov exponents for 1 cm^3 of the corresponding fluid.

Lyapunov exponents:

$$h(\mathscr{P}_\epsilon) \simeq \frac{D}{\epsilon^2} \leq \sum_{\lambda_n > 0} \lambda_n \tag{4.4}$$

Brownian motion can be observed with video cameras and optical microscopes which can reach resolutions of the order of $\epsilon \simeq 0.5\ \mu$m [47]. Recording the motion of colloidal particles, we can attempt to measure the ϵ-entropy per unit time. Table I contains several possible values for the ϵ-entropy corresponding to the preceding experimental conditions. Lower bounds on the order of 1–400 digits/sec may be expected. Although minute with respect to the theoretical value of Eq. (3.8), it nevertheless shows that the sum of positive Lyapunov exponents is nonvanishing.

A similar reasoning can be applied to other random processes, such as the erratic oscillations of a torsion balance in a dilute gas or to Nyquist electric noise of thermodynamic origin.

V. CONCLUSIONS

In this chapter, we developed in detail the argument that the dynamical randomness of Brownian motion gives us a lower bound on the sum of Lyapunov exponents for the microscopic dynamics in gases or liquids. The argument is based on several hypotheses, in particular, the existence of positive Lyapunov exponents in Hamiltonian systems like Eq. (4.1). Such an existence can be proved in hard-sphere systems and there is strong numerical evidence of the validity of such an assumption in Hamiltonians like Eq. (4.1) at high energies and low densities corresponding to room temperatures and pressures.

The measurement of dynamical randomness of Brownian motion as described in Section IV would give us experimental evidence that microscopic chaos is intimately related to a transport process like diffusion. Such a result would provide indirect support for the new theory described in the Introduction. Indeed, a positive entropy per unit time shows the existence of a mechanism of stretching and folding in the phase-space dynamics. As a consequence, the eigenprobabilities of the Liouvillian dynamics may acquire irregular properties which are reminiscent of fractal objects (as seen in Figs. 1 and 2).

In the title of this chapter, we asked whether microscopic chaos can be observed in the laboratory. In terms of a lower bound on the KS entropy, the answer to this question would be affirmative. Brownian motion and other thermodynamic noises provide evidence that microscopic dynamics have random time evolution. Our main result in this chapter has been to show that quantitative relations exist between apparently disconnected features. In particular, it is often thought that deterministic chaos has nothing to do with stochastic noises. In our discussion, we introduced the distinction between macroscopic and microscopic chaos. From this viewpoint, the previous statement should be rephrased in the sense that macroscopic chaos and stochastic noises are distinct phenomena (which may nevertheless appear to be combined in the same system [48]). On the contrary, stochastic noises and microscopic chaos are parts of the very same phenomenon, as we argued. The entropy per unit time is the concept that helps us to establish quantitatively the connection.

Nevertheless, we also note that microscopic chaos does not seem to be completely accessible to observation because Newtonian dynamics and ergodic theory predicts values of the Lyapunov exponents and the KS entropy which are extremely large with respect to what can practically be measured. In this regard, we may also wonder to what extent this extreme dynamical randomness is a property of the system or of the classical description. The answer to this question should come from a better understanding of dynamical randomness in a quantum description. We have summarized elsewhere several results to this problem [49–51].

We conclude with the remark that the property of chaos is concerned with the behavior of the n-time correlation functions, as we mentioned in Section III.4. Multitime properties have not been explored much until now via nonequilibrium statistical mechanics, which has mainly been focused on two-time correlation functions. Nevertheless, we think that modern multitime techniques in nuclear magnetic resonance and in nonlinear quantum optics may give access to general or particular n-time statistical properties. In this context, the exponential decay (3.4) could be used for more direct evaluations of the entropies per unit time.

ACKNOWLEDGMENTS

It is my pleasure to thank Professors G. Nicolis, I. Prigogine, and S. A. Rice for support and encouragement in this research. I am grateful to I. Antoniou, I. Prigogine, and S. Tasaki for asking me to contribute to this volume. The author is financially supported by the National Fund for Scientific Research (F. N. R. S. Belgium).

REFERENCES

1. J.-P. Eckmann and D. Ruelle, *Rev. Mod. Phys.* **57**, 617 (1985).
2. Hao Bai-Lin, Ed., *Chaos I & II: A Reprint Collection* (World Scientific, Singapore, 1984, 1990).
3. J. Lebowitz and O. Penrose, *Physics Today*, February 1973, p. 23.
4. N. N. Krylov, *Works on the Foundations of Statistical Mechanics* (Princeton University Press, Princeton, NJ, 1979); Ya. G. Sinai, *ibid.* p. 239; Ya. G. Sinai and N. I. Chernov, *Russ. Math. Surveys* **42**: 3, 181 (1987).
5. P. Gaspard and G. Nicolis, *Phys. Mag.* (*J. Belg. Phys. Soc.*) **7**, 151 (1985).
6. P. Gaspard, in *Solitons and Chaos*, I. Antoniou and F. Lambert (eds.) (Springer, Berlin, 1991), pp. 46–57.
7. R. Livi, A. Politi, and S. Ruffo, *J. Phys. A: Math. Gen.* **19**, 2033 (1986); H. A. Posch and W. G. Hoover, *Phys. Rev. A* **38**, 473 (1988); **39**, 2175 (1989).
8. P. Gaspard and S. A. Rice, *J. Chem. Phys.* **90**, 2225, 2242, 2255 (1989); **91**, E3279 (1989).
9. P. Gaspard and G. Nicolis, *Phys. Rev. Lett.* **65**, 1693 (1990).
10. P. Gaspard and F. Baras, in *Microscopic Simulations of Complex Hydrodynamic Phenomena*, M. Maréschal and B. L. Holian (eds.) (Plenum, London, 1991), pp. 301–322; P. Gaspard and F. Baras, *Phys. Rev. E* **51**, 5332 (1995).
11. P. Cvitanović, J.-P. Eckmann, and P. Gaspard, *Chaos, Solitons, and Fractals* **6**, 113 (1995); P. Cvitanović, P. Gaspard, and T. Schreiber, *Chaos* **2**, 85 (1992).
12. P. Gaspard, in *From Phase Transitions to Chaos*, G. Györgyi, I. Kondor, L. Sasvári, and T. Tél (eds.) (World Scientific, Singapore, 1992), pp. 322–334.
13. P. Gaspard, *J. Stat. Phys.* **68**, 673 (1992).
14. P. Gaspard and D. Alonso Ramirez, *Phys. Rev. A* **45**, 8383 (1992).
15. H. H. Hasegawa and W. C. Saphir. *Phys. Lett. A* **161**, 471, 477 (1992); H. H. Hasegawa and W. C. Saphir. *Phys. Rev. A* **46**, 7401 (1992); H. H. Hasegawa and D. J. Driebe, *Phys. Lett. A* **168**, 18 (1992); H. H. Hasegawa and D. J. Driebe, *Phys. Lett. A* **176**, 193 (1993).
16. P. Gaspard, *J. Phys. A: Math. Gen.* **25** L483 (1992); P. Gaspard, *Phys. Lett. A* **168**, 13 (1992).
17. S. Tasaki, I. Antoniou, and Z. Suchanecki, *Phys. Lett. A* **179**, 97 (1993); I. Antoniou and S. Tasaki, *J. Phys. A: Math. Gen.* **26**, 73 (1993); I. Antoniou and S. Tasaki, *Physica A* **190**, 303 (1992).
18. P. Gaspard, *Chaos* **3**, 427 (1993); *Phys. Rev. E* **53**, 4379 (1996).
19. S. A. Rice, P. Gaspard, and K. Nakamura, *Adv. Class. Traj. Meth.* **1**, 215 (1991).
20. P. Cvitanović and B. Eckhardt, *J. Phys. A: Math. Gen.* **24**, L237 (1991).
21. P. Gaspard, in *Quantum Chaos*, G. Casati, I. Guarneri, and U. Smilansky (eds.) (North-Holland, Amsterdam, 1993), pp. 307–383.
22. I. Prigogine, *Nonequilibrium Statistical Mechanics* (Wiley, New York, 1962); I. Prigogine, *From Being to Becoming* (Freeman, San Francisco, 1980).

23. K. F. Gauss, *J. Reine Angew. Math.* **IV**, 232 (1829).
24. S. Smale, in *The Mathematics of Time* (Springer, New York, 1980), p. 137.
25. S. Nosé, *J. Chem. Phys.* **81**, 511 (1984); *Mol. Phys.* **52** 255 (1984); W. G. Hoover, *Phys. Rev. A* **31**, 1695 (1985).
26. D. J. Evans, E. G. D. Cohen, and M. P. Morris, *Phys. Rev. A* **42**, 5990 (1990).
27. N. I. Chernov, G. L. Eyink, J. L. Lebowitz, and Ya. G. Sinai, *Phys. Rev. Lett.* **70**, 2209 (1993).
28. S. Tasaki and P. Gaspard, in *Toward the Harnessing of the Chaos*, M. Yamaguti (ed.) (Elsevier Science B. V., 1994), pp. 273–288; *J. Stat. Phys.* **81**, 935 (1995).
29. M. Hara and M. Yamaguti, *Jpn. J. Appl. Math.* **1**, 183 (1984).
30. P. Gaspard and X.-J. Wang, *Phys. Rep.* **235**, 321 (1993).
31. C. E. Shannon and W. Weaver, *The Mathematical Theory of Communication* (The University of Illinois Press, Urbana, 1949).
32. A. N. Kolmogorov, *Dokl. Acad. Sci. USSR* **124**:4, 754 (1959); Ya. G. Sinai, *ibid.* **124**:4, 768 (1959).
33. G. J. Chaitin, *Algorithmic Information Theory* (Cambridge University Press, Cambridge, 1987).
34. V. I. Arnold and A. Avez, *Ergodic Problems of Classical Mechanics* (Benjamin, New York, 1968); P. Billingsley, *Ergodic Theory and Information* (Wiley, New York, 1965); I. P. Cornfeld, S. V. Fomin, and Ya. G. Sinai, *Ergodic Theory* (Springer, Berlin, 1982).
35. M. S. Green, *J. Chem. Phys.* **22**, 398 (1954); R. Kubo, *J. Phys. Soc. Jpn.* **12**, 570 (1957).
36. Ya. B. Pesin, *Math. USSR Izv.* **10**(6), 1261 (1976); *Russ. Math. Surveys* **32**(4), 55 (1977).
37. H. Kantz and P. Grassberger, *Physica D* **17**, 75 (1985).
38 P. Szépfalusy and T. Tél, *Phys. Rev. A* **34**, 2520 (1986); **35**, 477 (1987).
39. D. Ruelle, *Ann. NY Acad. Sci.* **316**, 408 (1979); Ju. I. Kifer, *Math. SSSR Izv.* **8**, 1083 (1974).
40. A. N. Kolmogorov, *IRE Trans. Inf. Theory* **1**, 102 (1956); A. N. Kolmogorov and V. M. Tikhomirov, *Uspekhi Mat. Nauk* **14**, 3 (1959); V. M. Tikhomirov, *Russ. Math. Surveys* **18**, 51 (1963).
41. B. B. Mandelbrot and J. W. Van Ness, *SIAM Rev.* **10**, 422 (1968).
42. A. Cohen and I. Procaccia, *Phys. Rev. A* **31**, 1872 (1985).
43. N. G. Van Kampen, *Stochastic Processes in Physics and Chemistry* (North-Holland, Amsterdam, 1981), p. 358 and references therein.
44. P. Résibois and M. De Leener, *Classical Kinetic Theory of Fluids* (Wiley, New York, 1977).
45. J. L. Lebowitz and E. Rubin, *Phys. Rev.* **131**, 2381 (1963); P. Résibois and H. T. Davis, *Physica* **30**, 1077 (1964).
46. L. Landau and E. Lifshitz, *Fluid Mechanics* (Pergamon Press, New York, 1963).
47. J. Perrin, *Les atomes* (Presses Universitaires de France, Paris, 1970).
48. G. Nicolis and P. Gaspard, *Chaos, Solitons, and Fractals* **4**, 41 (1994).
49. P. Gaspard, in *Quantum Chaos—Quantum Measurement*, P. Cvitanović, I. Percival, and A. Wirzba (eds.) (Kluwer, The Netherlands, 1992), pp. 19–42.
50. P. Gaspard, in *Quantum Chaos*, H. A. Cerdeira, M. C. Gutzwiller, R. Ramaswamy, and G. Casati (eds.) (World Scientific, Singapore, 1991), p. 348; T. Hudetz and P. Gaspard, unpublished.
51. P. Gaspard, *Prog. Theor. Phys. Suppl.* **116**, 369 (1994).
52. J. R. Dorfman and P. Gaspard, *Phys. Rev. E* **51**, 28 (1995); P. Gaspard and J. R. Dorfman, *Phys. Rev. E* **52**, 3525 (1995).

PROTON NONLOCALITY AND DECOHERENCE IN CONDENSED MATTER—PREDICTIONS AND EXPERIMENTAL RESULTS

C. A. CHATZIDIMITRIOU-DREISMANN

Iwan N. Stranski Institute for Physical and Theoretical Chemistry, Technical University of Berlin, Berlin, Germany

CONTENTS

Advances in Chemical Physics, Volume XCIX, Edited by I. Prigogine and Stuart A. Rice.
ISBN 0-471-16526-3

I. INTRODUCTION

As is well known, the fundamental dynamical laws of classical and quantum mechanics are microscopically reversible because they are invariant with respect to time inversion. However, mesoscopic and macroscopic phenomena (like diffusion, energy dissipation, and transport processes in condensed matter) are irreversible. Therefore, the description of irreversibility at the microscopic level of natural laws represents a challenging problem of basic importance. During the last hundred years, diverse solutions to this problem have been suggested, some of them being of fundamental character (e.g., the Boltzmann equation of kinetic theory, the so-called Green–Kubo formulas of statistical mechanics, and the subdynamics and complex spectral decompositions of the Brussels school). The theoretical basis of this chapter is related to the complex spectral form of the second-order density operator of fermionic systems, which is obtained in the frame of the so-called complex scaling method (CSM) (for dilation analytic operators).

A criticism that usually is aimed at new theories, and especially at those that take novel approaches to fundamental physical problems, is that no *new experimentally verifiable consequences* appear to have emerged from the new approaches. Doubtless, this point is crucial, since a truly "new theory" must be able to make qualitatively new predictions that, furthermore, are experimentally testable [1]. In light of this remark, this chapter only parenthetically provides "mathematical proofs" and "derivations," but deals in detail with real experiments. Some of these experiments (see Section III) concern certain counterintuitive *predictions* (published 1989 [2]) of the underlying theory, which were experimentally verified afterward. It is also shown that the predicted new effects have no classical or conventional interpretation. Thus, it is hoped that the new experimental findings presented will stimulate further theoretical research and initiate a new collaboration between theoreticians and experimentalists.

A. Protons in Condensed Systems

The knowledge of proton-transfer quantum dynamics in condensed systems is indispensable for the understanding of many physical, chemical, and biological processes of basic importance. Here let us just mention some examples: proton-transfer reactions in water and ionic solutions; proton conductance and/or mobility in water, ice, and metallic hydrides; intramolecular proton transfer in organic molecules; proton dynamics in H-bonded organic crystals and polymers; proton dynamics in DNA and certain H-bonded structures of proteins (like α-helices and β-sheets); structure of H-bonded DNA-protein complexes. Accordingly, the scientific literature dealing with these topics is huge.

By considering the corresponding theories and/or models in some detail, however, one easily recognizes the following characteristics:

1. The majority of the existing investigations of chemical and biological systems do consider the protons as *classical* particles (typical example: quantum chemistry).
2. In most cases in which the quantum character of protonic (or H) motion is studied, the *Born–Oppenheimer approximation* is used as the starting point.
3. During the last decade, significant progress was achieved through the formulation of certain nonadiabatic theories (typical example: the famous Kondo theory with applications to metal–hydrogen systems at low temperatures). These ideas, however, have not been applied to chemical or biological topics until now.
4. Moreover, in *all* these theories and/or models, the dynamics of *single* protons (or H atoms) are studied, and possible quantum interference or quantum correlation effects between different protons are discarded completely.

These points were recently considered critically, in the frame of our general CSM treatment of quantum dynamics of condensed nonequilibrium systems [3,4]. It was argued that point 4 especially corresponds to a physical assumption which—in most realistic cases—cannot be justified physically. To illustrate, let us mention here that quantum correlations are intrinsically related with (partial) delocalization of, and interference between, the considered particles. Owing to their large *thermal de Broglie wavelength* λ_{dB}, protons may show short-time and spatially restricted quantum correlations even at relatively high temperatures. It holds that

$$\lambda_{dB} = h/\sqrt{2\pi k_B T m}$$

where k_B is Boltzmann's constant, T is temperature, and m is mass. Note that λ_{dB} is a bit smaller than the standard de Broglie wavelength h/p (p is momentum corresponding to the relation $p^2/2m = 3k_B T/2$). For a quasi-free proton at, say 300 K, one finds that $\lambda_{dB}(H) \approx 1.0$ Å, and for μ^+ under the same conditions, $\lambda_{dB}(\mu^+) \approx 3.0$ Å. Protons lying nearby in dense systems (like water) may therefore exhibit quantum interference (or correlation) effects within sufficiently short time intervals.

B. Quantum Correlations Between Protons

As is well known, the striking nonlocal character of quantum objects has been repeatedly demonstrated through interference effects appearing in Young's double-slit experiments of different kinds. In this context, the first successful experiments concerning quantum interference of *single* He and Na atoms are worth mentioning [5]. A counterintuitive quantum interference effect between *two* well-separated (by about 10,000 Å) Ca atoms emitting *one* photon has also been demonstrated [6]. Furthermore, the discovery of Bell's inequalities [7] and the corresponding experimental investigations proving their violation (see [8,9]) clearly disprove the existence of so-called local hidden variables and stress the reality as well as the physical significance of Einstein–Podolsky–Rosen (EPR) correlations [10,11]; see also [12,13]. A large number of theoretical investigations of new variants of Bell-type inequalities recently appeared, so that it was said that "nonlocality bursts into life" [14].

All of these theoretical and experimental investigiations, however, deal with quantum systems being (quasi-)isolated from their environment. In condensed systems, EPR (or quantum) correlations are considered to be of great importance, but their theoretical treatment as well as experimental measurement represent extremely difficult tasks. To deal with this problem, we recently applied, for the first time [3], the general theory of dilation analyticity (or CSM [15,16]) to the dynamics of large systems in nonequilibrium. This work emphasized the fundamental role that quantum correlations play in dynamical processes in condensed phases. A particularly perplexing result is the following. If certain specific quantization conditions (concerning the complex-energy eigenvalues of resonances; see Section II.D) are fulfilled, the thermalized and CSM-transformed second-order density matrix of fermionic systems, $\Gamma^{(2)}$, has no diagonal representation. In more technical terms, this means that $\Gamma^{(2)}$ contains *Jordan blocks* (of higher order). This unexpected result was further analyzed and shown to imply physically that there is—in the considered framework—no way to "transform away" quantum correlations [4]. Thus, entanglements between quantum systems, as expressed by quantum (or EPR) correlations, are now revealed to be intrinsic to a broad class of condensed systems. The possible connection of this finding with Yang's concept of off-diagonal long-range order (ODLRO) [17] was pointed out. These persisting correlations, as revealed by our CSM theory [3,4], constitute the short-lived and spatially restricted *coherent-dissipative structures.*

It has been discussed and demonstrated through many applications that coherent-dissipative structures represent a new form of spontaneously emerging self-organization in nonequilibrium systems on the microscopic level of description.

C. On Intrinsic Irreversibility and the Dynamics of Nonintegrable Systems

In the present context it is also interesting to point out, as Prigogine and co-workers have emphasized (cf. [18–22]), the entity "density operator" has to be considered as the primary concept in the dynamics of large systems, whereas the "wave function" should be considered as a quantity of approximative character, that is, as an idealization. In the physical context of condensed-matter processes, this may easily become clear—at least if one wants to go beyond the usual "coarse graining," "time smoothing," and other similar approximations. Additionally, in large systems and condensed matter, the physical meaning of the concept "stationary state" is lost [23].

The fundamental question concerning the existence of microscopic processes being intrinsically irreversible has been investigated extensively by Prigogine and co-workers during the last 30 years, (cf. [18–22]). These investigations clarified many properties of chaotic and/or nonintegrable dynamical systems, classical as well as quantal. It has been pointed out that quantum large Poincaré systems—which are of particular interest for our purposes too—are mixing and have an absolutely continuous spectrum for their Liouville superoperators L, which determine the time evolution of the associated density operators of the systems.

Recent work of the Brussels–Austin school succeeded in proving that L (or the Hamiltonian H, if it exists) may admit a *complex* spectral decomposition [19–22]. Although the underlying mathematical structure (i.e., the rigged Hilbert space formalism) differs from the CSM which underlies our work [3,4], one sees that complex energies, resonant states, decay states, and so on are not just "approximations," but fundamental entities of both theories.

Moreover, the subdynamical decomposition of the Liouvillian [19–22] revealed a feature which is essentially equivalent to the quantum correlated (coherent-dissipative) structures [4,3]. Namely, the non-hermitian intermediate operator Θ of the "reduced" Liouville–Von Neumann equation, which is associated with dissipative processes, may have a Jordan block structure—like the CSM-transformed and thermalized density operator $\Gamma^{(2)}$ in our formalism.

Most of the experimental results and predicted effects, which are presented below, are shown to have no conventional (classical or quantal) explanation. For this reason, we believe that they represent the first direct "fingerprints" of microscopic processes which are intrinsically irreversible [2,18–22].

D. Outline

In this chapter, the focus is mainly on the *experimental applicability* of the CSM theory of quantum correlations in the framework of dynamical processes in large and dense systems. We consider only quantum correlation effects between protons in condensed matter. (For applicability of the CSM theory to electronic quantum correlations, see [24–27] as well as [4].) In the following sections, after some remarks on quantum correlations, decoherence, and the CSM formalism, certain predictions [2,4] of the theory and proposed experiments to test them are presented. Eight different experimental topics are discussed. The diversity of the topics considered clearly demonstrates the predictive power of the general theory [3,4].

II. QUANTUM CORRELATIONS AND DECOHERENCE

A. On Quantum or EPR Correlations

The classic argument of Einstein, Podolsky, and Rosen [10] is nowadays well known and even treated in popular books. The reason is that, since the discovery of Bell's inequalities [7], an impressive number of novel experiments disproved these inequalities beyond any reasonable doubt, thus clearly confirming the existence of nonlocal quantum correlations (or entanglements) between systems that can even be spatially well separated. These correlations, which have no classical–mechanical analogue [17], are usually called EPR correlations (e.g., see [9], [11–13], [31], and references cited therein).

Let us consider these correlations a little more formally. Let S_A and S_B be two physical systems with the associated Hilbert spaces H_A and H_B. Then the density operator ρ_{A+B} of the complete system is generally given by

$$\rho_{A+B} = \sum_{k,\,l} c_k c_l^* \,|a_k b_k\rangle\langle a_l b_l| \tag{1}$$

(The state vectors $|a_k\rangle$ and $|b_k\rangle$ belong to H_A and H_B, respectively. Here, Schrödinger's "natural expansion" of correlated states is used; cf. [31].) Standard theory says that measurements on a quantum ensemble of S_A systems are described with the aid of the partial density operator

$$\rho_A \equiv \mathrm{Tr}_B(\rho_{A+B}) = \sum_k |c_k|^2 \,|a_k\rangle\langle a_k| \tag{2}$$

(Tr_B is the partial trace over the variables of S_B.) An analogous formula holds for S_B. As a result, the maximal physical information accessible by measurements on the subsystems S_A and S_B is given by

$$\rho_{\mathrm{unc}} = \rho_A \otimes \rho_B \tag{3}$$

Comparison with Eq. (1) shows that the quantal phase factors between the different (generally complex) amplitudes c_k are lost in this case (3). Thus, the physical information one can obtain from all possible measurements on the two subsystems does *not* allow the reconstruction of the complete density operator ρ_{A+B}, because $\rho_{A+B} \neq \rho_{\text{unc}}$. In this context, one often says that "the whole is qualitatively more than the combination of its parts."

The aforementioned phase factors represent physically the quantum correlations between the subsystems S_A and S_B. These phase factors can be determined only by appropriate measurements on both subsystems "at the same time." If this is the case, one says that the corresponding experiments measure the possible correlations between the subsystems; ρ_{unc} describes uncorrelated subsystems.

The formulas given above also demonstrate the EPR paradox easily: If a measurement on the system S_A is done and determines a specific state, say $|a_1\rangle$, it follows from Eq. (1) that the well-separated subsystem S_B—"without in any way disturbing it" [10]—will be in the state $|b_1\rangle$ with certainty. This fact represents the so-called generalized EPR phenomenon or EPR paradox [31].

These considerations show that the reason for the EPR paradox is given by the superposition principle of quantum mechanics. *All EPR correlation effects are quantum interference effects* [31]. Thus, it should be clear that the terms "quantum" and "EPR" correlations have essentially the same physical meaning.

In the following sections, the focus is on quantum correlation effects—rather than interaction mechanisms—of specific dynamical processes in condensed matter and at "high" (e.g., room) temperatures.

B. Decay of Correlations and Decoherence

As shown above, quantum or EPR correlations are mathematically represented by the off-diagonal matrix elements of the density operator under consideration (usually denoted by ρ or Γ). The physical terms "coherence" and "coherent superposition of states" refer to a specific kind of quantum correlation, which—illustratively speaking—may be regarded as the "maximal possible" or the "most intense" form of correlation. Quantum coherence is intrinsically related to the well-known superposition principle of quantum mechanics.

Decoherence, a term that became "modern" in the last decade, means the decay of quantum correlations between two or more quantum states. Thus, decoherence may mean, in specific cases, the disappearance of quantum interference between two quantum systems.

It is important to note that decoherence is also intrinsically connected with dissipation (or entropy production), the (still unsolved) "measurement problem" of quantum theory, and the "direction of time" problem of quantum dynamics of large systems (cf. [32] and [18]).

After these remarks, one may recognize that certain actual scientific topics and theories (e.g., dissipative quantum tunneling, subdynamics, microscopic irreversibility, off-diagonal long-range order, partial coherence, creation and destruction of correlations, and system–environment interactions) are strongly interrelated and also have essentially the "same" goals and aims.

As an explicit and illustrative example, let us consider here the very simple decoherence model of reference [33]. Let $\rho(x, x')$ be the density operator (or matrix) of a particle in the coordinate representation. In the high-temperature and weak-coupling limit (of the particle interactions with its environment), the master equation [33] for ρ may be written as

$$\frac{d\rho}{dt} = -\frac{i}{\hbar}[H, \rho] - r(x - x')\left(\frac{\partial\rho}{\partial x} - \frac{\partial\rho}{\partial x'}\right) - \frac{2mk_B T}{\hbar^2}(x - x')^2\rho \tag{4}$$

where H is the particle's effective (or renormalized) Hamiltonian and r is the relaxation rate. The first term is well known. The second term represents "friction" effects and causes dissipation. The third term takes into account fluctuations or "random kicks" due to the environment. The effect of the third term on quantum superpositions of states is of particular interest, because it is easily shown that it *destroys quantum coherence*, that is, it causes *decoherence*. In other words, this term causes the *elimination* of the off-diagonal terms $\rho(x, x')$ (with $x \neq x'$) of ρ which represent *quantum correlations* between the associated states being "centered" (in this example [33]) around x and x'.

Further reasoning shows [33] that the quantum coherence will disappear exponentially on a *decoherence time scale*

$$\tau_D \approx \tau_R \frac{\hbar^2}{2mk_B T(\Delta x)^2} = r^{-1}\left(\frac{\lambda_T}{\Delta x}\right)^2 \tag{5}$$

where $\tau_R = r^{-1}$ is the standard relaxation time (due to friction processes), $\Delta x = x - x'$, and $\lambda_T \equiv \hbar/\sqrt{2mk_B T}$. (Note that λ_T is not identical to the thermal de Broglie wavelength λ_{dB} mentioned above.)

It is well known that the decoherence time τ_D for macroscopic systems is extremely short. Omnès [32] mentions the decoherence process of the center of mass of the moon being due to the disturbance of the moon's motion by the photons coming from the sun, and he estimates that τ_D is about 10^{-35} s. Furthermore, for a system at room temperature with mass 1 g, and for a spatial separation $\Delta x = 1$ cm, Eq. (5) gives $\tau_D/\tau_R = 10^{-40}$. These (and similar) remarks result in a tendency to motivate the "esoteric character" of quantum (or EPR) correlations (and the associated delocalization) in *condensed* matter at elevated temperatures. In this context, one also says that the thermal motion "smears out" quantum (EPR) correlations.

For our purposes, however, it is important to realize that the estimates above do *not* apply to light particles. For example, for an electron, τ_D can be much larger than τ_R [33]. Here, let us consider the well-studied H–Nb system, which is discussed in detail in Section IV.B. For protons at $T = 100$ K, one finds $\lambda_T \approx 0.5$ Å. The neighboring tetrahedral interstitial sites in the bcc lattice of Nb (with lattice constant $a_{Nb} = 3.3$ Å), which can be occupied by a trapped proton, are separated by $\Delta x = a\sqrt{2}/4 \approx 1.2$ Å (see Section IV.B). Thus, according to Eq. (5), τ_D is not very different from τ_R, which may have direct experimental consequences (τ_R is associated with the well-known "friction" effects of a proton with phonons and electrons, which causes dissipation). For example, the trapped protons may exhibit a short-time delocalization (of the order of τ_D). This effect should not be confused with the usually considered time-independent delocalization of isolated systems over stationary states.

C. Some Remarks on the Complex-Scaling Theory of Thermally Activated Quantum Correlations

Recently, the general quantal theory of dilation analyticity (or CSM) was extended and applied to quantum dynamics of condensed nonequilibrium systems for the first time [3,4]. The possibility of novel, thermally activated EPR correlations, which follow from first quantum theoretical principles, was pointed out. The corresponding irreducible structures were called *coherent-dissipative* because they represent a short-lived and spatially restricted cooperative phenomenon within the microscopic level of description. The conceptual connection with standard quantum delocalization, EPR correlations, and Yang's concept of ODLRO was mentioned. For a recent review of the theory and its different applications, see [4] and references therein. We state here only some basic steps behind the proposal about coherent-dissipative structures:

1. Coleman [28,29] showed that many quantum correlation effects appear to be intrinsically connected with general properties of the second-order reduced-density matrix for *fermionic systems*, $\Gamma^{(2)}$. Namely, it was proved that this matrix has a universal decomposition:

$$\Gamma^{(2)} = \Gamma^{(2)}_{\mathrm{L}} + \Gamma^{(2)}_{\mathrm{S}} + \Gamma^{(2)}_{\mathrm{incoh}} \tag{6}$$

The "large-box" part, $\Gamma^{(2)}_{\mathrm{L}}$ represents the coherent part (i.e., it is associated with the macroscopic pair wave function of the "condensate" of paired fermions); $\Gamma^{(2)}_{\mathrm{incoh}}$ is an incoherent component. The so-called "small-box" part, $\Gamma^{(2)}_{\mathrm{S}}$, has a nondiagonal form and is associated with possible "short-range" quantum correlations between paired fermions.

2. In the recent work [3,4] on the extension of the CSM to quantum

dynamics of large systems and on thermally activated, short-lived quantum correlations in condensed matter, the crucial role of $\Gamma_S^{(2)}$ has been demonstrated, which usually is completely discarded. More concretely, the "part of the ensemble" correlations represented by $\Gamma_S^{(2)}$ has been subjected to the "thermalization transformation"

$$\gamma \equiv \frac{1}{Z} e^{-(\beta/2)H^c} \Gamma_S^{(2)c} e^{-(\beta/2)H^c} \tag{7}$$

where $\beta = 1/k_B T$ and H^c refers to the CSM-transformed second-order reduced Hamiltonian. Under specific quantization conditions [3,4,34,35], the density operator γ has a *Jordan block* [36] structure, which means that it is by no means (i.e., under no similarity transformation) diagonalizable. Thus, the well-known probabilistic interpretation of the diagonal elements of the density operator γ is here completely lost. This is a crucial finding, because it implies that the states constituting γ coalesce and act "cooperatively" as an indivisible unit, which has been called a *coherent-dissipative structure*. In more plain terms, one can say that all CSM-transformed states under consideration appear to be intrinsically connected, or entangled.

3. The minimal number $s_{\min}$ of states participating in a coherent-dissipative structure has been proved to be given by

$$s_{\min} = \frac{4\pi k_B T}{\hbar} \tau_{\text{rel}} \tag{8}$$

where τ_{rel} represents the characteristic relaxation time (lifetime or decoherence time) of the specific microdynamical process of a microsystem (and it is related to the imaginary energy parts of the exposed resonances of H^c). Because $s_{\min}$ is proportional to T, one has the unexpected result that temperature increases may favor the increase of the size (in Hilbert space) of these structures.

4. Coherent-dissipative structures are short lived and have a microscopically small size. For example, the lifetime of coherent-dissipative structures consisting of protons in water appears to be of the order of some picoseconds [4]; the spatial extension of magnetic coherent-dissipative structures in *paramagnetic* Gd seems to be some tens of angströms, and their lifetime is less than 1 ps [4,24].

5. In this context, it is also interesting to observe that the trace of γ representing coherent-dissipative structures vanishes identically, $\mathrm{Tr}\,\gamma = 0$ [2,4]. Current investigations indicate that this result may be of considerable importance in the physical context of laser light stattering (Raman, Rayleigh–Brillouin; see Sections III.D and E) and neutron scattering on water (cf. [2,4]).

6. The different applications of the general CSM theory make use of the following *ansatz* concerning the actual delocalization (in geometric space) of coherent-dissipative structures:

$$\Xi_X = F_X(\hat{H}_{\text{eff}}, \ldots) \cdot \lambda_{\text{dB}} \cdot s_{\min, X} \tag{9}$$

The symbols have the following meanings: Ξ may represent—depending on the specific application—(1) a transport coefficient or (2) the geometrical size of a coherent-dissipative structure; X is the specific microscopic quantum system; $s_{\min, X}$ is the "size" (in the space of state functions) of the structure; $\lambda_{\text{dB}, X}$ is the conventional thermal de Broglie wavelength of one quantum system X; and $F_X(\hat{H}_{\text{eff}}, \ldots)$ is a functional of the "effective" or "relevant" Hamiltonian $\hat{H}_{\text{eff}}$ being proper for the dynamics of X in the condensed system. The term F may also depend on some external parameters (like magnetic field, temperature, etc.). The specific "mechanism," $\hat{H}_{\text{eff}}$, describing the system for a sufficiently short-time interval, appears explicitly in this *ansatz*. Fortunately, its actual form is often not needed, as the different applications demonstrate.

III. PROTONS IN WATER

A. Anomalous Activation Energy of Proton Transfer in Water

One of the most important problems in the physical chemistry of water is the experimental and theoretical treatment of the rate constants characterizing the following processes:

$$H_3O^+ + H_2O \underset{}{\overset{k_1, E_1}{\rightleftharpoons}} H_2O + H_3O^+ \tag{10a}$$

$$OH^- + H_2O \underset{}{\overset{k_2, E_2}{\rightleftharpoons}} H_2O + OH^- \tag{10b}$$

The pioneering work of Meiboom [37,38] proved that one can measure these reaction rates, k_i, and "activation energies," E_i, (with $i = 1, 2$) by NMR spectroscopic methods (see below). The study of proton-transfer reactions in water is also important for the understanding of the excess (or anomalous) conductivities (or mobilities) of the hydronium (H_3O^+) and hydroxyl ions in water, aqueous solutions, and ice (see the classic work of Eigen [39,40]).

Our investigations of these reactions [41] were based on the following physical assumption or conjecture: From the quantum mechanical viewpoint, it appears that the protons forming the H^+ ions are indistinguishable from those belonging to the water molecules and being in the vicinity of the ions. This unconventional consideration may be motivated by the fact that the thermal de Broglie wavelength of a "quasi-free" proton, $\lambda_{dB}(H^+)$, at room temperature (cf. Section I.A) is large enough that one almost always will find water protons at a distance of the order of $\lambda_{dB}(H^+)$ around each H^+. This fact may lead to typical quantal delocalization and/or interference effects. The assumption of quantum effects between protons "belonging" to ions and water molecules, of course, represents a *working hypothesis*, that is, it must be tested experimentally. Fortunately, it appears that (at least) two important predictions, which follow straightforwardly from our physical assumption, *contradict* all thus far known conventional theories (or models).

The first prediction is given by a novel form [2,41] of the connection of (1) the proton transfer rates, k_i, of Eqs. (10a,b) with (2) the excess ionic conductivities of H^+ and OH^-, $\lambda^e_{H^+}$ and $\lambda^e_{OH^-}$, in water. The latter quantities are conventionally defined as

$$\lambda^e_{H^+} \equiv \lambda_{H^+} - \lambda_{X^+} \quad \text{and} \quad \lambda^e_{OH^-} \equiv \lambda_{OH^-} - \lambda_{Cl^-} \tag{11}$$

with $X^+ = K^+$ or Na^+, where λ_Y represents the experimentally measured ionic conductance of the ion Y in water [42].

The *conventional treatment* of the connection under consideration (see [4,37] for a short derivation) is based on the well-established equations of Nernst and Einstein:

$$\lambda^e = \frac{qD}{k_B T} \quad \text{and} \quad D = \frac{\langle x^2 \rangle}{6\tau_{rel}} \tag{12}$$

The notations are as follows: q is the elementary charge; D is the diffusion coefficient describing charge transport due to proton transfers; τ_{rel} is the average lifetime of a H_3O^+ (or OH^-) ion, and $\langle x^2 \rangle$ is the average of the square of the displacement of a proton. In a simple model, one may identify $\langle x \rangle$ with the mean distance between two oxygen atoms of water molecules [37]. From these equations and the standard relation for the "relaxation times" associated with the reaction (10a),

$$\frac{1}{\tau_{rel, H^+}} = k_1 \cdot [H_2O] \tag{13}$$

one obtains straightforwardly $\lambda^e_{H^+} T = C \cdot k_1$, where C is a temperature-independent constant and, therefore,

$$\frac{\lambda^e_{H^+}}{\lambda^e_{OH^-}} = \frac{k_1}{k_2} \tag{14}$$

With an Arrhenius-type *ansatz* for the rate constants of Eqs. (10a,b), $k_i = C_i \cdot \exp(-E_i/RT)$, $(i = 1, 2)$ it follows immediately that

$$\log\left(\frac{\lambda^e_{\mathrm{H}^+}}{\lambda^e_{\mathrm{OH}^-}}\right) = C - \frac{E_1 - E_2}{RT} \tag{15}$$

It will be shown that the result (14), or equivalently (15), is definitely in disagreement with the corresponding prediction of our CSM theory of quantum correlations.

With the aforementioned precise data of reference [42] for the ionic conductances in water, Eq. (15) yields the classically predicted value

$$E_1 - E_2 \approx -2.0 \text{ kJ/mol} \qquad \text{for } T = 15\text{–}55°\text{C} \tag{16}$$

This result is in accordance with the well-known [45] "traditional" treatment of the reactions (10a,b) which predicts a negative sign for $E_1 - E_2$. This result can be illustrated by the following argument [39]: Eq. (10a) describes a proton jump between two *neutral* water molecules, whereas Eq. (10b) describes a proton jump between two *charged* centers, that is, two OH^-. Therefore, the activation energy of the proton transfer in the latter case should be higher than in the former, that is, $E_1 < E_2$.

Interestingly, our CSM theory made quite the opposite prediction [2,4,41]. Namely, application of the general *ansatz* (9) to the transport coefficients $\lambda^e_{\mathrm{H}^+}$ and $\lambda^e_{\mathrm{OH}^-}$ yields

$$\lambda_X = F_X(\hat{H}_{\mathrm{eff}}, T) \cdot \lambda_{\mathrm{dB},X} \cdot s_{\mathrm{min},X} \qquad (\mathrm{X} = \mathrm{H}^+, \mathrm{OH}^-) \tag{17}$$

and thus

$$\frac{\lambda^e_{\mathrm{H}^+}}{\lambda^e_{\mathrm{OH}^-}} = \frac{F_{\mathrm{H}^+}}{F_{\mathrm{OH}^-}} \cdot \frac{\lambda_{\mathrm{dB,H}^+}}{\lambda_{\mathrm{dB,OH}^-}} \cdot \frac{\tau_{\mathrm{rel,H}^+}}{\tau_{\mathrm{rel,OH}^-}} \tag{18}$$

Further insertion of the explicit form of the thermal de Broglie wavelength $\lambda_{\mathrm{dB},X} = \hbar\sqrt{2\pi/m_X k_B T}$ and the standard relation (13) in Eq. (18) yields the result

$$\frac{\lambda^e_{\mathrm{H}^+}}{\lambda^e_{\mathrm{OH}^-}} = \frac{F_{\mathrm{H}^+}}{F_{\mathrm{OH}^-}} \cdot \sqrt{\frac{m_{\mathrm{OH}^-}}{m_{\mathrm{H}^+}}} \cdot \frac{k_2}{k_1} \tag{19}$$

(Regretfully, in the scientific literature the symbol λ is conventionally used for both "wavelength" and "ionic conductance".)

This formula is the main result of the CSM theory. To make it capable of experimental testing, we considered [2,41] the following slight simplification: In the present context we may assume for physical reasons that

$$F_{H^+} \approx F_{OH^-} \tag{20}$$

because in both cases (10a,b), the larger part of the system with Hamiltonian $\hat{H}_{eff}$ consists of water molecules, that is, of the same compound. Equation (20) represents a physical assumption, which is probably of *approximate* character. Nevertheless, its validity and/or physical significance is supported by the experiment; see below. Thus, we obtain

$$\frac{\lambda^e_{H^+}}{\lambda^e_{OH^-}} = \sqrt{\frac{m_{OH^-}}{m_{H^+}}} \cdot \frac{k_2}{k_1} \tag{21}$$

With an Arrhenius-type *ansatz* for the reaction rates k_i, as above, Eq. (21) yields

$$\log\left(\frac{\lambda^e_{H^+}}{\lambda^e_{OH^-}}\right) = C' + \frac{E_1 - E_2}{RT} \tag{22}$$

and with the aid of the conductivity data of reference [42], we obtain the predicted value [2]

$$E_1 - E_2 \approx +2.0 \text{ kJ/mol} \qquad \text{for } T = 15\text{–}55°\text{C} \tag{23}$$

It is important to observe that the classically and CSM-predicted numerical values of $E_1 - E_2$, Eqs. (16 and 23), differ *even by sign*!

Owing to the considerable scattering of the existing old (from the 1960s) NMR experimental data for this difference $E_1 - E_2$ (see [4,30] for original references), new high-precision experiments (utilizing the ^{1}H–NMR spin-echo techniques) have been carried out by Hertz and co-workers [43,44]. The NMR experimental data [43] are graphically presented in Fig. 1, together with the classical and CSM theoretical predictions based on the same conductivity data [42]. In our treatment of these data, we prefer to omit the experimental values of k_1 and k_2 at $T = 5.3$ C, because of the well-known "anomaly" that water exhibits at 4 C. For the temperature range $T = 10.7$–55°C, the experimental value

$$E_1 - E_2 = +(1.9 \pm 0.5) \text{ kJ/mol} \tag{24}$$

was obtained. Further measurements up to 85°C [44] confirmed our prediction as well. Thus, we may conclude that the experimental results

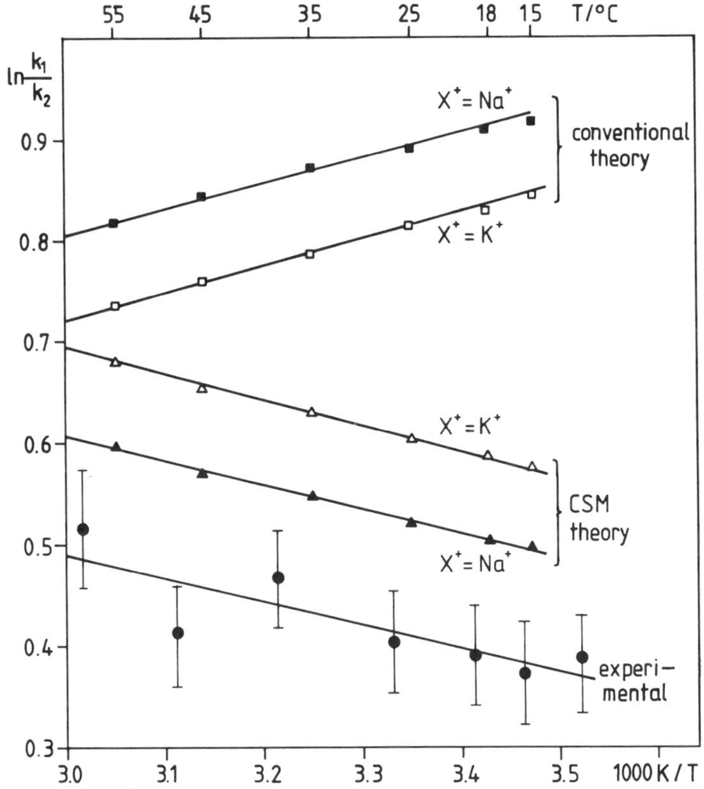

Figure 1. The temperature dependence of the logarithm of k_1/k_2. The predictions of conventional and CSM theory and the experimental results [43] are shown. The data points for the two theoretical predictions were calculated with the conductivity data. The reference cations (K^+ and Na^+) used to calculate the H^+-excess conductivity are indicated. (Reproduced from [30].)

confirm the predicted [2] *positive* sign of $E_1 - E_2$ definitely, and are in clear disagreement with the prediction of the classical theory.

The results above also allow us to test the relation (20), which was assumed to be approximately valid. From the data of Fig. 1, one obtains the relation $F_{H^+} \approx 1.16 \cdot F_{OH^-}$.

B. Anomalous Decrease of H^+/D^+ and OH^-/OD^- Conductances in H_2O/D_2O

1. Anomalous H^+/D^+ Conductance

Another prediction [2,4] of the general theory concerns an *anomalous decrease* of the H^+ (and probably also D^+) ionic conductivity in H_2O/D_2O

mixtures. This "anomaly" was shown to follow from the assumed short-time quantum correlations between protons in water and aqueous H^+ solutions. Of course, this physical picture is in contrast with the viewpoint taken by, say, quantum chemistry and molecular dynamics, where protons are considered as classical particles.

The following crucial point represents our *working hypothesis*. If the well-known high H^+ conductance, λ_{H^+}, in liquid water is caused by the assumed quantum interference effects, there must be an anomalous decrease of λ_{H^+} in H_2O/D_2O mixtures due to the so-called mass and spin superselection rules (cf. [31]). Namely, in these mixtures the possible quantum interference between appropriate protonic states becomes disrupted by deuterons "near" or "between" the considered protons. For exactly the same reasons, we also expect an anomalous decrease of the OH^- conductance in H_2O/D_2O (see Section III.B.2).

Similar reasoning may also apply to D^+ conductance in H_2O/D_2O. Owing to the *bosonic* character of deuterons, however, the general CSM theory of coherent-dissipative structures does not apply directly in this case (cf. [47] for the possible extension of the theory to the bosonic case).

It may be interesting to give a crude *estimate* of the expected decrease of λ_{H^+}, in light of the CSM theory of quantum correlations. The general theory speaks primarily of "delocalization" in "Hilbert space," therefore, the desired estimate should here be reached with an illustrative model of *classical–mechanical* character. First, let us consider the water molecules as classical bodies and a volume V of liquid water. Let V_O and V_H be the volume parts being occupied by oxygen and hydrogen atoms, respectively. Then one may write the equation $V = V_O + V_H$. Second, let us consider the 1 : 1 mixture of H_2O/D_2O where one has $V = V_O + V_H^* + V_D^*$ with $V_H^* = V_D^*$. The coherent-dissipative structure around an H^+ ion may be imagined as a sphere of radius r, in the case of pure H_2O, and r^*, in the mixture. The idea of the disruption of the protonic coherent-dissipative structures by some D atoms in the mixture leads immediately to the following simple relation:

$$\frac{r^*}{r} = \left(\frac{V_O + V_H^*}{V_O + V_H}\right)^{1/3} \equiv \left(\frac{V - V_D^*}{V}\right)^{1/3}$$

In light of the general CSM theory, the H^+ mobility is assumed to be proportional to the spatial size r or r^* of the coherent-dissipative structures [cf. Eq. (20)]. Thus, this model gives the following estimated values: Assuming that $V_O : V_H = 1 : 1$, one has $r^*/r \approx 0.91$ and thus an anomalous decrease of the H^+ mobility (or conductance) of about -9%. Correspondingly, $V_O : V_H = 4 : 1$ implies an "anomalous decrease" of about -3%.

To test the considered prediction experimentally, molar conductances, Λ, of different HCl/DCl and KCl solutions in H_2O/D_2O mixtures were measured [46]. The experimental results are summarized in Figs. 2 and 3. The conductances of KCl solutions depend almost *linearly* on the D-atom fraction, X_D, of the solvent (cf. Fig. 2). This result is as expected from standard (or classical) electrochemical theory (cf. [42]), because (1) it is experimentally well established that the fluidity (i.e., the inverse of viscosity) of the

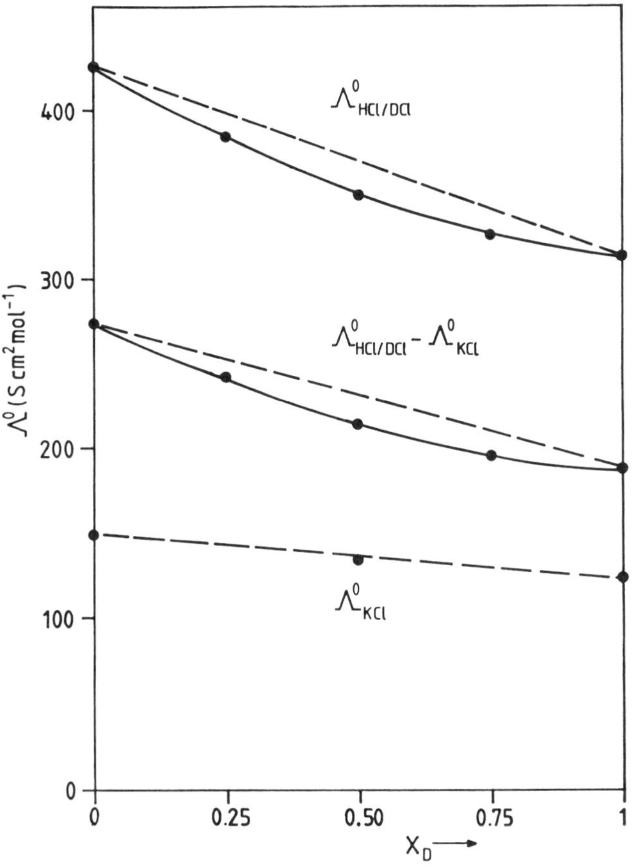

Figure 2. Molar conductances of HCl/DCl, $\Lambda^0_{HCl/DCl}$, and of KCl, Λ^0_{KCl}, in H_2O/D_2O mixtures at infinite dilution and $T = 25°C$ as a function of the mole fraction X_D of deuterium. Also shown is the excess conductance as determined from the difference between the data for HCl/DCl and KCl. Error bars are smaller than the size of each data point. (Reproduced from [46].)

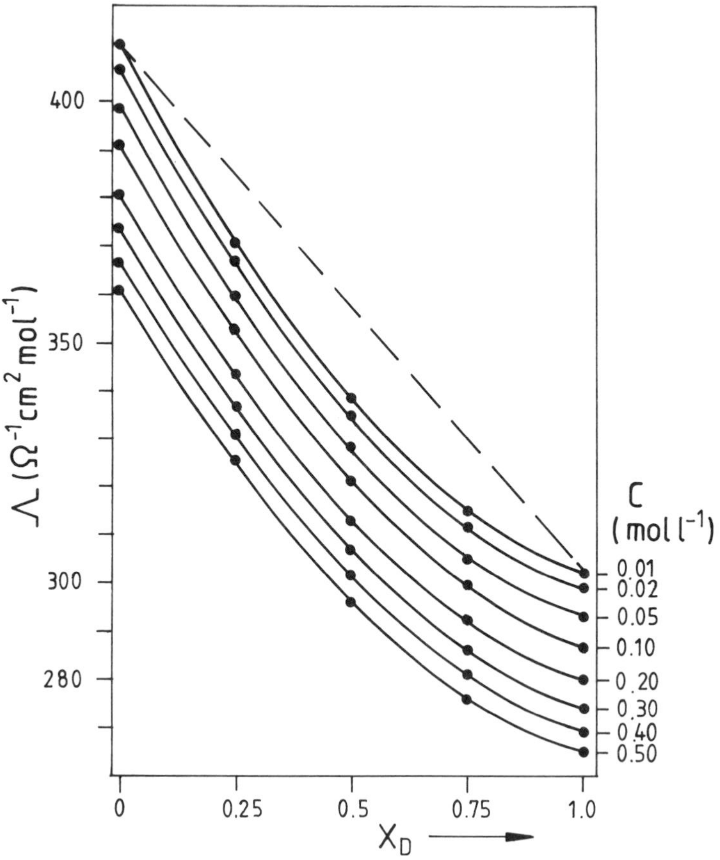

Figure 3. Graphical representation of the experimentally determined molar conductances of HCl/DCl in H_2O/D_2O mixtures at 25°C as a function of atom fraction X_D of deuterium and of acid concentration C. (Reproduced from [30].)

considered mixtures depends almost linearly on X_D and (2) ionic conductances are, to a very good approximation, directly proportional to the fluidity of the solvent (Walden's rule); see [46, 30] for details.

The conductivity data of HCl/DCl in H_2O/D_2O mixtures at infinite dilution, Λ^0, are shown in Fig. 2. At intermediate solvent compositions, the curve lies distinctly below the straight line connecting the limiting values in pure H_2O, where $X_D = 0$, and D_2O, where $X_D = 1$. For the quantitative treatment of the experimental results, we define the deviation of the measured conductance of a mixture of concentration C, $\Lambda(X_D, C)$, from that determined by the linear interpolation, $\Lambda_{lin}(X_D, C)$, between the values of the two pure solutions ($X_D = 0$ and 1):

$$\Delta\Lambda(X_D, C) = \Lambda(X_D, C) - \Lambda_{lin}(X_D, C)$$
$$= \Lambda(X_D, C) - [(1 - X_D) \cdot \Lambda(0, C) + X_D \cdot \Lambda(1, C)] \quad (25)$$

The corresponding percentage of relative deviation is

$$\Delta^{rel}(X_D, C) \equiv 100 \cdot \Delta\Lambda(X_D, C)/\Lambda_{lin}(X_D, C) \quad (26)$$

The relative deviation at $X_D = 0.5$ was found to be $\Delta^{rel}(0.5, C \to 0) \approx -5.1\%$.

Furthermore, the effect of interest is expected to be related to the *excess* conductance, which is defined by the difference of the data obtained for HCl/DCl and KCl. (This conventional definition is based on the fact that the main thermodynamic data of these two solutions are very similar [42].) Note that the excess conductance is usually associated with "quantum tunneling effects" of H^+ mobility in water (cf. [40]). The corresponding anomalous decrease of the excess conductance at $X_D = 0.5$ is now -7.7%.

It is interesting to note that the magnitude of the anomalous decrease under consideration is *independent of the acid concentration* (cf. Fig. 3). For full details, see [46].

Thus, the experiments show that the conductivities of KCl solutions in H_2O/D_2O mixtures are completely in accord with standard electrochemical theory [42]. This stresses that the predicted anomalous decrease $\Delta\Lambda^0$ of the (excess) molar conductivity of HCl/DCl is associated with some—classically unexpected—property of H_2O/D_2O. Therefore, it is hardly conceivable that the considered effect could be interpreted in terms of conventional theory.

2. *Anomalous OH^-/OD^- Conductance*

To investigate further the anomalous decrease of the protonic mobility in H_2O/D_2O, similar experiments have been carried out which concern the ionic conductivity of OH^-/OD^- [48]. These experiments deal with the measurement of ionic conductivities of several solutions of NaOH/NaOD in H_2O/D_2O mixtures. The experimental technique is equivalent to that mentioned in the previous section. The measured conductivities, at 25°C, always exhibit the predicted anomalous decrease, as in the case of H^+/D^+ presented above. The main experimental result [48] is presented graphically in Fig. 4, where the conductivities at infinite dilution are plotted against the mole fraction X_D. The anomalous decrease of the conductivity at the equimolar solvent composition is about -4.75%. The *excess conductance* of the solutions, however (which in the present case may be defined by the difference between the measured data for NaOH/NaOD and the corresponding

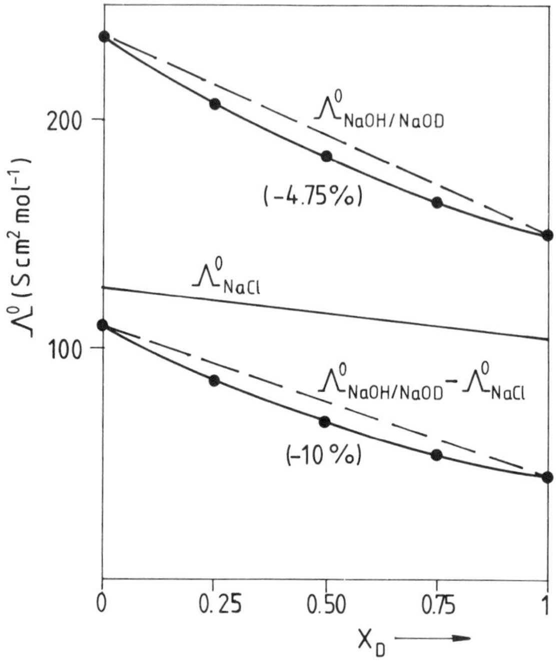

Figure 4. Graphical representation of the experimentally determined molar conductances of NaOH/NaOD and of NaCl in H_2O/D_2O mixtures at infinite dilution, at 25°C as a function of solvent composition X_D. Also shown is the excess conductance of OH^-/OD^- at infinite dilution. (Data taken from [48].)

conductance of NaCl), exhibits an anomalous decrease of about -10% at $X_D = 0.5$. Obviously, these experimental results represent an additional confirmation of our predictions [2,4].

C. Dielectric Relaxation Time in H_2O/D_2O

Recently Eigen noticed [49] that the anomalous decrease of the H^+ conductivity in H_2O/D_2O mixtures, as discussed above, could be due to another physical reason—an *increase* of the *lifetime* of the water clusters ("containing" the H^+ particle, such as $H_9O_4^+$ [39,40]). Note that the H^+ ions are not "free" particles in water, but are rather "captured" in proper water clusters, so that a possible increase of the mean lifetime of these clusters would also imply a decrease of the mobility of H^+. Additionally, it is well known that the lifetime of water clusters can be inferred from the knowledge of the dielectric relaxation time τ_{DR} in water. Therefore, measurement of the quantity τ_{DR} in H_2O/D_2O mixtures has been suggested [49].

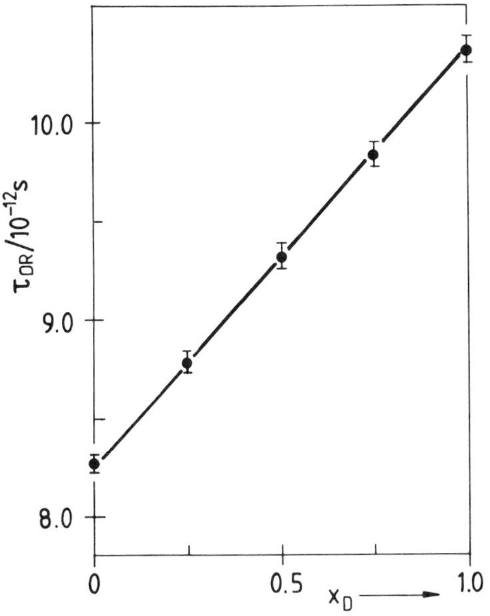

Figure 5. Experimentally determined values of the dielectric relaxation time in H_2O/D_2O as a function of the composition, expressed by the atom fraction X_D, of deuterium. (Data taken from [50].)

To test this assertion, precise dielectric relaxation measurements (in the spectral range between 1 MHz and 40 GHz) in H_2O, D_2O, and H_2O/D_2O mixtures have been carried out (by Kaatze, Göttingen) [50]. The results clearly exhibit a linear dependence of τ_{DR} on X_D (see Fig. 5). Thus, there is no anomalous increase of the lifetime of water clusters in H_2O/D_2O. Therefore, we may conclude that the dielectric relaxation results give further (although indirect) evidence of the aforementioned conjectured proton delocalization and EPR correlations between protons in water.

D. Anomalous Raman Light Scattering in H_2O/D_2O

The light-scattering experiment [51] discussed in this section was motivated by the following theoretical considerations. In the framework of our CSM theory of quantum correlations, a general theorem concerning scattering of external fields by coherent-dissipative structures has been proved [2,4]. This theorem states that Jordan blocks, those representing coherent-dissipative structures, have a vanishing trace, $\mathrm{Tr}\,\gamma = 0$; (cf. Section II.D). This formal result may have an important physical consequence. Namely, it implies that the measurable average

$$\langle \vec{E} \rangle = \frac{\mathrm{Tr}(\Gamma \vec{E})}{\mathrm{Tr}\,\Gamma} \tag{27}$$

(where Γ denotes here the full second-order density matrix, after proper thermalization and complex scaling) on an appropriate physical quantity $\vec{E}$ is expected to contain no contribution from the Jordan blocks γ, because (1) the latter are not normalizable and (2) it holds trivially that

$$\langle \vec{E} \rangle < \infty \tag{28}$$

Now let $\vec{E}$ represent the electric-field component of a continuous laser field acting on an appropriate H_2O/D_2O mixture. The physical meaning of Eq. (28) may then be that the laser light field must be "expelled" from the spatial domains being occupied by these structures. (Another possibility is, of course, that these structures become completely transparent.) Thus we should assume the validity of the equation

$$\mathrm{Tr}(\gamma \vec{E}) = 0 \tag{29}$$

in order to fulfill the necessary physical condition (28).

Now let us consider the Raman spectra of the OH- and OD- *stretching vibronic region* in liquid H_2O, D_2O, and H_2O/D_2O mixtures. (This is the spectral region between, say, 2000 and 4000 cm^{-1}.) As is well known, the broad range of these spectra is, plainly speaking, due to the continuous distribution of H-bond strengths and the short lifetime of these bonds (cf. [52–55]). Of particular importance for the present considerations is experimental evidence [53,54,56] that spatially correlated H-bonds may exist in liquid water. Quantum and/or collective dynamical properties of liquid water have been also proposed [57,58].

It is important to realize that the OH- and OD-vibrational stretching modes can represent fermionic and bosonic "systems" (or degrees of freedom), respectively. Then the application of the aforementioned physical arguments apply to the case of the stretching Raman spectra of H_2O/D_2O mixtures straightforwardly. Thus, we may predict an anomalous Raman-scattering intensity dependence of these stretching bands on X_D [51].

For the assessment of the Raman data presented below, knowledge of the most precise neutron scattering results [59,60] on the same systems is indispensable. In all cases, the data are fully consistent with the usual (see below) expectation that the microstructures of light and heavy water, as well as their mixtures, are almost equal. Possible differences—due to the quantum character of H and D—are below present measuring accuracies associated with all pair-correlation functions and/or partial structure factors [60–63]. This statement is further supported by a quantum mechanical study [63] which showed that nuclear quantum corrections to the structure of water, resulting from isotopic H/D substitution, are very small indeed (see also [62,61]). Of course, this finding is consistent with the well-known Born–Oppenheimer (BO) approximation.

Furthermore, it is crucial to stress the following two alternative points concerning H/D substitution:

1. If all molecules (i.e., H_2O, D_2O, and HDO) behave classically, it follows that all pair-correlation functions $g_{XX}(r)$ and $g_{XO}(r)$ ($X =$ H,D) remain unchanged.
2. If the molecules behave quantum mechanically [63], these equations may not hold strictly; however, the neutron experiments above imply that

$$g_{\mathrm{HH}}(r) \approx g_{\mathrm{DD}}(r) \approx g_{\mathrm{HD}}(r) \quad \text{and} \quad g_{\mathrm{HO}}(r) \approx g_{\mathrm{DO}}(r)$$

where the sign "$\approx$" here means "equal within measuring accuracy."

Thus, in both cases we observe a "constancy of the microstructure" of water. According to the BO approximation, we are then forced to conclude that H/D substitution leaves the electron-density distribution in liquid water (and/or the wave function of different clusters of water molecules [62]) essentially unchanged. Therefore, the same holds for all possible intra- and intermolecular interactions (e.g., dipole–dipole). This conclusion is confirmed by the most recent infrared [64] and dielectric relaxation [50] experiments (see Section III.C) on H_2O–D_2O as well.

The last conclusion implies straightforwardly that the total Raman intensity of the OH (or OD) spectral region must be strictly proportional to the number of the scattering OH (or OD) "particles" (or oscillators). It should be emphasized that this implication ought to be considered as unequivocal.

Surprisingly, the Raman data presented in the following are in clear contrast to the strict proportionality inferred above. Our experiment consists simply of determining the integral Raman scattering cross sections of the OH- and OD-stretching vibrational bands of H_2O–D_2O mixtures and their comparison with the corresponding cross sections of pure H_2O and pure D_2O. As is well known, the considerable width of these bands is caused by the H- and D-bond networks. The temperature was about 20°C. Some examples of the measured, individual Raman spectra are shown in Fig. 6. These graphs also demonstrate one important advantage—for our purposes—of the Raman effect as compared with the infrared absorption

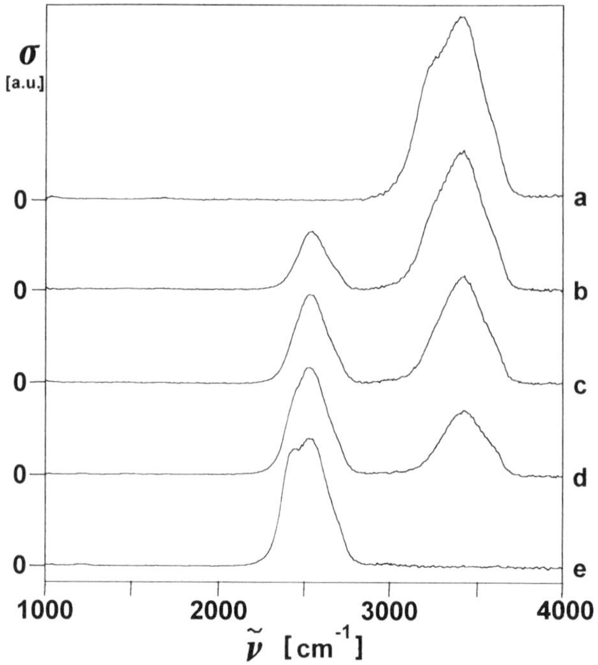

Figure 6. Some typical Raman scattering cross sections ($T = 20°C$) versus Stokes frequency shift for (a) pure water, (e) pure heavy water, and H_2O/D_2O mixtures with the following H : D atom compositions: (b) H : D = 7 : 3, (c) H : D = 5 : 5, and (d) H : D = 3 : 7. The cross sections are in arbitrary units. For convenience, spectra are parallel shifted along the ordinate. Spectra are presented as measured, that is, without any smoothing and/or baseline corrections [51].

[64]: the OH- and OD-spectral regimes are well separated (cf. [52,65,66]). As a consequence, our results are completely free of any data fitting and/or deconvolution procedure.

The experimental method used is described in detail in reference [51]. Here let us just mention that we used a spectrograph with an optical multichannel analyzer (OMA) detection system (SIT-Vidikon), which allowed simultaneous measurement of the complete spectrum from 1000 to 4000 cm^{-1}. Frequency reduction was performed in the standard way, that is, by multiplying the measured intensity $I(\nu)$ by the factor $\nu(1 - \exp[-h\nu/kT])/(\nu_L - \nu)^4$, where ν_L is the frequency of the laser light and ν is the frequency difference of the scattered light (i.e., the Raman shift) (see [52,66]). This transformation is necessary to obtain the intrinsic molar scattering activity or, equivalently, the scattering cross section for a Raman process.

The first feature of our effect is presented in Fig. 7. The quotient $Q = \sigma^H/\sigma^D$ gives the ratio between the integral Raman scattering cross sec-

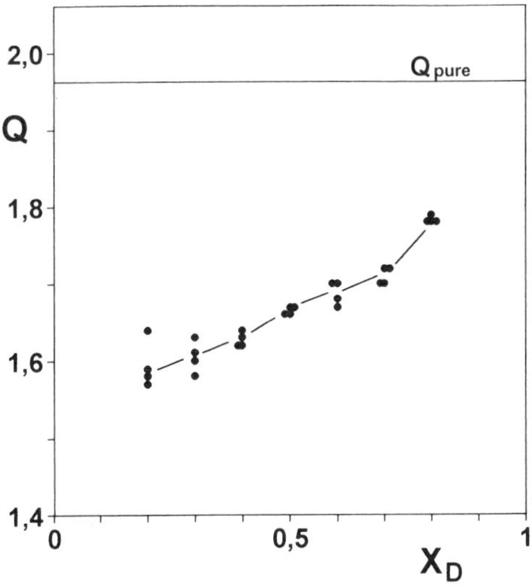

Figure 7. The quotient $Q = \sigma^H/\sigma^D$ of the integral Raman scattering cross sections of the OH- and OD-stretching vibration versus the mole fraction X_D of deuterium. Overlapping Q data points of the mixtures are slightly shifted along the abscissa to make them visible. The straight lines connecting the points are guides for the eyes. $Q_{pure} = 1.96$ (maximal error: ± 0.02) is the quotient Q defined above for pure H_2O and pure D_2O, and its magnitude is marked by the horizontal line. (Reproduced from [51].)

tions of the OH-stretching and that of the OD-stretching vibration. Repeated measurements of the corresponding cross sections of pure H_2O and pure D_2O gave for Q the value $Q_{pure} = 1.96$, as represented by the straight line in Fig. 7. This is in good agreement with the result in reference [66]. The comparison is based on Crawford's intensity sum rule [67]. The Q values for different H_2O–D_2O mixtures are given by the points, representing four independent experimental series. The predicted anomaly is represented here by the large difference between Q_{pure} and the Q values of all mixtures. Note also that Q varies with the H/D composition of the mixture.

The second feature of our effect (see Fig. 8) throws more light on this effect. Let $\Delta\sigma^H = (\sigma^H_{mix} - \sigma^H_{pure})/\sigma^H_{pure}$ represent the relative deviation of the considered OH vibronic integral cross section for a mixture, σ^H_{mix}, from that of pure water, σ^H_{pure}; $\Delta\sigma^D$ is defined analogously. The deviation $\Delta\sigma^H$ decreases with decreasing protonic concentration [H], that is, with increasing X_D, whereas, $\Delta\sigma^D$ increases with decreasing concentration [D]. This is the most surprising feature of the considered anomalous effect. Additionally, note also the substantial differences in magnitude, sign, and shape

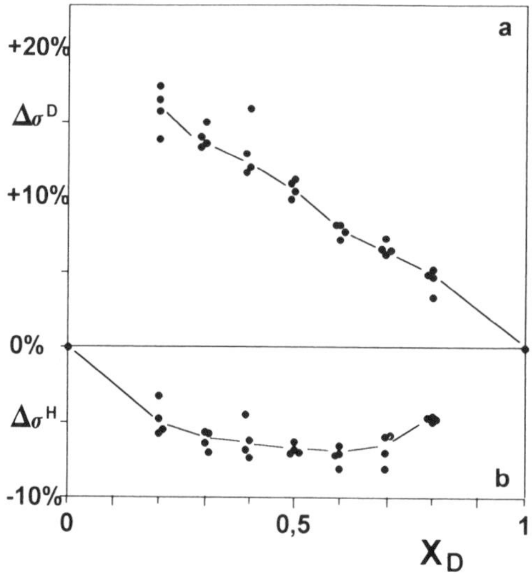

Figure 8. (*a*) Relative deviation of the OD-vibronic integral cross section of the mixtures from that of pure D_2O, that is, $\Delta\sigma^D = (\sigma^D_{mix} - \sigma^D_{pure})/\sigma^D_{pure}$ versus the D atom mole fraction of the mixtures. (*b*) Relative deviation of the OH-vibronic integral cross section of the mixtures from that of pure H_2O, that is, $\Delta\sigma^H = (\sigma^H_{mix} - \sigma^H_{pure})/\sigma^H_{pure}$ versus the D atom mole fraction of the mixtures. Overlapping data points are slightly shifted along the abscissa to make them visible. The straight lines connecting the points are guides for the eyes. (Reproduced from [51].)

of the functions $\Delta\sigma^H$ and $\Delta\sigma^D$. The quantitative treatment of these results is the subject of current investigations.

One should also emphasize the striking differences in the Raman scattering behavior of the mixtures having the same HDO, but different H_2O and D_2O concentrations (e.g., those with $X_D = 0.3$ and $X_D = 0.7$). Therefore, in contrast to other statements (e.g. [64,65]), we find that one cannot extract "individual" spectra of the "components" H_2O, D_2O, and HDO from the measured spectra of the mixtures.

Summarizing the considerations above, let us stress that if (1) the BO approximation is valid and (2) the constancy of the microstructure of water (see above) holds, one should have $\Delta\sigma^H = 0$ and $\Delta\sigma^D = 0$ for all X_D. These equations are expected to hold "within measuring accuracy," if one takes into account the quantization of hindered translations and librations of the different molecules, which, however, were treated as rigid bodies [63]. Thus, our experiments strongly affect at least one of the assumptions (1) and (2). Now one should observe that $|\Delta\sigma^H(0.5)| \leq |\Delta\sigma^D(0.5)|$ (see Fig. 8), although D is heavier than H. This indicates that our new experimental findings are not (mainly) caused by the missing adiabatic decoupling of electron and nuclear motion. Thus, it seems more probable that assumption (2) is violated, within the physical context of Raman scattering. But, because (2) cannot be violated in the case of classical nuclei, as mentioned above, we are necessarily leading to the conclusion that "large quantum effects" of protons and deuterons represent the more probable reason for the experimental results presented. Since quantum chemistry and molecular dynamics do treat the nuclei as classical particles, even the qualitative description of the revealed effect represents a considerable challenge for conventional theories.

According to our working hypothesis, however, a qualitative description of the rationale of the effect is straightforward. Namely, the spatial extension of the (possible) protonic quantum delocalization should be reduced in H_2O–D_2O mixtures [4]. The X_D-dependence of $\Delta\sigma^H$ simply shows that the "unreduced" microstructures being associated with protonic delocalization in pure water have effectively a larger Raman cross section than the reduced ones in the mixtures. Surprisingly, the bosonic OD-vibrational degrees of freedom reveal quite the opposite behavior. For a simple theoretical treatment, see [51].

E. Correlated Two-Photon Scattering in Water

The CSM theoretical relation (47) allows for another surprising prediction [2,4]: A sufficiently intense light pulse (e.g., from a nanosecond Nd–YAG laser, after second-harmonic generation) should exhibit an anomalous nonstochastic increase (also called bunching)—or decrease (called antibunching)—of the probability that two photons are simultaneously measured (i.e., the coincidence rate), if one detects scattered photons in nearly *antiparallel* directions. This novel predicted effect occurs because the external field must have a vanishing expectation value in the small spatial domains occupied (temporarily) by coherent-dissipative structures. Thus, illustratively speaking, the fields of the two photons may interfere destructively in the aforementioned spatial domains. Recent photoncounting/coincidence experimental results in our laboratory clearly confirm the existence of this new effect [68].

IV. PROTONS IN METALS AND ORGANIC CRYSTALS

A. Introductory Remarks

The quantum diffusion of hydrogen and other light particles, like D and the positive muon μ^+, in metals is a subject of great interest in basic as well as applied research areas (cf. [69,70]). In the following sections, two counterintuitive effects are presented and qualitatively explained from first theoretical principles. These effects are

1. The anomalous diffusion coefficient D_H of H in NbH_aD_b, with $a + b =$ constant (e.g., $=0.60$), which shows a striking nonlinear dependence of the decrease of D_H (up to one order of magnitude) with decreasing partial H concentration [69]
2. The anomalously high mobility of H, as compared to that of μ^+, in different metal hydrides [71]

Thus far, both effects lack any theoretical interpretation. In the framework of the CSM theory of thermally activated quantum correlations, however, they may find a natural explanation in terms of broken quantum correlations between protons when deuterons (or muons) are present in their environment [72,73]. Note that the *same* physical assumption (conjecture or working hypothesis) was applied to the proton-mobility effects treated in the preceding sections.

The aforementioned experimental findings (1) and (2) concern metal–hydrogen systems with a *large* H content and at relatively *high* temperatures (e.g., 100–500 K). (These experimental conditions appear to be convenient for the applicability of the CSM theory of quantum correlations [4].) To prevent confusion, let us emphasize that this physical context has nothing to do with the so-called *coherent quantum tunneling* of hydrogen in metals, where the focus is exclusively on the tunneling of *single* protons. As is well known, coherent quantum tunneling is studied in experiments with extremely *low* H concentration and temperature (usually <10 K) (see [74–77]). It should also be mentioned that the conventional coherent quantum tunneling process is considered to have an "infinite" lifetime (since it is treated with the aid of the time-*independent* Schrödinger equation), whereas the aforementioned thermally activated quantum correlations clearly represent *short-time* effects (of the order of the decoherence time; cf. Section II.C).

In Section IV.C, certain NMR experiments on an anomalous H/D transfer in benzoic acid crystals (which consist of hydrogen-bonded dimeric units) are reported [78]. It will be shown that these observations bear an unexpected resemblance to the aforementioned Fukai effect [69].

It might be helpful to mention that in the following—as in the scientific literature of metal–hydrogen systems—the terms "hydrogen" and "proton" are used synonymously. This is clearly justified, because there are no well-defined hydrogen atoms in metallic lattices.

B. Anomalous Decrease of H Diffusion Coefficient in Metal–H/D Systems

Hydrogen diffusion in NbH_aD_b has been studied extensively by Fukai [69], with the aid of the pulsed-field gradient H-NMR method. This experimental technique provides a unique possibility of observing diffusion of one particular isotope (H) in the presence of others (D). One striking observation in the case of NbH_aD_b with constant total concentration of hydrogen and deuteron, $a + b = 0.6$, is presented schematically in Fig. 9. With an increasing mole fraction of deuterium X_D ($=[D]/[H + D]$), the mobility of H remains constant for a large range of X_D, then suddenly drops off significantly. This unexpected effect is also found in samples with different total $[H] + [D]$ concentrations and in a wide temperature range. For experimental details and further references, see the review article of Fukai and Sugimoto [69] and Fukai's monograph [70].

It has been pointed out [69] that these experimental results provide clear evidence that the motions of H atoms are strongly correlated with each other. The physical nature of this correlation, however, remains unclear, and thus far, existing quantitative theoretical investigations to explain this effect have not been successful [69].

In light of the general theory of quantum correlations [4], however, the interpretation of this effect is straightforward [72,73]. In this context it is assumed that protons (and probably also deuterons) may be EPR correlated in metallic hydrides within sufficiently short time intervals (of the order τ_D or τ_R), even at the high temperatures considered. The observed mobility of H is then expected to be—at least in part—due to this quantum correlation effect and/or the associated delocalization of protons. But if these protonic EPR-correlated structures become disrupted by deuterons in their vicinity, the protonic mobility should decrease. (As is well known, delocalized particles appear to be more mobile than their localized counterparts.) As noted in preceding Sections III.B–D, this restriction is due to spin and/or mass *superselection rules* [4,31].

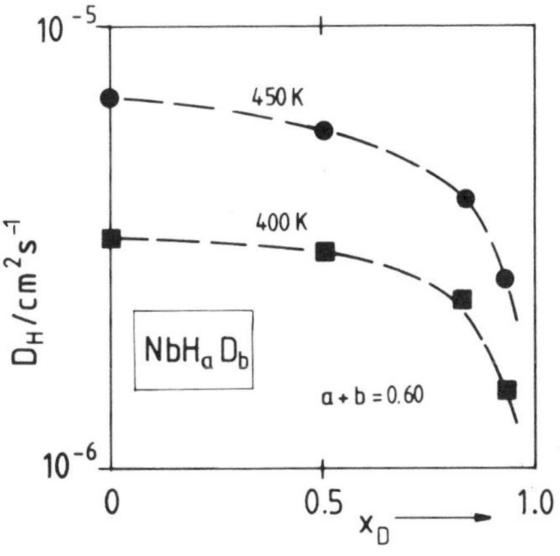

Figure 9. Schematic representation of the diffusion coefficient of H (as measured by NMR) in α-NbH_aD_b, with $a + b = 0.6$, for two different temperatures. (Data taken from [69].)

The considered effect of Fukai is now easy to understand [72,73]. Namely, increasing D content reduces the protonic EPR-correlated domains, and thus the protons become less mobile. An *estimate* of the size of these EPR-correlated domains can be inferred as follows. One sees from Fig. 9 that the protonic mobility decreases significantly (by half an order of magnitude) at $X_D \approx 0.9$. The corresponding concentration of protons (per lattice cell of Nb) is then

$$[\mathrm{H}] \approx 0.6 \times 2 \times (1 - 0.9) \approx 0.12 \tag{30}$$

which would typically be the smallest concentration at which the protonic EPR-correlated domains may still exist. The mean distance between protons at this characteristic concentration gives an estimate of the maximum extension of such a domain:

$$l_{\mathrm{EPR}} = [\mathrm{H}]^{-1/3} \cdot a_{\mathrm{Nb}} \approx 2 \cdot a_{\mathrm{Nb}}$$

where a_{Nb} is the lattice constant of the host metal Nb.

C. Anomalous H/D Transfer in H-Bonded Organic Crystals

Recently, a precise investigation of the spin-lattice relaxation rates of protons and deuterons in different partially deuterated benzoic acid crystals by Takeda [78] revealed a striking quenching of the transfer rate of an HD pair in the H-bonded dimeric units of carboxyl groups with increasing concentration of D (or, equivalently, decreasing H concentration). A similar effect was also observed in partially deuterated crystals of acetylenedicarboxylic acid. This remarkable effect is considered here, because it bears a clear resemblance to the finding of Fukai mentioned above.

The so-called "double proton transfer" in H-bonded dimers of carboxyl groups in carboxylic acid crystals (and other related systems) has been studied extensively. In these systems, quantum tunneling is believed to play a dominant role in the "cooperative" transfer of two protons (HH pair) and two deuterons (DD pair) in dimeric units of carboxyl groups, even at elevated temperatures (e.g., 200 K). As the temperature increases, additional thermally activated processes also become important [79,80], and related coupling effects between protonic (and deuteronic) motion and other degrees of freedom have been discussed as a key mechanism for clarifying the temperature dependence of the proton transfer in H bonds.

In addition to the conventional knowledge, Takeda [78] recently presented a novel result showing the aforementioned remarkable quenching of

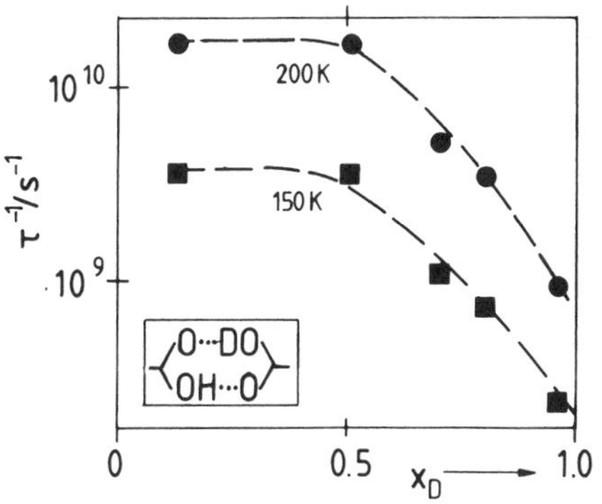

Figure 10. Transfer rate of an HD pair (measured by H- and D-NMR) in hydrogen-bonded dimers of carboxyl groups in benzoic acid-crystals as a function of the mole fraction X_D. (Data taken from [78].)

the transfer rate of an HD pair with increasing mole fraction of D in the hydrogen bond of benzoic acid crystals. The experiment consists of measuring the transfer rate of an HD pair (in the dimeric units described above) as a function of the H/D fraction of the hydrogen bonds. (The phenyl groups of the molecules are always fully protonated. It is well established that these protons do not exchange, contrary to the protons or deuterons of the carboxylic groups.)

The HD transfer rate under consideration was determined by measuring the spin-lattice relaxation rate T_1^{-1} of H and D nuclear magnetic resonances (NMR). The experimental results are as follows.

First, it is quite remarkable that the H and D transfer rates in an HD pair are similar, despite the different masses of protons and deuterons. The most striking feature of the results, however, is the "sudden" decrease of the transfer rate of the HD pair with decreasing H content in the surrounding hydrogen bonds of below 50% (see Fig. 10). The similarity of this effect to Fukai's findings becomes obvious if one compares Fig. 9 with Fig. 10. These experiments were analyzed in detail, in order to test possible quantum correlations between H-bonding protons of these crystals [78].

This effect has, hitherto, no conventional interpretation [78]. However, according to our working hypothesis, the protons of the H bonds may be EPR correlated. Therefore, deuterons "between" the protons may break these correlations, thus causing a "localization" of protons (and correspondingly of deuterons), which is tantamount to a decrease of the transfer rate of a HD pair.

D. "Slow" Muons and "Fast" Protons in Metals

A prominent method for the determination of hydrogen diffusion constants in metals is the method of *incoherent quasi-elastic neutron scattering* (QNS). The main advantages are: (1) the simultaneous study of the spatial and temporal development of the H diffusion process, and (2) the sensitivity of this method owing to the high value of the neutron scattering cross section of H (which is larger by an order of magnitude than that of other elements).

Here we give some details of the QNS experiment (which are used below). The double-differential incoherent neutron cross section for diffusive H motion at small (momentum transfer) wave vectors Q is a single Lorentzian (cf. [71,70] and references therein), that is,

$$\frac{\partial\sigma^2}{\partial\omega\,\partial\Omega} = \frac{\sigma_{\text{inc}}^{\text{H}}}{4\pi} N_{\text{H}} \exp[-2W(Q)] \frac{1}{\pi} \frac{\hbar D_{\text{H}} Q^2}{(\hbar D_{\text{H}} Q^2)^2 + (\hbar\omega)^2} \tag{31}$$

where $\sigma_{\text{inc}}^{\text{H}}$ is the incoherent cross section of H, $\exp[-2W(Q)]$ is the Debye–Waller factor, and N_{H} is the total number of H atoms "seen" by the neutron beam. In the type of experiments considered (i.e., QNS), one measures H diffusion over distances of the order of $2\pi/Q \approx 30$–40 Å. In view of the large number of jump directions for H in intermetallic hydrides, a liquid-like isotropic arrangement of interstitial sites appears to be a reasonable approximation. Then the linewidth of the Lorentzian in Eq. (31) is given by [71]

$$\Gamma(Q, T) = \frac{6\hbar D_{\text{H}}(T)}{l^2}\left(1 - \frac{\sin(Ql)}{Ql}\right) \tag{32}$$

where l is an "effective" jump length which is determined by the specific lattice of the host metal (and the interstitial sites where the protons can be placed). It turned out that this equation represents a sufficiently accurate approximation for small Q values, even with respect to the well-known additional broadening caused by multiple scattering with a larger Q (cf. [71]). From knowledge of the parameter l and the measurement of the Lorentzian line width for different values of Q and T, and using Eq. (32), one determines the hydrogen diffusion coefficient D_{H}.

To compare the proton mobility with that of μ^+, one calculates the so-called *mean hydrogen residence time* $\tau_{\text{H}}(\text{QNS})$, (where the index QNS refers to the QNS experimental situation). The characteristic time $\tau_{\text{H}}(\text{QNS})$ fulfills the relation

$$\tau_{\text{H}}(\text{QNS}) = \frac{l^2}{6D_{\text{H}}(\text{QNS})} \tag{33}$$

The proton mobility in different metal–hydrogen systems has also been studied with the aid of the μSR method (which is not described here). In this case, one measures directly the fluctuation rate (or relaxation time) of the μ^+–H dipolar interactions.

In metallic hydrides with *high* H content, one easily sees that the H and μ^+ residence times should be almost the same, because a muon can change site only when a proton jumps to leave a place for the μ^+ motion. During these investigations, however, the following astonishing observation has been made repeatedly: The measured muon residence time τ_{μ^+} is larger than the proton residence time τ_H, as measured by QNS, even though μ^+ is lighter than H. Or, in more concrete terms,

$$\tau_H(\mu SR) \approx \tau_{\mu^+} \approx 10 \cdot \tau_H(QNS) \tag{34}$$

where $\tau_H(\mu SR)$ is the proton residence time being deduced from the μSR experiments; see Fig. 11. This "paradoxical" [81] finding holds for a large temperature range [71] and for a considerable number of different metal–hydrogen systems (see [71,81] and references cited therein). It is also striking that the QNS results for protons are in agreement with the corresponding results of PAC (perturbed angular correlation) and NMR experiments [71,81].

We now consider in some detail the experimental methods used in these experiments [72,73]. This appears to be crucial since this relation is based on the results of two *different* experimental methods being applied to hydrogen–metal systems: muon spin rotation (μSR) and neutron scattering. According to our working hypothesis (see Section IV.A): In the case of the μSR experiment, the presence of μ^+ may break quantum correlations between protons in its neighborhood. This causes the restriction of the spatial extension of protonic coherent-dissipative structures, which is tantamount to decrease of the mobility of the muon and the protons surrounding it. In this context, the large de Broglie wavelength of the muon (see Section I.A) should be remembered. Because *more* (less) extended EPR-correlated coherent-dissipative structures—and the corresponding protonic spatial delocalization—correspond to an *increased* (decreased) mobility of the participating protons, the restriction above "caused by" the appearance of μ^+ also causes a certain decrease of the mobility (i.e., the diffusion coefficient D_H) of those protons, which interact with the considered μ^+. In the μSR experiment one measures the μ^+–H dipolar interactions. Therefore, the quantity $\tau_H(\mu)$ is thought to be associated with the motion of "single" (i.e., less-correlated) protons.

In the case of QNS, there is no breaking of the protonic coherent-dissipative structures. Thus, in an illustrative way one can say that, during

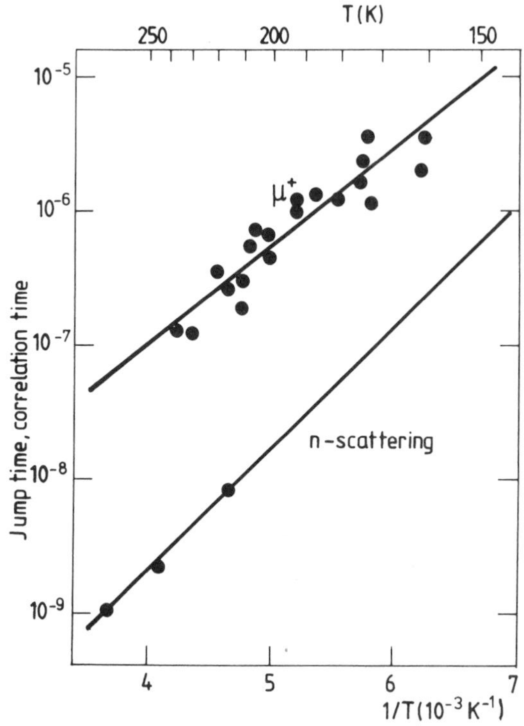

Figure 11. Schematic representation of the "correlation time" of μ^+ (measured by μ^+SR) and the "jump time" of protons (measured by QNS) in different ZrV_2H_a, with $3.25 < a < 4.0$. (Data taken from [71].)

the scattering, the neutrons "see" several quantum-correlated protons "at the same time," which is tantamount to "single and more mobile" protons.

These considerations of the physical conditions underlying the two experimental methods imply that the "experimentally established" relation (33) should be reconsidered by carefully taking into account what is *directly measured*—rather than inferred on the basis of "well-established" and/or tacit assumptions. This analysis is as follows.

First, the actual values of the mean hydrogen residence time according to Eq. (33) are based on the tacit assumptions that:

1. The neutrons are scattered by "single" protons
2. The assumed value and/or physical relevance of the quantity l are correct

According to our working hypothesis, however, the neutrons are scattered by EPR-correlated structures of linear dimension l_{EPR} which may "contain" several protons, and therefore, l_{EPR} should replace l in Eq. (33). Furthermore, the aforementioned experimental finding is tantamount to

$$10 \cdot D_H(\mu SR) \approx D_H(QNS)$$

(with obvious notation), as follows from assumption (2). For the residence time of a "single" proton, one thus obtains

$$l_{EPR}^2/D_H(QNS) \approx l^2/D_H(\mu SR) \approx 10 \cdot l^2/D_H(QNS) \tag{35}$$

and therefore $l_{EPR} \approx 3.3l$. For jumps between tetrahedral sites in Nb one has $l = 1.2$ Å, which gives the estimate $l_{EPR} \approx 4$ Å, which is of the same order as the corresponding one following from the analysis of Fukai's experiments (Section IV.B).

V. ADDITIONAL REMARKS

When the scientific community realized that EPR correlations are of fundamental physical character, it was often conjectured that they may play a significant role in many dynamical processes in dense systems. In this chapter, various predicted new effects and/or experimental results are described which give strong evidence for the physical significance of quantum correlations in condensed-matter dynamics. In connection with each topic, the specific reasons have been given which demonstrate that the considered effects have no classical or conventional explanation. These effects clearly stress the predictive power and experimental relevance of the new CSM theory of thermally activated quantum correlations in condensed matter.

To investigate the novel time-dependent EPR correlations between protons in more detail, we recently launched a series of inelastic and elastic *neutron* scattering experiments in water and metal–hydrogen systems.

ACKNOWLEDGMENTS

Fruitful collaboration and many insightful discussions with E. Brändas and E. Karlsson (Uppsala) as well as G. Hertz and H. Weingärtner (Karlsruhe) are gratefully acknowledged. M. Eigen (Göttingen) and G. Ertl (Berlin) are thanked for their suggestions about carrying out the experiments presented in Sections III.C and B, respectively, and U. Kaatze (Göttigen) for the performance of the experiment reported in Section III.C. This work was supported, in part, by the Commission of the European Communities (Science program) and the Fonds der Chemischen Industrie (Frankfurt/Main).

REFERENCES

1. T. S. Kuhn, *The Structure of Scientific Revolutions*, 2nd ed. (University of Chicago Press, Chicago, 1970).
2. C. A. Chatzidimitriou-Dreismann, *Int. J. Quantum Chem. Symp.* **23**, 153 (1989).
3. E. J. Brändas and C. A. Chatzidimitriou-Dreismann, *Lect. Notes Phys.* **325**, 485 (1989).
4. C. A. Chatzidimitriou-Dreismann, *Adv. Chem. Phys.* **80**, 201 (1991).
5. O. Carnal and J. Mlynek, *Phys. Rev. Lett.* **66**, 2689 (1991); D. W. Keith, C. R. Ekstrom, Q. A. Turchette, and D. E. Pritchard, *Phys. Rev. Lett.* **66**, 2693 (1991).
6. P. Grangier and A. Aspect, *Phys. Rev. Lett.* **54**, 418 (1985).
7. J. S. Bell, *Physics* **1**, 195 (1964).
8. A. Aspect, J. Dalibard, and G. Roger, *Phys. Rev. Lett.* **49**, 1804 (1982).
9. J. A. Wheeler and W. H. Zurek (eds.), *Quantum Theory and Measurement* (Princeton University Press, Princeton, NJ, 1983).
10. A. Einstein, B. Podolsky, and N. Rosen, *Phys. Rev.* **47**, 777 (1935).
11. N. Bohr, *Phys. Rev.* **48**, 696 (1935).
12. B. d'Espagnat, *Sci. Am.* **241**, 128 (1979); *Phys. Rep.* **110**, 201 (1984).
13. B. d'Espagnat, *Reality and the Physicist—Knowledge, Duration and the Quantum World* (Cambridge University Press, Cambridge, 1989).
14. J. Maddox, *Nature* **352**, 277 (1991).
15. E. Balslev and J. M. Combes, *Commun. Math. Phys.* **22**, 280 (1971).
16. J. Aguilar and J. M. Combes, *Commun. Math. Phys.* **22**, 269 (1971).
17. C. N. Yang, *Rev. Mod. Phys.* **34**, 694 (1962).
18. I. Prigogine, *From Being To Becoming* (Freeman, San Francisco, CA, 1980).
19. T. Petrosky and I. Prigogine, *Physica A* **175**, 146 (1991).
20. T. Petrosky, I. Prigogine, and S. Tasaki, *Physica A* **173**, 175 (1991).
21. I. E. Antoniou and I. Prigogine, *Physica A* **192**, 443 (1993).
22. I. E. Antoniou and S. Tasaki, *Int. J. Quantum Chem.* **46**, 425 (1993).
23. L. D. Landau and E. M. Lifshitz, *Statistical Physics—Part I* (Pergamon Press, Oxford, 1980).
24. C. A. Chatzidimitriou-Dreismann, E. J. Brändas, and E. Karlsson, *Phys. Rev. B* (*Rapid Commun.*) **42**, 2704 (1990).
25. E. Karlsson, E. J. Brändas, and C. A. Chatzidimitriou-Dreismann, *Phys. Scripta* **44**, 77 (1991).
26. C. A. Chatzidimitriou-Dreismann and E. J. Brändas, *Physica C* **201**, 340 (1992).
27. C. A. Chatzidimitriou-Dreismann, *Physica C* **219**, 420 (1994).
28. A. J. Coleman, *Rev. Mod. Phys.* **35**, 668 (1963).
29. A. J. Coleman, *J. Math. Phys.* **6**, 1425 (1965).
30. C. A. Chatzidimitriou-Dreismann and E. J. Brändas, *Ber. Bunsenges. Phys. Chem.* **95**, 263 (1991).
31. H. Primas, *Chemistry, Quantum Mechanics and Reductionism* (Springer-Verlag, Berlin, 1983).
32. R. Omnès, *Rev. Mod. Phys.* **64**, 339 (1992) and references therein.
33. W. H. Zurek, *Phys. Today* **44**, 36 (1991) and references therein.

34. C. E. Reid and E. J. Brändas, *Lect. Notes Phys.* **325**, 475 (1989).
35. H. Lehr and C. A. Chatzidimitriou-Dreismann, *Phys. Rev. A* **52**, 2935 (1995).
36. F. R. Gantmacher, *The Theory of Matrices*, Vols. I and II (Chelsea, New York, 1974).
37. S. Meiboom, *J. Chem. Phys.* **34**, 375 (1961).
38. Z. Luz and S. Meiboom, *J. Am. Chem. Soc.* **86**, 4768 (1964).
39. M. Eigen and L. de Maeyer, *Proc. Roy. Soc.* (*London*) **A247**, 505 (1958).
40. M. Eigen, *Angew. Chem. Int. Ed.* **3**, 1 (1964).
41. C. A. Chatzidimitriou-Dreismann and E. J. Brändas, *Int. J. Quantum Chem.* **37**, 155 (1990).
42. R. A. Robinson and R. H. Stokes, *Electrolyte Solutions* (Butterworths, London, 1970).
43. R. Pfeifer and H. G. Hertz, *Ber. Bunsenges. Phys. Chem.* **94**, 1349 (1991).
44. R. Pfeifer, PhD thesis, University of Karlsruhe, 1991.
45. A. Gierer and K. Wirtz, *Ann. Phys. Lpz.* (6. *Folge*) **6**, 257 (1949).
46. H. Weingärtner and C. A. Chatzidimitriou-Dreismann, *Nature* **346**, 548 (1990).
47. E. J. Brändas and C. A. Chatzidimitriou-Dreismann, *Int. J. Quantum Chem.* **40**, 649 (1991).
48. S. Gluth, Diploma thesis, University of Karlsruhe, 1990.
49. M. Eigen, personal communication.
50. U. Kaatze, *Chem. Phys. Lett.* **203**, 1 (1993).
51. C. A. Chatzidimitriou-Dreismann, U. K. Krieger, A. Möller, and M. Stern, *Phys. Rev. Lett.* **75**, 3008 (1995).
52. M. H. Brooker, G. Hancock, B. C. Rice, and J. Shapter, *J. Raman Spectrosc.* **20**, 683 (1989).
53. J. L. Green, A. R. Lacey, and M. G. Sceats, *J. Chem. Phys.* **90**, 3958 (1986).
54. J. L. Green, A. R. Lacey, M. G. Sceats, S. J. Henderson, and R. J. Speedy, *J. Chem. Phys.* **91**, 1684 (1987).
55. S.-H. Chen and J. Teixeira, *Adv. Chem. Phys.* **64**, 1 (1986).
56. J. L. Green, A. R. Lacey, and M. G. Sceats, *Chem. Phys. Lett.* **130**, 67 (1986).
57. S. A. Rice and M. G. Sceats, *J. Phys. Chem.* **85**, 1108 (1981).
58. M. A. Ricci, D. Rocca, G. Ruocco, and R. Vallauri, *Phys. Rev. Lett.* **61**, 1958 (1988).
59. W. E. Thiessen and A. H. Narten, *J. Chem. Phys.* **77**, 2656 (1982).
60. A. K. Soper and M. G. Phillips, *Chem. Phys.* **107**, 47 (1986).
61. A. K. Soper and J. Turner, *Int. J. Mod. Phys.* **B7**, 3049 (1993).
62. K. Laasonen, M. Sprik, M. Parrinello, and R. Car, *J. Chem. Phys.* **99**, 9080 (1993).
63. R. A. Kuharski and P. J. Rossky, *J. Chem. Phys.* **82**, 5164 (1985).
64. Y. Maréchal, *J. Chem. Phys.* **95**, 5565 (1991).
65. G. E. Walrafen, in *Water—A Comprehensive Treatise*, Vol. 1, F. Franks (ed.) (Plenum Press, New York, 1972), pp. 151–214.
66. J. R. Scherer, M. K. Go, and S. Kint, *J. Phys. Chem.* **78**, 1304 (1974); *ibid.* **77**, 2108 (1973).
67. B. Crawford, *J. Chem. Phys.* **20**, 977 (1952).
68. U. K. Krieger, F. Aurich, and C. A. Chatzidimitriou-Dreismann, *Phys. Rev. A* (*Rapid Commun.*) **52**, 1827 (1995).
69. Y. Fukai and H. Sugimoto, *Adv. Phys.* **34**, 263 (1985).
70. Y. Fukai, *The Metal–Hydrogen System* (Springer-Verlag, Berlin, 1993).
71. R. Hempelmann, D. Richter, O. Hartmann, E. Karlsson, and R. Wäppling, *J. Chem. Phys.* **90**, 1935 (1989).

72. C. A. Chatzidimitriou-Dreismann and E. Karlsson, in "Proceedings of the International Symposium on Hydrogen–Metal Systems (Uppsala, June 1992)," *Z. Phys. Chem.* **181**, 165 (1993).
73. C. A. Chatzidimitriou-Dreismann, *Ber. Bunsenges. Phys. Chem.* **96**, 1742 (1992).
74. H. Grabert and H. Wipf, *Festkörperprobleme/Adv. Solid State Phys.* **30**, 1 (1990).
75. Y. Kagan and A. J. Leggett, *Quantum Tunnelling in Condensed Media* (North-Holland, Amsterdam, 1992).
76. U. Weiss, *Quantum Dissipative Systems* (World Scientific, Singapore, 1993).
77. H. Wipf, *Ber. Bunsenges. Phys. Chem.* **95**, 438 (1991).
78. S. Takeda, A. Tsuzumitani, and C. A. Chatzidimitriou-Dreismann, *Chem. Phys. Lett.* **198**, 316 (1992).
79. A. Stäckli, B. H. Meier, R. Kreis, R. Meyer, and R. R. Ernst, *J. Chem. Phys.* **93**, 1502 (1990).
80. A. Heuer and U. J. Haeberlen, *J. Chem. Phys.* **95**, 4201 (1991).
81. A. Baudry, P. Boyer, A. Chikdene, S. W. Harris, and S. F. J. Cox, *Hyp. Int.* **64**, 657 (1990).

AUTHOR INDEX

Numbers in parentheses are reference numbers and indicate that the author's work is referred to although his name is not mentioned in the text. Numbers in *italic* show the pages on which the complete references are listed.

SUBJECT INDEX